# Negative Differential Resistance and Instabilities in 2-D Semiconductors

# NATO ASI Series

## Advanced Science Institutes Series

*A series presenting the results of activities sponsored by the NATO Science Committee, which aims at the dissemination of advanced scientific and technological knowledge, with a view to strengthening links between scientific communities.*

The series is published by an international board of publishers in conjunction with the NATO Scientific Affairs Division

| | | |
|---|---|---|
| **A** | **Life Sciences** | Plenum Publishing Corporation |
| **B** | **Physics** | New York and London |
| **C** | **Mathematical and Physical Sciences** | Kluwer Academic Publishers |
| **D** | **Behavioral and Social Sciences** | Dordrecht, Boston, and London |
| **E** | **Applied Sciences** | |
| **F** | **Computer and Systems Sciences** | Springer-Verlag |
| **G** | **Ecological Sciences** | Berlin, Heidelberg, New York, London, |
| **H** | **Cell Biology** | Paris, Tokyo, Hong Kong, and Barcelona |
| **I** | **Global Environmental Change** | |

### *Recent Volumes in this Series*

*Volume 302*—Microwave Discharges: Fundamentals and Applications
    edited by Carlos M. Ferreira and Michel Moisan

*Volume 303*—Particle Production in Highly Excited Matter
    edited by Hans H. Gutbrod and Johann Rafelski

*Volume 304*—Growth Patterns in Physical Sciences and Biology
    edited by Juan-Manuel Garcia-Ruiz, Enrique Louis, Paul Meakin,
    and Leonard M. Sander

*Volume 305*—Chemical Physics of Intercalation II
    edited by Patrick Bernier, John E. Fischer, Siegmar Roth,
    and Stuart A. Solin

*Volume 306*—Ionization of Solids by Heavy Particles
    edited by Raúl A. Baragiola

*Volume 307*—Negative Differential Resistance and Instabilities in 2-D
    Semiconductors
    edited by N. Balkan, B. K. Ridley, and A. J. Vickers

*Volume 308*—Photonic Band Gaps and Localization
    edited by C. M. Soukoulis

*Series B: Physics*

# Negative Differential Resistance and Instabilities in 2-D Semiconductors

Edited by

## N. Balkan
## B. K. Ridley
## A. J. Vickers

University of Essex
Colchester, United Kingdom

Springer Science+Business Media, LLC

Proceedings of a NATO Advanced Research Workshop on
Negative Differential Resistance and Instabilities in 2-D Semiconductors,
held September 20–25, 1992,
in Il Ciocco, Lucca, Italy

**NATO-PCO-DATA BASE**

The electronic index to the NATO ASI Series provides full bibliographical references (with keywords and/or abstracts) to more than 30,000 contributions from international scientists published in all sections of the NATO ASI Series. Access to the NATO-PCO-DATA BASE is possible in two ways:

—via online FILE 128 (NATO-PCO-DATA BASE) hosted by ESRIN, Via Galileo Galilei, I-00044 Frascati, Italy

—via CD-ROM "NATO-PCO-DATA BASE" with user-friendly retrieval software in English, French, and German ( ©WTV GmbH and DATAWARE Technologies, Inc. 1989)

The CD-ROM can be ordered through any member of the Board of Publishers or through NATO-PCO, Overijse, Belgium.

Library of Congress Cataloging-in-Publication Data

---

Negative differential resistance and instabilities in 2-D
  semiconductors / edited by N. Balkan, B.K. Ridley, and A.J. Vickers.
      p.    cm. -- (NATO ASI series. Series B, Physics ; vol. 307)
    "Published in cooperation with NATO Scientific Affairs Division."
    "Proceedings of a NATO Advanced Research Workshop on Negative
  Differential Resistance and Instablities in 2-D Semiconductors, held
  September 20-25, 1992, in Il Ciocco, Lucca, Italy"--T.p. verso.
    Includes bibliographical references and index.
    ISBN 978-1-4613-6220-3      ISBN 978-1-4615-2822-7 (eBook)
    DOI 10.1007/978-1-4615-2822-7
    1. Hot carriers--Congresses.  2. Semiconductors--Congresses.
  3. Transport theory--Congresses.  4. Electric resistance-
  -Congresses.  5. Gunn effect--Congresses.   I. Balkan, N.
  II. Ridley, B. K.   III. Vickers, A. J.   IV. North Atlantic Treaty
  Organization.  Scientific Affairs Division.   V. NATO Advanced
  Research Workshop on Negative Differential Resistance and
  Instabilities in 2-D Semiconductors (1992 : Il Ciocco, Italy)
  VI. Series: NATO ASI series.  Series B, Physics ; v. 307.
  QC611.6.H67N44   1993
  537.6'226--dc20                                      93-15664
                                                         CIP

---

Additional material to this book can be downloaded from http://extra.springer.com.

©1993 Springer Science+Business Media New York
Originally published by Plenum Press, New York in 1993
Softcover reprint of the hardcover 1st edition 1993

## PREFACE

Instabilities associated with hot electrons in semiconductors have been investigated from the beginning of transistor physics in the 1940s. The study of NDR and impact ionization in bulk material led to devices like the Gunn diode and the avalanche-photo-diode. In layered semiconductors domain formation in HEMTs can lead to excess gate leakage and to excess noise. The studies of hot electron transport parallel to the layers in heterostructures, single and multiple, have shown abundant evidence of electrical instability and there has been no shortage of suggestions concerning novel NDR mechanisms, such as real space transfer, scattering induced NDR, inter-subband transfer, percolation effects etc. Real space transfer has been exploited in negative-resistance FETs (NERFETs) and in the charge-injection transistor (CHINT) and in light emitting logic devices, but far too little is known and understood about other NDR mechanisms with which quantum well material appears to be particularly well-endowed, for these to be similarly exploited. The aim of this book is therefore to collate what is known and what is not known about NDR instabilities, and to identify promising approaches and techniques which will increase our understanding of the origin of these instabilities which have been observed during the last decade of investigations into high-field longitudinal transport in layered semiconductors.

The book covers the fundamental properties of hot carrier transport and the associated instabilities and light emission in 2-dimensional semiconductors dealing with both theory and experiment. The material is organised into subject areas that can be classified broadly into the following groups: review of negative differential mechanisms; hot electron transport; the effects of size quantization and band structure on hot carrier-phonon interactions, real space transfer and impact ionization; space charge waves; novel techniques for the study of instabilities; and devices. The chapters have been grouped together according to these classifications as closely as possible. However, although there is much overlap of ideas, each chapter is essentially independent of others.

The book is based on the contributions to the NATO, ARW entitled 'NDR and Instabilities in 2-D Semiconductors' which was held at Il Ciocco Conference Centre, Lucca, Italy between 20th and 25th September 1992.

We would like to acknowledge the sponsorship of NATO Scientific Affairs Division. Also, we would like to thank Prof. J.H. Wolter and Prof. E. Schöll for their help with the organisation of the workshop.

N. Balkan
B.K. Ridley
A.J. Vickers

December 1992

# CONTENTS

Negative Differential Resistance:
A Brief History and Review ........................................................ 1
B.K. Ridley

Electronic Transport in Semiconductors at High Energies:
Effects of the Energy Band Structure ............................................. 23
K. Hess

Theory of Oscillatory Instabilities in Parallel and Perpendicular Transport
in Heterostructures .................................................................. 37
E. Schöll

Light Emitting Logic Devices Based on Real Space Transfer in Complementary
InGaAs/InAlAs Heterostructures ........................................... 53
S. Luryi and M. Mastrapasqua

Electron Mobility in Delta-Doped Quantum Well Structures ............................. 83
W.T. Masselink

Application of a New Multi-Scale Approach to Transport in a
GaAs/AlAs Heterojunction Structure ............................................ 99
P.D. Yoder and K. Hess

Negative Differential Resistance, Instabilities and Current Filamentation
in $GaAs/Al_xGa_{1-x}As$ Heterojunctions ............................................. 109
J.H. Wolter, J.E.M. Haverkort, P. Hendriks and E.A.E. Zwaal

Hot Electron Instabilities in QWs: Acoustoelectric Effect and Two-Stream
Plasma Instability ............................................................... 127
R. Gupta, N. Balkan and B.K. Ridley

Hybrid Optical Phonons in Lower Dimensional Systems and their Interaction
with Hot Electrons ............................................................. 141
N.C. Constantinou and B.K. Ridley

Negative Differential Resistance in Superlattice and Heterojunction Channel
Conduction Devices ............................................................ 153
S.W. Kirchoefer

Electronic Transport in a Laterally Patterned Resonant Structure ........................ 171
M.C. Yalabik

Travelling Domains in Modulation-Doped GaAs/AlGaAs Heterostructures ............. 179
R. Döttling and E. Schöll

Negative Differential Resistance and Domain Formation in
    Semiconductor Superlattices ............................................................ 189
H.T. Grahn

High Frequency DC Induced Oscillations in 2D ............................................. 203
A.J. Vickers, E.S.-M. Tsui and A. Straw

Hot Electron Induced Impact Ionization and Light Emission in GaAs based
    MESFET's, HEMT's, PM-HEMT's, and HBT's ......................................... 215
C. Canali, C. Tedesco, E. Zanoni, M. Manfredi and A. Paccagnella

Negative Differential Resistance, High Field Domains and Microwave Emission
    in GaAs Multi-Quantum Wells .......................................................... 251
A. Straw, A. Da Cunha, N. Balkan and B.K. Ridley

On Negative Differential Resistance and Spontaneous Dissipative Structure
    Formation in the Electric Breakdown of p-Ge at Low Temperatures ........... 261
J. Peinke, W. Clauss, A. Kittel, J. Parisi, U. Rau and R. Richter

Dissipative Structures in Bistable Electronic and Optoelectronic
    Semiconductor Devices ................................................................. 269
R. Symanczyk, S. Knigge and D. Jäger

NDR, Hot Electron Instabilities and Light Emission in LDS ............................. 283
A. Da Cunha, A. Straw and N. Balkan

Light Emission and Domain Formation in Real-Space Transfer Devices ............... 305
T.K. Higman, J. Chen, M.S. Hagedorn and R.T. Fayfield

Hot Electrons in δ-Doped GaAs ............................................................. 317
M. Asche

Optical Phonon Modes in Semiconductor Quantum Wells and Superlattices ........... 335
M.P. Chamberlain

Plasmons on Laterally Drifting 2DEGs ....................................................... 351
H.P. Hughes, R.E. Tyson, L.C. O'Súilleabhán and R.J. Stuart

Temperature-Dependent Screening Calculation of Hot-Electron Scattering
    in Heavily Doped Semiconductors ..................................................... 361
K.O. Jensen, J.M. Rorison and A.B. Walker

Optical Measurements of Carrier Mobilities in Semiconductors using Ultrafast
    Photoconductivity ....................................................................... 373
V. Brückner

A Time-Dependent Approach for the Evaluation of Conductance
    in Two Dimensions ..................................................................... 385
K. Stratford and J.L.Beeby

Hierarchy of Current Instabilities in the Impact Ionization Avalanche
    in Semiconductors ...................................................................... 393
K. Aoki

Impact Ionization of Excitons and Donors in Center-Doped $Al_{0.3}Ga_{0.7}As$/GaAs
    Quantum Wells ........................................................................ 409
H. Weman

Hot Exciton Luminescence in Quantum Wells as a Spectroscopic Tool ................... 421
    F. Calle, C. López, F. Meseguer, L. Viña, J.M. Calleja and C. Tejedor

Combined Quantum Mechanical-Classical Modelling of Double Barrier
    Resonant Tunneling Diodes ........................................................... 431
T.G. Van de Roer

Index ........................................................................................ 439

# NEGATIVE DIFFERENTIAL RESISTANCE:
# A BRIEF HISTORY AND REVIEW

B.K. Ridley

Department of Physics
University of Essex
Colchester  CO4 3SQ
U.K.

## 1.  HISTORY OF N.D.R.

The simplest considerations concerning power consumption in an electrical circuit soon leads to the idea that if the circuit contained an element with a negative differential resistance (NDR) it would be possible to create a.c. power rather than consume it.  Such an element would have highly non-linear electrical properties, but it was well known from the study of dielectric breakdown that large non-linearities were common in insulators at high fields. The cause of breakdown was seen to have its origin in either a thermal runaway through excessive Joule heading or in purely electrical effects.  The latter were infinitely more interesting since thermal breakdown was often synonymous with melting.  The first theory by Zener in 1934[1] described breakdown in terms of quantum-mechanical tunnelling between valence and conduction bands, but the critical fields predicted proved to be significantly larger than those observed in the alkali halides.  It took the invention of the Zener diode for this theory to come into its own, though, ironically in its role as a switch, the breakdown, in practice, is frequently a consequence of avalanche rather than tunnelling.  To compound the irony, pure Zener tunnelling has been exploited in the "tunnel diode" which is famous for its NDR.  Other mechanisms of breakdown emphasized impact ionization and subsequent avalanche effects, phenomena familiar from the studies of breakdown in gases.  Theories of how electrons in a polar solid could pick up more energy from the field than they could  dissipate in a single collision were developed by von Hippel[2] and Fröhlich[3,4] and subsequently elaborated by Callen[5] and Fröhlich and Paranjape.[6]  This was really the beginning of the study of hot electrons, and a parallel study by Davydov[7,8] and others in the USSR on the effect of a field on the distributed function of an electron and its mobility in a semiconductor helped to lay the theoretical foundations of the topic.

*Negative Differential Resistance and Instabilities in 2-D Semiconductors*
Edited by N. Balkan *et al.*, Plenum Press, New York, 1993

1

The advent of the transistor in the mid-forties triggered off research into the properties of hot electrons[9] which has continued to the present day. The first experiments by Ryder[10] demonstrated an interesting fall in mobility of electrons in Ge with increasing field (Fig.1). It was Krömer[11,12] who advanced the first idea for an NDR mechanism. This was based on the form of the degenerate valence band structure of Ge and allied semiconductors which suggested that at high fields holes would be driven into regions where their effective mass was negative, and above a critical field the differential mobility would be negative. Unfortunately, it turned out that the number of holes which could be induced into occupying negative-mass states was always far too small and the differential mobility remained obstinately positive. Nevertheless, the general idea of an NDR led Shockley[13] to examine the behaviour of NDR devices and he concluded that because of electron injection at the cathode, necessary to preserve current continuity, a two-terminal system would always exhibit a d.c. positive resistance, and only at high frequencies would the properties be favourable. He also considered transit-time NDR, of which more later.

What was needed was a viable mechanism for getting an NDR. In 1959 Krömer visited the Mullard Research Laboratory at Redhill in England and talked about his negative-mass amplifier. Tom Watkins and I were in the audience and were stimulated to come up with two mechanisms for NDR[14,15] both of which have since been observed – one involving the intervalley transfer of electrons, the other involving the field-induced capture of electrons by negatively-charged impurities. Experimental verification of the electron-transfer mechanism was attempted by Tweedale and Watkins[16] in strained p-type Ge, since Ge was the only high-quality semiconductor available at the time, but this was unsuccessful. But Hilsum[17] presented a simple analysis which showed that n-type GaAs would exhibit an electron-transfer NDR at fields above 3kV cm$^{-1}$. The capture NDR was successfully observed in Sb-compensated Au-doped Ge by Ridley and Pratt.[18-20] In addition to the field-induced capture of electrons directly observable by the current decay following the onset of a voltage pulse, there appeared an instability in the form of a slow-moving high-field domain.

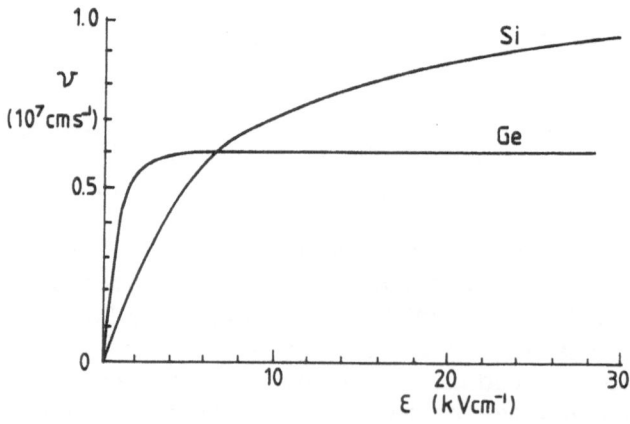

Fig.1  Velocity saturation of n-type Ge and Si

Krömer had already seen that an NDR would promote an electrical instability, a feature of NDR I first became aware of through conversations on hot electrons with Reik, and Adawi had also noted this.[21] Fortunately I was not aware of Shockley's conclusion concerning the unobservability of NDR at zero frequency - ignorance being bliss in this case - and deduced that the form of instability would be a propagating high-field domain in the case of a voltage-controlled NDR and a high-current filament in the case of a current-controlled NDR,[22] (Fig.2). At the time Pratt and I believed that the domains in Ge: Au: Sb were the first to be observed, but at the Semiconductor Conference in Paris in 1964 we discovered that Böer and co-workers had been observing stationary domains in CdS for a number of years,[23,24,25] which they attributed to a field-quenching mechanism involving several band-impurity transitions. Furthermore, slow domains had been observed in semi-insulating GaAs by Barraud[26] and by Northrup and co-workers[27], but all of this paled into technological insignificance in the face of Ian Gunn's discovery of high-frequency current oscillations in n-type GaAs.[28] I remember trying to persuade Ian that this was caused by the electron-transfer mechanism plus propagating high-field domains, but he was convinced that the effect was not associated with hot electrons for the apparently very good reason that his measurement of noise temperature had yielded the room-temperature value. It was only when the field structure of a high-field domain was understood - namely that most of the specimen was in a low-field region for most of the time - that the apparent discrepancy between the theory of the electron-transfer mechanism and the results of noise measurements was resolved. Krömer showed conclusively that the Gunn effect in GaAs was caused by the electron-transfer mechanism[29] as Hilsum[17] had predicted, and he went on to demonstrate and to resolve, thereby, another conflict, that for propagating high-field domains to form the product of electron density and specimen length - the so-called $n\ell$ product - had to exceed roughly $10^{12} \text{cm}^{-2}$, otherwise the field-distribution became that predicted by Shockley - a stationary domain at the anode. Modes of operation at high frequencies with limited space-charge accumulation were shown to be possible by Copeland.[30]

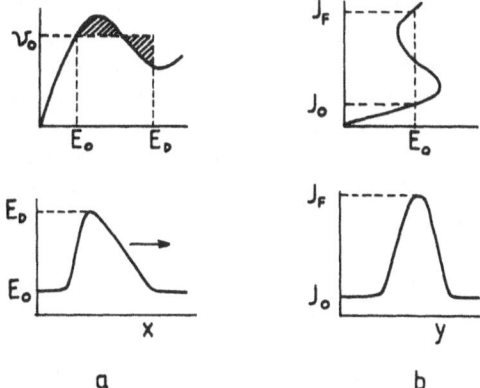

Fig.2    The two types of NDR and associated product of instability.
(a) Field-controlled NDR and domain formation - the shaded areas are approximately equal.
(b) Current-controlled NDR and filament formation

The transferred-electron device (TED) is now well established. Operating frequencies reach to over 100 GHz, pulse powers attain 6 kW, with efficiencies as high as 30%, and they are in wide use in microwave systems. An analysis of TED materials GaAs, InP, GaSb, InAs and InSb, shows that, surprisingly, the truly important material parameter is the mass ratio of the two ions in the binary system since this determines the ratio of zone-centre LO and zone-boundary LA phonon energies.[31] A large LO phonon energy enhances the scattering rate and thus allows high frequency operation, and a small LA phonon energy means that a large phonon population is available at room temperature to keep the mobility of electrons in the upper valleys low. High ionicity also helps to enhance the scattering rate in the central valley, but variations in band structure are relatively unimportant. Thus InP emerges as the favourite material, being the most ionic and having the largest mass ratios of well-researched materials.

Low frequency voltage-controlled NDR was also commonly observed in amorphous oxides as Hickmott reported in 1962 but here the picture is much darker.[32,33] The oxide film is "formed" by applying a voltage above some critical value. Before being formed it is an insulator, afterwards it switches to a conducting state which exhibits an NDR and electroluminescence.[34,35] Models to explain this type of NDR involve the creation of space-charge with conduction via impurity states or via a combination of electronic and ionic motion. Another model involves ohmic conduction along filaments physically different from the host matrix which can fracture. These models are reviewed by Dearnley, Stoneham and Morgan.[36] They warn, correctly, that there are as many models as authors, but their model involving filaments has proved reasonably viable.

Space-charge was a common ingredient of models which describe current-controlled NDR associates with electrical breakdown in insulators. Electrons are provided at the cathode by field-emission, usually described by Fowler-Nordheim tunnelling theory[37-39] taking into account field-enhancement via cathode projections.[40] In the mechanism suggested by O'Dwyer[41,42] impact ionization produces mobile holes which drift towards the cathode and enhance the field, thereby inducing a runaway avalanche. In another mechanism, mobile positive ions (protons, hydrated protons or sodium ions, for example) are thermally activated by the injected current and drift towards the cathode enhancing the field there.[43] Impact ionization of impurities in insulators can also induce a current-controlled NDR under the right circumstances, even though the resultant positive charges are immobile. In this case impact ionization is spatially localized in planes more or less equidistantly spaced, analagous to the striations produced in the breakdown of a gas. Current-controlled NDR appears if the distance between cathode and anode satisfies quasi-periodic conditions, set by the ionization wave.[44,45]

The mechanisms for NDR in insulators are usually not applicable to bulk semiconductors. They have been mentioned in this account for two reasons. One is that heterostructures contain elements which are semi-insulating and it is therefore prudent to be aware of theories of instabilities in insulators in the interpretation of instabilities of quantum-well structures. The other reason is that it demonstrates that two different categories of NDR exist, namely, **intensive** NDR, in which the NDR arises as a bulk effect in a uniform electric field, and **extensive** NDR, in which space-charge and its interaction with electrodes create non-uniform conditions without which the NDR would not occur. Intensive NDR is essentially field-controlled, whereas extensive NDR is voltage-controlled and dependent on the dimensions of the specimen.

In what follows the focus will be on intensive NDR produced electronically rather than thermally. As we have seen the physics of intensive NDR began in the 'fifties with the research of Krömer. We will mostly be concerned with NDR which operates from d.c. to high frequencies, but we will also mention the transit-time mechanism which only works at high frequencies.

The principal landmarks in the development of the study of NDR and instabilities in bulk semiconductors are listed in Table 1 (see end of paper).

## 2. MECHANISMS FOR INTENSIVE NDR IN BULK SEMICONDUCTORS

Here we review very briefly the principal mechanisms giving rise to intensive NDR in bulk semiconductors, dealing first with **field-specific** NDR in which the current-density/field characteristic is N-shaped, and then with **current-density-specific** NDR in which the current-density/field characteristic is S-shaped (Fig.3). The order will be roughly in decreasing generality, rather than historical.

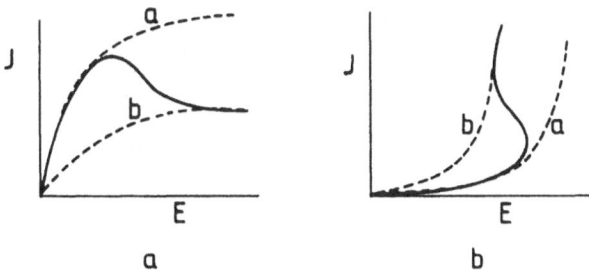

Fig.3    General pattern of NDR mechanism: (a) field-controlled  (b) current-controlled. A field-induced transition occurs between states a and b

### 2.1  *Scattering-induced NDR*

NDR can occur in some cases when there is more than one scattering mechanism. Hilsum and Welborn[46] pointed out that mixed polar LO and acoustic phonon scattering could give rise to an NDR in a parabolic band. Above a critical field polar LO scattering cannot maintain an energy balance and a transfer to deformation-potential scattering via acoustic modes occurs. If the relative strengths of polar and deformation-potential coupling lie within certain limits the transfer from dominant LO to acoustic-phonon scattering results in an NDR.[47]

## 2.2 *Non-parabolicity*

Increase of effective mass with energy occurs quite generally and tends to reduce the mobility of electrons (we will usually speak of electrons but our remarks will be applicable to holes) through both the increase in inertia and the increased scattering rate consequent on the increased density of states per unit energy. In non-polar semiconductors the drift velocity of electrons is predicted

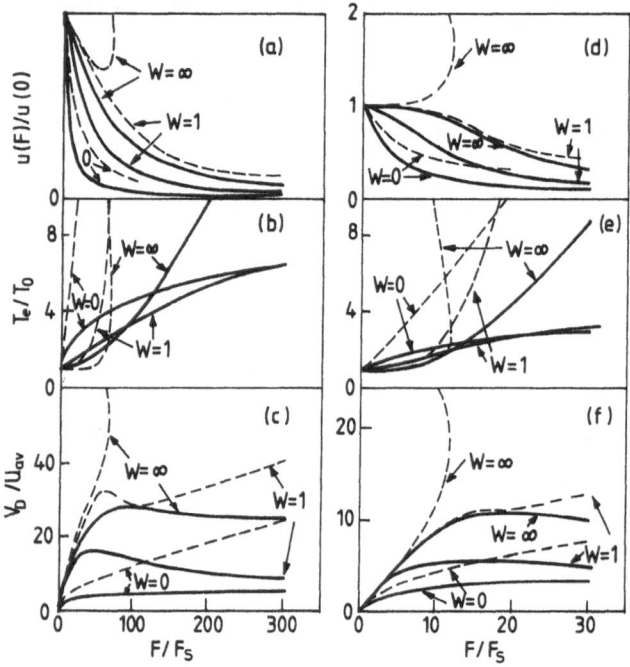

Fig.4  Variation with applied field $F/F_S$ of (a,d) mobility $\mu(F)/\mu(0)$, (b,e) electron temperature $T_e/T_0$, (c,f) drift velocity $v_0/u_{av}$ ($u_{av}$ is the average sound velocity) for different combinations of acoustic and polar optical scattering, as given by $W = \mu_{po}^o/\mu_{ac}^o$. Figs. (a), (b), (c): $\lambda_0 = \hbar\omega/k_B T_0 = 2$; (d), (e), (f) $\lambda_0 = 0.5$. The dashed lines are for the parabolic case, the continous for the non-parabolic case with $k_B T_0/E_g = 0.1$. A Maxwellian distribution is assumed (Harris and Ridley[87])

to saturate in a parabolic band as a consequence of optical-phonon emission, and clearly this saturation will give way to a reduction with field when the band is non-parabolic, and hence an NDR. This also can occur in the case of polar LO scattering,[47,48] so that non-parabolicity of quite modest proportions can prevent runaway, though only at the expense of an NDR instability (Fig.4).

## 2.3 *Electron-transfer*

This mechanism relies on the electron transferring when hot from a light-mass valley to a heavy-mass valley[14] (Fig.5a). In a sense it is a special case of non-parabolicity NDR in which there is a sudden increase in the density of states at a certain energy. Favourable band-structures abound among the III-V and II-VI semiconductors where the effect is well-known and needs no further elaboration here except to remark that it masks, or perhaps incorporates is a better description, the NDRs predicted as a consequence of mixed scattering and single-valley non-parabolicity. In Ge the effect involves the transfer from the

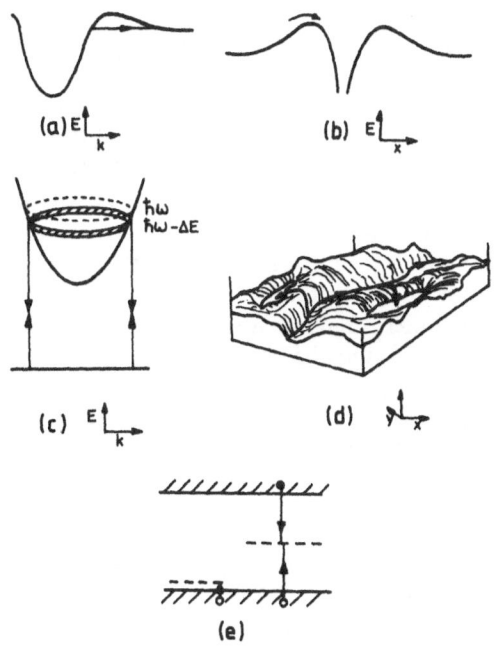

Fig.5    Mechanisms for field-controlled NDR.
(a) Intervalley transfer  (b) Impurity barrier  (c) Delta-function excitation
(d) Percolation    (e) Field-emission of holes

set of 4 <111> valleys to the upper-lying set of 6 <100> valleys where there are not only more states to scatter into but also more intervalley scattering. If a suitable band-structure does not exist naturally it can sometimes be created by applying stress to the material. Thus a uniaxial stress applied to Ge splits the degenerate light and heavy hole bands into two mutually orthogonal spheroidal bands one lying above the other in energy. Applying the field along the light-mass direction of the lower valley causes transfer of holes to the upper valley where their conductivity mass is greater.

### 2.4 *Negative mass*

One of the most intriguing aspects of solid-state physics stemming from the quantum properties of an electron is the natural tendency of an electron in a periodic potential to execute spatial oscillations under the accelerating influence of a d.c. electric field. These so-called Bloch oscillations stem from the presence of Bragg reflections and are associated with regions of energy in a conduction band where the effective mass of an electron is negative. Such regions of negative mass occur in the heavy-hole band near the zone-centre in certain crystallographic directions, and if a sufficient density of holes could be induced to inhabit these states at high fields an NDR would result.[12] Unfortunately scattering is so strong, in general, that an NDR based on Bloch oscillations is difficult to achieve.

### 2.5 *Optically-induced NDR*

Optical excitation of electrons from a well-defined level such as that provided by an impurity to a correspondingly well-defined ring of levels in a conduction band by the action of monochromatic light can produce a quasi-delta-function distribution provided that recapture occurred much faster than scattering (Fig.5c). If the level to which the electrons were excited lay just below the threshold for optical phonon emission the result of applying a field would be, effectively, to switch on a potent scattering mechanism as accelerated electrons reach that threshold. Barker and Hearn[49] predicted that under the right conditions the result would be an absolute negative resistance. Unfortunately the rate of capture is usually small compared with that of scattering and the distribution function produced is far from the required form. However, using two intense laser lines does produce in theory markedly non-Maxwellian distribution functions in which population inversion within the band occurs leading to instability.[50]

### 2.6 *Magnetically-induced NDR*

In a quantizing magnetic field such that all electrons reside in the lowest Landau level, transport occurs via scattering between spatially separated magnetic states. Because scattering rates are proportional to the density of final states and because the density of states in this quasi-1D system decreases with increasing energy, a single carrier tends to migrate "upstream" towards higher voltages in the presence of a field[51] (Fig.6). This negative mobility effect is usually eliminated when the difference between the occupation probabilities of the initial and final states is taken into account, but for quasi-delta-function distributions this would not be the case.

### 2.7 *Percolation NDR*

In semi-insulating material the presence of substantial potential fluctuations, due to impurity clustering for instance, which in conducting material would be screened out, can reduce transport to a few percolation paths. In the case of thermionic emission across such material it is notorious that a substantial difference is usually found between the theoretical and observed value of Richardson's constant, and this may be explained by the conducting area of the barrier being smaller than the geometric area. At high fields it is expected that the higher energy of the carriers will allow percolation paths to expand and conductivity to rise. But it is also possible for more carriers to be excited into non-conducting regions where they become effectively immobile. If the latter effect dominates an NDR would ocur[52] (Fig.5d).

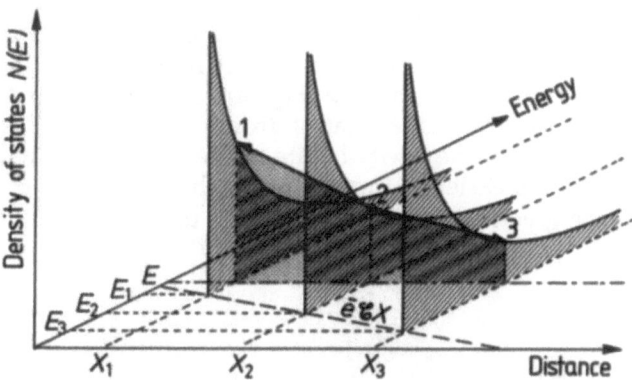

Fig.6   Density of states in a quantizing magnetic field. In a field an electron tends to hop upstream where the density of states is larger

### 2.8   *Field-enhanced capture*

Several mechanisms exist under this heading. The impurity barrier effect[15] is associated with any potential barrier to capture surrounding an impurity such as the Coulomb barrier of an oppositely charged centre, or the deformation barrier of a centre whose radius is larger (or smaller, depending on the sign of the deformation potential) than the host atom (Fig.5b). At zero field the population of mobile electrons derived by thermal ionization of the impurity levels in question is in thermal equilibrium with the population in the levels. Heating the electrons (but not the lattice) enhances the capture rate over the generation rate (provided the field is not so high as to cause impact ionization of the levels) and thus reduces the population of mobile electrons. Previously described mechanisms were all to do with reducing the mobility of electrons. In field-enhanced capture the mobility change, if it occurs, is not central to the effect, but, instead, the reduction in mobile electron density.

Another conceivable mechanism for field-enhanced capture is associated with the parity of the electron in the band relative to that in the impurity, in a radiative transition. An example of a parity barrier to radiative recombination is an electron in a $\Gamma_1$-valley with s-orbital parity captured by the s-orbital ground state of a donor. Increasing the energy of the electron means moving the electron into states with more p-orbital character through k.p mixing and therefore enhancing the capture probability. This can happen more potently in an intervalley transfer, from $\Gamma$ to $X$, for example, where the final state has more p-orbital characters.[53]

When electrons and holes are generated by light a quite different mechanism can occur which enhances the capture of electrons at high fields.

This is the field-emission of holes. For this to work holes must be captured very rapidly, and the rate at which electrons recombine has then to be crucially determined by the emission rate of holes. If the hole traps are not too deep the emission rate can be increased by field-emission - either via tunnelling or impact ionization - and hence the photoconductivity is reduced[25] (Fig.5e).

## 2.9 *Two-level impurity NDR*

Several mechanisms involving impact ionization of impurities exist for a current-controlled NDR. An example is that involving an impurity centre with a ground state and an excited state (Fig.7a). Impact ionization and capture into the excited state plus impact excitation from the ground to the excited state increases the population of the excited state at the expense of the ground state. Since the excited state is easier to impact ionize than the ground state the field required is less for a given current, corresponding to an S-type NDR.[54]

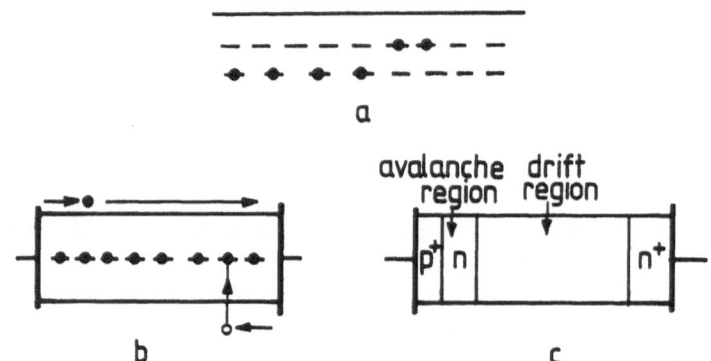

Fig.7    Mechanisms for current-controlled NDR and transit-time NDR.
(a) Two-level impact ionization (b) Double injection (c) Read diode

## 2.10 *Double injection*

When the system is such that the current is essentially determined by the injection of electrons at the cathode and injection of holes at the anode - and this can be realized if the semiconductor is semi-insulating - a current-controlled NDR can arise if there exists a recombination centre completely occupied with electrons at equilibrium (Fig.7b). In such a case injected electrons cannot be trapped, but holes are trapped very rapidly. At sufficiently high currents the injected holes can be present in high enough concentration so that the occupation of the recombination centre is changed to make electron and hole lifetimes equal. Provided that the cross-section for capturing holes, $\sigma_p$, is much greater than that for electrons, $\sigma_n$, the high-injection case corresponds to the recombination centre being virtually full of holes, and increasing the current can proceed with much less voltage.[55]    Although space-charge is involved it is not crucial to the NDR.

The basic cause of NDR in this case is the transition from what is effectively one-carrier injection to two-carrier injection. This can also occur when, for example, the injection of one type of carrier is blocked until the field at the blocking contact is high enough either to induce tunnelling or to break down the material via impact ionization near the contact. The I/V characteristic then jumps from the single to the two-carrier curve, giving a current-controlled NDR. Light emission at a contact can also lead to NDR.[56]

### 2.11 *Transit-time NDR*

The mechanisms for NDR so far discussed are operative at zero frequency, but there exist a number of devices such as the IMPATT (impact ionization avalanche transit time) diode, that operate via an NDR which appears at microwave frequencies, and which provide the most powerful solid-state sources of microwave power. These devices are all variants of the diode proposed by Read[57] in 1958, following Shockley's consideration[13] of transit-time devices in 1954. Basically the diode consists of a high-field region in which injected carriers produce an avalanche by impact ionization, followed by a low-doped drift space with sufficiently high field to cause the electrons generated by impact ionization to drift at their saturation-drift velocity (Fig.7c). There are two phase delays of importance: one is that associated with injection into the drift region, the other is associated with the transit time. With suitable design these phase delays can be engineered to make the real part of the impedance negative.

### 3. INSTABILITIES

As mentioned previously a material exhibiting NDR is electrically unstable. Instability, however, is not restricted to the presence of an NDR, and it is often useful to be aware of low-field mechanisms which can give rise to phenomena which may be wrongly attributed to the presence of an NDR. Here we mention the principal instabilities which can arise in bulk semiconductors.

### 3.1 *NDR instabilities*

Voltage-controlled NDR manifests itself in the formation of a high-field domain, which can be propagating or stationary (Fig.8). When transverse drift and diffusion is unimportant a propagating domain can form near the cathode (assuming electrons to be the majority carrier) provided sufficient space charge can be accumulated i.e.. provided Krömer's $n\ell$ product is exceeded, otherwise a stationary domain forms at the anode (following Shockley's description). A propagating domain generates a current spike every time it vanishes at the anode and thus the period of the current oscillations is determined by the transit time, which can be influenced by trapping[58] (Fig. 9). A special case can occur when the domain velocity, reduced by trapping, approaches the velocity of a piezoelectrically active acoustic wave in the direction of propagation. In this situation it is predicted that the domain induces large acoustic strains.[59] When trapping is important the criterion for domain formation involves an

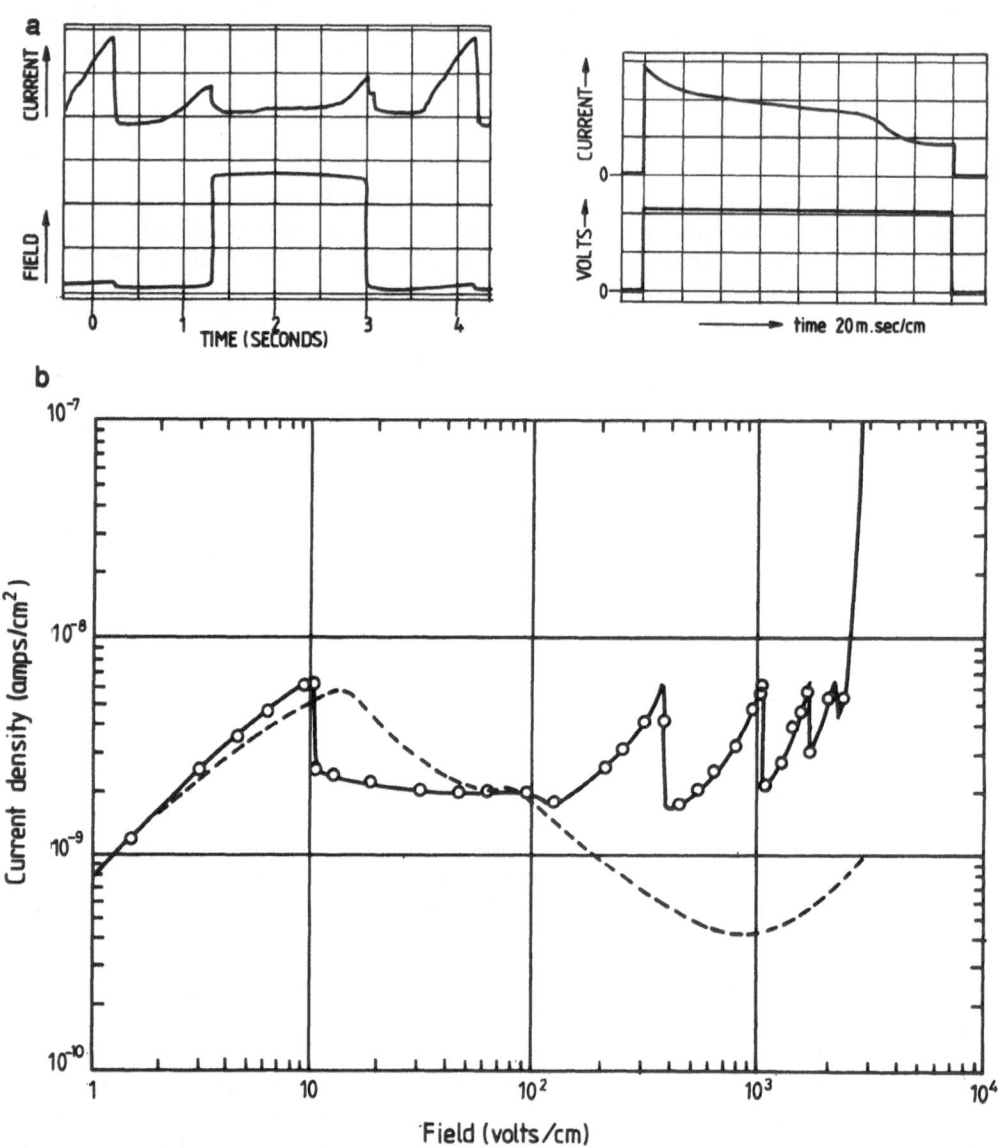

Fig.8    Impurity-barrier domains:  (a) Static- the dashed line is the measured initial current
reduced by the ratio of capture times. $\tau$ (F)/$\tau$(0) ;  (b) Time-dependence of current and
voltage between two probes in the presence of a propagating domain (Ridley and
Pratt[88])

N$\ell$ product, where N is the density of free and trapped carriers. Domain formation is also dependent on the transverse dimensions. Kino and Robson[60] showed that for a thin slab of thickness d the growth rate of a space-charge wave of wavevector $k_x$ in the plane was reduced by a factor $[1 + (2/k_xd)]^{-1}$ as a consequence of field spreading, which led to an nd product for GaAs of $1.6 \times 10^{11} cm^{-2}$ that had to be exceeded before domains could form. A simple model of a propagating domain was advanced by Butcher[61] who showed that the domain travelled with the drift velocity of electrons at the field outside the domain, and that this field was related to the maximum field inside the domain via an equal-areas rule applied to the current-density/field characteristic. An equal-areas rule for filaments associated with current-controlled NDR has been given by Schöll.[62]

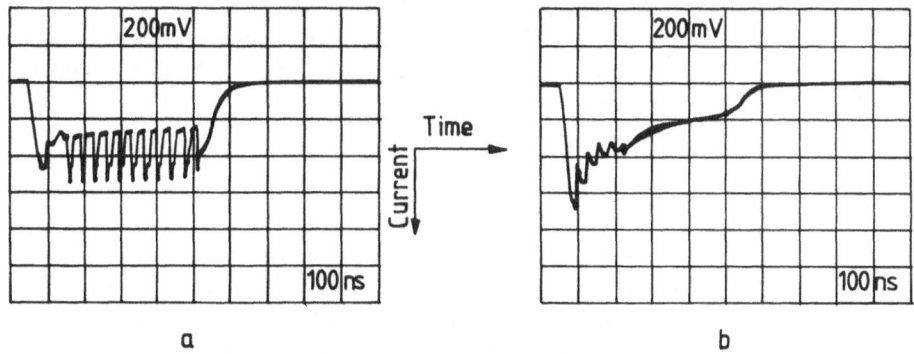

Fig.9    Fast and slow domains in GaAs (T = 160 K).
(a) Gunn oscillations in dark  (b) Under illumination Gunn domains propagate during the incubation of a slow domain. (Leach and Ridley[89])

## 3.2   Recombination waves

When there is a recombination level deep in the forbidden band of a semiconductor such that the trapping rates of electrons and holes are independent of one another, and when the concentration of minority carriers is high enough, a recombination instability can develop in the presence of a modest applied field.[63,64] The effect depends on the ratios $p_0/n_0$ are the equilibrium densities of holes and electrons) and $\upsilon_n/\upsilon_p$ (the ratio of capture rates for electrons and holes). In n-type material slow waves with velocity $v_p(p_0/n_0)$, where $v_p$ is the drift velocity of holes, exist if $(\upsilon_n/\upsilon_p) < (p_0/n_0) \ll 1$ and the field is approximately high enough for the drift length and diffusion length of the shorter lived carrier to be equal. Fast waves with velocity $v_p$ occur for $(\upsilon_p/\upsilon_n) < (p_0/n_0) \ll 1$. Recombination instabilities were reported in gold-compensated n-type Si.[65]

## 3.3   Acoustoelectric domains

At fields sufficiently high to produce supersonic drift velocities carriers begin to excite acoustic waves and an instability develops when the creation rate exceeds the loss rate in the cavity. The loss rate for acoustic phonons rises with frequency, so low frequencies are favoured. The deformation-potential interaction, however, is strong only at high frequencies and it is easier to achieve amplication via the piezoelectric interaction, which is intrinsically strong at low frequencies. Free-carrier screening becomes important, therefore, since the interaction with very low frequencies is screened out. There exists a frequency for a given carrier concentration at which the gain is maximum,[66,67] lying typically between 100 MHz and 10 GHz. When conditions are right the sample becomes the microsonic analogue of the laser.[68] As acoustic flux builds up the electrical resistance of the sample increases, which tends to limit the build up. One possible end point is the appearance of an acoustoelectric domain - a region of broad-band incoherent acoustic flux parametrically down-converted in frequency,[69] which travels at the velocity of sound and drops an appreciable fraction of the voltage. The carrier density must lie between a lower and upper limit. If too low the gain is too small for acoustoelectric effects to materialize in a single transit; if too high, the frequency for maximum gain is so high that lattice losses become very large, necessitating high applied fields which move the system into hot-electron regimes.

### 3.4 *Other instabilities*

The acoustoelectric effect is an example of energy being transferred from a drifting electron to an excitation of the crystal when the drift velocity exceeds the phase velocity of the excitation. In principle, optical phonons can be excited in the same way[70] but, typically, drift velocities around $10^7 cm^{-1}$ are required, and hot electron effects often intervene. The same is true of plasmons, though in the case of two-stream instabilities acoustic-like plasma waves occur and can more easily be excited.

### 4. NDR IN MICROSTRUCTURES

All mechanisms for bulk NDR are applicable to microstructures. Indeed, the study of superlattices was first motivated by the possibility of observing Bloch oscillations in transport perpendicular to the layers i.e. "vertical" transport.[71] One modification regarding transport parallel to the layers i.e. "parallel" transport, is the influence of field-spreading on the development of a domain. In the bulk, the growth rate (neglecting diffusion) is given by $|\sigma|/\varepsilon$ where $\sigma$ is the (negative) differential conductivity and $\varepsilon$ is the permittivity, whereas in 2D it is given by $|\sigma_s| k_x/2\varepsilon$,[72,73] where $\sigma_s$ is the (negative) differential surface conductivity and $k_x$ is the in-plane wavevector, (Fig.10). (We note in passing that with $|\sigma_s| = |\sigma| d$, where d is the thickness of the layer, the 2D rate is exactly the reduced bulk rate obtained by Kino and Robson[60] in the limit d→0.)

New mechanisms for NDR in parallel transport include the following (Fig.11). An example of oscillations which have been observed is given in Fig.12.

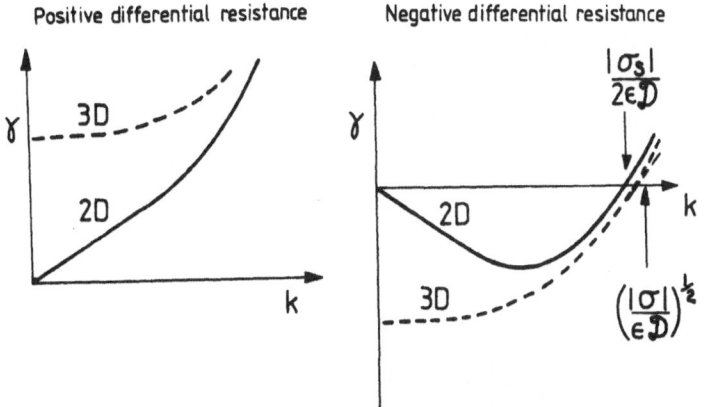

Positive differential resistance        Negative differential resistance

Fig.10  Attenuation rate of space-charge waves with wavevector k in 3D and 2D

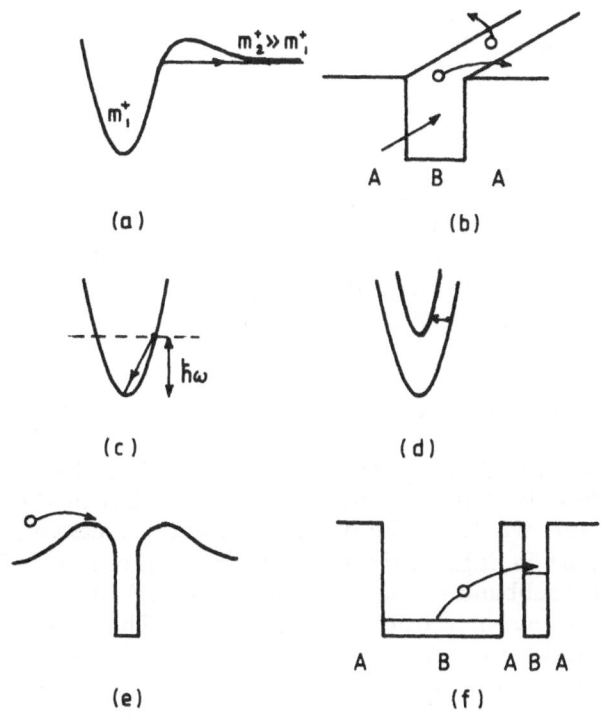

(a)

(b)

(c)

(d)

(e)

(f)

Fig.11  Some mechanisms for NDR in microstructures.
(a) Intervalley transfer  (b) Real-space transfer  (c) Delta-function excitation
(d) Intersubband scattering  (e) Impurity-barrier capture  (f) Tunnelling real-space
transfer

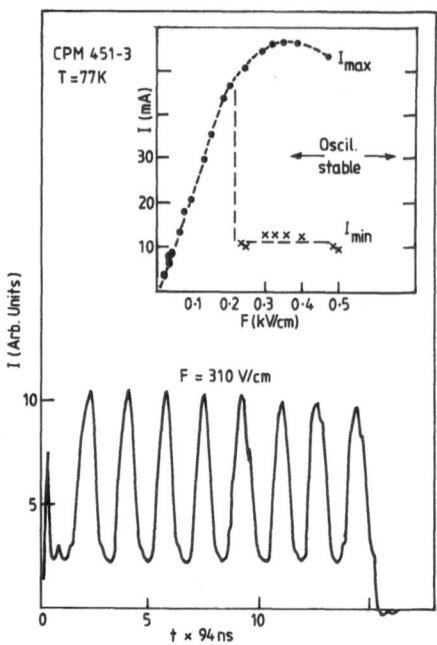

Fig.12 Current oscillations in a heavily-doped GaAs/Al$_{.45}$Ga$_{.55}$As single quantum well sample (Balkan and Ridley[90])

## 4.1 *Real-space transfer*

Thermionic emission of hot electrons over the quantum-well barrier will produce an NDR if the mobility in the well is much greater than that in the barrier.[74] Emission through the barrier from a wide well to a narrow well can also lead to NDR under the right circumstances.[75]

## 4.2 *Scattering-induced NDR*

In bulk semiconductors stable transport of hot electrons in a parabolic band is possible only if polar scattering is absent, but in the 2D case no such stability is possible whatever the phonon scattering mechanism. Thus, 2D hot electron transport is inherently unstable.[76] A more specific mechanism is provided by intersubband scattering. Electrons heat up in the ground state in which the scattering is purely intrasubband. The onset of intersubband scattering and more intense scattering via charged impurities in the upper subband can reduce the mobility sufficiently to give an NDR.[77] Another specific mechanism exploits the finite density of states at the bottom of the lowest subband, which leads to an abrupt scattering threshold for optical phonon emission. This leads to an NDR only if the electron density is strong enough to randomize the energy below the threshold but not above.[78,79] In GaAs this implies a carrier density of no more than about $10^{11}$ cm$^{-2}$. Scattering-induced NDR is intrinsically a high-speed effect with a characteristic time-constant of 1 ps or less.

## 4.3  *Parallel conduction in barriers*

The more complex connection between the contacts and the wells and barriers provides new possibilities for instability.[80]  The key phenomenon is the opening of a conducting channel in the barrier material running parallel to the quantum well.  Thus trapping of injected carriers and impact ionization of donors in the barrier become possible, and these processes can affect the electron density in the well via space-charge effects.

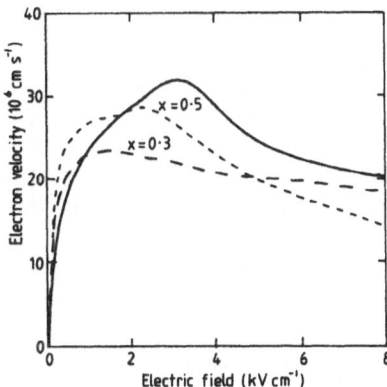

Fig.13  Velocity-field curves at 77 K in GaAs/$Al_xGa_{1-x}As$ measured at 35 GHz.  The continuous line is for bulk GaAs, the dashed lines for two single-heterojunction structures with x = 0.3 (n = 4.1 x $10^{11}$ cm$^{-2}$) and x = 0.5 (n = 3.2 x $10^{11}$ cm$^{-2}$) (Masselink et al.[84])

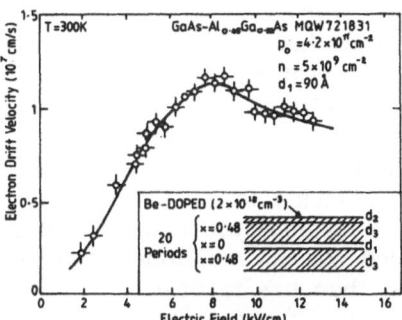

Fig.14  Velocity-field curve at 300 K for minority electrons in a GaAs (90Å)/$Al_{0.48}Ga_{0.52}As$ MQW (p = 4 x $10^{11}$ cm$^{-2}$,  n = 5 x $10^9$ cm$^{-2}$).  (Höpfel et al[85])

Only real-space transfer has been observed and exploited.[81]  In real-space-transfer transistors the electrical behaviour has proved to be remarkably complex, but there have been many reports of oscillations in the frequency range 100 MHz – 1 GHz which are largely unexplained.  Some evidence for acoustoelectric effects have been reported on electron densities in GaAs/AlGaAs around $10^{11}$ cm$^{-2}$;  which is near the theoretical cut-off.[83] Negative differential mobilities have been observed in a GaAs/$Al_xGa_{1-x}As$ single heterojunction with x = 0.3 at 35 GHz by Masselink et al[84] above

2 kV cm$^{-1}$ (Fig.13), and with x = 0.48 in a 90Å well at fields above 8 kV cm$^{-1}$ for minority carrier electrons by Höpfel et al[85] (Fig.14) but, again, the origins are not evident. In some cases there is copious band-gap light emission,[86] and while there is evidence of field inhomogeneities developing at high fields there is none that suggest the formation and propagation of domains. It is to be expected that real-space transfer, in allowing a degree of transverse diffusion, will act to inhibit the formation of domains.

The study of instabilities in microstructures has attracted the interest of chaotologists, and there is an increasing number of papers which view the non-linear phenomena which lie at the base of the topic in the general terms of chaos theory.

As regards the basic physics, there are several elements which arise from the confinement of excitations and the idiosyncrasies of microstructures which make the situation different from the bulk. Apart from the obvious quantization of the electronic motion, there is the affect of confinement on the vibration spectrum. This causes the acoustic phonon spectrum to be folded and the optical branch to exhibit confinement. In addition, the optical modes become hybridized through the necessity to satisfy elastic and electromagnetic boundary conditions at every interface, and the envelope functions which describe the vibration patterns arise from the triple coherent combination of LO, TO and interface polaritons (IP) in polar heterostructures, or the double coherent combination of LO and TO modes in non-polar heterostructures. The presence of interface polaritons in polar material means that an electromagnetic interaction is added to the usual Fröhlich interaction, the consequences of which are currently being researched. While confinement favours a weaker Fröhlich interaction, the proximity of interfaces suggests at first sight, a stronger interaction with interface polaritons, but this turns out not to be true in all cases. Specifically, it is not true when there is a large mechanical mismatch and sufficient LO bandwidth to hybridize with IP modes of all frequencies between those of the zone-centre LO and TO modes. Hybridization also affects the lifetime and scattering of optical waves and hence the rôle of hot phonons (one should say hot hybridons) in transport. Another common feature is the high density of carriers in most experiments - usually necessary in order to carry out the experiment, given the small volumes of typical microstructures. Plasma modes have to be taken into account in determing the energy and momentum relaxation of carriers by transforming the LO modes into coupled LO/plasma modes which are then hybridized with TO and IP modes. One then has to deal with quadruple hybrids in polar microstructures, triple hybrids in non-polar microstructures. Again, these are current topics of research.

It is clear there are many things we do not understand about NDR and other instabilities connected with parallel transport in microstructures. One would like answers to the following questions.

1. Why are there no travelling domains?
2. Does scattering-induced NDR exist?
3. Is there an acoustoelectric effect?
4. How does optical-phonon hybridization affect hot-electron transport?

These and other questions are likely to occupy out attention for some time to come.

**Table 1.** NDR Chronology 1951-1964

| Date | Name | Topic |
|------|------|-------|
| 1951 | Shockley | Hot electrons in Ge, theory. |
| 1953 | Ryder | Hot electrons in Ge, experiment. |
|      | Krömer | Negative-mass mechanism. |
| 1954 | Shockley | Transit-time NDR and stationary domains. |
| 1958 | Read | Transit-time NDR diode. |
| 1959 | Böer | Field-quenching NDR in Cds. |
|      | Stafeev | Double injection. |
| 1961 | Ridley and Watkins | Intervalley transfer and impurity barrier NDRs. |
| 1962 | Hutson and White | Acoustoelectric effect. |
|      | Hickmott | Low-frequency NDR in oxides. |
|      | Hilsum | Transferred-electron effect in GaAs. |
| 1963 | Ridley and Pratt | NDR domains in Au:Sb:Ge. |
|      | Ridley | Domains and filaments. |
|      | Gunn | Gunn effect. |
|      | Barraud | NDR in semi-insulating GaAs. |
| 1964 | Krömer | Gunn effect = intervalley transfer, $n\ell$ product. |
|      | Northrop,Thornton and Trezise | Slow domains in semi-insulating GaAs. |
|      | Dumke | Single to double injection NDR. |
|      | Konstantinov and Perel | Recombination instability. |

## ACKNOWLEDGMENTS

I am grateful to my colleagues at Essex – Drs. Naci Balkan, Rita Gupta, Anthony Vickers and Andrew Straw – for their stimulating contributions, experimental and theoretical, in this area of research. I am also indebted to SERC and the U.S. Office of Naval Research for supporting work on this topic.

## REFERENCES

1.  C. Zener, *Proc.Roy.Soc.* 145, 523 (1934).
2.  A. von Hippel, *J.Appl.Phys.* 8, 815 (1937).
3.  H. Fröhlich, *Proc.Roy.Soc.* 160, 230 (1937).
4.  H. Fröhlich, *Proc.Roy.Soc.* A188, 521 (1947).
5.  H.B. Callen, *Phys.Rev.* 76, 1394 (1948).
6.  H. Fröhlich and B.V. Paranjape, *Proc.Phys.Soc.* B69, 21 (1956).
7.  B. Davydov, *Physik Z.Sov.* 9, 433 (1936).
8.  B. Davydov, *Physick Z.Sov.* 12, 269 (1937).

9.  W. Shockley, *Bell Syst.Tech.J.* 30,990 (1951).
10. E.J. Ryder, *Phys.Rev.* 90, 766 (1953).
11. H. Krömer, *Zeits. f. Phys.* 134, 435 (1953).
12. H. Krömer, *Phys.Rev.* 109, 1856 (1958).
13. W. Shockley, *Bell Syst.Tech.J.* 33, 799 (1954).
14. B.K. Ridley and T.B. Watkins, *Proc.Phys.Soc.* 78, 293 (1961).
15. B.K. Ridley and T.B. Watkins, *Proc.Phys.Soc.* 78, 710 (1961).
16. K. Tweedale and T.B. Watkins, M.R.L. Report (1959).
17. C. Hilsum, *Proc. IRE* 50, 185 (1962).
18. B.K. Ridley and R.G. Pratt, *Physics Lett.* 4, 300 (1963).
19. R.G. Pratt and B.K. Ridley, *Proc.Phys.Soc.* 81, 996 (1963).
20. B.K. Ridley and R.G. Pratt, *J.Phys.Chem.Solids* 26, 21 (1965).
21. I. Adawi, *J.Appl.Phys.* 32, 1101 (1961).
22. B.K. Ridley, *Proc.Phys.Soc.* 82, 954 (1963).
23. K.W. Böer, H.J. Hänsch and U. Kümmel, *Z.Physik* 155, 170 (1959).
24. K.W. Böer, *Z.Phys.* 155, 184 (1959).
25. K.W. Böer and W.E. Wilhelm, *Phys.Stat.Sol.* 3, 1718 (1963).
26. A. Barrand, *C.R. Acad.Sci, Paris* 256, 3632 (1963).
27. D.C. Northrop, P.R. Thornton and K.E. Trezise, *Solid St.Electro* 7, 17 (1964).
28. J.B. Gunn, *Solid State Comm.* 1, 88 (1963).
29. H. Krömer, *Proc.IEEE* 52, 1736 (1964).
30. J.A. Copeland, *Proc.IEEE* 54, 1479 (1966).
31. B.K. Ridley, *J.Appl.PPhys.* 48, 754 (1977).
32. T.W. Hickmott, *J.Appl. Phys.* 33, 2669 (1962).
33. T.W. Hickmott, *J. Appl. Phys.* 35, 2679 (1964).
34. T.W. Hickmott, *J. Appl. Phys.* 36, 1885 (1965).
35. T.W. Hickmott, *J. Appl. Phys.* 37, 4380 (1966).
36. G. Dearnley, A.M. Stoneham and D.V. Morgan, *Rep.Progr.Phys.* 33, 1129 (1970).
37. R.H. Fowler and L.W. Nordheim, *Proc.Roy.Soc.* 119, 173 (1928).
38. L.W. Nordheim, *Proc.Roy.Soc.* A121, 626 (1928).
39. E.L. Murphy and R.G. Good, *Phys.Res.* 102, 1464 (1956).
40. T.J. Lewis, *J. Appl.Phys.* 26, 1405 (1955).
41. J.J. O'Dwyer, *J.Appl.Phys.* 39, 4356 (1968).
42. J.J. O'Dwyer, *J. Appl.Phys.* 40, 3887 (1969).
43. B.K. Ridley, *J.Appl.Phys.* 46, 998 (1975).
44. B.K. Ridley and F.A. El-Ela, *Solid State Electron* 32, 1393 (1989).
45. B.K. Ridley, *J.Phys.Condens.Matt.* 2, 2941 (1990).
46. C. Hilsum and J. Welborn, *J.Phys.Soc.Jap.* 21, 532 (1966).
47. J.J. Harns and B.K. Ridley, *J.Phys.Chem.Solids* 34, 197 (1973).
48. G. Persky and D.J. Bartelink, *I.B.M. J. Res.Dev.* 13, 607 (1969).
49. J.R. Barker and C.J. Hearn, *J.PPhys.C.* 6, 3097 (1973).
50. B.K. Ridley and J.J. Harris, *J.Phys.C.* 9, 991 (1976).
51. B.K. Ridley, *J.Phys.C.* 16, 2261 (1983).
52. B.K. Ridley, *Solid St.Electron* 33, 859 (1990).
53. B.K. Ridley, *J.Phys.C.* 7, 1169 (1974).
54. E. Schöll, *J.Physique* 42, C7-57 (1981).
55. V.I. Stafeer, *Sov.Phys. Solid St.* 1, 763 (1959).
56. W.P. Dumke, *Proc. ICPS (7th Paris)* p.611 (Dumod, Paris) (1964).
57. W.T. Read, *Bell System.Tech.J.* 37, 401 (1958).
58. B.K. Ridley, *Phys.Lett.* 16, 105 (1965).

59. B.K. Ridley, *J.Phys.D.* 7,1555 (1974).
60. G.S. Kino and P.N. Robson, *Proc.IEEE* 56, 2056 (1968).
61. P.N. Butcher, *Phys.Lett.* 19, 546 (1965).
62. E. Schöll, *Solid-State Electron* 29, 687 (1986).
63. O.V. Konstantinov and V.I. Perel, F.T.T. 6, 3364 (*Sov.Phys.-Solid State* 6, 2691) (1964).
64. O.V. Konstantinov, V.I. Perel and G.V. Tzarenkov. F.T.T. 9 1761 (*Sov.Phys.-Solid State* 9, 1381( (1967).
65. J.S. Moore, C.M. Penchina, N. Holonyak, M.D. Sirkis and T. Yamara, *J.Appl.Phys.* 37, 2009 (1966).
66. A.R. Hutson and D.L. White, *J.Appl.Phys.* 33, 40 (1962).
67. D.L. White, *J.Appl.Phys.* 3, 2547 (1962).
68. B.K. Ridley, R.S. Sussman and J.D. Allan, *Solid St.Commun.* 10, 713 (1972).
69. N.I. Meyer and M.H. Jørgenson, *Adv.Solid St.Phys.* 10, 21 (1970).
70. P. Kocevar, *J.Phys.C.* 5, 3349 (1972).
71. L. Esaki and R. Tsu, *I.B.M. Res.Dev.* 14, 61 (1970).
72. T. Kurosawa and S. Nagahashi, *J.Phys.Soc.Jap.* 45, 707 (1978).
73. B.K. Ridley, *Semicond.Sci.Technol.* 3, 542 (1988).
74. K. Hess, H. Morkoc. H. Shichijo and B.G. Streetman, *Appl.Phys.Lett.* 35, 469 (1979).
75. S.W. Kirchoeffer, R. Magno and J. Comas, *Appl.Phys.Lett.* 44, 1054 (1984).
76. B.K. Ridley, *Rep.Progr.Phys.* 54, 169 (1991).
77. K. Tsubaki, A. Livingstone, M. Kawashima, H. Okamoto and K. Kumabe, *Solid St.Comm.* 46, 517 (1983).
78. B.K. Ridley, *J. Phys.C.* 17, 5357 (1984).
79. M.A.R. Al-Mudares and B.K. Ridley, *J.Phys.C.* 19, 3179 (1986).
80. P. Hendriks, E.A.E. Zwaal, J.E.M. Haverkost and J.H. Wolter, *S.P.I.E.* 1362, (Aachen) 217 (1990).
81. S. Lurgi and A. Kastalsky, *Physica* 134B/C 453 (1985).
82. S. Lurgi and M.R. Pinto, *Phys.Res.Lett.* 67, 2351 (1991).
83. N. Balkan, R. Gupta, B.K. Ridley, M. Emeny, J. Roberts and I. Goodridge, *Solid St.Electron* 32, 1641 (1989).
84. W.I. Masselink, N. Breslan, D. La Tulipe, N.I. Wang and S.L. Wright, *Solid St.Electron* 31, 337 (1988).
85. R.A. Höpfel, J. Shah, A.C. Gossard and W. Wiegmann, *Physica* 134B, 509 (1985).
86. N. Balkan and B.K. Ridley, "Properties of impurity states in superlattice semiconductors" ed. C.Y. Fong, I.P. Batra and S.C. Ciraci (Brussels: NATO) ASI B183, 229 (1988).
87. J.J. Harris and B.K. Ridley, *J. Phys.Chem.Solids* 34, 197 (1973).
88. B.K. Ridley and R.G. Pratt, *J.Phys.Chem.Solids* 26, 21 (1965).
89. M.F. Leach and B.K. Ridley, *J.Phys.C.* 11, 2265 (1978).
90. N. Balkan and B.K. Ridley, *Semicond.Sci.Technol.* 3, 507 (1988).

# ELECTRONIC TRANSPORT IN SEMICONDUCTORS AT HIGH ENERGIES: EFFECTS OF THE ENERGY BAND STRUCTURE

Karl Hess

Beckman Institute, Coordinated Science Laboratory and
Department of Electrical and Computer Engineering
University of Illinois, Urbana-Champaign

A number of hot electron effects in semiconductors, related to device reliability, involve scattering and accelerations at very high energies and therefore are dominated by bandstructure effects. To describe these effects quantitatively it is necessary to go beyond the usual inclusion of non-parabolicity and to use a full bandstructure as calculated, for example, from the empirical pseudopotential method. Current Monte Carlo simulations of these effects develop in two major directions. Attempts are currently being made to avoid the uncertainties of a large number of deformation potential constants for the electron phonon interaction and to treat electron phonon interaction and bandstructure within a single framework. In addition efforts are continuing to include complex quantum effects such as collisional broadening and the intracollisional field effect. These developments are reviewed using the example of the theory of impact ionization as developed by Bude which is representative in many respects for general processes that exhibit a high energy threshold.

## INTRODUCTION

This is a brief review of the progress that has been made in the understanding of electronic transport in semiconductors at high energies due to the introduction of so called full band models [1].

Transport theory, after the invention of the transistor, was mostly confined to the band edges, although the possible importance of hot electrons was early recognized [2]. In particular, investigations of the phenomena of impact ionization were made by Wolff (electron temperature model) and Shockley (lucky electron model) and involved electron energies of the order of several electron volts. However, detailed knowledge of transport properties was, as Conwell stated in her well-known book [3], confined to the band edges especially with regard to an inclusion of complex bandstructures in the motion and scattering of electrons.

The basis of the most sophisticated early transport investigations was the Boltzmann equation for the electrons (holes) and the golden rule of Fermi to describe scattering events [4]. The possible importance of specific bandstructure features for the occurrence of transport phenomena such as negative differential resistance was recognized by Ridley and Watkins [5]

*Negative Differential Resistance and Instabilities in 2-D Semiconductors*
Edited by N. Balkan *et al.*, Plenum Press, New York, 1993

23

and by Hilsum [6] in the two celebrated papers that preceded the discovery of the Gunn effect [7]. These developments brought forth a variety of papers investigating multi-valley transport far away from equilibrium, particularly by use of Monte Carlo methods to obtain a solution of the Boltzmann transport equation.

Subsequently improvements were made in two directions. Quantum effects such as the intracollisional field effect were considered and Quantum transport equations were tested [8]. It was also recognized that the multi-valley effective mass treatment and obvious extensions including nonparabolicity needed and could be replaced by one band approximations that invoke a full bandstructure [9], as calculated, for example, by the empirical pseudopotential method. These developments turned out to be particularly important for the physical explanation of hot electron device reliability related effects such as impact ionization and emission of hot electrons from silicon into silicon dioxide. They have led to an increased understanding of transport at high energies and are the main subject of this review. The experimental work on high energy transport has recently received considerable expansion by Femto-second spectroscopy experiments [10] and goes now well beyond the previously available experimental material from current-voltage characteristics [3] (including negative differential resistance), impact ionization [9] and transport over heterojunctions [11] including real space transfer [12].

The recent theoretical efforts have attempted to unify electron motion and scattering by deriving them from the same bandstructure model and by eliminating the multitude of empirically determined deformation potential constants [13]. Other recent developments attempt to include quantum effects into the Monte Carlo simulations and will also be briefly reviewed.

The inclusion of heterojunction transport in such complete transport descriptions is just gaining momentum. A detailed account on these developments is given by P. D. Yoder later in these proceedings.

## FULL BAND MONTE CARLO AND EXPERIMENTAL RESULTS

The first Monte Carlo code that included a full bandstructure beyond the inclusion of nonparabolicity was developed in [9]. This numerical approach is equivalent to the solution of the Boltzmann equation within the one band approximation, i.e., it solves the Boltzmann equation recognizing that the velocity $\vec{v}$ of electrons is given by

$$\vec{v} = \frac{1}{\hbar} \, \nabla_{\vec{k}} \, E(\vec{k}) \tag{1}$$

where $E(\vec{k})$ is the Energy as a function of wave vector $\vec{k}$ in a given band, $\vec{k}$ being determined in an electric field $\vec{F}$ from

$$\hbar \, \dot{\vec{k}} = e\vec{F} \tag{2}$$

The transition probabilities $S(\vec{k},\vec{k}')$ due to scattering, which appear in the Boltzmann Equation, are determined from the Golden rule and a number of scattering mechanisms such as electron-phonon and electron-impurity interactions is usually included.

It is important to notice that in this way a large number of quantum effects are taken into account in the otherwise classical Boltzmann equation. Scattering is treated properly by the Golden rule as long as collisional broadening and the intracollisional field effect are of lesser importance. Bandstructure effects are included in both the scattering terms and the acceleration and streaming terms. It is now recognized that the inclusion of a full bandstructure is particularly important for the scattering rates. These rates are roughly proportional to the density of states which is given by the well-known expression:

$$\frac{2}{(2\pi)^3} \int \frac{dS}{\nabla_k E\ (k)}$$ (3)

where $dS$ designates the integration over a surface of constant energy E. This density of states has very large maxima typically one to two electron volts above the conduction band edge of many important semiconductors. Accordingly the scattering rate increases steeply at these energies and cannot be approximated by a constant (a constant mean free path is also not permissible). Fig. 1 shows the electron-phonon scattering rate for conduction band electrons in silicon as it is obtained using the proportionality to the density of states. The energy distribution function of electrons (holes) is also strongly influenced by this enormous increase in scattering and becomes extremely non-Maxwellian. This has profound consequences on effects that have a high energy threshold and depend on the high energy tail of the distribution function. Therefore impact ionization, emission of hot electrons into $SiO_2$ and similar reliability related hot electron effects cannot be understood, not even qualitatively, without an inclusion of these bandstructure effects. This is the precise reason for the gaining importance of Monte Carlo simulations in devices. Energy balance equations that deal with a single average over the complex energy distribution cannot satisfactorily deal with the intricacies of the distribution function at high energies. Illustrations of this fact are provided throughout this paper.

The full band Monte Carlo approach as currently practiced has two major drawbacks. First, the method consumes an extraordinary amount of cpu-time even on the largest and fastest current computers. This fact is mainly due to the necessity of representing bandstructure and scattering rates on a large number of mesh points and due to the necessity of interpolations. Some researchers have tried to circumvent this problem by using simplified bandstructures [14]. However, from the viewpoints of accuracy it is more appropriate to attempt to gain speed by employing modern methods of numerical mathematics [15]. I assume that this problem will be alleviated also by the increasing capabilities of computers. The second problem is related to the large parameter space that is opened with increasing energy, particularly with respect to the parameters entering the electron-phonon interactions. The most important electron-phonon interactions at high energies occur via the deformation potential and include acoustical and optical phonons. The interaction strength is usually characterized by deformation potential constants that are adjusted to achieve agreement with a number of experimental results, most prominent the current voltage characteristic. The current, however, depends on many other parameters (e.g., impurity scattering, electron-electron interactions) and consequently, a complete determination of these constants is difficult, even at the band edges where the number of eligible deformation potentials is more limited. The number of deformation potentials constants increases vastly at higher energies. Consider, for example, electrons in GaAs located at the $\Gamma$ minimum for low electric fields. As the field approaches the Gunn threshold the L and X valleys start to become populated and $\Gamma$–$L$, $\Gamma$-X and X-L intervalley transitions become important. For each of these scatterings there are a number of possible contributory phonons, determined from energy-momentum conservation and symmetry, and corresponding different coupling constants. It therefore comes as no surprise that our knowledge of the values of these constants and even the contributory phonon types is incomplete.

The contributions of a low energy acoustic phonon was, for example, only recently recognized [16]. Experimental results on the Gunn effect, on impact ionization and real space transfer effects have contributed significantly to the body of knowledge of the electron phonon interaction in many semiconductors. It has also been shown that effects with high energy

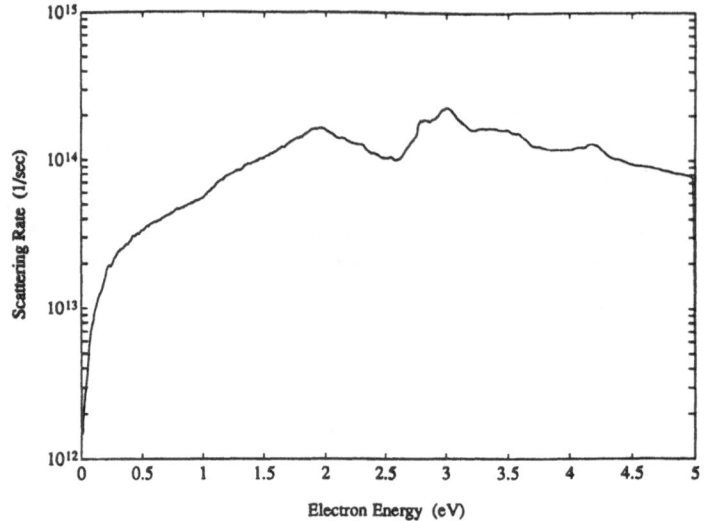

Figure 1. Approximate total electron-phonon scattering rate in silicon.

Figure 2. Monte Carlo simulation showing carrier density as a function of energy and time in the Γ valley of GaAs (after Bailey et al. [10]).

thresholds add significantly to the possibilities of narrowing down the number of eligible deformation potential constants [17]. Powerful support has come recently from Femto-second spectroscopy experiments. These studies explore the sensitivity of photon absorption to the electron distribution. Consider, for example, the generation of a high electron density in the $\Gamma$ valley of GaAs by a pump pulse of ~ 2 e V energy. The electrons are generated at various energies exceeding also the energy of the L and X-minima. A second probe pulse will then exhibit absorption dependent on the existing carrier density (due to the Pauli principle). Fig. 2 shows the electron density as a function of time and energy in the $\Gamma$ valley of GaAs as obtained from Monte Carlo simulations. It can clearly be seen that the highest energy peak decays extremely fast due to $\Gamma$-X and $\Gamma$-L scattering. Careful comparisons to experiments [10] permit then to deduce deformation potential constants. Although these deductions are still not entirely precise, due to uncertainties introduced by other scattering mechanisms such as electron-electron interactions, it has been convincingly shown that the deformation potential for scattering from $\Gamma$ to L in GaAs is lower than previously assumed and of the order of $D_{\Gamma-L} = 5 \times 10^8$ e V instead of around $10^9$ e V. The analysis of experimental data from Femto second spectroscopy is in general not straightforward and actually requires the most sophisticated Monte Carlo simulations. This is true for all high energy transport effects that investigate situations far from equilibrium and therefore require a solution of the Boltzmann equation. It is beyond the scope of this paper to describe the intricacies of evaluation of the large body of experimental materials. I, therefore, concentrate here on a single effect, impact ionization, that has a long history of theoretical investigations and demonstrates clearly the complexities involved in understanding high energy transport.

## IMPACT IONIZATION

The phenomenon of impact ionization provides an excellent example of the complexities of high energy transport and of the value of large scale computational models which are necessary to include the correct physics and obtain results that can be compared to experiments.

Impact ionization is the exact inverse of the Auger process [18]. An electron is accelerated in the conduction band to high energies and looses its energy by ionization, i.e., causing a valence band electron to transition to the conduction band. The ionization rate $S_I$ corresponding to this process is difficult to calculate. First estimates were performed by Keldysh [19] and have been used subsequently by many researchers. The Keldysh formula, however, does not apply to ionization by electrons located at arbitrary points of the Brillouin zone as it assumes parabolic bands with minima and maxima at $\Gamma$. A much more general estimate for $S_I$ has been given by Kane [20] for silicon and by Bude and Hess [21] (using the Kane theory) for several III-V compound materials. They employed Monte Carlo integration to obtain $S_I$. The main difference between the Keldysh and Kane results is that Keldysh predicts a relatively hard threshold for the ionization process while Kane calculated a very soft threshold and much lower rates close to the threshold.

The experimentally determined ionization coefficient $\alpha$ is related to the ionization rate $S_I$ by:

$$\alpha \approx \frac{\Sigma_{nc} \int_{BZ} d\vec{k}\, f_{nc}(\vec{k}) \langle S_I \rangle}{\Sigma_{nc} \int_{BZ} d\vec{k}\, f_{nc}(\vec{k})\, v_{nc}} \tag{4}$$

Here $\langle S_I \rangle$ is the average ionization rate at a given energy E, f is the electron energy distribution and $v$ the group velocity as given by Eq (1). The subscript $nc$ denotes the specific conduction band in consideration. Usually two conduction bands will be sufficient to obtain accurate $\alpha$.

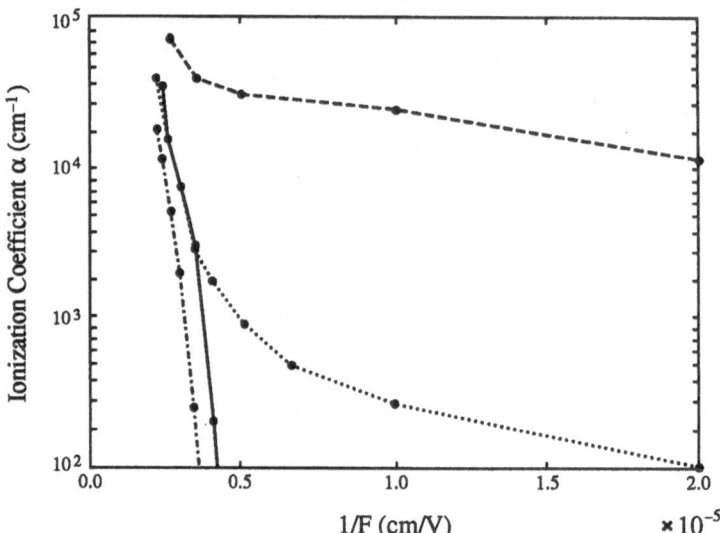

Figure 4. Calculated impact ionization coefficient $\alpha$ as a function of inverse field, $1/F$. Solid line, GaAs; dashed line, InAs; dot-dashed line, InP; and dotted line, $Ga_{0.43}In_{0.57}As$ (after [21]).

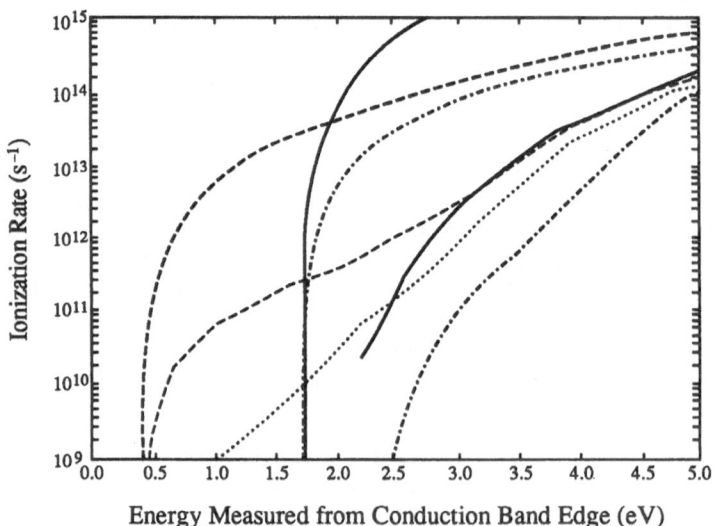

Figure 3. Ionization rates: upper set of three curves, Keldysh formula; lower set of four curves (Kane like theory); solid line, GaAs; dashed line, InAs; dot-dashed line, InP; and dotted line, $Ga_{0.43}In_{0.57}As$ (after [21]).

The exact form of the distribution function is extremely important to obtain precise values of $\alpha$ since $S_I$ has a threshold and only the tail of the distribution contributes to $\alpha$. This tail of the distribution has been subject of many investigations including the lucky electron model of Shockley [2], the parametrized solution of the Boltzmann equation by Baraff [4] and the full band Monte Carlo simulations by Shichijo and Hess [9] as well as the lucky drift model of Ridley et al. [22]. All of these models have certain validity ranges and are useful to explain impact ionization phenomena. Here we concentrate on the full band approach since it leads to a virtually exact solution, at least within the semiclassical framework of the one band approximation. Before proceeding to discuss some of the results, I would like to discuss a basic property of the Boltzmann equation that has led to misinterpretation of impact ionization phenomena. Any scattering mechanism which is both strongly inelastic and has a threshold will influence the distribution function in such a way that f becomes depleted above the threshold. The depletion will be the stronger the stronger the scattering rate. One can easily find a general proof of this fact [23]. It is also well-known in other fields (e.g., the possibility to decreasing revenue by increasing taxes). As a consequence, the value of $\alpha$ in equation (4) can both decrease and increase if $<S_I>$ is increased because of the possibility of a strong decrease of the tail of $f_{nc}$ with increasing $<S_I>$. In large parameter-ranges of deformation potentials and bandstructure, $\alpha$ becomes insensitive to the value of $<S_I>$ above threshold because $f_{nc}$ and $<S_I>$ counteract each other so that the product stays about constant. This is the precise reason why it was possible to obtain agreement with the experimental results of $\alpha$ by using extremely large $<S_I>$ and the Keldysh formula (which is inappropriate to be used) and also agreement with experimental $\alpha$ by using the much smaller values of $<S_I>$ from Kane's correct theory. These facts were first recognized in ref [17] and to a fuller extent in [21]. Fig. (3) shows the results for $< S_I >$ from the Keldysh formula and from Bude et al. [21]. Although the $< S_I >$ are vastly different, they lead to comparable $\alpha$. One might then ask why then precise $< S_I >$ are needed. The answer is not only that a correct theory is, of course, more desirable, but also that only for steady state calculations $\alpha$ is given by Eq (4). For transients as they occur in devices $\alpha$ needs to be calculated differently. Then the exact form of $< S_I >$ plays indeed a major role. One can easily see this considering for example the so called dead-space for impact ionization which sensitively depends on the average ionization energy [24] and therefore on the precise form of $< S_I >$. The same arguments apply to rapid spacial changes of the electric field as they are encountered in submicron devices such as MOSFET's.

Even under steady state conditions, the form of $< S_I >$ can give rise to very different functional forms of $\alpha$ vs the electric field or as usually plotted vs the inverse electric field. An example is given in Fig. (4)which shows $\alpha$ for GaAs, InP and also In GaAs. The In GaAs field-dependence is clearly different and deviates from the exp (const/F) form that is obtained by the use of the Keldysh formula. $< S_I >$ and $\alpha$ are in this case strongly influenced by the relatively high energy of the X and L minima (measured from the $\Gamma$ minimum) and therefore by density of states considerations [21].

Figs. [5-6] show equal energy surfaces in $\vec{k}$-space for the conduction band in silicon as well as ionization rates. It is clear from the complex shape of these graphs and eqs (1)-(3) that a quantitative understanding of high energy transport takes a quantitative full band approach.


## QUANTUM EFFECTS BEYOND BAND STRUCTURE AND THE GOLDEN RULE

From a quantum point of view, a distribution function f $(\vec{r}, \vec{k}, t)$ cannot be defined except if the uncertainty relation for position $(\vec{r})$ and momentum ($\vec{k}$) coordinate permits. If the device or structure in question lacks geometrical features that give rise to tunneling, propagation across heterointerfaces, or quantum interference, only two quantum effects are not captured by the full band Monte Carlo method. Both of these effects, the intracollisional field effect and collision broadening are well-known and have been considered in detail in the literature [25].

Figure 5. Equal energy surfaces for the silicon conduction bands (in $\vec{k}$ space, for E = 1.63 eV). Shading indicates the impact ionization rate (courtesy F. Bodine, U. Ravaioli).

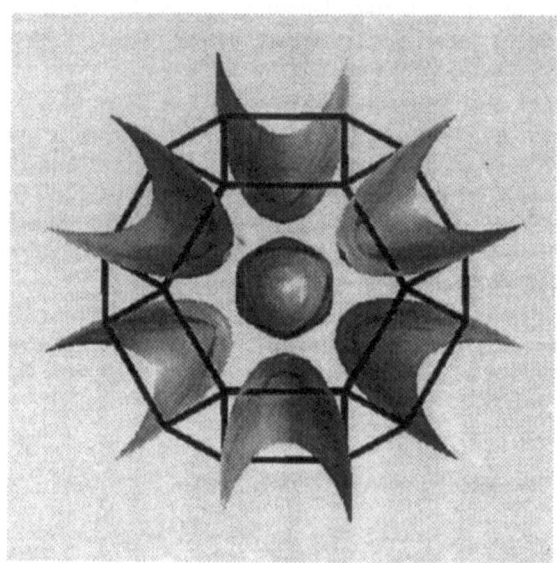

Figure 6. Same as Fig. 5, but in the ⟨1,1,1⟩ direction (courtesy F. Bodine, U. Ravaioli).

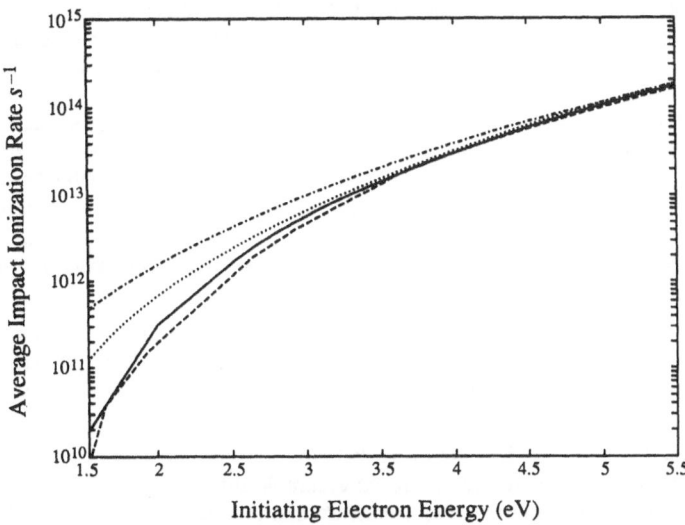

Figure 7. Impact ionization rates for Si averaged over all initial electron states with a given energy measured from the bottom of the conduction band. Dashed curve, Kane's result; solid curve, zero-field, no collision broadening; dotted curve, collision broadening; dashed-dot curve, $F = 5 \times 10^5$ V/cm and collision broadening (after Bude et al. [25]).

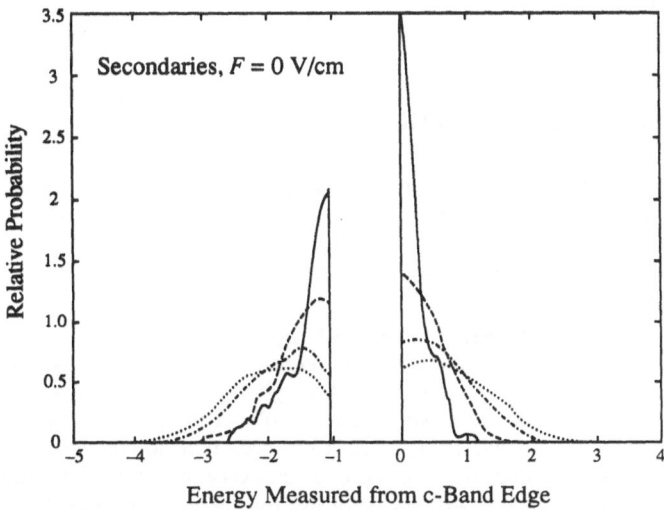

Figure 8. Secondaries produced by impact ionizing electrons in the conduction band for no field. Solid curve, initial electron of 1.5 eV; dashed curve, 2.5 eV; dashed-dot, 3.5 eV; dotted, 4.5 eV. The two sets of curves correspond to holes for the $E < -E_G$, and electrons, for $E > 0$. Energies are referenced from the bottom of the conduction band (after Bude et al. [25]).

It turns out that for weakly energy dependent scattering mechanisms, the intracollisional field effect is negligible [26] and collision broadening is without bigger consequence. It has been shown by Bude et al. [26], however, that effects that involve thresholds and sensitively depend on threshold are also quite sensitive to these two quantum corrections. The reason is, of course, that both the intracollisional effect and collision broadening smear-out the thresholds. Semiclassical impact ionization as well as the other reliability related hot electron effects do, of course, exhibit a threshold. For impact ionization, most textbooks report that the threshold can be calculated within the effective mass approximation from momentum and energy conservation. This is, however, completely wrong and misleading. Reliable thresholds can only be obtained from *three dimensional* calculations using, for example, the Anderson Crowell criteria and full bands [21].

The inclusion of the quantum corrections changes the threshold behavior considerably. This is a consequence of the fact that the electron system by itself is not Hamiltonian and energy is only conserved in the whole system of both electrons and phonons (and maybe photons).

Fig (7) shows the ionization rate $S_I$ vs initialing electron energy with and without quantum effects. It can be seen that there are considerable differences at low energies (close to threshold). Fig. (8) shows the electron secondary distribution for two values of electric field strength. Note that even for zero electric field, there exists a considerable broadening of final? energies. The actual computation of all of these effects is extremely cpu time consuming. The simplest accurate $S_I$ can be obtained, according to our experience, from a Kane-type theory with a certain broadening included in the Monte Carlo integration [25].

For rough estimates one can put the threshold to the value of the energy of the band gap $E_g$ above the conduction band edge, and assume a Kane-like slowly increasing $S_I$. The actual maximum of ionization events is often far above threshold as shown for a specific electric field and silicon in Fig. (9).

## RECENT DEVELOPMENTS AND OUTLOOK

The full band methods described above have demonstrated considerable success, in particular for those processes for which the non Maxwellian form of the distribution function is of importance. In some instances these methods have demonstrated predictive power and in many cases they have been instrumental to highlight the complex physics involved in certain types of measurement and device function.

However, with all its success, the full band approach remains unsatisfactory as long as it treats the motion of electrons and the scattering on different footage. To be sure the scattering rates include some of the bandstructure effects through the density of states. However, from a more basic point of view, one also would want to obtain the phenomenological deformation potential constants from the same treatment that has been used to compute the bandstructures. It is also clear that the deformation potential constants vary throughout the Brilloiun zone and the constants at the band-minima will not be of use for points of arbitrary symmetry and location in $\vec{k}$ space. Similarly the contributory phonon modes, transversal and longitudinal will vary and be different at different symmetry points. Finally, one also would like to include the dispersion of the phonon spectrum more precisely than is usually done (no dispersion for optical linear for acoustical phonons). An ambitious project with these goals has been undertaken by Higman and Fischetti [13] and Yoder et al. [27] using the empirical pseudopotential approach to calculate both bandstructure and scattering rates. This method still contains (highly accurate) approximations such as the rigid pseudo-ion approximation. It also relies on an interpolation of the pseudopotential which is not unique.

It is therefore important that this method be further extended and put on a more solid theoretical footage. This has recently been attempted by Yoder et al. [28] using the Harris functional approach for both bandstructure calculation and analysis of the electron-phonon

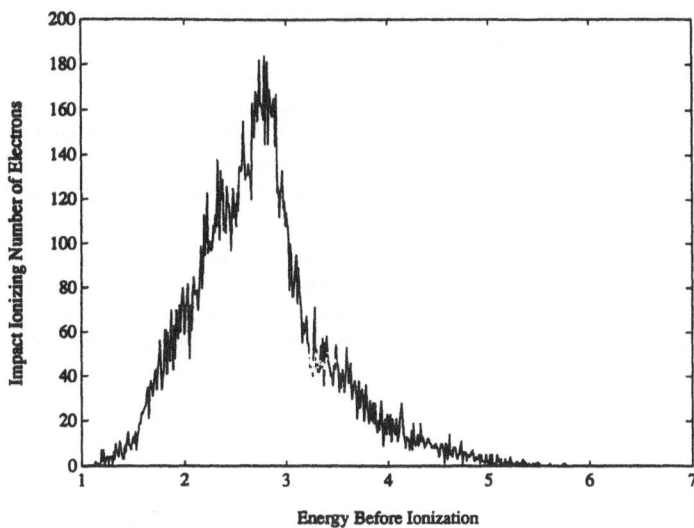

Figure 9. Energy distribution of impact ionizing electrons in silicon for an electric field F = 400 kV/cm (courtesy Doug Yoder).

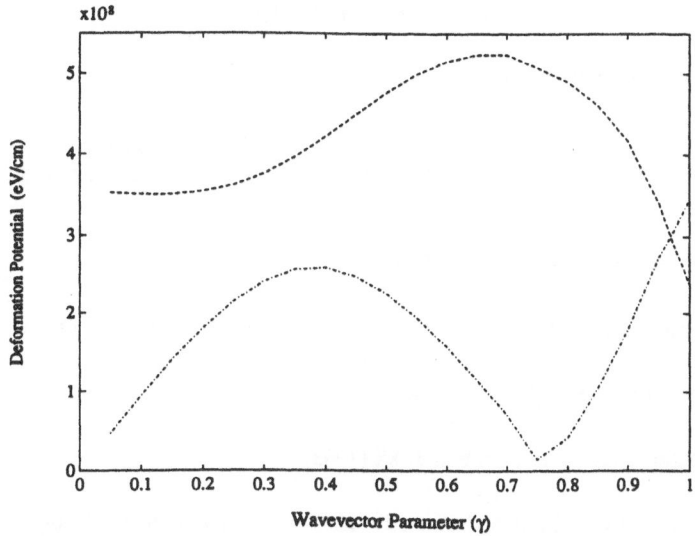

Figure 10. Deformation potential "constants" for interaction with transverse optical modes (dashed line) and longitudinal acoustic modes (dot-dashed lines) vs. wavevector parameter $\gamma$ [$\vec{k}$ = ($\gamma$,0,0)] (after P.D. Yoder et al. [28]).

interaction in silicon. The results of this research agree roughly with the results obtained from empirical pseudopotentials. Figure (10) shows the deformation potentials (now a function of $\vec{k}$) for a set of parameterized transitions corresponding to "f-type" scattering in silicon. The particular transitions chosen are from initial states at various distances along the (1,0,0) axis to final states at corresponding distances along one of the four axes transverse to the initial axis. Selection rules dictate that only two modes may contribute strongly to the electron-phonon matrix element for this set of transitions, depending critically on the phonon polarization. In fact, these modes are the ones with $\Sigma_1$ symmetry, which for an N-process are the longitudinal acoustic and one of the transverse optical modes. As the initial and final electronic states grow further apart, the phonon modes become mixed to some degree, and those which contribute to the electronic transition are no longer purely optical or acoustical. Nonetheless, the contributing modes will always be found to have the same symmetry and we see this reflected in a continuous deformation potential.

There exists therefore the distinct possibility that electronic transport at high energies may be accurately simulated and understood from the basic input of a few parameters such as pseudopotential Fourier coefficients. The large demand on cpu time of such approaches is well compensated by their fundamental accuracy, descriptive power and logically satisfactory design.

## ACKNOWLEDGEMENT

The work was supported by the Army Research Office and the Office of Naval Research.

## REFERENCES

1.  Monte Carlo Device Simulation: Full Band and Beyond, edited by K.Hess, Kluwer Academic Publishers, Boston/Dordrecht/London (1991)

2.  W. Schockley, Solid-State Electron. 2, 35 (1961)

3.  E. M. Conwell, High Field Transport in Semiconductors, Supplement 9 Solid-State Physics, F. Seitz, D. Turnbull and H. Ehrenreich, Editors, Academic Press, New York/London (1967)

4.  G. A. Baraff, Phys. Rev. 128, 2507 (1962)

5.  B. K. Ridley and T. B. Watkins, Proc. Phys, Soc. (London) 78, 293 (1961)

6.  C. Hilsum Proc. IRE 50, 185 (1962)

7.  J. B. Gunn, Solid State Commun 1, 88 (1963)

8.  See Chapter 15 in Semiconductors, D. K. Ferry, Macmillan Publishing Company, New York (1991)

9.  H. Shichijo and K. Hess, Phys. Rev. B 23, 4197-4207 (1981)

10. See the references in D. W. Bailey, C. J. Stanton and K. Hess, Phys. Rev. B 42, 3423 (1990)

11. See for example IBM Journal of Research and Development Volume $\underline{34}$, 4, (1990)

12. I. C. Kizilyalli and K. Hess, J. Appl. Phys. $\underline{65}$ 2005 (1989)

13. M. V. Fischetti and J. M. Higman, see Chapter 5 in ref above

14. A recent review and references to this work have been given by J. Frey, FED Journal Vol. 2, 12 (1992)

15. Milestone progress has recently been achieved by J. Bude, E. Grosse and R. K. Smith by developing a phase-space simplex Monte Carlo approach. J. Bude and R. Kent Smith, private communication

16. S. Zollner, S. Gopalan and M. Cardona, J. Appl. Phys. $\underline{68}$, 1682 (1990)

17. J. Y. Tang and K. Hess, J. Appl. Phys. $\underline{54}$, 5145 (1983)

18. See, e.g., B. K. Ridley, Quantum Processes in Semiconductors Clarendon Press, Oxford (1982) Chapter 6.9

19. L. V. Keldysh, Sov. Phys. JETP $\underline{21}$, 1135 (1965)

20. E. O. Kane, Phys. Rev. 159, 624 (1967)

21. J. Bude and K. Hess, J. Appl. Phys. 7, 8, October (1992)

22. B. K. Ridley, Semicond. Sci. Technol. $\underline{2}$, 116 (1986)

23. See the treatment of the warm electron distribution in K. Hess and C. T. Sak, Phys. Rev. B $\underline{10}$, 3375 (1974)

24. J. Bude, Private communication

25. J. Bude, K. Hess and G. J. Iafrate, Phys. Rev. B. $\underline{45}$, 10958, (1992)

26. P. Liopavsky, F. S.Khan, F. Abddsalami and J. W. Wilkins, Phys, Rev.B $\underline{43}$, 4885 (1991)

27. P. D. Yoder, J. M. Higman, J. Bude and K. Hess, Semicond. Sci. Technol. $\underline{7}$, B357 (1992)

28. P. D. Yoder, V. Natoli and R. Martin, to be published

# THEORY OF OSCILLATORY INSTABILITIES IN PARALLEL AND PERPENDICULAR TRANSPORT IN HETEROSTRUCTURES

Eckehard Schöll

Institut für Theoretische Physik
Technische Universität Berlin
Hardenbergstr.36
D-1000 Berlin 12
Germany

## ABSTRACT

The theory of dynamic hot-electron transport instabilities in low-dimensional semiconductor structures is reviewed with an emphasis on the following aspects:
(i) General conditions for self-generated oscillatory instabilities in semiconductor nonequilibrium transport are elaborated in the framework of nonlinear dynamic systems.
(ii) Real space transfer in modulation-doped heterostructures is investigated. In single and double heterojunctions, and in coupled quantum wires different bifurcation scenarios, including symmetric and nonsymmetric oscillation modes well above the 100 GHz regime, and asymptotic and transient chaos, are predicted. Electrically and optically induced real-space transfer are compared. (iii) Vertical hot-electron transport across a single heterojunction is considered. Bistability between a tunneling and a thermionic emission state, and different self-organized spatio-temporal oscillation modes are found.

## INTRODUCTION

Nonlinear transport instabilities of hot electrons are widespread in semiconductors driven far from thermodynamic equilibrium (Schöll, 1987; Abe, 1989; Thomas, 1992; Shaw et al., 1992). Apart from the occurrence of negative differential conductivity (NDC) in the static current density-versus-field characteristic, such instabilities are often associated with self-generated temporal current or voltage oscillations, and self-organized spatio-temporal pattern formation like current filamentation, field domain formation, or more complex spatio-temporal turbulent dynamics. Although the microscopic mechanism inducing such behavior may be very different, the observed macroscopic nonlinear phenomena like bifurcations, nonequilibrium phase transitions,

*Negative Differential Resistance and Instabilities in 2-D Semiconductors*
Edited by N. Balkan *et al.*, Plenum Press, New York, 1993

37

periodic and chaotic oscillatory instabilities, and self-organization are often quite similar, and are germane to a variety of different dissipative dynamic systems ocurring in physics, chemistry, biology and other sciences (Haken, 1983; Nicolis and Prigogine, 1977; Ebeling and Feistel, 1986).

Recently, hot carrier transport instabilities in low-dimensional semiconductor structures have received increasing attention (Ridley, 1991). The miniaturization of semiconductor microstructures results in high electric fields and strong carrier heating combined with quantum size effects, which can induce novel instabilities not normally accessible in bulk materials. Examples, such as electric-field domains in superlattices, Wannier-Stark ladders, nonresonant and resonant tunnelling, and real space transfer, are treated in separate articles in this volume. In the present paper, recent progress in the theoretical description of oscillatory instabilities associated with hot carrier transport parallel and perpendicular to the layers of a semiconductor heterostructure is reviewed. After a general outline of conditions for oscillatory instabilities, real space transfer in modulation-doped semiconductor structures giving rise to NNDC (N-shaped NDC), and perpendicular transport in a heterostructure hot electron diode leading to SNDC (S-shaped NDC) are treated in detail.

## OSCILLATORY INSTABILITIES IN NONEQUILIBRIUM TRANSPORT

A frequently used approach to hot carrier transport theory is based upon hydrodynamic balance equations for slow macroscopic observables like the carrier density $n(\mathbf{r}, t)$, the mean momentum per carrier $\mathbf{p}(\mathbf{r}, t)$, and the mean energy per carrier $E(\mathbf{r}, t)$. These can be derived by a moment expansion of the Boltzmann transport equation for the semiclassical carrier distribution function $f(\mathbf{r}, \mathbf{k}, t)$, where $\mathbf{r}$ is the spatial coordinate and $\hbar \mathbf{k}$ is the crystal momentum:

$$\frac{\partial f}{\partial t} + \mathbf{v_g} \cdot \nabla_r f + q\hbar^{-1}\mathcal{E} \cdot \nabla_k f = (\frac{\partial f}{\partial t})_{coll} \qquad (1)$$

Here $\mathbf{v_g}$ is the group velocity, $\mathcal{E}$ is the electric field, and the subscript *coll* denotes the collision integral, which includes all dissipative processes like phonon and impurity scattering, carrier-carrier scattering and impact ionization. The carrier charge is $q = \pm e$ for holes or electrons, respectively.

Multiplying (1) with appropriate powers of $\mathbf{k}$ and integrating over the first Brillouin zone, one obtains a closed system of moment equations

$$\dot{n} + \nabla \cdot (n\mathbf{v}) = \Phi_0(n, E) \qquad (2)$$

$$\dot{\mathbf{p}} + (\mathbf{v} \cdot \nabla)\mathbf{p} + \frac{1}{n}\nabla(nk_BT_e) - q\mathcal{E} = -\mathbf{p}\Phi_0/n - \mathbf{p}/\tau_m(E) \qquad (3)$$

$$\dot{E} + (\mathbf{v} \cdot \nabla)E + \frac{1}{n}\nabla(nk_BT_e\mathbf{v}) - \frac{\kappa}{n}\Delta T_e - q\mathbf{v} \cdot \mathcal{E} = -E\Phi_0/n - (E - E_0)/\tau_e(E) \qquad (4)$$

where for a nondegenerate, isotropic, parabolic band with an effective mass $m^*$,

$$\mathbf{p}(\mathbf{r}, t) = m^*\mathbf{v}(\mathbf{r}, t) \qquad (5)$$

$$E(\mathbf{r}, t) = \frac{m^*}{2}\mathbf{v}^2 + \frac{3}{2}k_BT_e \qquad (6)$$

and the following assumptions are used: (i) The carrier temperature $T_e$ is scalar, (ii) The energy-flow density is

$$\mathbf{j}_E = n\mathbf{v}E + n\mathbf{v}k_BT_e + \mathbf{j}_Q$$

where the heat flux is approximated phenomenologically by

$$\mathbf{j}_Q = \kappa \nabla T_e$$

with thermal conductivity $\kappa$

(iii) The collision integrals are approximately evaluated yielding the generation-recombination (g-r) rate

$$\Phi_0(n, E) = \int (\frac{\partial f}{\partial t})_{coll} \, z \, d^3k \, , \, z = 2(2\pi)^{-3} \tag{7}$$

the momentum relaxation rate

$$-n\frac{\mathbf{p}}{\tau_m(E)} = \int \hbar \mathbf{k} (\frac{\partial f}{\partial t})_{coll} \, z \, d^3k \tag{8}$$

and the energy relaxation rate

$$-n\frac{E - E_0}{\tau_e(E)} = \int \frac{\hbar^2 \mathbf{k}^2}{2m^*} (\frac{\partial f}{\partial t})_{coll} \, z \, d^3k \tag{9}$$

with mean energy-dependent momentum and energy relaxation times $\tau_m$, $\tau_e$, respectively, and $E_0 = (3/2)k_B T_L$. This is equivalent to a self-consistent parametrization of the nonequilibrium distribution function by $n$, $\mathbf{p}$ and $E$. Eqs (2,3,4) with (5,6) constitute a closed set of nonlinear partial differential equations for $n, \mathbf{p}, E$. They are coupled to the local electric field $\mathcal{E}$ by Maxwell's equations

$$\epsilon_0 \epsilon_s \nabla \cdot \mathcal{E} = \rho \tag{10}$$

with the local charge density $\rho = e(N_D - N_A) + q(n + n_t)$ ($N_D$ and $N_A$ are donor and acceptor densities, $n_t$ is the concentration of trapped electrons, $\epsilon_0$ and $\epsilon_s$ are absolute and relative permittivity, respectively), and

$$\nabla \times \mathbf{H} = \epsilon_0 \epsilon_s \dot{\mathcal{E}} + \mathbf{j} \equiv \mathbf{j}_0 \tag{11}$$

with conduction current density $\mathbf{j} = qn\mathbf{v}$ and magnetic field $\mathbf{H}$. We shall confine attention to the important case that $\mathbf{H}$ is not a dynamic variable, but can be regarded as an external control parameter. From (11) it follows by applying $\nabla$ that the total external current density $\mathbf{j}_0$ satisfies $\nabla \cdot \mathbf{j}_0 = 0$, and may be regarded as an additional control parameter.

The time-scale on which the variables $n, \mathbf{p}, E, \mathcal{E}$ occurring in (2,3,4,11) change is determined by the g-r lifetime, the momentum relaxation time, the energy relaxation time, and Maxwell's dielectric relaxation time, respectively. Since macroscopic dynamics is dominated by the slow variables, in particular at the onset of an instability, it depends on the relations between these time-scales which of the above variables must be considered as relevant dynamic variables in a particular system. The fast variables are enslaved by the slow ones and may be eliminated adiabatically from the dynamic equations by neglecting their time derivatives.

Momentum relaxation often occurs faster than other processes, so that $\mathbf{p}$ can be eliminated adiabatically by setting $d\mathbf{p}/dt \equiv \dot{\mathbf{p}} + (\mathbf{v}\nabla)\mathbf{p} = 0$. Introducing the mobility

$$\mu(E) \equiv \frac{e}{m^*}\tau'_m(E), \quad \frac{1}{\tau'_m} \equiv \frac{1}{\tau_m} + \frac{\Phi_0}{n} \tag{12}$$

the drift-diffusion equation

$$\mathbf{j} = qn\mathbf{v} = en\mu(E)\mathcal{E} - qD(E)\nabla n - \frac{q}{e}\mu(E)nk_B\nabla T_e \tag{13}$$

39

is recovered if the diffusion coefficient $D$ is defined via the Einstein relation $eD = \mu k_B T_e$.

Current instabilities can occur only if the constitutive dynamic equations contain sufficiently strong nonlinearities. These may be manifest in the mobility, in the electron temperature, or in the generation-recombination rate. Accordingly, the physical mechanisms can be classified as drift instabilities, electron overheating instabilities, or g-r instabilities (Schöll, 1992a). A widely studied example of the third class involves impact ionization of carriers from impurity levels (low temperature impurity breakdown) or across the bandgap (avalanching). If a free carrier is strongly heated by an applied electric field, it may transfer its kinetic energy to a bound carrier, thereby generating an additional free carrier, which may, in turn, impact ionize other carriers. Such a positive feedback ("autocatalysis") leads to a rapid increase of the free carrier density. This simple g-r mechanism can explain self-generated oscillatory instabilities including period-doubling routes to chaos, and intermittency, if coupled with dielectric relaxation of the electric field (Schöll, 1987), possibly in the presence of an additional magnetic field (Hüpper and Schöll,1991).

The different oscillation mechanisms may be discussed from a unified viewpoint (Schöll, 1989) by analyzing the bifurcations of the underlying dynamic system (2), (4), (11). To this purpose we restrict ourselves to the simplest, spatially homogeneous form, although extensions to include current filamentation are possible. If the semiconductor is operated in a resistive external circuit described by Kirchhoff's law $U_0 = IR_L + V$, where $U_0$ is the applied bias voltage, $R_L$ is the load resistance, $I$ and $V$ are the sample current and voltage, respectively, eq. (11) with (13) can be cast into the form

$$\dot{\mathcal{E}} = [J_0 - (\sigma_L + \sigma(\mathbf{x}, \mathcal{E}))\mathcal{E}]/(\epsilon_0 \epsilon_s) \tag{14}$$

where $\sigma \equiv ne\mu$ is the conductivity which depends on the field $\mathcal{E}$ and additional dynamic variables $\mathbf{x} \equiv (x_1, x_2 \ldots, x_N)$, such as carrier densities and mean energy, governed by balance equations of the type (2),(4), and

$$J_0 \equiv U_0/(R_L A) \text{ and } \sigma_L \equiv L/(R_L A) \tag{15}$$

are control parameters, $A$ is the cross-section of the current flow, and L is the contact distance.

The static differential conductivity is given by

$$\sigma_{diff} = \frac{d}{d\mathcal{E}}[\sigma(\mathbf{x}(\mathcal{E}^*), \mathcal{E}^*)\mathcal{E}^*] \tag{16}$$

where $*$ denotes the steady-state values obtained from $\dot{\mathbf{x}} = 0$, $\dot{\mathcal{E}} = 0$. The stability of the steady state with respect to small fluctuations $(\delta\mathbf{x}, \delta\mathcal{E}) \sim exp(\lambda t)$ follows from linearizing (14) and the additional transport equations for $\dot{\mathbf{x}}$ around $(\mathbf{x}^*, \mathcal{E}^*)$, which yields a secular equation of degree $N + 1$ for the eigenvalues $\lambda_1, \ldots \lambda_{N+1}$. This stability analysis can easily be extended to include space-dependent fluctuations, in which case a family of linear modes $\lambda(\mathbf{k})$ is obtained (Schöll,1987).

The simplest oscillatory instability is the Hopf bifurcation of a limit cycle. It occurs when a pair of complex conjugate eigenvalues crosses the imaginary axis. At the bifurcation point, $\lambda$ is purely imaginary. A periodic oscillatory solution (i.e., a closed cycle in phase space) bifurcates from the steady state solution, which loses its stability thereby. The generated limit cycle may be stable (*supercritical* Hopf bifurcation) or unstable (*subcritical*). Its amplitude grows from zero with a square-root law as a function of the control parameter. Here we shall not consider the general case (Schöll 1989), but restrict ourselves to $N = 1$. The condition for a Hopf bifurcation can then be stated as

$$\lambda_1 + \lambda_2 = 0, \ \lambda_1\lambda_2 > 0. \tag{17}$$

One can prove that this is equivalent to

$$\tilde{\lambda} \equiv \frac{\partial \Phi_0}{\partial n} = \tilde{\nu} \quad , \quad \tilde{\lambda}(\sigma_L + \sigma_{diff}) < 0 \qquad (18)$$

where

$$\tilde{\nu} \equiv (\sigma_L + \frac{\partial}{\partial \mathcal{E}}[\sigma(\mathbf{x}, \mathcal{E})\mathcal{E}])/(\epsilon_0 \epsilon_s) \qquad (19)$$

For $\sigma_L = 0$, $\tilde{\nu}$ is proportional to the differential mobility $dv/d\mathcal{E}$. Then (18) can be satisfied by either

$$\sigma_{diff} < 0 \quad \text{and} \quad \tilde{\lambda} = \tilde{\nu} > 0$$

or

$$\sigma_{diff} > 0 \quad \text{and} \quad \tilde{\lambda} = \tilde{\nu} < 0$$

This demonstrates that oscillatory instabilities are possible for either *negative* differential conductivity (and *positive* differential mobility), or *positive* differential conductivity (and *negative* differential mobility). With $N > 1$, more complex situations are possible, but $\sigma_{diff} < 0$ and $\sigma_{diff} > 0$ may still occur (Schöll 1989). Note that a Hopf bifurcation is not the only way to create self-generated oscillations; global bifurcations are another possibility (Thompson and Stewart 1986, Döttling and Schöll 1991).

## REAL SPACE TRANSFER

The occurrence of NNDC due to electron transfer in real space in a modulation-doped $GaAs/Al_xGa_{1-x}As$ heterostructure has been demonstrated theoretically (Gribnikov, 1973; Pacha and Paschke, 1978; Hess et al., 1979; Sakamoto et al., 1980; Aoki et al., 1989) and experimentally (Keever et al., 1981). Fig. 1 shows the conduction band diagram of such a structure. The AlGaAs layer is heavily n-doped with donor density $N_D$, while the GaAs layer is undoped. In thermodynamic equilibrium the electrons fall into the GaAs quantum well, where they experience strongly reduced impurity scattering. Thus the mobility $\mu_1$ in the GaAs layer is high. Application of an electric field, $\mathcal{E}_\parallel$, parallel to the layers induces carrier heating. As a result, the electrons are thermionically emitted into the AlGaAs, where their mobility, $\mu_2$, is much lower due to strongly enhanced impurity scattering. This real-space transfer from the high-mobility to the low-mobility layer at fields of the order of 2 $kV/cm$ causes NNDC, in analogy with the intervalley transfer in momentum space in the Gunn effect (Shaw et al., 1992). This effect has been exploited in devices like the charge injection transistor (CHINT), see Luryi (1990).

A novel mechanism for self-generated oscillatory instabilities under dc conditions in the real-space transfer regime has been proposed by Schöll and Aoki (1991). It is based upon the coupled nonlinear dynamics of the real-space electron transfer and of the space-charge in the doped AlGaAs layer, which controls the interface potential barrier $\Phi_B$ (Fig. 1a). Real-space transfer of the electrons in the GaAs layer leads to an increase of the carrier density in the AlGaAs, which diminishes the positive space charge controlling the band bending. Subsequently, the potential barrier $\Phi_B$ decreases with some delay due to the finite dielectric relaxation time. This leads to an increased backward thermionic emission current which decreases the carrier density in the AlGaAs. Hence the space charge and $\Phi_B$ are increased. This, in turn, decreases the backward thermionic emission current, which completes the cycle.

For a single heterojunction this oscillation mechanism can be modelled in terms of balance equations for the spatially averaged carrier densities in the GaAs layer and in the AlGaAs layer, respectively.

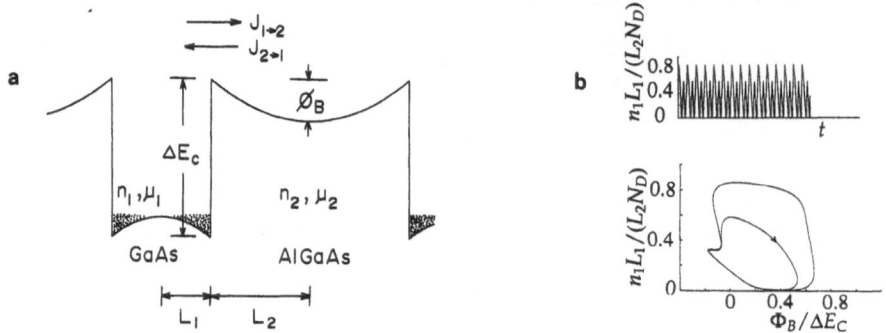

**Figure 1.** (a) Schematic energy band diagram of a GaAs/AlGaAs heterostructure. (b) Simulated self-generated oscillations. The phase portrait of $n_1$ in units of $N_D L_2/L_1$ versus the potential barrier $\Phi_B/\Delta E_c$, and the time series $n_1(t)$ is shown (after Schöll and Aoki, 1991)

$$n_1 = \frac{1}{L_1} \int_{-L_1}^{0} n(x,t)\, dx \quad , \quad n_2 = \frac{1}{L_2} \int_{0}^{L_2} n(x,t)\, dx,$$

supplemented by dielectric relaxation equations for the parallel electric field $\mathcal{E}_\parallel$ and the interface potential barrier $\Phi_B$. If the spatial variation along the y,z-direction is neglected, one obtains

$$\dot{n}_1 = (j_{1\to 2} - j_{2\to 1})/(eL_1) \ , \ \dot{n}_2 = (j_{2\to 1} - j_{1\to 2})/(eL_2) \tag{20}$$

$$\epsilon_0 \epsilon_s \dot{\Phi}_B = -e\mu_2 N_D \Phi_B + e^3 \mu_2 (N_D - n_2)^2 L_2^2/(2\epsilon_0 \epsilon_s) + e^2 L_2^2 \dot{n}_2 \tag{21}$$

$$\epsilon_0 \epsilon_s \dot{\mathcal{E}}_\parallel = J_0 - (\sigma_L + \frac{en_1 \mu_1 L_1 + en_2 \mu_2 L_2}{L_1 + L_2}) \mathcal{E}_\parallel \tag{22}$$

with the thermionic emission current densities

$$j_{1\to 2} = -en_1 (\frac{T_1}{2\pi m_1^*})^{1/2} exp(\frac{-\Delta E_c}{k_B T_1}) \tag{23}$$

$$j_{2\to 1} = -en_2 (\frac{T_2}{2\pi m_2^*})^{1/2} exp(\frac{-\Phi_B}{k_B T_2}) \tag{24}$$

The mean energies , i.e., the electron temperatures $T_1, T_2$ in the two layers can be adiabatically eliminated

$$T_2 \approx T_L \ , \ T_1 \approx T_L + \frac{2}{3k_B} \tau_E e\mu_1 \mathcal{E}_\parallel^2$$

Local charge neutrality requires

$$n_1 L_1 + n_2 L_2 = N_D L_2$$

so that one of the concentrations $n_1, n_2$ can be eliminated as a dependent variable. Eq. (22) for the space charge dynamics of $\Phi_B = -e \int_0^{L_2} \mathcal{E}_\perp dx$ is derived by spatially averaging eq. (11) for the internal perpendicular field $\mathcal{E}_\perp$, using (10).

The dynamic system (20),(21),(22) displays a Hopf bifurcation of a limit cycle, period-doubled oscillations (Fig. 1b) and chaos for certain values of the control parameters (Schöll and Aoki, 1991). If $R_L$ and $U_0$ are chosen such that three steady state operating points coexist, more complex bifurcation scenarios are found, including global bifurcations of limit cycles by condensation of paths or from a separatrix, and bistability of oscillatory and steady states (Döttling and Schöll, 1992). The latter leads to hysteretic switching transitions between self-generated oscillations and a steady state. For typical numerical parameters, there is a distinct time-scale separation between fast real space transfer and slow dielectric relaxation. In such systems not only bifurcations, but also sudden *quantitative* changes of the phase portrait in exponentially small ranges of the control parameter can occur. This results in *excitability* of the medium in the following sense: If the system is perturbed beyond a threshold excitation, it makes an extensive excursion in phase space prior to returning to its stable steady state (Döttling et al., 1992). The formation of travelling field domains is treated in a separate article in this volume (Döttling and Schöll, 1993).

## DOUBLE HETEROSTRUCTURES

Here we consider a thin, undoped $GaAs$ layer of width $2L_1$ spaced between two identical, n-doped $Al_xGa_{1-x}As$ layers of widths $L_2$. Quantum size effects and the coupled real space transfer into *both* AlGaAs barriers then lead to qualitatively new nonlinear dynamic features as compared to a single-heterojunction real-space-transfer oscillator (Reznik and Schöll, 1993). The nonlinear transport processes are modelled by a set of dynamic equations for the carrier densities $n_2$ and $n_2'$, and the potential barriers $\Phi_B$ and $\Phi_B'$ in each of the two AlGaAs barriers, respectively, identical to (20) and (21), but with the different current densities

$$j_{1\to 2} = -en_1 e^{E_1/k_B T_1}(L_1\frac{k_B T_1}{2\pi\hbar} + \frac{2m_1^*}{\pi\hbar^2}(\sqrt{E_1} + \sqrt{E_2})D_1(\Phi_B))$$

$$j_{2\to 1} = -en_2((8\pi k_B T_L m_2^*)^{-1/2}D_2(\Phi_B) + (\frac{k_B T_L}{2\pi m_2^*})^{1/2}e^{-\Phi_B/k_B T_L})$$

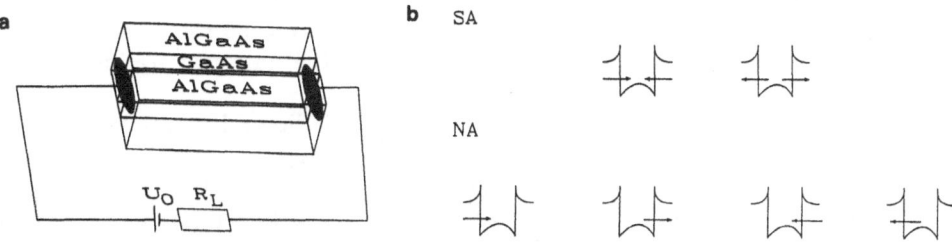

**Figure 2.**(a) Double heterostructure (schematic). (b) Symmetric (SA) and nonsymmetric (NA) oscillation mode of the double heterostructure (schematic). The arrows show different phases of the electron motion during one oscillation cycle, respectively.

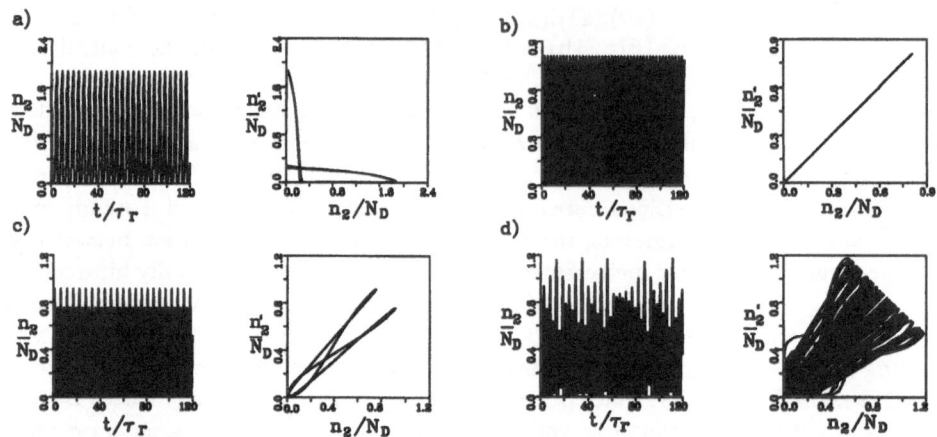

**Figure 3.** Time series (left) and phase portraits (right) for (a) nonsymmetric, (b) symmetric, (c) period-doubled, (d) chaotic oscillations of a double heterostructure with $\tau_r = 1.4ps$, $N_D = 10^{17}\ cm^{-3}$ (After Reznik and Schöll, 1993).

incorporating classical thermionic emission and nonresonant tunneling from the 2D subbands $E = E_i + \hbar^2(k_y^2 + k_z^2)/2m_1^*$ $(i = 1, 2)$ of the GaAs quantum well with tunneling coefficients $D_1$, $D_2$ (Reznik, 1992). In the equations for $\dot{n}_2'$ and $\dot{\Phi}_B'$ we have to replace $n_2, \Phi_B$ by $n_2', \Phi_B'$. The local neutrality condition is now given by

$$2n_1 L_1 + (n_2 + n_2')L_2 = 2N_D L_2 \ .$$

The parallel field $\mathcal{E}_\parallel$ relaxes fast and can be eliminated adiabatically. The solution of the dynamic equations shows that in the static current-field characteristic negative differential conductivity is quenched as a result of the modified 2D density of states, but oscillatory instabilities still occur. With increasing dc bias voltage $U_0$ one first finds a Hopf bifurcation of a nonsymmetric (NA, Fig. 3b) and then of a symmetric (SA, Fig. 3a) nonlinear self-generated oscillation mode, in which the electrons cycle asynchronously or synchronously, respectively, between the two barriers and the well (Fig. 2b). Thereafter there is bistability of these two modes, and the selection can be made, e.g., by changing the initial distribution of the electrons over the three layers. Upon further increase of bias, SA performs one period doubling (Fig.3c) and a transition to chaos (Fig. 3d) via quasiperiodicity. This is followed by a crisis at $\mathcal{E}_c$ whereby the chaotic attractor is suddenly transformed into a chaotic repeller (Fig. 4a). The latter manifests itself by long chaotic transients of the current, before the nonsymmetric attractor NA is eventually approached for $t \to \infty$. The escape rate $\kappa = 1/T$ of the repeller, where $T$ is the mean transient time, obeys the universal scaling law $\kappa \sim (\mathcal{E}_0 - \mathcal{E}_c)$.

The repeller contains a dense set of unstable periodic orbits, which can be stabilized by a delayed feedback control as proposed by Pyragas (1992). In the semiconductor structure this feedback might be realized by coupling the barrier potential $\Phi_B$ back to the dynamic equations for $\dot{\Phi}_B$ and $\dot{\Phi}_B'$ with a certain time delay $\tau$, i.e., by adding a term $K \cdot (\Phi_B(t - \tau) - \Phi_B(t))$. Such a transient chaos control may provide a powerful method of widely tuning the frequency and the oscillation mode of a real-space-transfer oscillator by selecting a suitable unstable periodic orbit to be stabilized, as shown, e.g., in Fig. 4b.

## COUPLED QUANTUM WIRES

A further reduction in dimensionality occurs if an array of lateral, mesa etched quantum wires is considered. In Fig. 5 two modulation-doped quantum wires A and B are shown, representing two real-space transfer oscillators coupled via the load of the external circuit to which they are connected (Utecht and Schöll, 1993). The two sets of model equations (20),(21) for A and B, respectively, are modified to take account of the quasi-1D electronic structure and density of states of the laterally confined electron gas in the $GaAs$-layer. The two quantum wires are coupled via the dielectric relaxation

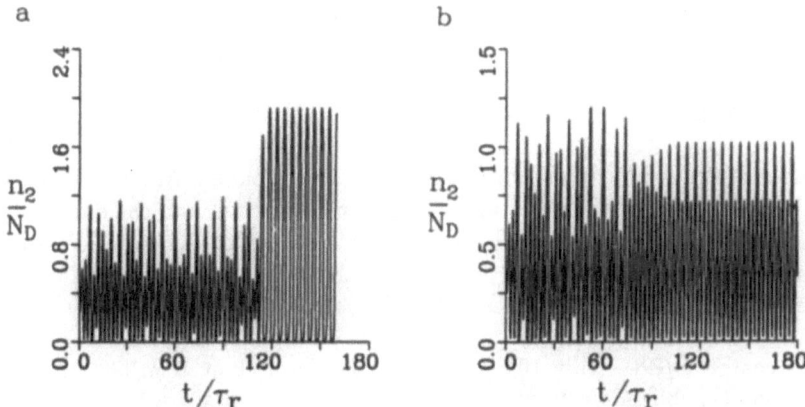

**Figure 4.** Transient chaos control by delayed feedback. The carrier density in the barrier 2 is plotted versus time ($\tau_r = 1.4ps$). (a) no control, (b) at $t = 75\tau_r$ the control is switched on. (After Reznik and Schöll, 1993).

equation analogous to (22). As a consequence, for symmetric quantum wires only the symmetric oscillation mode is stable (Fig. 5a) while the nonsymmetric oscillation is transient, representing a periodic repeller (Fig. 5c). The stability character of the symmetric and the nonsymmetric mode can be interchanged by connecting an external capacitor $C_{ext}$ parallel to the quantum wires (Fig. 5b). In this case, the phase lag of $\pi$ of the two wires leads to frequency-doubled current oscillations. It should be noted that the predicted frequencies are generally much higher than in the 2D and 3D case and are estimated of the order of several hunderd GHz.

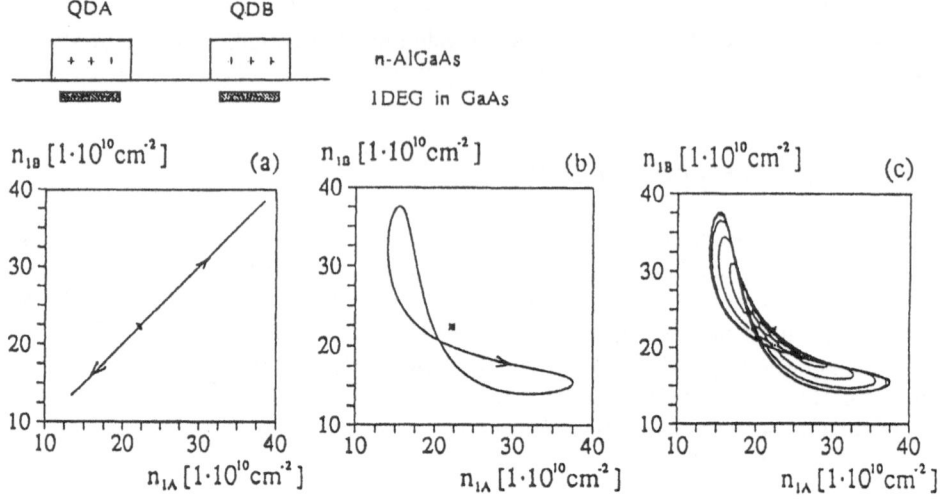

**Figure 5.** Phase portraits of carrier density in quantum wire B versus that in wire A (see top): (a) symmetric, (b) nonsymmetric, (c) nonsymmetric transient oscillations for two symmetric quantum wires of width 50 Å (After Utecht and Schöll, 1993).

## OPTICALLY INDUCED REAL SPACE TRANSFER

Real space transfer can also be induced by intersubband infrared (IR) excitation of carriers in the GaAs quantum well and subsequent tunneling into the adjacent AlGaAs layer, instead of field-induced carrier heating. A system of dynamic equations including the generation-recombination processes in the GaAs layer as well as resonant and nonresonant tunneling has been derived and analyzed with the methods of nonlinear dynamics by Kunz and Schöll (1993). For typical sample geometries ($2L_1 = 65$ Å, $2L_2$ = 500 Å center-doped within 100 Å, as used by Heinrich et al., 1990) the GaAs well contains two 2D subbands, where the higher level may or may not be in resonance with the localized level in the space-charge potential well of the AlGaAs layer. Therefore, large resonant or much smaller nonresonant tunnelling currents, respectively, are the dominant coupling mechanism between the two layers. Thermionic emission across the barrier can be neglected at the relatively small fields $\mathcal{E}_\parallel$ applied parallel to the layers. The dynamic equations (20),(21),(22) describing *electrically* induced real space transfer are modified accordingly. In particular, the GaAs intersubband dynamics for electrons in the lower subband, $n_1^{(1)}$, and in the higher subband $n_1^{(2)}$, has to be taken into account by including IR excitation and radiative and nonradiative relaxation. Further, unlike for a homogeneously doped AlGaAs layer, the barrier height $\Phi_B$ does not suffice for a consistent description of the space charge dynamics in the center doped case, but has to be supplemented by a dynamic equation for the potential loss in the doped region $\Phi_1 = -e \int_{L_u}^{L_2} \mathcal{E}_\perp \, dx$, where $L_u$ is the width of the undoped boundary stripes of the AlGaAs layer. Thus the system is described by the 5 dynamic variables $n_1^{(1)}$, $n_1^{(2)}$, $\Phi_B$, $\Phi_1$ and $\mathcal{E}_\parallel$.

It can be shown under quite general conditions (Kunz and Schöll, 1993) that oscillatory instabilities do not occur, but optoelectronic bistability is predicted. It involves the existence of two locally stable steady states at a given electric field, corresponding to two different types of tunnelling prevalent, and associated with different photoconductive responses: the regime of dominant resonant tunnelling shows negative photoconductivity as a function of the IR intensity $I_0$, whereas the nonresonant tunneling state is practically invariant with respect to changes of $I_0$.

## VERTICAL HOT-ELECTRON TRANSPORT

In the last part of this review we shall now consider vertical transport in a $GaAs/Al_xGa_{1-x}As$ heterostructure, i.e., the electric field is applied perpendicular to the heterolayers. Our model system is the heterostructure hot-electron diode (HHED), a two-terminal device proposed by Hess et al. (1986). This device consists of an $n^- - GaAs$ and an $n^- - Al_xGa_{1-x}As$ layer, spaced between two $n^+ - GaAs$ buffer layers. This device shows bistability between a low tunnelling current and a high thermionic emission current over the energy barrier in the conduction band as shown in Fig. 6, leading to SNDC. Oscillatory behaviour (Kolodzey et al., 1988) has been found, too. A similar device consisting of a multiple structure has been investigated by Belyantsev et al. (1986). SNDC had been predicted for such structures earlier by Gribnikov and Melnikov (1966).

A model which explains the measured SNDC characteristics and predicts an oscillatory instability was developed by Wacker and Schöll (1991). Assuming that the electric fields $\mathcal{E}_i$ and the current densities $j_i$ should be homogeneous in the x-direction (perpendicular to the heterojunction) for each layer $i = 1, 2$, one can write the following formulae for the total voltage drop $U$ accross the sample and the surface charge density $\rho_s$ between the two layers

$$U = L_1\mathcal{E}_1 + L_2\mathcal{E}_2 \tag{25}$$

$$\rho_s = \epsilon_0(\epsilon_2\mathcal{E}_2 - \epsilon_1\mathcal{E}_1) \tag{26}$$

If carriers are injected only from layer 1 into layer 2, which is correct as long as $|\mathcal{E}_2| \gg |\mathcal{E}_1|$, one obtains $j_1 = j_1(\mathcal{E}_1)$ and $j_2 = j_2(\mathcal{E}_1, \mathcal{E}_2)$. Irrespectively of the explicit forms of $j_1$ and $j_2$, conditions for negative differential conductance $dU/dI < 0$ and for oscillatory instabilities via a Hopf bifurcation can be derived (Wacker and Schöll 1991,1992). Since in systems with SNDC filamentation is likely to occur ( this has in fact been reported in a GaAs/AlGaAs structure by Higman et al., 1988), spatial inhomogeneities in the y,z-direction parallel to the layers has to be considered. For Ohmic planar contacts, however, $U$ is not space-dependent; hence it is convenient to use $U(t)$ and $\rho_s(y, z, t)$ as dynamic variables rather than $\mathcal{E}_1$ and $\mathcal{E}_2$.

The dynamic equation for $\rho_s$ can be obtained from the continuity equation

$$\dot{\rho}_s = j^- - \nabla \cdot \mathbf{j}_\| \quad with \quad j^-(\rho_s, U) = j_1 - j_2$$

where $\mathbf{j}_\|$ is the 2D current density in the 2D electron gas at the heterojunction parallel to the interface. In the drift-diffusion approximation

$$\dot{\rho}_s = j^-(\rho_s, U) + \frac{\mu}{\epsilon_0(\epsilon_1/L_1 + \epsilon_2/L_2)}\nabla \cdot (|\rho_s|\nabla\rho_s) \tag{27}$$

The dynamics of $U$ follows from Maxwell's eq. (11):

$$\dot{U} = (I - \int_A dydz \, j^+)/(C + C_{ext}) \quad with \quad j^+(\rho_s, U) = C(j_1/C_1 + j_2/C_2) \tag{28}$$

where $C_i = A\epsilon_0\epsilon_i/L_i$ are the intrinsic capacitances of the single layers and $C^{-1} = C_1^{-1} + C_2^{-1}$. A linear stability analysis of the homogeneous steady state

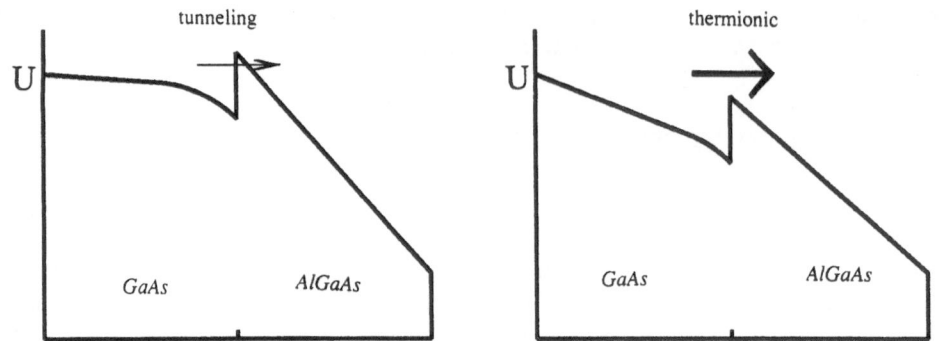

**Figure 6.** Schematic conduction band structure of a *GaAs/AlGaAs* heterostructure with a perpendicular electric field. The two possible conduction states at a given voltage U in the bistability regime are shown.

$j_1 = j_2 = I/A$ yields a Hopf bifurcation if

$$\frac{\partial j^-}{\partial \rho_s} = \frac{A}{C_{ext}} \left( \epsilon_2 \frac{\partial j_1}{\partial \mathcal{E}_1} + \epsilon_1 \frac{\partial j_2}{\partial \mathcal{E}_2} \right) / (\epsilon_1 L_2 + \epsilon_2 L_1)$$

which occurs on the NDC branch of the I(V) characteristic for sufficiently large $C_{ext}$. The resulting spatially homogeneous self-generated oscillation is shown in Fig. 7 for a specific, simple model of the current densities

$$j_1 = e\mathcal{E}_1(n_{\Gamma 1}\mu_\Gamma + n_{L1}\mu_L), \quad j_2 = j_\Gamma + j_L + j_T$$

where the conduction electrons are distributed in the $\Gamma$- and the L-valley of the *GaAs* layer according to a common electron temperature $T_1$, and $j_2$ is composed of tunnelling and thermionic emission currents from the $\Gamma$-valley, $j_\Gamma$, from the L-valley, $j_L$, and tunnelling from the bound state of the 2D electron gas (representing the interface charge density $\rho_s$), $j_T$, respectively. The electron temperature $T_1$ is enslaved by the field $\mathcal{E}_1$ and can be eliminated adiabatically. The oscillation cycle is visualized in the $(\mathcal{E}_1, \mathcal{E}_2)$ phase portrait in Fig. 7(b) as follows. For small fields $\mathcal{E}_1, \mathcal{E}_2$ the tunneling current is small and $j_2 < j_1$. Hence $\rho_s$ increases and so does $\mathcal{E}_2$ while $\mathcal{E}_1$ remains almost constant. For large $\mathcal{E}_2$ the tunneling current becomes quite large due to a reduced barrier width (Fig. 6). Thus $j_1 < j_2$, and $\rho_s$ decreases while $\mathcal{E}_1$ increases, enhancing thermionic emission over the barrier. Subsequently $\mathcal{E}_2$ decreases until $j_2 < j_1$ is reached again, which completes the cycle.

## SELF-ORGANIZED SPATIO-TEMPORAL DYNAMICS

The linear stability of (27),(28) reveals that the uniform steady state can also become unstable against the formation of an *inhomogeneous* distribution of $\rho_s(y)$ leading to a filamentary current flow $j_1(y), j_2(y)$. Under current bias different instabilities can occur (Wacker and Schöll, 1992), depending upon the control parameters and initial conditions chosen: (i) spatially homogeneous oscillations as shown in Fig. 7, (ii) stationary current filaments, (iii) a novel type of spatio-temporal oscillations where an

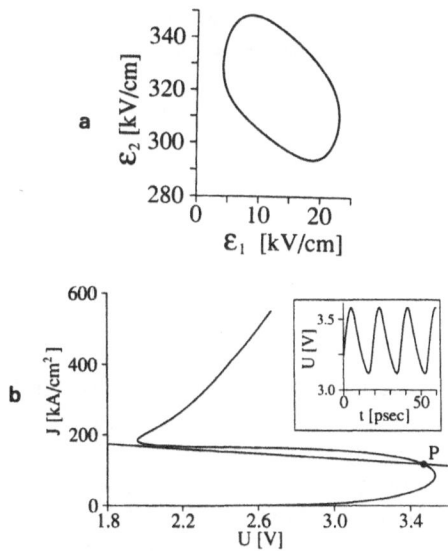

**Figure 7.** Spatially homogeneous oscillation in the heterostructure hot electron diode. In (a) the static current-voltage characteristic is shown together with the load-line. The inset depicts voltage oscillations U(t). (b) Phase portraits of the electric fields $\mathcal{E}_2$ versus $\mathcal{E}_1$. (After Wacker and Schöll, 1991).

inhomogeneous filamentary distribution $\rho_s(y)$, $j_1(y)$, $j_2(y)$ is formed and quenched periodically, see Fig. 8. This periodicity is very sensitive to stochastic fluctuations, since $\rho_s$ and $U$ grow very slowly after the extinction of the filamentary spikes (see Fig. 8 b and Fig. 8a). Thus the time intervals between two spikes are governed by noise, resulting in an irregular sequence of voltage spikes. These fluctuations can be shown to be particularly pronounced at the onset of the instabilities. The spatio-temporal dynamics can be interpreted in the general context of domain wall motion in bistable systems (Schimansky-Geier et al., 1991). It should be noted that filament switching oscillations of type (iii) have been observed in Si pin diodes (Symanczyk et al., 1991).

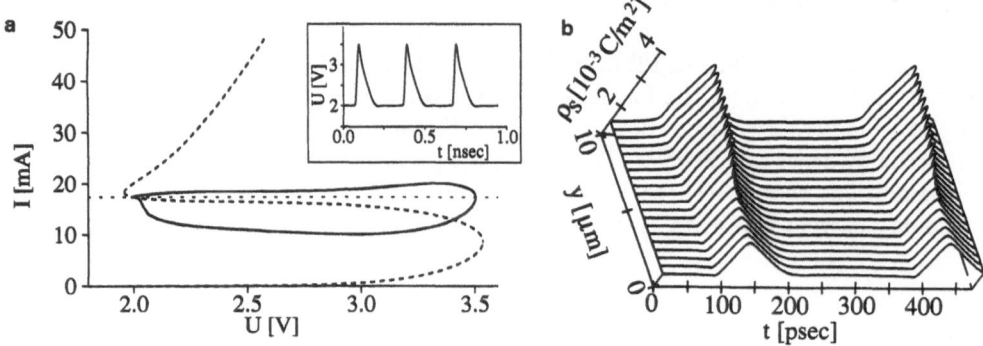

**Figure 8.** Spatio-temporal oscillation in the heterostructure hot-electron-diode. In (a) the static current-voltage characteristic (dashed) and the current bias (dotted horizontal line) are shown. The time-dependent solution is shown as the full line in the I-U phase portrait and as U(t) in the inset. The associated spatio-temporal dynamics of the surface charge density $\rho_s$ is plotted in (b). (After Wacker and Schöll, 1992)

## CONCLUSION

Theoretical approaches to oscillatory hot-electron instabilities in different semiconductor heterostructures, quantum wells, and quantum wires have been reviewed from the unified viewpoint of nonlinear dynamic systems. It has been demonstrated that a variety of complex temporal and spatio-temporal dynamic behaviour and bifurcation scenarios can arise in such nonlinear transport models. The underlying physical mechanism can be roughly divided into two classes: (i) Parallel transport in modulation-doped heterostructures associated with NNDC and real-space transfer. (ii) Vertical transport across a heterostructure associated with SNDC and bistability.

Experimentally, oscillations in 2D heterolayer systems under dc conditions have been observed by several authors (e.g., Balkan and Ridley, 1989; Vickers et al., 1989; Hendriks et al., 1991; see also the review by Ridley, 1991), but the physical origin is not yet completely clear. Future research should be directed to clarify these instabilities by developing more sophisticated physical models, and analyzing them with the tools of nonlinear dynamics as outlined for some exemplary cases in this review.

### Acknowledgement

I am indebted to K.Aoki, M.Asche, R.Döttling, O.Hess, G.Hüpper, R.Kunz, K.Pyragas, W.Quade, D.Reznik, R.Schmolke, C.Utecht, A.Wacker for their valuable discussions and contributions. I am grateful to D.Reznik for technical assistance in preparing this manusript.

# REFERENCES

Abe, Y. (ed.), 1989, Special Issue on Nonlinear and Chaotic Transport in Semiconductors, Appl. Phys. A48, 93.

Aoki, K., K. Yamamoto, N. Mugibayashi, and E. Schöll, 1989, Sol. State Electron. 32, 1149.

Balkan, N. and B.K. Ridley, 1989, Semicond. Sci. Technol. 3, 507; Superlatt. Microstruct. 5, 539.

Belyantsev, A.M., A.A. Ignatov, V.I. Piskarev, M.A. Sinitsyn, V.I. Shashkin, B.S. Yavich, and M.L. Yakovlev, 1986, JETP Letters 43, 437.

Döttling, R. and E. Schöll, 1992, Phys. Rev. B 45, 1935.

Döttling, R. and E. Schöll, 1993, see this volume.

Döttling, R., E. Schöll, and D. Reznik 1992, Proceedings of the NATO ASI on Chaotic Dynamics: Theory and Practice, ed. T. Bountis, Plenum, N.Y.

Ebeling, W., and R. Feistel, 1986, "Physik der Selbstorganisation und Evolution" Akademie-Verlag, Berlin.

Gribnikov, Z.S., and M.I. Mel'nikov, 1966, Sov. Phys. Solid State 7, 2364.

Gribnikov, Z.S., 1973, Sov. Phys. Semicond. 6, 1204.

Haken, H., 1983, "Synergetics, An Introduction", 3rd ed., Springer-Verlag, Berlin.

Heinrich, R., R. Zachai, M. Besson, T. Egeler, G. Abstreiter, W. Schlapp and G. Weimann, 1990, Surface Science 228, 465.

Hendriks, P., E.A.E. Zwaal, J.G.A. Dubois, F.A.P. Blom, and J.H. Wolter, 1991, J. Appl. Phys. 69, 302.

Hess, K., H. Morkoc, H. Shichijo, and B.G. Streetman, 1979, Appl. Phys. Lett. 35, 469.

Hess, K., T.K. Higman, M.A. Emanuel and J.J. Coleman, 1986, J. Appl. Phys. 60, 3775.

Higman, T.K., L.M. Miller, M.E. Favaro, M.A. Emanuel, K. Hess and J.J. Coleman, 1988, Appl. Phys. Lett. 53, 1623.

Hüpper, G., and E. Schöll, 1991, Phys. Rev. Lett. 66, 2372.

Keever, M., H. Shichijo, K. Hess, S. Banerjee, L. Witkowski, and H. Morkoc, 1981, Appl. Phys. Lett. 38, 36.

Kolodzey, J., J. Laskar, T.K. Higman, M.A. Emanuel, J.J. Coleman, and K. Hess, 1988, IEEE EDL-9, 272.

Kunz, R. and E. Schöll, 1993, Phys. Rev. B 47.

Luryi, S., 1990, Superlatt. Microstr. 8, 395.

Nicolis, G., and I. Prigogine, 1977, "Self-Organisation in Non-Equilibrium Systems", Wiley, N.Y..

Pacha, F., and F. Paschke, 1978, Electron. Commun. 32, 235.

Pyragas, K., 1992, Phys. Lett. A 170, 421.

Reznik, D., 1992, Diplom Thesis, Technical University of Berlin.

Reznik, D., and E. Schöll, 1993, Z. Phys. B,

Ridley, B.K., 1991, Rep. Prog. Phys. 54, 169.

Sakamoto, R., K. Akai, and M. Inoue, 1989, IEEE Trans. Electron. Devices 36, 2344.

Schimansky-Geier, L., Ch. Zülicke, and E. Schöll, 1991, Z. Phys. B84, 433.

Schöll, E., 1987, "Nonequilibrium Phase Transitions in Semiconductors", Springer-Verlag, Berlin.

Schöll, E., 1989, Physica Scripta T29, 152.

Schöll, E., 1992a, in: "Nonlinear Dynamics in Solids", H. Thomas, ed., Springer-Verlag, Berlin.

Schöll, E., 1992b, in: "Handbook on Semiconductors," P.T. Landsberg, ed., North Holland, Amsterdam, 2nd ed.

Schöll, E., and K. Aoki, 1991, Appl. Phys. Lett. 58, 1277.

Shaw, M.P., V.V. Mitin, E. Schöll, and H.L. Grubin, 1992, "The Physics of Instabilities in Solid State Electron Devices", Plenum, N.Y.

Symanczyk, R., S. Gaelings, and D. Jäger, 1991, Phys. Lett. A160, 397.

Thomas, H. (ed.), 1992, "Nonlinear Dynamics in Solids", Springer-Verlag, Berlin.

Thompson, J.M.T., and H.B. Stewart, 1986, "Nonlinear Dynamics and Chaos", Wiley, N.Y.

Utecht, C. and E. Schöll, 1993, to be published.

Vickers, A.J., A. Straw, and J.S. Roberts, 1989, Semicond. Sci. Technol. 4, 743.

Wacker, A., and E. Schöll, 1991, Appl. Phys. Lett. 59, 1702.

Wacker, A., and E. Schöll, 1992, Semicond. Science Techn. 7, 1456.

# LIGHT EMITTING LOGIC DEVICES BASED ON REAL SPACE TRANSFER IN COMPLEMENTARY InGaAs/InAlAs HETEROSTRUCTURES

Serge Luryi and Marco Mastrapasqua

AT&T Bell Laboratories
Murray Hill, NJ 07974

## ABSTRACT

We discuss the principle and the implementation of novel light-emitting logic devices based on the real-space transfer (RST) of hot electrons between complementary conducting layers. One of these layers, the emitter, is doped $n$-type and has two or more contacts for applying the lateral electric field. Heated by this field, electrons are injected into a complementary $p$-type collector layer, contacted independently. The injection current is accompanied by a luminescence signal arising from the recombination of the transferred electrons in a specially designed collector active region.

The peculiar symmetry of charge injection by RST enables one to implement functional logic gates. The simplest structure with two emitter contacts acts as an exclusive OR gate in the dependence of both the collector current and the output light on the input voltages. The multiterminal device performs such functions as OR and NAND and is electrically reprogrammable between these functions. These powerful logic operations are demonstrated at room temperature.

We also review recent theoretical studies of the symmetry properties of RST transistors. These studies, based on continuation modeling and transient device simulation, reveal a variety of instabilities and a striking novelty of *multiply-connected* current-voltage characteristics. A RST transistor can support anomalous steady states in which hot-electron injection occurs in the absence of any voltage between the emitter electrodes. Some of these states break the reflection symmetry in the plane normal to the channel at midpoint. In the anomalous states, the electron heating is due to a fringing field from the collector electrode. The formation of hot-electron domains in real-space transfer represents a transition to such a collector-controlled state.

## 1. INTRODUCTION

An important direction in the microelectronics research is the development of new *functional devices*, which can perform logic tasks that would normally require an assembly of several transistors. The charge injection transistor (CHINT) offers interesting opportunities in this context.[1] The CHINT concept refers to a class of devices based on the principle[2,3] of real-space transfer (RST) of hot electrons between independently contacted conducting layers. Several functional devices employing this principle have been discussed in the literature.[4-8]

The generic CHINT structure is illustrated in Fig. 1a. One of the conducting layers, the emitter, plays the role of a hot-electron cathode, with the heating voltage applied to the contacts $S$ and $D$. The other conducting layer, the collector, is separated by a heterostructure barrier. The RST effect manifests itself in the increase of the collector current $I_C$ at a constant positive collector bias $V_C$, when a sufficiently high heating bias $V_{DS}$ is applied.

*Negative Differential Resistance and Instabilities in 2-D Semiconductors*
Edited by N. Balkan *et al.*, Plenum Press, New York, 1993

53

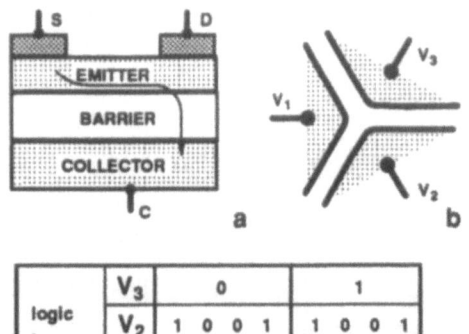

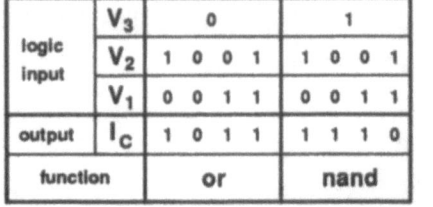

| logic input | $V_3$ | 0 | | | | 1 | | | |
|---|---|---|---|---|---|---|---|---|---|
| | $V_2$ | 1 | 0 | 0 | 1 | 1 | 0 | 0 | 1 |
| | $V_1$ | 0 | 0 | 1 | 1 | 0 | 0 | 1 | 1 |
| output | $I_C$ | 1 | 0 | 1 | 1 | 1 | 1 | 1 | 0 |
| function | | or | | | | nand | | | |

c

**Fig. 1.** Charge injection transistor principle and the RST logic.

(a) Schematic diagram of a CHINT. Emitter electrons, heated by the field applied between electrodes $S$ and and $D$, undergo RST into the collector layer, as indicated by the arrow.

(b) ORNAND logic gate. Input terminals arranged with a 3-fold cyclic symmetry define three channels 1-2, 2-3, and 3-1.

(c) Truth table of $I_C(V_1, V_2, V_3)$. The RST current is *off* in two states when $V_1 = V_2 = V_3$ and *on* in the other six states. By symmetry, every *on* state has the same $I_C$.

A fundamental property of RST transistors, essential for our present discussion, is the *symmetry equivalence*[9,10] between the internal states $S[V_D, V_C]$ of the device at different external bias configurations:

$$S[V_D, V_C] \rightleftarrows S[-V_D, (V_C - V_D)]. \tag{1}$$

This correspondence follows from the reflection symmetry in the plane normal to the source-drain direction which cuts the channel in the middle. Although a similar relation exists between internal states $S[V_D, V_G]$ in a field-effect transistor, there is an important difference in that the CHINT collector is the output terminal and the symmetry expressed by Eq. (1) implies that the output current is invariant under an interchange of the input voltages $V_S$ and $V_D$. Thus the device exhibits an exclusive-OR (**xor**) dependence of the collector current on the input voltages, regarded as binary logic signals.

Even more powerful logic functionality is obtained in a RST device with three input terminals.[6] This device, which we shall refer to as the ORNAND gate, has a cyclic 3-fold symmetry, Fig. 1b. Its truth table (Fig. 1c) corresponds to **ornand**$(\{V_j\})$ = **or**$(V_1, V_2)$ when $V_3$ is low, and **ornand**$(\{V_j\})$ = **nand**$(V_1, V_2)$ when $V_3$ is high.

The invention of charge injection logic[6] has focused attention on the symmetry properties of RST transistors. Their potential applications are likely to be based on the peculiar symmetry with respect to the heating field polarity. The same symmetry shows up in the analysis of the *instabilities* associated with the hot-electron injection and the domain formation in RST transistors. An unexpected recent discovery[9] is the existence of a number of anomalous states at $V_D = 0$, some of which are not only stationary but also *stable* with respect to small perturbations. In these states, the electron heating is due to the fringing field from the collector electrode. Because of the relation (1), states at $V_D = 0$ must either be symmetric with respect to reflections in the midplane normal to channel or possess broken-symmetry partners. By theoretical modeling of the CHINT operation it has been found[9] that the device possesses intrinsically complicated – often multiply-connected – $IV$ characteristics. Application of a sufficiently high $V_D$ at a fixed $V_C > 0$ forces a switching transition, accompanied by the formation of a hot electron domain. Physically, the domains form when the finite supply rate of electrons to a "hot spot" is exceeded by the RST flux from that spot. The depleted domains unscreen the fringing field ("normally" screened by channel electrons) and the RST becomes collector controlled.

Until recently, all of the work on CHINT and related logic devices involved heterostructures with the *same type of conductivity* in both the emitter channel and the collector. The review[10] cites a number of general references on unipolar real-space transfer transistors. Recently, one of us discussed[11] the possibility of using the RST of carriers into a *complementary* collector layer to implement light emitting devices endowed with a logic functionality with respect to electrical input. Significant progress in this program has been made in our laboratories during the last year. The first complementary CHINT was implemented in a InGaAs/InAlAs heterostructure and demonstrated the **xor** operation.[12,13] Subsequently, the multiterminal ORNAND gate was implemented,[14] demonstrating the **or** and **nand** functions – both in the output light and the collector current – as well as the electrical switching between these functions.

The present work reviews the recent contributions.[12–14, 9] The epitaxial structures used and the fabrication sequence are discussed in the next section. Electrical characterization of the complementary CHINT structures are described in Sect. 3, their optical characterization in Sect. 4, and the logic operation in Sect. 5. Results of the numerical simulation[9] of hot-electron transport in the simplest CHINT structure (Fig. 1a) are reviewed in Sect. 6. Our conclusions are summarized in Sect. 7.

## 2. STRUCTURES

Figures 2 show schematic cross-sections of the logic device structures discussed below. Structure of the XOR gate is illustrated in Fig. 2a and ORNAND gate in Fig. 2b. The $In_{0.53}Ga_{0.47}As/In_{0.52}Al_{0.48}As$ heterostructures have been grown by molecular beam epitaxy on InP substrate. The design is based on the epitaxial contact scheme, first used in the fabrication of CHINT by Mensz et al.[15] In this scheme, the channel length is defined by a trench etched in the $n^+$ cap layer (*a*), while the remaining portions of the cap layer make ohmic contacts to the source and drain metal. The 25 Å InAlAs layer (*b*) is used as an etch stop in the selective etching of the cap layer[16] All patterns, including the trenches, were defined by standard optical contact lithography.

In the ORNAND gate, instead of the triangular electrode arrangement of Fig. 1b, the required cyclic symmetry is obtained with four electrodes 1, 2, 3, 3̃ – two of which, 3 and 3̃, being logically identical ($V_{\tilde{3}} \equiv V_3$) though physically split.

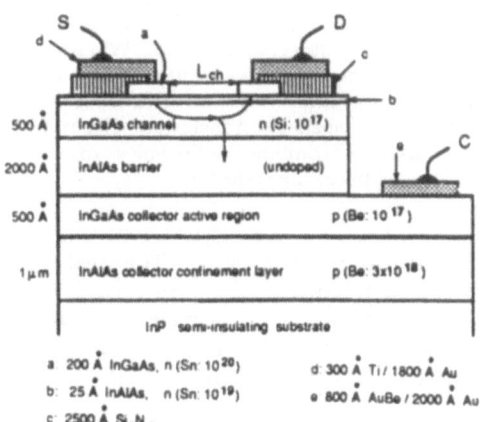

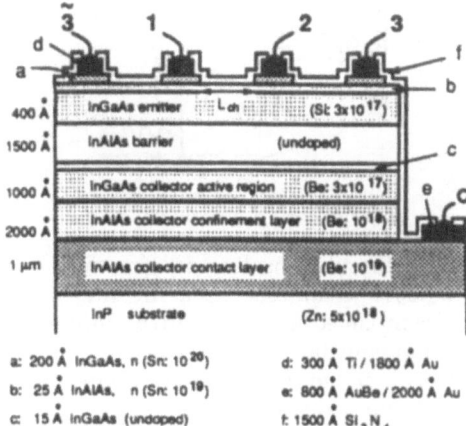

**Fig. 2a.** Cross section of the first complementary charge injection transistor.[12,13] The emitter channel length, defined by the trench etched in the cap layer (*a*), is $L_{CH} = 3\,\mu m$ and the width $W = 50\,\mu m$.

**Fig. 2b.** The ORNAND gate.[14] Each of the three channels, defined by trenches in the cap layer (*a*), has length $L_{CH} = 1\,\mu m$ and width $W = 40\,\mu m$. Cyclic symmetry results from the periodic boundary condition $V_{\tilde{3}} \equiv V_3$.

The epitaxial structures, shown in Figs. 2a and 2b are generally similar, but there are several noteworthy differences. The ORNAND structure is grown on a $p$-type substrate and uses a more heavily doped collector contact layer to reduce the parasitic series resistance in the collector circuit. The emitter channel is also more heavily doped to avoid its complete depletion by the surface potential and ensure a "normally-on" channel, which conducts even in the absence of a positive $V_C$.

Figure 3 shows a schematic energy-band diagram of the XOR device (Fig. 2a) in a cross-section under the trench. The corresponding diagrams for the ORNAND structure are quite similar, the main difference being that the emitter channel is not depleted ("normally-on") in equilibrium, due to heavier doping. In our XOR device in equilibrium, Fig. 3a, the channel is entirely depleted by the surface potential and the $pn$ junction, so that the channel conduction requires a positive collector voltage to induce a two-dimensional electron gas at the interface with the barrier. In the "flat-band" condition, illustrated in Fig. 3b, the collector bias $V_C = V_{FB} \approx 0.6\,\mathrm{eV}$ corresponds to the work-function difference between $n$-InGaAs in the channel and $p$-InGaAs in the collector. In the operating regime, Fig. 3c, the $pn$ junction is forward biased and the main obstacle to current is due to the band discontinuities at the two interfaces with the InAlAs barrier. Inasmuch as the valence band discontinuity in the InGaAs/InAlAs heterosystem is smaller than the conduction band discontinuity[17] ($\Delta E_V = 0.20\,\mathrm{eV}$ < $\Delta E_C = 0.50\,\mathrm{eV}$) we can expect that, in the absence of RST, the collector current will be dominated by holes.

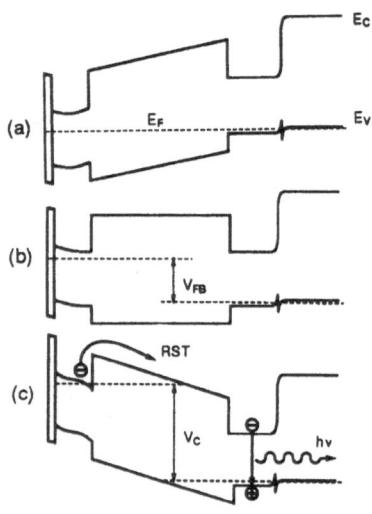

Fig. 3. Schematic energy-band diagram in a cross-section under the trench of the complementary CHINT of Fig. 2a.

(a) Equilibrium. The collector and the emitter-channel layers form a $pn$ junction, separated by a wide-gap barrier. The channel is depleted by the surface potential.

(b) Flat-band condition. The flat-band voltage $V_C = V_{FB}$ is undetermined because of an uncertainty in the surface potential. The calculated flat-band voltage in a cross-section under the contacts is $V_{FB} = 0.63\,\mathrm{V}$.

(c) Operating regime. The channel charge is induced by the collector field. When the heating electric field is applied along the channel, the latter becomes a hot-electron emitter. Real-space transferred electrons radiatively recombine with holes in the active region of the collector.

Electron heating is generated by a drain-to-source bias $V_{DS}$. The real-space transfer manifests itself in the increasing collector current $I_C$. As usual, it is accompanied by a negative differential resistance in the drain current $I_D$. In our complementary structures, the RST injection is accompanied by a 1.6 μm luminescence signal, arising from the recombination of the injected electrons with holes in the InGaAs collector active region.

The purpose of the wide-gap InAlAs layer in the collector is to spatially confine electrons injected over the barrier and, at the same time, to provide a low-resistance path for the collector current. Both requirements are very important. Confinement is necessary for the radiative efficiency (otherwise most of the injected electrons would reach the collector contact prior to recombination) and low collector resistance is necessary to minimize a spurious back-gating effect, which would otherwise severely limit the available RST current. If the doping level $N_A$ in the confining barrier is uniform, then a balance must be struck between the requirements of a low resistance (minimized by higher $N_A$) and a low recombination velocity at the interface between the active and confinement layers (minimized by lower $N_A$). A better solution, well-known in the art of designing heterostructure lasers, is to heavily dope most of the confinement layer but leave a thin low-doped sublayer immediately adjacent the active region. This approach was used (though by no means optimized) in the design of our more recent structure, Fig. 2b.

## 3. Electrical Characterization

Electrical properties of a complementary CHINT are quite different from those of unipolar devices.[4,10] Most of the differences are rooted in the fact that the collector and the emitter form a *pn* junction with a wide-gap barrier in between. The device operating regime corresponds to a forward bias of the *pn* junction, hence we must rely exclusively on the band-gap discontinuities to block the unwelcome leakage current. In what follows we shall refer to as the leakage that part of the collector current which flows independently of the electron temperature $T_e$ controlled by the lateral field in the emitter channel. Identification of the leakage mechanism is an important part of the device characterization.

### 3.1 Leakage Current

In a unipolar RST transistor, the leakage can be strongly suppressed by choosing a heterostructure with a large conduction band discontinuity. Excellent results have been obtained[16] in the InGaAs/InAlAs system which has $\Delta E_C = 0.50$ eV. In a complementary CHINT, we have to worry about the flux of holes from the collector into the emitter channel.

Figures 4 and 5a show the collector leakage current characteristics of the Fig. 2 devices with all emitter contacts grounded. The measured values of $I_C$ are plotted against the collector bias $V_C$ for different temperatures. At high temperatures, $T \geq 200$ K, and relatively low bias the current obeys the thermionic model

$$I_C = I_C^{SAT} e^{qV_C/n\,kT} \,, \tag{2a}$$

$$I_C^{SAT} = S\,A^*T^2\,e^{-\Phi/kT} \,, \tag{2b}$$

where $n$ is the diode ideality factor, $I_C^{SAT}$ the saturation current, $S$ is the total emitter area including contacts, $\Phi$ is the barrier height at zero bias (separation between the hole Fermi level in the collector and the valence band edge in the barrier at the channel interface), and $A^*$ the effective Richardson constant. From the slopes of the $I_C(V_C)$ curves at low bias, indicated by the dashed lines in Figs. 4 and 5a we determine $n \approx 1.4$. Extrapolating to $V_C = 0$, we determine $I_C^{SAT}$ and find that it can indeed be fitted to the form (2b).

The inset to Fig. 4 and Fig. 5b show the Arrhenius plots of $I_C^{SAT}/T^2$ versus $T^{-1}$ in the high-temperature range. The slopes of these plots give $\Phi = 0.90$ eV for the device in Fig. 2a and $\Phi = 0.84$ eV for the ORNAND device, Fig. 2b. These values are in reasonable agreement with calculated values $\Phi^h$ of the barrier height for holes relative to their Fermi level in the collector, based on the room-temperature energy gaps $E_G^{In_{0.53}Ga_{0.47}As} = 0.75$ eV, $E_G^{In_{0.52}Al_{0.48}As} = 1.45$ eV, the band discontinuities $\Delta E_C = 0.50$ eV, $\Delta E_V = 0.20$ eV, and given doping levels. The calculated values for the devices in Figs. 2a and 2b are, respectively, $\Phi^h = 0.89$ eV and $\Phi^h = 0.92$ eV. This is to be contrasted with the estimated zero-bias barrier heights for emitter electrons, $\Phi^e = 1.10$ eV and $\Phi^e = 1.13$ eV, respectively. The slight discrepancy between the measured and calculated values of $\Phi^h$ in the multiterminal device may be explained by a minor deviation of the actual doping levels from nominal or perhaps by thermally-assisted tunneling effects.

At higher collector biases, exceeding the flat-band condition $V_C > V_{FB}$, the top of the barrier for holes is at the collector interface. This is the operating regime of a complementary CHINT. Further increase of the collector current with $V_C$ occurs primarily because of the accumulation of holes (increasing the Fermi level) and also,

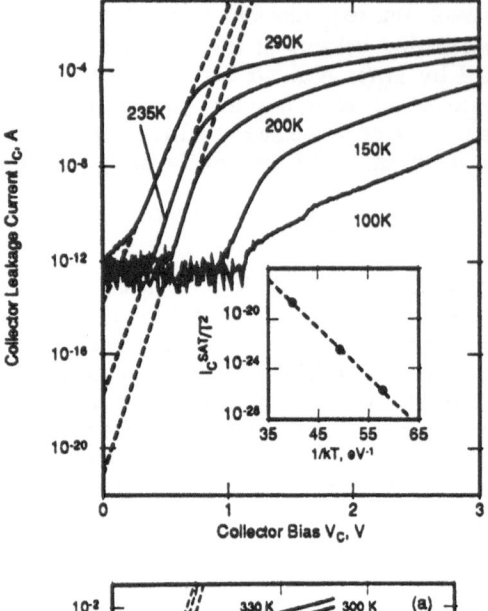

Fig. 4. Collector diode characteristics of the Fig. 2a device at different temperatures. Source and drain are connected together and grounded. Dashed lines indicate the linear extrapolation of $I_C$ to zero forward bias. The intercepts of the dashed lines with the ordinate axis define $I_C^{SAT}$. Inset shows an Arrhenius plot of $I_C^{SAT}/T^2$. The measured activation energy $\Phi = 0.90\,eV$ agrees with $\Phi^h = 0.89\,eV$, calculated taking $E_F^e = 12\,meV$ and $E_F^h = 105\,meV$ for the equilibrium Fermi level separations from the conduction band in the emitter layer under the source/drain contacts and from the valence band in the collector active layer, respectively.

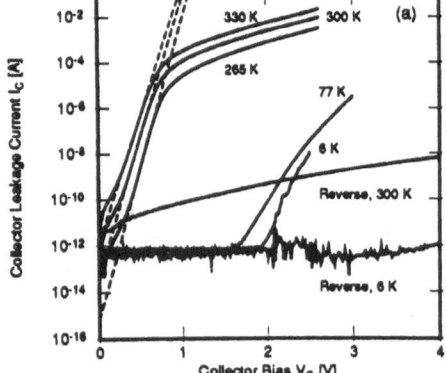

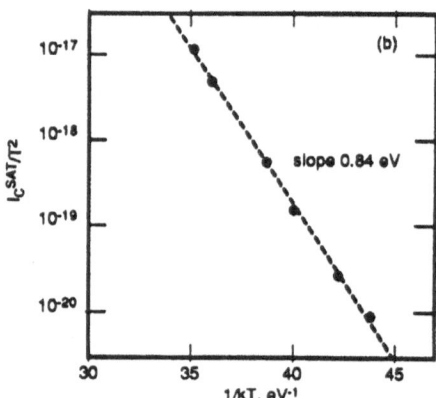

Fig. 5. Collector leakage current characteristics of the ORNAND device (Fig. 2b) at several temperatures. Curves $I_C(V_C)$ were measured with all four emitter electrodes grounded. Forward-biased collector-emitter $pn$ junction corresponds to $V_C > 0$. The leakage current under reverse bias is of opposite polarity and the plotted curves represent $|I_C|$. Dashed lines in (a) indicate a linear extrapolation of $\log I_C$ to zero forward bias. From the Arrhenius plot (b) of the extrapolated value $I_C^{SAT}/T^2$ the activation energy is $0.84\,eV$. The calculated barrier height relative to the hole Fermi level is $\Phi_h = 0.92\,eV$, assuming $E_F^e = -21\,meV$ (degenerate) and $E_F^h = 82\,meV$ estimated from the given doping levels.

because of the increasing role of hole tunneling which lowers the effective barrier. Hence the leakage curve departs from Eq. (2) when $V_C > V_{FB}$. As is evident from Figs. 4 and 5a, in the operating regime the slope of $\log I_C(V_C)$ is relatively gentle. At low temperatures, $T \leq 150\,K$, thermally assisted tunneling of holes is the dominant leakage mechanism.

## 3.2 Transistor Characteristics

The usual way of presenting current-voltage characteristics of a charge injection transistor is to plot the family of $I_C(V_D)$ and $I_D(V_D)$ curves at constant values of $V_C$. Figures 6(a)-(c) show such characteristics of the complementary CHINT at $T = 290\,K$, $T = 235\,K$, and $T = 100\,K$. The low $V_D$ part of the characteristics is similar to that in a "normally-off" field-effect transistor (FET), with the collector playing the role of a gate.

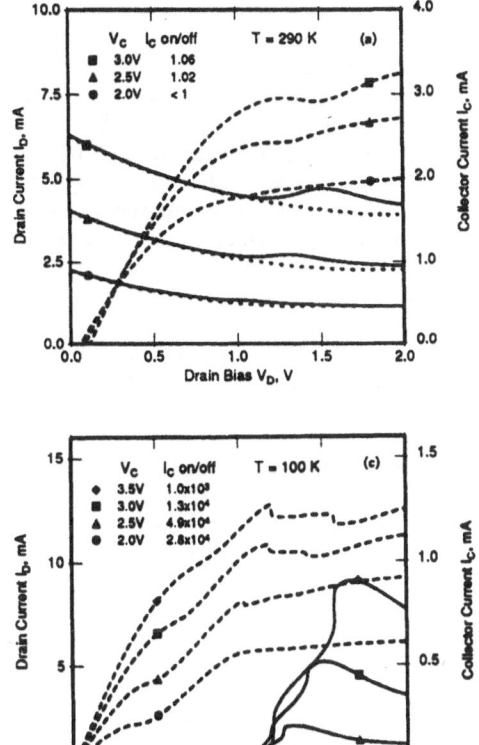

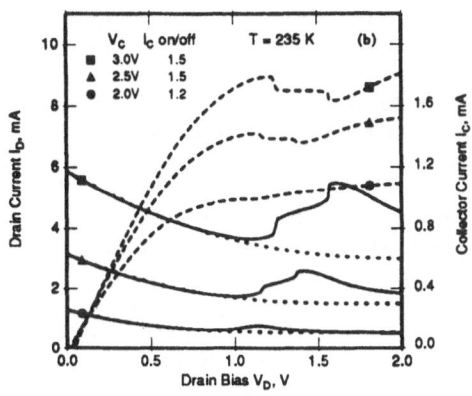

**Fig. 6.** Current-voltage characteristics at different temperatures and collector biases $V_C$. Drain current $I_D$ is shown by the dashed lines and collector current $I_C$ by the solid lines. The dotted lines show the expected leakage behavior of $I_C$ in the absence of RST, calculated from Fig. 4. The collector current on/off ratio is calculated by dividing the (relative) maximum value of $I_C$ by its minimum value before the onset of real-space transfer.
(a) $T = 290$ K; (b) $T = 235$ K; (c) $T = 100$ K.

At a fixed $V_C$ and increasing the heating voltage $V_D$, the channel current first ($V_D \leq 0.5$ V) increases linearly, then sublinearly with a tendency for saturation. The saturation value of the current increases with $V_C$, as it would in a FET. When the heating voltage is high enough to establish a significant RST, the drain current shows a negative differential resistance (NDR). The NDR effect is enhanced at lower temperatures, where it manifests itself by an abrupt drop in $I_D$. Simultaneously, the collector current rapidly increases from its leakage value.

Note that the increase in $I_C$ does not fully compensate for diminishing $I_D$, which means, by Kirchhoff's law, that the source current also decreases. This behavior can be explained by two mechanisms, both due to the backgating effect of the collector field. The first mechanism, which might be termed intrinsic, is due to the charge dynamically stored in the barrier layer in transit toward the collector.[4] The negative dynamic charge partially screens the collector field and reduces the electron concentration in the emitter channel. The second mechanism is due to a negative feedback of the collector current in the presence of a parasitic series resistance in the collector path. When the RST current begins to flow, the voltage drop on that series resistance reduces the collector potential, effectively lowering the $V_C$. Both effects result in decreasing the source current. When the collector contact is not properly alloyed, the series resistance may dominate.† In our device the dynamic space-charge effect may account for most of the decrease in the source current, as can be ascertained by simple estimates.[4] Nevertheless, a residual collector series resistance ($R_C \leq 0.5$ kΩ at 290 K but higher at lower temperatures, see below) is likely to be present in our structure; without $R_C$ we find it hard to understand certain features in the characteristics at high $V_D$ and still higher $V_C$.

---

† Such structures usually show a giant NDR accompanying a relatively little RST current. This anomaly goes away upon annealing the alloyed contact.

Indeed, as seen from Figs. 6, the collector current shows a well-defined maximum as a function of $V_D$. This behavior is faithfully tracked by the electroluminescence signal, which, as discussed in Sect. 4, indicates that the maximum is associated with a decrease of the electron RST current at high $V_D$. We believe, that the existence of a peak in the collector current and the decrease of $I_C$ at higher $V_D$ is related to the increase in the potential of the "hot spot" (i.e. the high-field region in the channel where most of the RST occurs) relative to the collector. Reversal of the field in the barrier at the hot spot exponentially suppresses the RST. No peak in $I_C$ is observed when $V_D$ is increased *simultaneously* with increasing the collector voltage, which corresponds to varying the source voltage at fixed $V_D$ and $V_C$. Field reversal at the hot spot naturally explains our data in Fig. 6 at lower $V_C \leq 2.5\,V$. However, a close examination of Fig. 6c shows that it does not work directly for higher values of $V_C$. Thus, for $V_C = 3.5\,V$, the maximum occurs near $V_D = 1.7\,V$. In the absence of a collector resistance, this would suggest that the entire emitter channel is in the operating regime and no field reversal takes place at the hot spot. To make the ends meet, it is sufficient to assume a series resistance $R_C \approx 1\,k\Omega$ in the collector circuit at low temperatures.

Consideration of the parasitic series resistances is also necessary to disentangle the RST from cold carrier leakage, which is important for estimating the internal radiative efficiency, especially at higher temperatures (see Sect. 4). In the higher-temperature plots, Fig. 6a and 6b, the collector current is seen to decrease prior to the onset of the RST. This is due to the diminishing leakage current with decreasing collector-to-drain bias. The same behavior occurs at low temperatures, but on the scale of Fig. 6c the leakage is not resolved. In order to estimate the RST current at higher temperatures, it is necessary to subtract the leakage contribution from the measured collector current. Ideally, when the junction in the drain region is near flat bands, the leakage from that region is effectively shut off and the overall leakage must diminish roughly by a factor of 2. However, as discussed below, inclusion of the parasitic resistances somewhat complicates the analysis.

We would like to distinguish between two contributions to the collector current:

$$I_C = I_C^{LKG} + I_C^{RST} , \tag{3}$$

where $I_C^{RST}$ is the real-space transfer current of hot electrons and $I_C^{LKG}$ is the collector leakage, defined as the current which would flow at a given voltage configuration $(V_C, V_D)$ if the electron heating phenomenon were absent. Both $I_C^{LKG}$ and $I_C^{RST}$ depend on the voltages $V_C$ and $V_D$. To estimate the behavior of the leakage component, we use an approximate equivalent circuit, shown in Fig. 7a. The FET symbol represents the "intrinsic" portion of the device under the trench; contribution of this area to the leakage current is assumed negligible. The diode symbols correspond to the *pn* junctions in the source and drain contact areas (it should be noted here that our "diode" is not a conventional *pn* junction, but has a heterostructure barrier separating the *p* and the *n* regions). For the purpose of calculating the leakage, we neglect the RST. The resistors $R_S$, $R_D$, and $R_C$ represent parasitic series resistances in the source, drain, and collector, respectively.

It would be nice to be able to determine the equivalent circuit parameters by a least square fit, but our data are insufficient for that purpose. Figure 7b illustrates the situation. The solid line represents the $I_C(V_D)$ curve at $V_C = 2.5\,V$ and $T = 235\,K$. We are interested in establishing the leakage component near the peak of the RST current at $V_D \approx 1.4\,V$ by extrapolating the pre-RST behavior corresponding to $V_D \leq 1\,V$. We use the circuit of Fig. 7a, assuming a perfect symmetry between the source and the drain and taking the single diode characteristic from the data of Fig. 4a. Dashed curve (a) corresponds to letting all the series resistances vanish. Adding a collector resistance shifts the curve upwards (b). A source resistance shifts it downwards (d,c) – in large part because of the voltage drop associated with the source-to-drain current. A good fit can be achieved with a continuous range of combinations of assumed $R_S$ and $R_C$, which means that the values of these resistances cannot be determined in this way. The use of our procedure derives from the fact that the asymptotic value of the leakage is

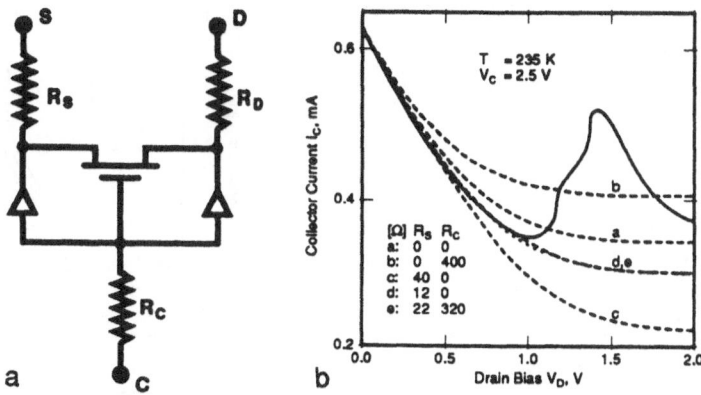

a       b

**Fig. 7.** Illustration of the procedure for estimation of the leakage current.

(a) Circuit model. The FET symbol represents the emitter channel assuming no RST current. The diodes correspond to *pn* heterostructure under the source and drain contacts and the resistors $R_S$, $R_D$, and $R_C$ to parasitic series resistances in the source, drain, and collector current paths, respectively.

(b) Estimation of the leakage current at a particular temperature and $V_C$. Characteristic of the FET symbol (the channel current) is taken from Fig. 6b. Diode characteristics are taken from Fig. 4. Series resistances are assumed symmetric, $R_D = R_S$.

insensitive to the actual values of $R_S$ and $R_C$, so long as they give a good fit to the region $V_D \leq 1\,V$. If we tried to use a simple spline fitting instead, we would need to specify a value for the horizontal asymptote – which cannot be determined lacking a physical basis.

The leakage curves, $I_C^{LKG}(V_D)$, determined in the indicated fashion for each value of $V_C$, are plotted by the dotted lines in Figs. 6a and 6b. Subtracting the leakage curve, we obtain a reasonable approximation to the RST current $I_C^{RST}$ at a given bias. Thus calculated curves $I_C^{RST}(V_D)$ are used below in Sect. 4.

### 3.3 Characteristics of a Single ORNAND Channel

The multi-terminal logic device actually forms three channels: $1-2$, $2-3$, and $\tilde{3}-1$. In the description of this device we shall use an arrow to indicate the source $\rightarrow$ drain direction in a particular measurement. Figure 8 shows the typical room-temperature characteristics, measured in the channel $\tilde{3} \rightarrow 1$. Characteristics of different channels coincide quite closely, but there is a slight asymmetry under the source-drain interchange in a single channel, due to an unavoidable off-center misalignment of the trenches. Thus the $\tilde{3} \rightarrow 1$ characteristics are slightly different from those measured in the $1 \rightarrow \tilde{3}$ configuration. As shown in Sect. 5, this systematic asymmetry cancels out in all the 8 states of the ORNAND gate.

Single-channel characteristics of the multiterminal device are significantly improved over those in Fig. 6a for the three-terminal CHINT of Fig. 2a. The key improvement lies in the larger $I_C^{RST}$, which is moreover achieved at lower $V_C$. This improvement can be attributed to a reduction of parasitic resistances. The present device, because of its multiterminal nature, naturally lends itself to a measurement of the series resistance $R$ in electrodes 1 and 2. We have found $R = (10 \pm 1)\,\Omega$, dominated by the contact resistance. The outer electrodes can be expected to have slightly lower $R$.

The $I_C^{LKG}$ contribution, calculated using the diode characteristics of Fig. 4a, is plotted in Fig. 8 by the bold dotted line. No parasitic resistance was assumed in this case, but we included the asymmetry in the emitter area associated with the source (electrode $\tilde{3}$) and the drain (electrode 1) when the other two contacts are floating.

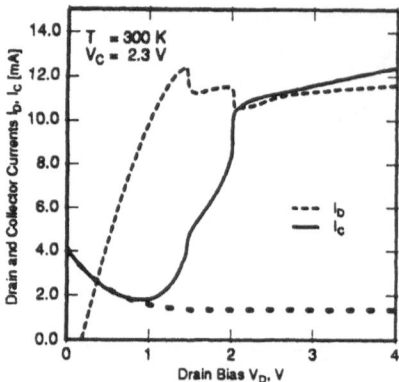

**Fig. 8.** Room temperature current-voltage $[I_C(V_D)$ and $I_D(V_D)]$ characteristics of a single-channel $\tilde{3} \to 1$ at a fixed collector bias $V_C = 2.3\,V$. Electrode $\tilde{3}$ is grounded and 1 acts as the drain. Electrodes 2 and 3 are kept floating. The bold dotted line indicates the leakage component $I_C^{LKG}$, calculated from the data in Fig. 5a.

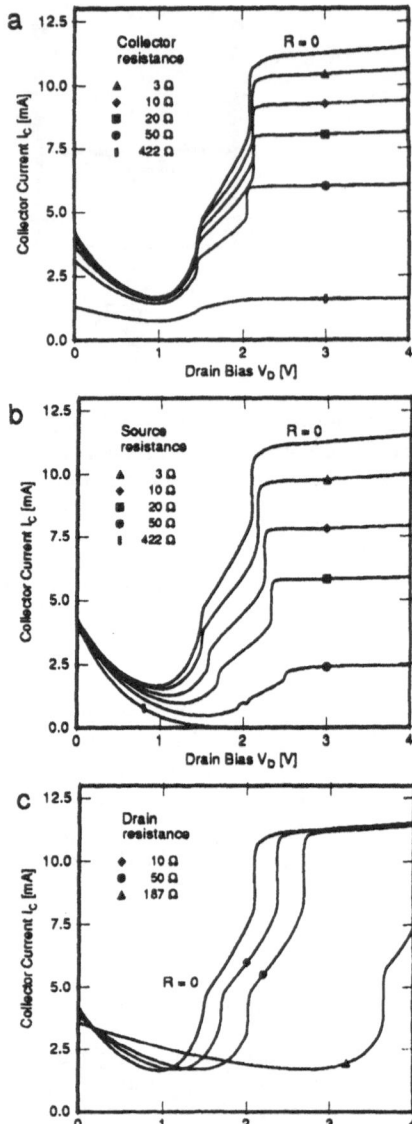

**Fig. 9.** The effect of series resistances. Current voltage characteristics $I_C(V_D)$ of a single channel $\tilde{3} \to 1$ measured under the same conditions as in Fig. 8 (though in a different device) with series resistances $R$ added in ($a$) the collector, ($b$) the source, and ($c$) the drain circuits.

Note the absence of a peak in the collector characteristic of Fig. 8, to be contrasted with the data in Fig. 6. At this point, we attribute this difference to a heavier channel doping in the multiterminal structure; a conclusive analysis requires an accurate two-dimensional simulation of the complementary device that we have not yet carried out.

The effect of adding external series resistances is illustrated in Fig. 9. Resistors in series with the source or the collector have a similar effect: its voltage drop effectively reduces the collector-to-source bias. The latter affects the normal electric field in the barrier and the sheet carrier concentration in the channel. A drain series resistance, on the other hand, reduces only the lateral channel field with the effect that the same $I_C$ is reached at a higher $V_D$.

Figure 10 reports a new and interesting finding. It shows the current-voltage characteristics for a grounded collector, both at 300 K and 6 K. The built-in electric field at $V_C = 0$ opposes the RST of channel electrons and increases the effective barrier height, cf. Fig. 3a. For $V_D > 0$, the device cross-section near the drain is a reverse-biased *pn* heterojunction. One would expect the characteristics to be similar to those in Fig. 5a for the reverse collector biases $V_C < 0$. However, at a certain drain bias, the curves $I_C(V_D)$ in Fig. 10 depart from the reverse-bias leakage curves, with the current increasing by several orders of magnitude. Such a regime has not been observed in the structure of Fig. 2a, where the emitter doping is lower and the channel is "normally off", so that the transport properties at $V_C = 0$ can not be properly investigated. Channel characteristics of the ORNAND structure closely resemble those in a normally-on field-effect transistor and show no particular structure in the anomalous region.

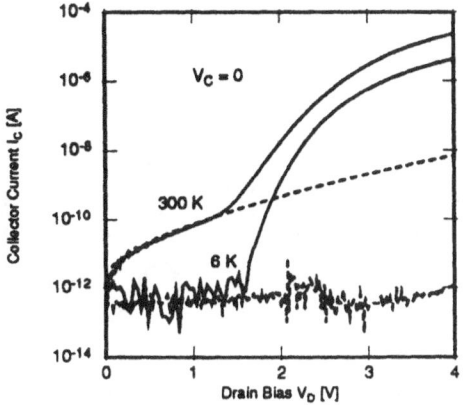

Fig. 10. Anomalous current-voltage characteristics of a single channel $3 \rightarrow 1$ at $V_C = 0$ and two different temperatures. Electrodes 2 and 3 are floating. Solid lines show $I_C(V_D)$ at $V_C = 0$. Polarity of $I_C$ corresponds to the flow of holes from the channel into the collector. Dashed lines represent the reverse-bias collector leakage current from Fig. 5a. The "soft breakdown" near $V_D \approx 1.5\,V$ is attributed to the RST of impact-ionized holes.

We believe the sharp rise of the reverse collector current (polarity corresponding to electrons flowing from the collector into the channel) at high values of $V_D \geq 1.5\,V$ can be imputed to the *impact ionization* by hot electrons in the channel. Holes, released by this process near the drain, drift toward the source and undergo a substantial heating along the way. The observed increase of $|I_C|$ is thus interpreted as a RST of *minority holes* from the channel. Note that the same electric field that suppresses the RST of electrons aids the RST of channel holes. Injection of *majority* carriers into the collector does not give rise to a luminescence – which is in accordance with our experimental observations. At cryogenic temperatures, the ionization threshold in InGaAs increases due to the band-gap widening and so does the critical value of $V_D$ at which the $I_C(V_D)$ curves in Fig. 10 depart from the corresponding leakage curves. Evidently, the complementary structure is not essential, although it simplifies observation of the minority RST effect by reducing the leakage current under reverse bias. The RST of minority carriers can be a powerful tool for studying impact ionization phenomena in field-effect transistor channels.

## 4. OPTICAL CHARACTERIZATION

Electroluminescence of the complementary RST devices was detected from the back of the polished substrate using a liquid nitrogen cooled Ge detector and a 0.75 m spectrometer. We observed a spectrum peaked at the photon energy corresponding to the band-to-band recombination in InGaAs. As the bandgap shrinks with increasing temperature, the peak wavelength increases from 1.56 μm at $T = 100\,K$ to 1.59 μm and 1.60 μm at 235 K and 290 K, respectively. Moreover, at low temperatures ($\leq 100\,K$) the position of the peak slightly shifts towards higher energy as the RST current increases. We attribute this shift to a band-filling effect due to the increasing electron injection. To measure the total emitted power, we used a broad-area Ge photodiode and suitable focusing optics.

Figures 11 show the dependence of the measured light power $P_L$ on the electron heating voltage $V_D$ at selected temperatures and different values of $V_C$. On the same graphs we show the "net" RST curves $I_C^{RST}(V_D)$, calculated as described in Sect. 3.2.

The first notable observation is that the $P_L(V_D)$ curves are to a good degree proportional to $I_C^{RST}(V_D)$, rather than to the total $I_C(V_D)$ of Fig. 6, as one could naively expect. Accordingly, the measured optical on/off ratio (Figs. 11) far exceeds the on/off ratio in the collector current, the difference being particularly striking at low temperatures. Even at $T = 290\,K$, where the RST structure in $I_C(V_D)$ is barely visible against the leakage background, the corresponding modulation of the optical signal is by an order of magnitude: for $V_C = 3.0\,V$ the on/off ratio is 9 in $P_L$ and 1.1 in $I_C$. The origin of this remarkable performance must be clearly understood, since the depth of modulation is of paramount importance for room-temperature logic applications of the light-emitting RST devices.

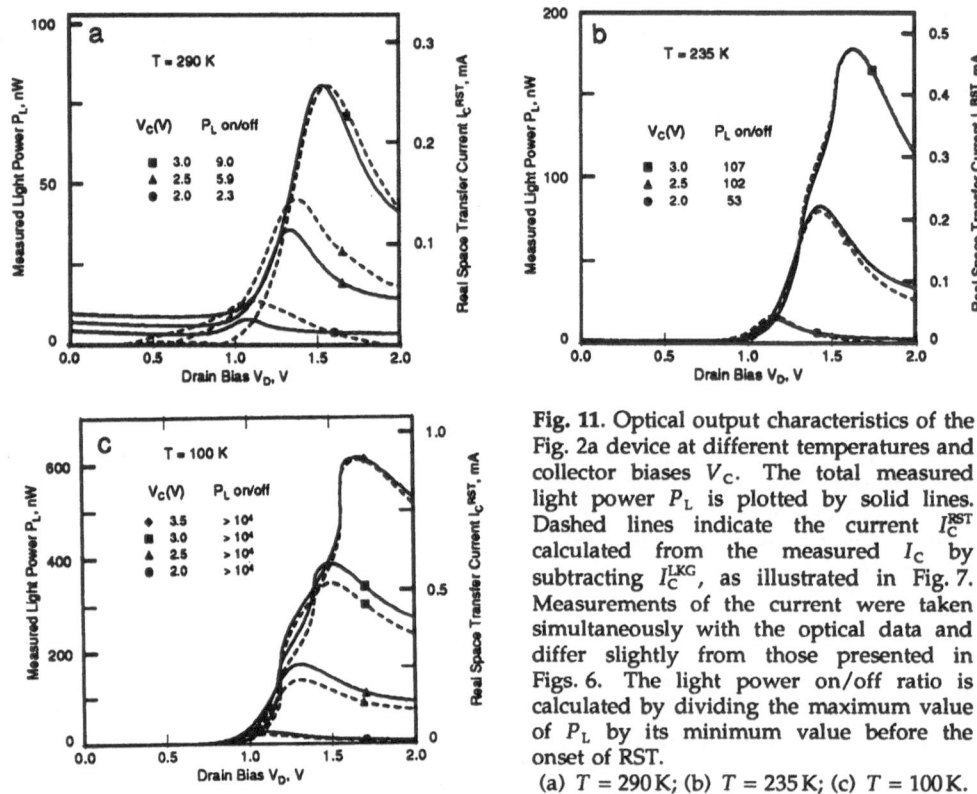

Fig. 11. Optical output characteristics of the Fig. 2a device at different temperatures and collector biases $V_C$. The total measured light power $P_L$ is plotted by solid lines. Dashed lines indicate the current $I_C^{RST}$ calculated from the measured $I_C$ by subtracting $I_C^{LKG}$, as illustrated in Fig. 7. Measurements of the current were taken simultaneously with the optical data and differ slightly from those presented in Figs. 6. The light power on/off ratio is calculated by dividing the maximum value of $P_L$ by its minimum value before the onset of RST.
(a) $T = 290\,K$; (b) $T = 235\,K$; (c) $T = 100\,K$.

By studying the temperature dependence of $I_C(V_C)$ at $V_D = 0$, we had concluded in Sect. 3.1 that the collector leakage is mostly due to the injection of holes into the channel. Now, the data of Fig. 11 clearly indicate that this injection does not contribute to the optical signal. It is significant that the leakage and the RST are caused by different types of carriers. The injected holes have a vastly lower radiative efficiency because they are likely to reach the source or drain contact before they recombine radiatively with electrons. The $In_{0.52}Al_{0.48}As$ etch-stop layer $b$ (Fig. 2) is insufficiently thick to confine minority holes in the emitter channel (the estimated hole escape time by tunneling across this layer is much shorter than the radiative recombination time). Most of the nonradiative recombination is likely to occur in the layers heavily doped with tin. In contrast, electrons injected into the collector are confined in the active region. Virtually all of the optical output can be attributed to the recombination of injected electrons in the collector active layer. This implies an important design consideration: to maximize the optical on/off ratio it is essential to suppress the leakage of electrons into the active region; the oppositely directed flux of holes can be tolerated.

At low temperatures, Fig. 11c, the light signal related to the leakage is below the noise level and the on/off ratio in $P_L$ is above $10^4$. At higher temperatures, the small electron component of the leakage current increases and consequently the on/off ratio disparity between the light-power output and the collector current is less pronounced. From the measured light power $P_L(V_D, V_C)$ we can estimate the internal radiative efficiency of the RST electrons as follows:

$$\eta_q = \frac{1}{h\nu\,\eta_c\,t_o}\;\frac{q\,\Delta P_L}{I_C^{RST}}\,, \qquad (4)$$

where $\eta_c \approx 0.48\%$ is the collection efficiency (estimated assuming an isotropic emission over the $2\pi$ solid angle, due to reflection from the surface metallization), $t_o \approx 88\%$ is the combined transmission of the optical components (lens and windows) between the detector and the device, and $h\nu$ is the photon energy at the peak of the luminescence spectrum. To account for the light power due to the leakage current, the quantity $\Delta P_L$ is taken equal to $P_L(V_D, V_C)$ minus the optical power measured at the same $V_C$ but lower $V_D$, prior to the onset of the RST. At lower temperatures, one has $\Delta P_L = P_L$.

The fact that our collection efficiency $\eta_c$ is low, is accounted for by the total internal reflection, Fresnel loss, and the collection solid angle. We have not used an anti-reflection coating. The relatively low numerical aperture ($N_A \approx 0.34$) has been forced by the use of a dewar for low-temperature measurements. Moreover, an error of $\pm 5\%$ in $N_A$ from the collection lens being slightly out of focus, produces an error of $\pm 10\%$ in $\eta_c$. Nevertheless, the systematic error in the determination of the efficiency was the same for all measurements, since the lens system was not adjusted during the whole set of measurements. The *relative* efficiency behavior at different temperatures and biases can be considered quite reliable.

Figure 12 shows the basic trends in the behavior of the radiative efficiency $\eta_q$ under varying temperature and collector bias. We found that near and above the peak of the RST current the efficiency is not modulated by varying $V_D$. This means that $\eta_q$ is practically independent of the injection current at a given $T$ and $V_C$. The data in Fig. 12 is taken near the RST peak; the dependence on $V_D$ in a wider range is discussed below. As is evident from the figure, the radiative efficiency decreases with increasing temperature. This dependence is likely to be explained, as in conventional light-emitting devices, by the increasing role at higher $T$ of non-radiative processes, such as Auger recombination.[18]

It is instructive to consider the radiative efficiency of the leakage current itself. This can be characterized by a parameter $\eta_q^{LKG}$, defined as in Eq. (4) but with $\Delta P_L$ replaced by $P_L(V_D=0)$ and $I_C^{RST}$ by the total $I_C$ at zero heating bias. For $V_C = 3.0\,\mathrm{V}$ and $T = 290\,\mathrm{K}$, $235\,\mathrm{K}$, and $200\,\mathrm{K}$, we find $\eta_q^{LKG} = 0.12\%$, $0.06\%$, and $0.04\%$, respectively. This temperature dependence, "opposite" to that in Fig. 12, gives a further evidence to the fact that the leakage light is due not to a small fraction of holes that recombine radiatively in the emitter channel but to a small electron contribution to the leakage current.

Returning to Fig. 12, the efficiency is also seen to decrease with increasing collector bias. This effect is prominent only at low $T \leq 150\,\mathrm{K}$ (no significant variation of $\eta_q$ with $V_C$ is observed at higher temperatures; the apparent dependence at $T \geq 200\,\mathrm{K}$ in Fig. 12 is not meaningful since the low-$V_C$ points are within an experimental error introduced by the uncertainty in the subtraction of the leakage component). One possible explanation for the $V_C$ dependence may be related to the injection of hot electrons into the wide-gap collector confinement layer, where they recombine nonradiatively.† Within this model, weakening of the $V_C$ dependence of $\eta_q$ at higher temperatures can be explained by a shorter hot-electron mean-free path due to higher rate of optical-phonon scattering. The fact that a sizable fraction of injected electrons

---

† Even if there were a radiative component from the confinement layer, it would not be detected experimentally, since all of the InAlAs light output would be re-absorbed in the InP substate. Moreover, as a matter of principle, one would not expect a significant radiation from the wide-gap layer, since electrons there are no longer confined and their radiative lifetime is longer than travel time to the contact.

can retain enough energy (especially at lower temperatures) to clear the second barrier, is well known from studies of ballistic transistors (see, for example, Ref. 19 and the literature cited therein). This interpretation does not contradict the fact that our measured spectra do not show hot-electron tails, because the electron cooling rates in InGaAs are known[20] to be faster than recombination.

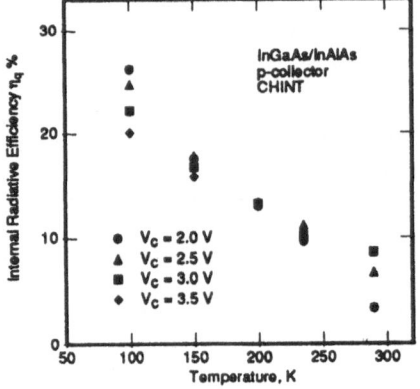

**Fig. 12.** Dependence of the internal radiative efficiency $\eta_q$ on the lattice temperature for different collector biases.

Nevertheless, we have difficulties with this interpretation. First, it is inconsistent with our understanding of the peak in the $I_C(V_D)$ dependence, which we had attributed in Sect. 3.2 to a field reversal in the "hot-spot" region of the channel. Clearly, one would expect a suppression of the ballistic-transistor effect in this case. This contradicts our data in Fig. 13a, which shows the dependence of the calculated $\eta_q$ on the heating voltage $V_D$ for several values of $V_C$ at $T = 100\,K$. No enhancement of the $\eta_q$ occurs at the heating voltages $V_D$ above the $I_C$ peak.

Next, consider the most striking feature of Fig. 13a, namely the dramatic downward steps in the efficiency, accompanying steps in the collector current. The low-bias values of the efficiency are relatively high and largely *independent* of the collector voltage.† As the heating bias reaches the location of the first current step, the efficiency suddenly drops to a lower plateau value. Moreover, curves corresponding to lower collector biases ($V_C = 2.0$ and $2.5\,V$), at which there are no steps in the current-voltage characteristics, show no significant decrease in efficiency either. For $V_C > 2.5\,V$, something evidently happens near $V_D \approx 1.2\,V$ which forces the step-like switching of the current and is accompanied with the efficiency drop. Steps in the current-voltage characteristics of the CHINT are usually explained[4,9,10,15,21,22] by an instability associated with the formation of hot-electron domains in the channel. Both the field and the electron temperature $T_e$ rise dramatically in the domains and that, of course, may lead to a higher injection into the confinement layer, because of the higher average energy of injected electrons. However, if the injection energy is the deciding factor, then we should expect the radiative efficiency go down strongly with increasing $V_C$ at all values of $V_D$. The trouble with the above ballistic-transistor interpretation of the $\eta_q(V_C)$ dependence is not that we cannot stretch it to explain the efficiency drop (*that we can do!*) but that it is clearly inconsistent with the fact that low-$V_D$ values of $\eta_q$ are practically *independent* of $V_C$. We are, therefore, led to consider alternative interpretations of the efficiency data.

---

† The apparent decrease in $\eta_q$ at lowest heating biases (near the sensitivity limit of our light detector) is probably related to nonradiative recombination with defects. The rate of such processes scales linearly with the minority concentration and at low carrier injection it exceeds the radiative recombination rate.

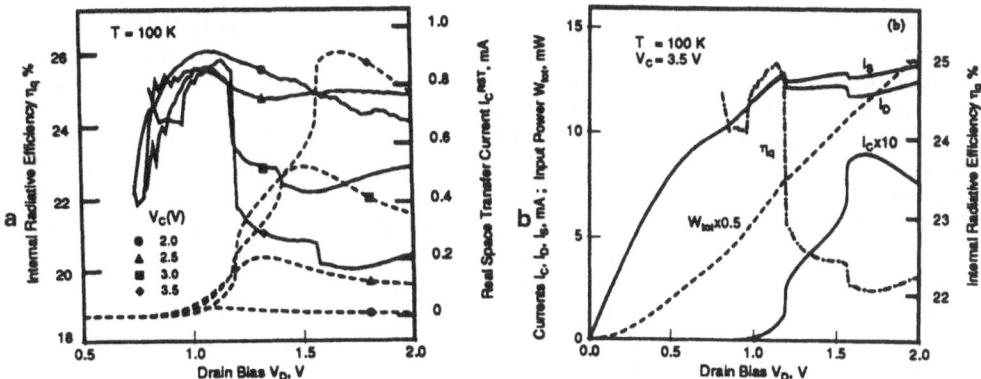

**Fig. 13.** Dependence of the internal radiative efficiency $\eta_q$ on the heating voltage $V_D$.
(a) Plots at different collector biases. Dashed lines indicate $I_C$ from Fig. 6c.
(b) Simultaneous plots of the radiative efficiency ($\eta_q$), the drain, collector, and source currents ($I_D$, $I_C$, and $I_S$, respectively), and the total electrical input power ($W_{tot}$).

In our opinion, a possible explanation involves a nonlocal heating of the lattice by hot electrons. In this picture, the lattice temperature in the active layer may substantially deviate from the ambient temperature, due to the power $W_e$ dissipated by hot-electrons in the channel via optical-phonon emission. This power can be estimated as

$$W_e = \mathcal{A}\, \frac{n\, kT_e}{\tau_e} \,, \tag{5}$$

where $n$ is the sheet carrier concentration in the emitter channel at the hot spot, $\mathcal{A}$ the hot-spot area, and $\tau_e\,(T_e)$ the energy relaxation time of hot electrons at temperature $T_e$. At low $V_D$ the heating field in the channel is low and relatively uniform, hence $T_e$ is low and so is $W_e$. The formation of a hot-electron domain shrinks $\mathcal{A}$ but at the same time strongly increases $T_e$ and therefore $W_e$ goes up. Higher phonon emission in the channel leads to a higher lattice temperature in the collector active layer and hence a lower radiative efficiency. From the data of Fig. 12, the required temperature increase at $T = 100\,\mathrm{K}$ is by less than $50\,\mathrm{K}$. The dependence on $V_C$ results from the field-effect gating action of the collector, expressed in Eq. (5) by the factor $n\,(V_C)$. After the domain is formed, the dependence on $V_D$ is weak.

We would like to stress that the lattice heating by hot carriers is an essential part of this picture. A simple Joule heating of the whole device is ruled out by the following observation. Consider the total input power $W_{tot}$ into the transistor, which in the common-source configuration is given by

$$W_{tot} = I_C\, V_C + I_D\, V_D \,. \tag{6}$$

To our surprise, we found that $W_{tot}$ is *continuous* at the steps in the current-voltage characteristics.§ This is illustrated in Fig. 13b, which shows the behavior of one of the efficiency curves of Fig. 13a together with $I_D$, $I_C$, $I_S$, and $W_{tot}$. We see that while all the terminal currents experience a discontinuity at $V_D \approx 1.2\,\mathrm{V}$, the total power input remains continuous.

---

§ In addition to our present device, we have re-analyzed the available data measured in previously reported unipolar charge injection transistors. To our surprise, we found that most (but not all) steps in $I_D$, $I_C$, and $I_S = I_D + I_C$ are accompanied by a continuous variation in the total input power. Typically, only the slope $W_{tot}\,(V_D)$ changes at the transition. We believe this phenomenon deserves further study, both experimental and theoretical. It may shed light on the nature of switching transitions between different branches of intrinsically complicated trajectories[23] in the device phase space, cf. below: Figs. 22 and 26-30.

To avoid misunderstanding, let us remark that $W_e$ is only a part of $W_{tot}$. Most of $W_{tot}$, however, is dissipated by acoustic phonons and distributed over a large volume. In contrast, the power $W_e$, dissipated by electrons with $T_e > \sim 1\,000\,K$, goes into the generation of relatively immobile optical phonons at the hot spot.

We conclude this Section by displaying in Fig. 14 the room-temperature light-voltage characteristics of a single-channel $(3 \rightarrow 1)$ of the multiterminal device. These characteristics were taken under the bias conditions identical to those in Fig. 8. For comparison, the figure also shows the $I_C^{RST}$ curve, obtained by subtracting $I_C^{LKG}$ (shown in Fig. 8 by the bold dotted line). We see that even though the injection current increases monotonically, the $P_L$ curve has a maximum — like the analogous curves in Fig. 11 — although the drop after the maximum is not so steep. At this time, we do not have a clear understanding why the two curves shown in Fig. 14 have a different behavior at $V \geq 2\,V$, cf. our comment below Fig. 8.

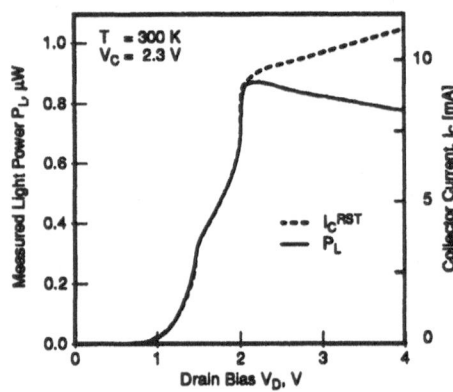

Fig. 14. Room-temperature light-voltage characteristics of a single-channel $\tilde{3} \rightarrow 1$ at a fixed collector bias $V_C = 2.3\,V$. Electrode 3 is grounded and 1 acts as the drain. Electrodes 2 and 3 are kept floating. The dashed line indicates the $I_C^{RST}$ curve, obtained by subtracting $I_C^{LKG}$ from $I_C$ in Fig. 8.

## 5. REAL SPACE TRANSFER LOGIC

The symmetry equivalence (1) between the internal states $S[V_D, V_C]$ of a CHINT device implies that its output optical power, which is proportional to $I_C^{RST}$ and is controlled by the heating bias $V_{DS}$, is invariant if the potentials on the source and drain terminals are interchanged. Thus the device exhibits an exclusive-OR dependence of the emitted light power on the input voltages $V_S$ and $V_D$, regarded as binary (high/low) logic signals.

Figures 15 demonstrate this logic operation at three different temperatures. The collector bias is fixed at $V_C = 3.0\,V$ and the input voltages are varied from $low = 0$ to $high = 1.5\,V$. Operation of an optical **xor** gate is illustrated in Fig. 15a. The data shown were obtained in DC measurements; no attempt was made to characterize the frequency response, as our device had not been designed for a fast operation. We see that the light-output power $P_L$ obeys $P_L = \text{xor}(V_S, V_D)$ at all temperatures, including room temperature. The slight asymmetry between the *on* states $(1, 0)$ and $(0, 1)$ indicates that our device is not perfectly symmetric due to processing variations. On the other hand, the difference between the *off* states $(0, 0)$ and $(1, 1)$ at high temperatures is owing to a nonvanishing radiation associated with the leakage current.

It should be stressed that the low radiative efficiency associated with the leakage current is crucial for a successful **xor** operation of our device. For comparison, Fig. 15b shows the electrical behavior of our device at similar temperatures and biasing conditions. While the approximate symmetry between the *on* states $(1, 0)$ and $(0, 1)$ is well maintained, the function $I_C = \text{xor}(V_S, V_D)$ obtains only at cryogenic temperatures. At higher $T$, the leakage of holes in the state $(0, 0)$ makes it effectively an *on* state and the resulting function looks more like a **nand** than a **xor**.

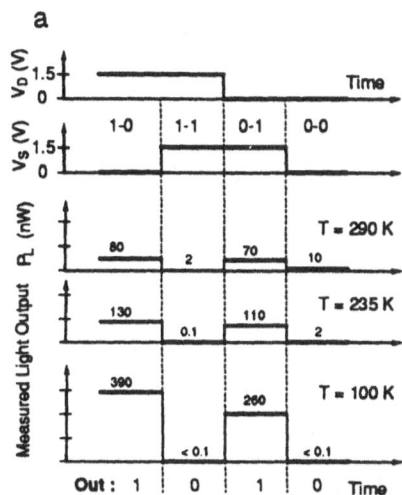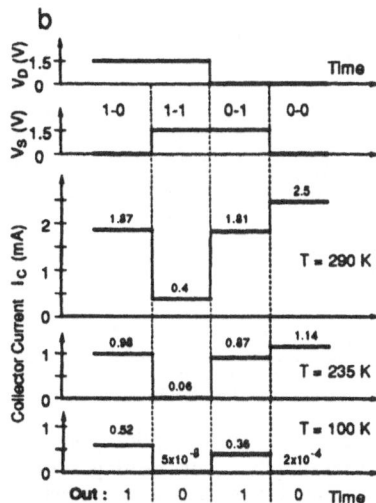

**Fig. 15.** Comparison of the optical and electrical logic operation of the complementary CHINT at different temperatures $T$. Four logic states $(1-0)$, $(1-1)$, $(0-1)$, and $(0-0)$ correspond to the electrical inputs $(V_S - V_D)$ taking the values $V = 0$ (logic-0) and $V = 1.5\,V$ (logic-1).
(a) Optical output, $P_L$, corresponds to $\mathbf{xor}(V_S, V_D)$ at all $T$, including room temperature.
(b) Electrical output, $I_C$, is $\mathbf{xor}(V_S, V_D)$ only at cryogenic temperatures.

Let us briefly discuss another possible application of the light-emitting CHINT: an optical frequency doubler of the electrical input signal.† Consider the device response to a sinusoidal variation of the voltage $V_D$ on one of the emitter electrodes, $D$, keeping the other electrode, $S$, grounded, $V_S = 0$. The light output for $V_D$ varying in the range $-2\,V < V_D < 2\,V$ and several fixed values of $V_C$, is shown in Fig. 16a, and the response to a sinusoidal variation of $V_D$ is illustrated in Fig. 16b. Note that in this situation, the $D$ electrode during the half-period when $V_D < 0$ is actually the source so that the effective collector-to-source voltage is varied. Because of the gating action of the collector field, the characteristics are asymmetric. The output comes twice per period in uneven pulses. No small-signal operation of this device is possible, because of the finite magnitude of the heating voltage required to initiate a tangible RST.

The threshold nature of the RST frustrates small-signal applications of the CHINT symmetry (1). Nevertheless, Mensz et al.[8] reported a unipolar operation of the CHINT as an electric frequency doubler. By intentionally driving up the leakage level with a high collector bias, they found a regime where the total $I_C (V_D)$ characteristic appears parabolic. A natural question arises as regards to the possibility of obtaining an optical analog of such a small-signal frequency doubler. In our present structure this is impossible precisely because of the low radiative efficiency associated with the leakage of holes. Perhaps, such an operation can be achieved using heterostructures in which both the leakage and the RST would be mostly due to the same carrier type (e.g. in $n$-channel complementary CHINT implemented in InGaAs/InP, where $\Delta E_V > \Delta E_C$). Of course, in such a structure the *on/off* ratio would be the same in both light and current. To recover superiority of the optical over the electrical logic operation in a InGaAs/InP complementary CHINT, one would have to use a $p$-type emitter.

---

† Such an application had been suggested to one of us (SL) by Professor K. Iga in a private communication (1990).

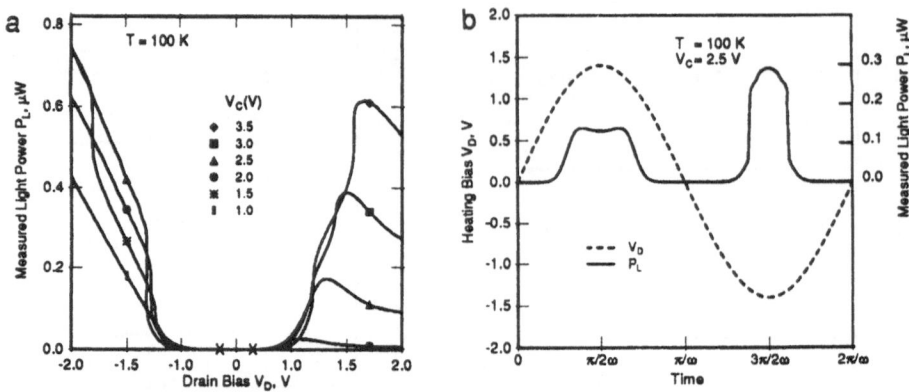

**Fig. 16.** Double-frequency optical pulse formation, illustrated at $T = 100\,\mathrm{K}$.
(a) Optical output for the heating voltage varying in the range $-2.0 < V_D < +2.0\,\mathrm{V}$ and several fixed collector biases. For a grounded $S$ electrode and negative $V_D$, it is the $D$ electrode that plays the role of a source. Note the broken voltage scale near $V_D = \pm 0.5\,\mathrm{V}$.
(b) Quasi-stationary evolution of the optical signal $P_L$ (solid line) in response to a sinusoidal heating voltage $V_D$ (dashed line).

## 5.1 ORNAND Logic

Since we set the periodic boundary conditions $\mathbf{V_{\bar{3}}} = \mathbf{V_3}$, our multi-terminal device has effectively three input terminals. One of these can be viewed as a control electrode which determines which of the two logic functions **or** or **nand** is executed on the other two inputs. Choice of the control electrode is arbitrary. The collector current $I_C$ and the light power $P_L$ represent the logic output. We shall refer to the states with *high* and *low* values of the output as *on* and *off* states, respectively.

Figure 17 demonstrates the room-temperature logic operation of the ORNAND gate. The light signal has been detected from the back of the polished substrate using a microscope objective to collect the light. The substrate thickness has been polished down to $35\,\mu\mathrm{m}$ reducing the free carrier absorption in the substrate. The collector bias is fixed at $V_C = 2.4\,\mathrm{V}$ and the input signals $\mathbf{V_1}$ and $\mathbf{V_2}$ are varied between $low = 0$ and $high = 3\,\mathrm{V}$, while the split electrode, chosen as the control, is fixed either at the *low* value $\mathbf{V_3} = 0$ for the **or** function or at the *high* value $\mathbf{V_3} = 3\,\mathrm{V}$ for the **nand** function.

As seen from Fig. 17, the symmetry between different *on* states is well maintained, both for $I_C$ ($\pm 8\,\%$) and $P_L$ ($\pm 6\,\%$). The larger variation in $I_C$ is due to the variation in the leakage current of holes. Different *on* states correspond to different areas $A_S$ of the

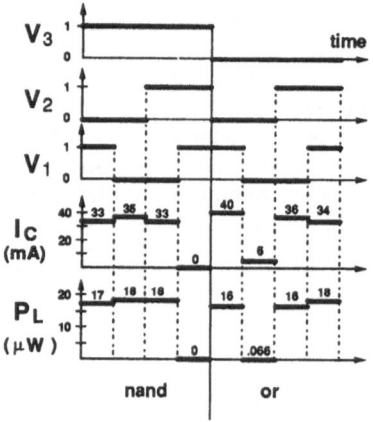

**Fig. 17.** Optical and electrical logic operation of the ORNAND gate, obtained in a quasi-stationary measurement at room temperature and $V_C = 2.4\,\mathrm{V}$. Electrodes 3 and $\bar{3}$ are tied together. The binary values "logic-0" and "logic-1" of the input signals $\mathbf{V_1}$, $\mathbf{V_2}$, and $\mathbf{V_3}$ correspond to 0 and $3\,\mathrm{V}$, respectively. The particular grouping of the states into OR and NAND reflects the choice of $\mathbf{V_3}$ as the "control" electrode.

source contact: $A_S$ is 25 % of the total emitter area for **nand**$(0,1)$ and **nand**$(1,0)$, while $A_S = 75$ % for **or**$(0,1)$ and **or**$(1,0)$, and $A_S = 50$ % for **nand**$(0,0)$ and **or**$(1,1)$. The larger the source area the larger is the leakage current which is added to the RST current. Since the radiative efficiency of $I_C^{LKG}$ is negligible compared to that of $I_C^{RST}$, the *on* states are more homogeneous in the measured light output. The low radiative efficiency of $I_C^{LKG}$ also explains the fact that the *off* state of the **or** function is less than perfect electrically (the ratio *on/off* $\approx 7$ in $I_C$), while more than satisfactory optically (*on/off* $\geq 200$ in $P_L$). In the *off* state there is no RST and most of leakage results from holes injected in the emitter layer.

Variation in the effective source area is not the only cause for violation of the basic symmetry Eq. (1). Another way to break this symmetry is to displace the channel-defining trench in the $n^+$ cap layer from its nominal position exactly in the middle between neighboring electrodes. Figure 18 shows the single-channel $I_C^{RST}(V_D)$ characteristics for two different channels and two different source → drain orientations.

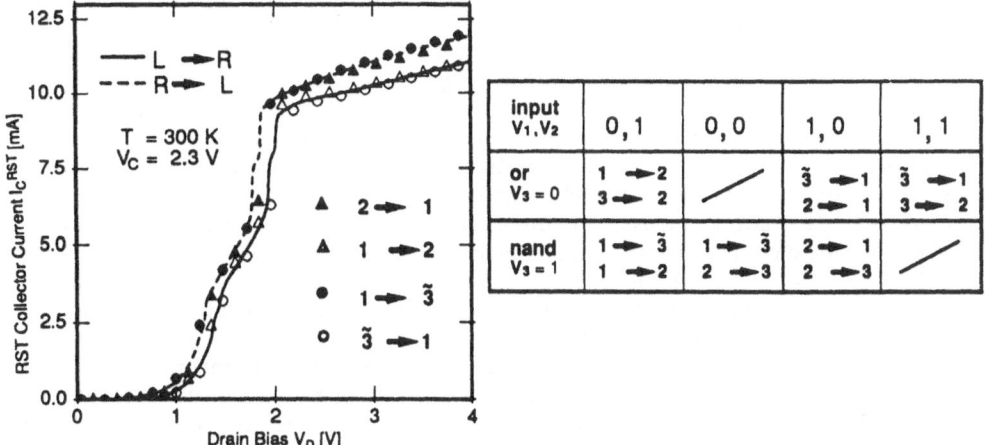

| input $V_1, V_2$ | 0, 1 | 0, 0 | 1, 0 | 1, 1 |
|---|---|---|---|---|
| or $V_3 = 0$ | 1 → 2   3 → 2 | ⟋ | 3̃ → 1   2 → 1 | 3̃ → 1   3 → 2 |
| nand $V_3 = 1$ | 1 → 3̃   1 → 2 | 1 → 3̃   2 → 3 | 2 → 1   2 → 3 | ⟋ |

**Fig. 18.** Asymmetry due to the misalignment of trenches in the ORNAND gate of Fig. 2b. The arrow indicates the source → drain orientation and its position the trench misalignment. Thus, the symbol $S \to D$ indicates that the trench is closer to the drain ($D$) electrode. The figure shows a comparison of the RST for two opposite orientations of the channel field in two different channels. The leakage current $I_C^{LKG}$ is subtracted, as in Fig. 14. The table on the right indicates the working channels, i.e. those channels that are under the heating field in all logic states of the gate.

We see that the curves corresponding to the same orientation in two different channels coincide more closely than those for the same channel but different orientations. This observation holds true for all devices examined. We remark that the data in the present report have been collected from a wafer with an exceptionally lucky alignment of the trench and the source/drain metal masks. In other wafers, the difference between the RST in opposite orientations of the same channel is even larger. Nevertheless, this has no effect on the symmetry of *on* states in the ORNAND gate. Indeed, the symmetry break due to a trench misalignment has to be identical in all the three channels, cf. Fig. 2b. Noting that each of the six *on* states has two channels under field – one of each orientation – we see that the systematic asymmetry cancels out in the ORNAND logic operation, cf. the Table in Fig. 18.

## 6. HOT-ELECTRON INSTABILITIES AND COLLECTOR-CONTROLLED STATES

As we have seen above, the $IV$ characteristics of charge injection transistors are extremely nonlinear, including a strong negative differential resistance (NDR) in the $I_D(V_D)$ dependence with sharp steps, cf., e.g., Fig. 6 or 8. Monte Carlo (MC) simulations[21,22] of the CHINT, demonstrate internal switching and the formation of high-field domains. These instabilities arise due to a positive feedback between the RST and the heating electric field in the emitter channel. Both experimental measurements and MC calculations have been limited in analyzing RST effects due to the restriction of tracing $IV$ characteristics exclusively in voltage increments. Significant progress in the understanding of the RST instabilities was achieved[9,23] with the help of continuation modeling and transient device simulation.

The study of hot-electron domains in RST transistors is important for understanding device limitations. Switching of the electron-heating control to the collector restricts the range where CHINT can be used as a linear amplifier. The frequency performance of CHINT in its usual mode is believed to be limited only by the time of flight of hot electrons over high-field regions of the device, i.e., over distances of order the barrier-layer thickness. The cutoff frequencies, extrapolated from the microwave measurements[24] of scattering parameters in a unipolar CHINT structure with a 2,000 Å -thick barrier, were 40 Ghz for both the current and the power gain. the transit time can be estimated to be of order 2-3 ps, which sets the upper limit for unity-gain cutoff frequencies at around 50 GHz. The ultimate speed performance can be achieved with an inverted CHINT structure in which the collector is the top layer.[25–27] This allows a reduction of the parasitic drain-collector capacitance and therefore the use of narrower barriers. With sufficiently narrow barriers, the CHINT can be expected to outperform a field-effect transistor of similar geometry (the RST collector corresponding to the FET gate, etc.) because the small-signal performance of CHINT is not limited[1] by the time of flight between the source and the drain. This advantage was recently demonstrated experimentally.[26] However, the domain formation itself seems to be a longer, FET-like, process, cf. Figs. 20 and 25 below.

Results of the numerical simulation[9,23] of the hot-electron transport in a three-terminal RST structure are reviewed below. A number of two-dimensional simulations of a unipolar CHINT have been performed, varying the device geometry (channel length $L_{CH}$ and barrier thickness $d_B$), transport parameters, and the external bias conditions. Both stationary and time-dependent problems have been investigated.

### 6.1 Mathematical Model

Any model which attempts to account for real-space transfer effects must directly include terms for carrier heating. Further, the ability to specify some form of Neumann boundary conditions is essential for tracing arbitrary, multivalued $IV$ characteristics. Both of these requirements are met by the general-purpose device simulator PADRE[28] which solves partial differential equations derived from moments of the Boltzmann equation. Simulations described below employed an energy balance system[29, 30], defined in terms of the electrostatic potential $\psi$, the electron density $n$, and the electron temperature $T_e$. Precise equations used are listed elsewhere.[31]

Through an element-based data structure, PADRE decomposes a device domain into arbitrary, nonplanar configurations of regions; for instance, any number of heterointerfaces can terminate abruptly at a single location (node). Data at the vertices of each element can have local material and model dependencies. The impurity concentration and solution variables are allowed to change abruptly across any heterostructure interface. In this analysis, the quasi-Fermi level and $T_e$ are assumed continuous, thus introducing a $T_e$ dependence in the interface condition on local electron density $n$; this means that the RST current density at energy barriers is thermionic and included self-consistently.

Models for $\mu$ and $\tau_e$ as a function of $T_e$ are typically derived from user-defined (e.g. MC) velocity-field $v(F)$ and temperature-field $T_e(F)$ relations for homogeneous slabs. To avoid confusion with negative differential mobility effects arising from momentum space transfer, a single population of electrons has been used with a monotonic $v(F) = \mu_0 F [1 + (\mu_0 F / v_{sat})^2]^{-1/2}$ relation[32] together with the $T_e(F)$ implied by a $T_e$-independent diffusivity.[33] All the results remain qualitatively similar for more realistic $v(F)$ and $T_e(F)$. For similar reasons, impact ionization and tunneling effects were excluded. Other parameters were selected to match the InGaAs/InAlAs system as exactly as possible.

In order to trace the complex *IV* curves shown below, it is necessary to apply mixed current/voltage boundary conditions. The predictor-corrector continuation method[34] based on a pseudo-arclength $\sigma$ was used for this purpose. Computationally, the continuation method requires the addition of a single algebraic auxiliary equation, typically written in terms of the voltage and current $(V_j, I_j)$ and the unit tangent $(\dot{V}_j, \dot{I}_j)$ at a known point on the curve $j$. The unit tangent can also be used to detect limit points (e.g. where $\dot{V} = 0$ or $\dot{I} = 0$) and to predict an initial guess for the subsequent bias point $j+1$. Figure 19 illustrates the continuation method and associated auxiliary condition on the pseudo-arclength used here.

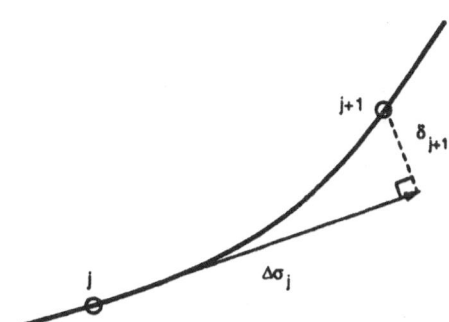

Fig. 19. Predictor-corrector continuation[34] applied to step between consecutive points $j$ and $j+1$ on an *IV* curve. The corresponding auxiliary equation is
$$\dot{I}_j (I - I_j) + \dot{V}_j (V - V_j) - \Delta\sigma_j = 0$$
where the next pseudo-arclength step $\Delta\sigma_{j+1}$ is automatically controlled by a user-specified tolerance on the error of the tangential projection $\delta_j$.

## 6.2 Broken Symmetry and Formation of RST Domains

*Rapid Ramping.* Figure 20 illustrates a time-dependent simulation of a device with $L_{CH} = 5\,\mu m$ and $d_B = 0.2\,\mu m$. Both $S$ and $D$ electrodes are kept grounded, while $V_C$ is linearly ramped from 0 to $V_C = 2\,V$. Depending on the ramping time $\tau$ the device settles in one of two states: for $\tau > \tau_{cr}$ it is the normal† state, whereas for $\tau < \tau_{cr}$ the steady state carries a large RST current (Fig. 20a). The critical ramping speed is determined by the rate at which the increasing fringing field (Fig. 20b) is screened by channel electrons. The value of $\tau_{cr} \approx 32.3\,ps$ roughly corresponds to the time of electron travel from $S$ and $D$ contacts to the middle of the channel.

The anomalous state at $V_D = 0$ exists only for a sufficiently high $V_C$, and the value of $V_C^{cr}$, below which this state disappears, depends on the device geometry and $v_{sat}$. The value of $V_C^{cr}$ is sharply defined, and for the ramp end voltage $V$ above $V_C^{cr}$, one has $\tau_{cr} \propto V$ to a good approximation, cf. Fig. 21a. This indicates that the relevant critical parameter is the displacement current ($\propto dV/dt$), which should be compared to a transient current ($\propto v_{sat}$) associated with electron screening processes in the emitter.

---

† In a "normal" state of the device, for $V_{DS} = 0$, the collector draws only a minimal current, determined by the barrier height and the temperature. In this state a variation of $V_C$ has the sole effect of changing capacitively, as in a field-effect transistor, the electron concentration in the channel.

The values of $\tau_{cr}$, determined to within 0.1 ps, are plotted in Fig. 21b against the emitter channel length $L_{CH}$ for different assumed values of $v_{sat}$, barrier thicknesses, and ramp end voltages $V$. For short $L_{CH}$ the dependence $\tau_{cr}(L_{CH})$ is approximately quadratic, and for long $L_{CH}$ the dependence is linear.

*Stationary Characteristics.* Starting from the two stationary states at $V_D = 0$, it is possible to determine the characteristics $I_D(V_D)$ and $I_C(V_D)$ at a fixed collector voltage $V_C = 2\,V$ relative to the $S$ electrode. With the predictor-corrector continuation method[34], one can trace arbitrarily shaped, connected components of the characteristic, starting from any established state within each component. The curves in Fig. 22 correspond to the locus of points in the $(V_D, I_D)$ plane for which the device has a steady state at a given $V_C$. To our knowledge, the displayed $I_D(V_D)$ dependence represents the first example of a multiply connected current-voltage characteristic. Any transition between disconnected components of the graph requires a global redistribution of the state fields

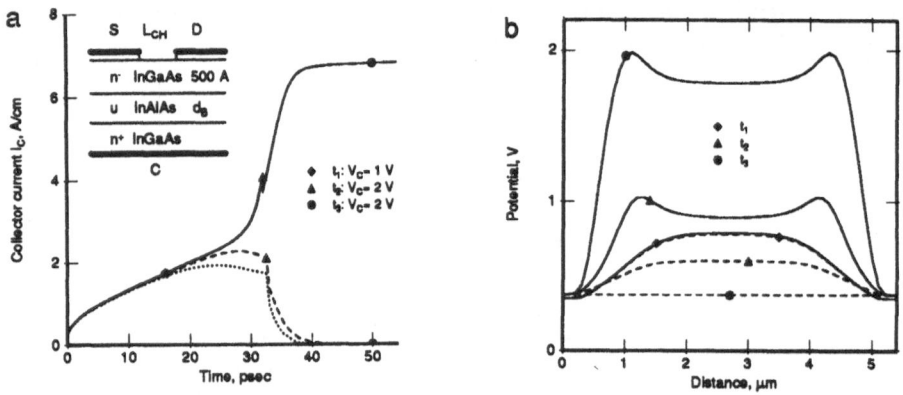

**Fig. 20.** Time-dependent simulation of a real-space transfer transistor with $L_{CH} = 5\,\mu m$, $d_B = 0.2\,\mu m$, and $v_{sat} = 10^7\,cm/s$. The collector voltage is ramped linearly from $V_C = 0$ at $t = 0$ to $V = 2\,V$ at $t = \tau$. The results are plotted for two situations: $\tau = 32.0\,ps < \tau_{cr}$ (solid lines) and $\tau = 32.5\,ps > \tau_{cr}$ (dashed lines).

**(a)** *left figure:* Collector current $I_C(t)$; inset shows cross-section of the device structure. Dotted curve corresponds to the absence of a RST (pure displacement current); it is obtained by artificially increasing the barrier height.

**(b)** *right figure:* Potential distribution $V(x)$ along the channel at selected times.

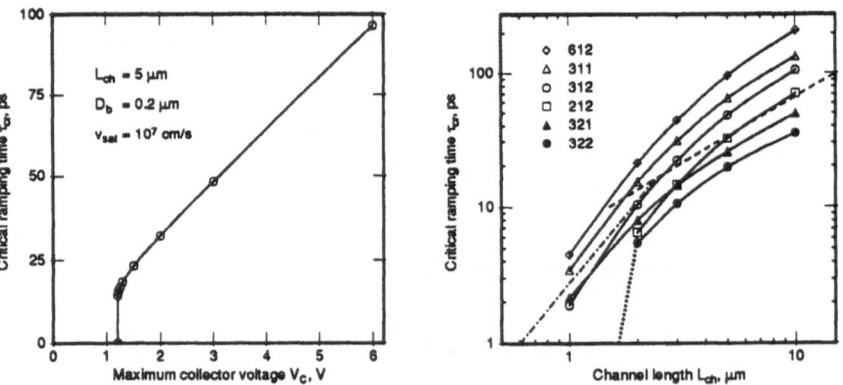

**Fig. 21.** Critical ramping speed for the formation of a stable anomalous stable state at $V_D = 0$. The collector bias is ramped linearly $V_C = 0 \rightarrow V$ in the time interval $\tau$. For $\tau > \tau_{cr}$ the device settles in the normal state, for $\tau < \tau_{cr}$ in the anomalous state (d in the notation of Fig. 22).

*Left figure:* Dependence of $\tau_{cr}$ on the endramp voltage for a $5\,\mu m$ device. $V_C^g = 1.21\,V$.

*Right figure:* Curve labels *VSD* indicate the endramp voltage ($V$) in volts, the saturated velocity ($S$) in $10^7\,cm/s$ and the barrier thickness ($D$) in 1000 A. The dashed and stipple lines indicate linear and quadratic dependences, respectively.

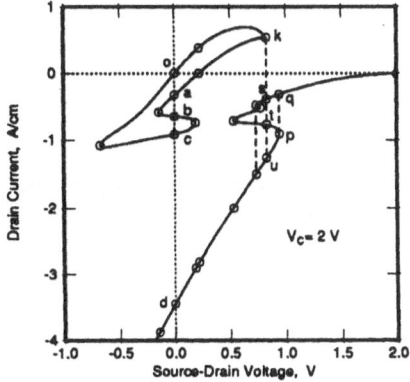

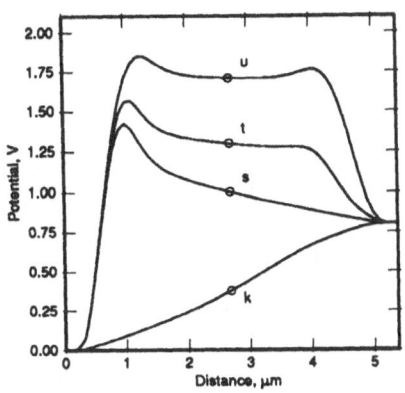

**Fig. 22.** Current-voltage characteristics obtained by the continuation method.

**Fig. 23.** The channel potential profile $V(x)$ in different CHINT states at the same external bias ($V_C = 2\,V$, $V_D = 0.825\,V$).

corresponding to the formation or repositioning of high-field, high-temperature domains in the structure. Such a redistribution, reminiscent of a phase transition, is forced as $V_D$ increases beyond the rightmost point (**k**) of the bounded graph component.

The potential profiles $V(x)$ along the channel – before and after the transition – are shown in Fig. 23. Of the three collector-controlled states, **s**, **t**, and **u**, corresponding to the same ($V_D \approx 0.82\,V$) external bias as the state **k**, two (**s** and **u**) are stable. The actual transition occurs into the state **u** which has the highest value of the collector current. This has been ascertained by a time-dependent simulation in which the initial state **k** was perturbed by a small step $V_D(\mathbf{k}) \rightarrow V_D(\mathbf{k}) + \delta V_D$. The hot-electron domains in the state **u** are characterized by a strong field concentration, accompanied by a dramatic rise in $T_e$. The electron concentration in the domain is depleted so that the collector field remains unscreened and the local potential goes below that of the drain, resulting in a negative $I_D$. All states on the $\mathbf{p} - \mathbf{u} - \mathbf{d}$ branch of the collector-controlled component are similar to **u** and perfectly stable.

This indicates that state **d** (which we had first found in rapid ramping of $V_C$ at $V_D = 0$, cf. Fig. 20) is experimentally accessible by a quasi-static variation of $V_D$ at fixed $V_C$. The existence of a stable anomalous state **d** is obviously a necessary (though insufficient) condition for the multiply-connected topology of the $I_D(V_D)$ characteristic. As discussed above, it results when the competition between RST and screening of the fringing collector field is resolved in favor of RST. Precisely when this happens depends on the transport parameters assumed and the device geometry.

*Unstable anomalous states.* In addition to the normal state **o** and the anomalous state **d**, Fig. 22 reveals three other anomalous states (**a**, **b**, and **c**) at $V_D = 0$. The profiles $V(x)$ along the channel in these states are shown in Fig. 24.

It is clear that because of the non-linear nature of the problem, the actual stationary states may not transform according to irreducible representations of the symmetry group of the equations governing the device behavior at $V_D = 0$. The distribution of internal fields in the anomalous states is either fully symmetric or these states form a set of partners and transform into one another under the symmetry operations. Thus, states **a** and **c** under reflection transform into each other, even though the group has only one-dimensional linear representations. On the other hand, states **o**, **b**, and **d** are symmetric. Biasing the $D$ electrode with respect to $S$ breaks the reflection symmetry and allows a continuous transformation between states of different symmetry on the loop.

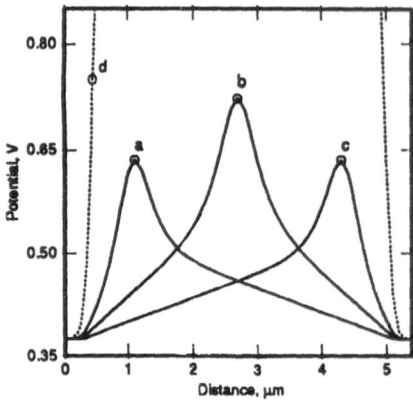

Fig. 24. The channel potential in the four anomalous states at $V_D = 0$.

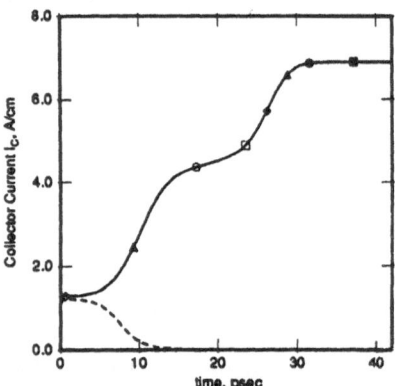 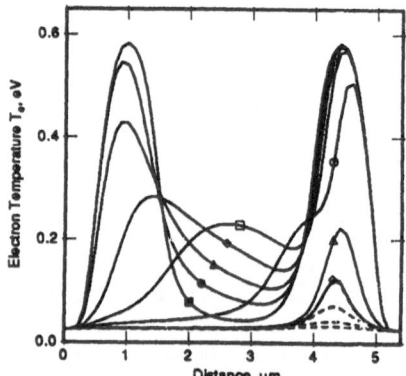

Fig. 25. Evolution of the non-stationary states $c_+$ (solid lines) and $c_-$ (dashed lines) at $V_D = 0$. Time dependence of the injection current is shown in *the left figure*; symbols mark the selected times in the evolution, at which the electron temperature profiles are plotted in *the right figure*. The "plateau" in the $I_C(t)$ dependence near $t = 20$ ps evidently corresponds to the situation when a fully-developed hot-electron domain exists already near $D$ but not yet near the $S$ electrode.

Of the five states at $V_D = 0$ and $V_C = 2$ V, only two (**o** and **d**) are stable with respect to small perturbations. This has been ascertained by following the evolution of states in the vicinity of the steady states at $V_D = 0$. In these simulations, Fig. 25, the initial states $a_\pm$, $b_\pm$, and $c_\pm$ have been assumed to coincide with a state on the loop displaced from **a**, **b**, and **c**, respectively, by an infinitesimal voltage $\delta V_D = \pm 10$ mV. Even though these states are virtually indistinguishable from the corresponding stationary ones, we found that $a_+$, $b_-$, and $c_-$ evolved into **o**, while $a_-$, $b_+$, and $c_+$ into **d**. All these instabilities develop on a rapid time scale, corresponding to the electron travel over the distances of the order of the domain size. They result in either the formation (repositioning) of a hot-electron domain, or its complete quench due to the screening by channel electrons.

Note that the question of stability of a given state can usually be ascertained *without* costly time-dependent simulations – by inspecting the phase diagrams $I_D(V_C, V_D)$ and $I_C(V_C, V_D)$ discussed below. The instability of states **a** and **c** is associated with a NDR in the $I_C(V_C)$ dependence, $\partial I_C / \partial V_C < 0$, and that of **b** with both $\partial I_C / \partial V_C < 0$ and $\partial I_D / \partial V_D < 0$. No counterexample to this rule has yet been found.

*DC mappings of CHINT current-voltage characteristics.* Fig. 26 shows $I_D(V_{DS})$ characteristics for a single device at $T = 300$ K using a series of fixed collector voltages ($V_{CS}$). Close examination of the characteristics shows numerous topological transformations. Beginning as an accumulation mode FET at low $V_{CS}$ (<1.0V), the onset of RST initiates the formation of a slight NDR region in the $+V_{DS}$ direction ($\approx 1.0$V). At higher $V_{CS}$, separate folds begin to appear for both $V_{DS} > 0$ and $V_{DS} < 0$, although the curves remain singly connected. At $V_{CS} \approx 1.2$V, a disconnected loop begins to appear in the left-hand plane, corresponding to a surface bounded by a minimum $V_{CS}$ in the 3D space in Fig. 26. As shown in Fig. 27 ($V_{CS} = 1.5$V), this closed loop and the needle-like fold emanating from the bottom of the left-hand plane both continue to open, and the "S-shaped" notch in the right-hand plane (the knee reached by tracing backwards from $V_{DS} = +\infty$) moves leftward as $V_{CS}$ increases. By $V_{CS} = 1.6$V, the characteristic is transformed into a loop which includes the origin, and a singly connected component which is multivalued but has no folds or intersections with the loop. These two components maintain essentially the same topology for larger $V_{CS}$ although their separation in $(I, V)$ increases, cf. Fig. 22.

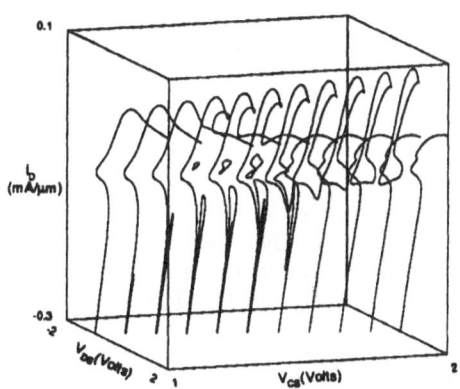

Fig. 26. CHINT $I_D$-$V_{DS}$ characteristics for $1.0$V $\leq V_{CS} \leq 2.0$V ($L_{CH} = 5\mu$m, $d_B = 0.2\mu$m, $v_{sat} = 1\times10^7$cm/s). Curves represent simulations at constant $V_{CS}$ separated by $\Delta V_{CS} = 0.1$V.

Note the multiplicity of anomalous $V_{DS} = 0$ states. In general, at $V_{DS} = 0$, we can expect an odd number $m_S$ of symmetric states and an even number $m_A$ of asymmetric ones, because there should always be one and only one unbounded path in the $(V_{DS}, I_D)$ plane, while, by symmetry, asymmetric states come in pairs. Varying $V_{CS}$, we have been able to realize cases with $(m_S, m_A) = (1, 0)$, $(3, 0)$, $(3, 2)$, $(3, 4)$, and $(5, 4)$. In a continuous variation of $V_{CS}$ one can, of course, arrive at a situation when the $I_D(V_{DS})$ curve only touches the $V_{DS} = 0$ axis without crossing. At this singular point there is an accidental degeneracy of two symmetric states, and the total number of distinct symmetric states becomes even.

Interestingly the symmetry partners may belong to topologically disconnected branches of the $I_D$–$V_{DS}$ characteristic. It is in fact the existence of unpaired asymmetric states along the singly connected, outer curve in Fig. 27 that has led[23] to the discovery of the loop, disconnected from the origin.

Figure 28 shows a map of the $I_C$-$V_{DS}$, corresponding to the $I_D$-$V_{DS}$ plot in Figs. 27. Together, they completely define the device state since $I_C = I_S + I_D$. Although slices of $I_C$-$V_{DS}$ space for a given $V_{CS}$ are not symmetric, there is symmetry in the 3D space, expressed by Eq. (1). Since the collector current is invariant under the symmetry transformation, plots of $I_C$-$V_{DS}$ have many self-intersections, cf. Fig. 28. Points defined by a single intersection with the $V_{DS} = 0$ axis correspond to symmetric states ($I_C = 2I_D = 2I_S$); points defined by two coincident intersections represent reflective, asymmetric pairs.

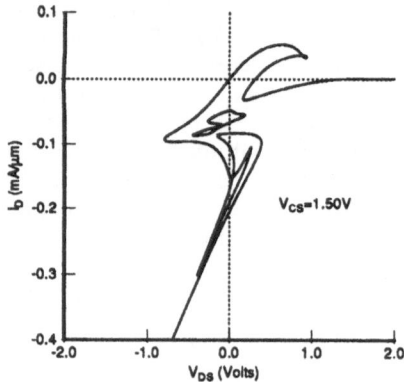

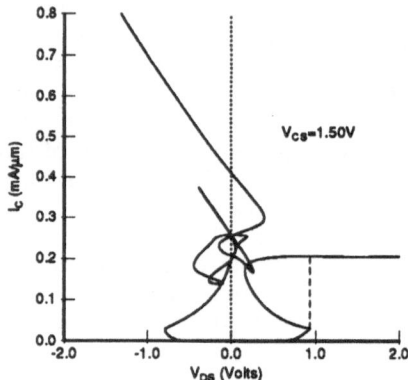

**Fig. 27.** Single CHINT $I_D$-$V_{DS}$ characteristic from Fig. 26 for $V_{CS} = 1.50\,V$ which contains both a self-intersecting component and a disconnected loop.

**Fig. 28.** Single CHINT $I_C$-$V_{DS}$ characteristic for $V_{CS} = 1.50\,V$ corresponding to the $I_D$-$V_{DS}$ in Fig. 27.

Phase space mappings, like those represented by curves in Figs. 27 and 28, have the property that except in the vicinity of self-intersections, their infinitesimally close points correspond to infinitesimally close state vectors $z \in R$ in the multidimensional space $R$, describing the state of the device [i.e. all the fields $n\,(x,y)$, $T_e\,(x,y)$, $\psi\,(x,y)$, etc.]. The converse is always true: points separated by a finite distance on a $(V, I)$ plane correspond to macroscopically distinct states $z$. Continuation in the pseudo-arclength produces a completely smooth evolution of the device state. In contrast, experimental measurements (and MC simulations) force abrupt transitions at limit points, for instance $k \rightarrow u$ in Figs. 22, as $V_{DS}$ is increased from 0, corresponding to the formation or repositioning of a high-field, high-temperature domain. The resultant negative $I_D$ at $u$ arises as the potential in the hot electron domain is lower than that of the drain, due to the unscreened collector field.

Fig. 29 shows the results of continuation simulations started from the origin for $V_{DS} > 0$, using a higher saturation velocity $v_{sat}$. Both the $V_{CS}$ and $V_{DS}$ thresholds for causing folds or limit points in $IV$ increase with $v_{sat}$. The thresholds are reduced for smaller $L_{CH}$, see Fig. 30. These dependencies are similar to those described above in connection with the critical ramp speed of a transient $V_{CS}$ excitation required to induce the stable $d$ state at $V_{DS} = 0$.

In spite of the somewhat artificial $v(F)$ and $T_e(F)$ assumed in the simulation, it can be safely concluded that the cause of the nonlinear steps observed past the RST threshold in experiments are the loops and folds predicted by continuation. Transient simulations corresponding to measurement procedures indicate that folds (e.g. in Fig. 29 for $V_{CS} = 3.0\,V$) can be followed to some length. Predictions of where the state transition will occur can be extracted accurately from complete dc $IV$ maps, but this analysis must include consideration of NDR effects at all terminals as well as external circuit configurations.†

---

† It would be very interesting to understand what is the physical basis for the continuity of $W_{tot}$ noted empirically at many transitions in experimental CHINT characteristics, see Fig. 13b and the footnote below Eq. (6).

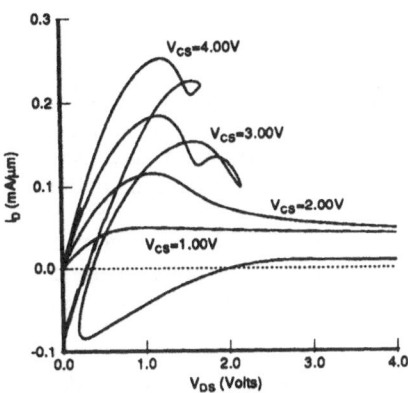

**Fig. 29.** CHINT $I_D$-$V_{DS}$ characteristics as a function of $V_{CS}$ ($L_{CH} = 5\,\mu m$, $d_B = 0.2\,\mu m$, $v_{sat} = 2 \times 10^7\,cm/s$). Results are shown only for the $V_{DS} > 0$ branches initiated from the origin.

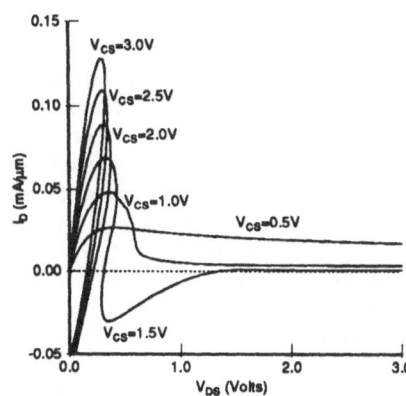

**Fig. 30.** CHINT $I_D$-$V_{DS}$ characteristics as a function of $V_{CS}$ ($L_{CH} = 2\,\mu m$, $d_B = 0.2\,\mu m$, $v_{sat} = 1 \times 10^7\,cm/s$). Results are shown only for the $V_{DS} > 0$ branches initiated from the origin.

### 6.3  Summary of the Simulation Results

The charge injection transistors possess complicated – often multiply-connected – IV characteristics. Application of a sufficiently high $V_{DS}$ at a fixed $V_{CS} > 0$ forces a switching transition, accompanied by the formation of a hot electron domain. Physically, the domains form when the finite supply rate of electrons to a "hot spot" is exceeded by the RST flux from that spot. The depleted domains unscreen the fringing field ("normally" screened by channel electrons) and the RST becomes collector controlled.

Potential applications of RST devices are likely to be based on their peculiar symmetry with respect to the heating field polarity. The same symmetry shows up in the analysis of the hot-electron domain formation. States of a multiterminal RST device under general bias are "adiabatically" connected to the anomalous states of the symmetric configuration at $V_D = 0$. The symmetry analysis is likely to prove very potent with devices of more complicated geometry, such as the ORNAND gate of Fig. 2b, whose symmetry group is $C_{3v}$. Phenomena that occur at $V_{DS} = 0$ capture the essential physics associated with the RST domains in general.

The simulations described above have been carried out using a transport model which eliminates the usual instabilities resulting from a negative differential mobility (the latter has been artificially excluded). The results have been verified[23] to remain qualitatively valid with a more realistic velocity-field model appropriate for InGaAs. For a quantitative simulation of a particular RST device structure, like those in Fig. 2, one would clearly need a realistic transport model. However, the novel anomalies discussed in Sect. 6, can be qualitatively reproduced in *any* model that allows channel electrons to be heated and self-consistently includes the RST flux.

An essential feature of the above analysis is the use of continuation methods which allow the inspection of the full device phase space. Slicing the $I_D(V_D, V_C)$ surface at different $V_C$ results in $I_D(V_D)$ curves of different topologies. The internal fields evolve smoothly along a connected $I_D(V_D)$ trajectory and do not signal the approach of a switching transition. Phase-space mappings successfully give the global type of information. Moreover, they usually give an unerring guess as to the *stability* of a given state, subsequently supported by costlier transient simulations. The understanding gained in this fashion will be invaluable in the application of RST transistors to the design of high-performance systems.

## 7. CONCLUSION

We have discussed the principle and the implementation of novel leight-emitting logic devices, based on the real-space transfer of hot-electrons between independently contacted complementary layers. A monolithic multiterminal logic device, that functions both optically and electrically as an ORNAND gate, has been described. The device, implemented in an InGaAs/InAlAs/InGaAs heterostructure, exhibits both the **or** and the **nand** functions of any two of the three input terminals. in either the output current or the optical output power. These functions are not fixed by the layout but are interchangeable by the voltage on the third ("control") terminal. Choice of the control electrode is, moreover, arbitrary. This is a unique logic device with such powerful capabilities; its function is electrically reprogrammable in the course of a circuit operation.

In the heterostructure used the valence band discontinuity is much smaller than that in the conduction band and most of the parasitic non-RST current is due to the injection of holes from the $p$-type collector layer into the $n$-type emitter. Due to a specially designed collector structure with a wide-gap InAlAs layer confining the active InGaAs layer, the radiative efficiency of minority carriers in the collector is much higher than that in the emitter. This makes the leakage current relatively non-radiative and substantially enhances the optical logic performance at room temperature.

The demonstrated device is an incoherent light-emitting source, similar to a conventional LED. In order to increase its output power and improve the frequency performance, we must take advantage of the stimulated emission in a resonant optical cavity. The high injection current density, $J_C \geq 25\,\mathrm{kA/cm^2}$, obtained in the present device and its high internal radiative efficiency, comparable to that in conventional long wavelength LED's, are promising for the future implementation of a real-space transfer logic laser.

Both electrically and optically, the RST logic devices show a complicated nonlinear behavior, including a variety of novel instabilities, in the internal state dependences on the input voltages. Much work is required, both experimental and theoretical, to use these instabilities in a controlled fashion for the implementation of fast functional devices.

## ACKNOWLEDGEMENTS

We wish to thank our collaborators G. L. Belenky, F. Capasso, A. Y. Cho, P. A. Garbinski, A. L. Hutchinson, M. R. Pinto, and D. L. Sivco for their contributions to the work reviewed here.

## REFERENCES

1. S. Luryi, A. Kastalsky, A. C. Gossard, and R. H. Hendel, "Charge Injection Transistor Based on Real-Space Hot-Electron Transfer", *IEEE Trans. Electron Devices* **ED-31**:832 (1984).

2. Z. S. Gribnikov, "Negative differential conductivity in a multilayer heterostructure," *Fiz. Tekh. Poluprovodn.* 6:1380 (1972) [*Sov. Phys. - Semicond.* 6:1204 (1973)].

3. K. Hess, H. Morkoç, H. Shichijo, and B. G. Streetman, "Negative differential resistance through real-space electron transfer," *Appl. Phys. Lett.* 35:469 (1979).

4. S. Luryi and A. Kastalsky, "Hot electron injection devices", *Superlatt. Microstr.* **1**:389 (1985).

5. A. Kastalsky, "Novel real-space transfer devices", in *High-speed Electronics*, ed. by B. Kallback and H. Beneking (Springer-Verlag, Berlin, 1986) pp. 62-71.

6. S. Luryi, P. Mensz, M. Pinto, P. A. Garbinski, A. Y. Cho, and D. L. Sivco, "Charge injection logic", *Appl. Phys. Lett.* **57**:1787 (1990).

7. H. Tian, K. W. Kim, and M. A. Littlejohn, "Novel heterojunction real-space transfer logic transistor structures: a model-based investigation", *IEEE Trans. Electron Devices* **ED-39**:2189 (1992).

8. P. M. Mensz, A. Y. Cho, and D. L. Sivco, "Charge injection frequency multiplier", *Appl. Phys. Lett.* **61**:934 (1992).

9. S. Luryi and M. R. Pinto, "Broken symmetry and the formation of hot-electron domains in real-space transfer transistors", *Phys. Rev. Lett.* **67**:2351 (1991); "Symmetry of the real-space transfer and collector-controlled states in charge injection transistors", *Semicond. Sci. Tech.* **7**:B520 (1992).

10. S. Luryi, "Charge injection transistors and logic circuits", *Superlatt. Microstr.* **8**:395 (1990).

11. Serge Luryi, "Light emitting devices based on the real-space-transfer of hot electrons", *Appl. Phys. Lett.* **58**:1727 (1991).

12. M. Mastrapasqua, F. Capasso, S. Luryi, A. L. Hutchinson, D. L. Sivco, and A. Y. Cho, "Light emitting charge injection transistor with *p*-type collector", *Appl. Phys. Lett.* **60**:2415 (1992).

13. M. Mastrapasqua, S. Luryi, F. Capasso, A. L. Hutchinson, D. L. Sivco, and A. Y. Cho, "Light emitting transistor based on real-space transfer: electrical and optical properties", To appear in *IEEE Trans. Electr. Dev.* (February, 1993).

14. M. Mastrapasqua, S. Luryi, G. L. Belenky, P. A. Garbinski, D. L. Sivco, and A. Y. Cho, "Multi-terminal light emitting logic device electrically reprogrammable between OR and NAND functions", 1992-IEDM *Tech. Digest* (1992) and submitted to *IEEE Trans. Electr. Dev.*.

15. P. M. Mensz, S. Luryi, A. Y. Cho, D. L. Sivco, and F. Ren, "Real space transfer in three-terminal InGaAs/InAlAs/InGaAs heterostructure devices", *Appl. Phys. Lett.* **56**:2563 (1990).

16. P. M. Mensz, P. A. Garbinski, A. Y. Cho, D. L. Sivco, and S. Luryi, "High transconductance and large peak-to-valley ratio of negative differential conductance in three-terminal InGaAs/InAlAs real-space transfer devices", *Appl. Phys. Lett.* **57**:2558 (1990).

17. R. People, K. Wecht, K. Alavi, and A. Y. Cho, "Measurement of the conduction-band discontinuity of molecular beam epitaxial grown $In_{0.52}Al_{0.48}As/In_{0.53}Ga_{0.47}As$ *N-n* heterojunction by *C-V* profiling", *Appl. Phys. Lett.* **43**:118 (1983).

18. G. P. Agrawal and N. K. Dutta, *Long-wavelength Semiconductor Lasers*, (Van Nostrand Reinhold, New York 1986).

19. S. Luryi, "Hot-electron transistors", in *High-Speed Semiconductor Devices*, S. M. Sze, Ed. (Wiley-Interscience, New York, 1990) Chap. 7.

20. K. Kash and J. Shah, "Carrier energy relaxation in $In_{0.53}Ga_{0.47}As$ determined from picosecond luminescence studies", *Appl. Phys. Lett.* **45**:401 (1984).

21. I. C. Kizilyalli, K. Hess, T. Higman, M. Emanuel, and J. J. Coleman, "Ensemble Monte Carlo simulation of real space transfer (NERFET/CHINT) devices", *Solid-St. Electron.* **31**:355 (1988).

22. I. C. Kizillyalli and K. Hess, "Physics of Real-Space Transfer Transistors", *J. Appl. Phys.* **65**:2005 (1989).

23. M. R. Pinto and S. Luryi, "Simulation of multiply connected current-voltage characteristics in charge injection transistors", 1991-IEDM *Tech. Digest*, pp. 507-510 (1991).

24. P. M. Mensz, H. Schumacher, P. A. Garbinski, A. Y. Cho, D. L. Sivco, and S. Luryi, "Microwave operation of charge injection transistors", 1990-IEDM *Tech. Digest*, pp. 323-326 (1990).

25. M. R. Hueschen, N. Moll, and A. Fischer-Colbrie, "Improved microwave performance in transistors based on real-space electron transfer", *Appl. Phys. Lett.* **57**:386 (1990).

26. K. Maezawa and T. Mizutani, "High-frequency characteristics of charge-injection transistor-mode operation in AlGaAs/InGaAs/GaAs metal-insulator-semiconductor field-effect transistors", *Jpn. J. Appl. Phys.* **30**:1190 (1991).

27. G. L. Belenky, P. A. Garbinski, S. Luryi, M. Mastrapasqua, A. Y. Cho, R. A. Hamm, T. R. Hayes, E. J. Laskowski, D. L. Sivco, and P. R. Smith, "Collector-up light-emitting charge injection transistors in n-InGaAs/InAlAs/p-InGaAs and n-InGaAs/InP/p-InGaAs Heterostructures", submitted to *J. Appl. Phys.* (1992).

28. M. R. Pinto, "Simulation of ULSI Device Effects", in *1991 ULSI Science and Technology*, J. Andrews and G. K. Cellar, eds., *Electrochem. Soc. Proc.* **91-11** (1991).

29. R. Stratton, "Diffusion of hot and cold electrons in semiconductor barriers", *Phys. Rev.* **126**:2002 (1962).

30. K. Bløtekjaer, "Transport equations for electrons in two-valley semiconductors", *IEEE Trans. Electron Dev.* **ED-17**:38 (1970).

31. S. Luryi and M. Pinto, "Collector-controlled states in charge injection transistors", in *Physics and Simulation of Optoelectronic Devices*, ed. by D. Yevick, *Proc. SPIE* **1679**, pp. 54-65 (1992).

32. D. M. Caughey and R. E. Thomas, "Carrier mobilities in silicon empirically related to doping and field", *Proc. IEEE* **55**:2192 (1967).

33. G. Baccarani and M. R. Wordeman, "An investigation of steady-state velocity overshoot in silicon", *Solid-St. Electron.* **28**:407 (1985).

34. W. M. Coughran, Jr., M. R. Pinto, R. K. Smith, "Continuation methods in semiconductor device simulation", *J. Comp. and Appl. Math.* **26**:47 (1989).

# ELECTRON MOBILITY IN DELTA-DOPED QUANTUM WELL STRUCTURES

**W. Ted Masselink**

IBM Research Division,
T.J. Watson Research Center,
Yorktown Heights, New York 10598 USA

**Abstract:**    Hall effect measurements of the mobility of electrons confined in doped GaAs/AlGaAs quantum wells indicate that the ionized impurity scattering of a quasi-two-dimensional electron gas immersed in the identical concentration of ionized impurities is greater than that of a bulk electron gas of the same density. This enhancement arises from fundamental differences between 2-D and 3-D scattering of charged carriers with ionized impurities. The enhanced scattering rate may be further increased by confining the dopant ions to a delta-like doping profile in the center of the well because of the further overlap of the electronic wavefunction with the impurities. Although electrons in center delta-doped AlGaAs/GaAs quantum wells have lower low-field mobilities than do electrons in uniformly doped quantum wells, at electric fields between 2 and 4 kV/cm the differential mobility in delta-doped quantum wells rises dramatically. This large increase in differential mobility may be a result of the heating of the electrons out of the symmetric ground state into the anti-symmetric first excited state.

## 1. Introduction

Electron transport in a quasi-two-dimensional (Q2D) system has provided both a fertile ground for fundamental physical studies as well as a basis for much of today's electronic device technology. Already in 1966 Fowler, Fang, Howard, and Stiles demonstrated that the silicon inversion layer in a MOS system is a two-dimensional electron gas [1]. With the invention of modulation-doping of single-crystalline semiconductor heterojunctions, high mobility 2DEG was demonstrated in 1979 [2]. Today the modulation-doped AlGaAs/GaAs system has been improved to the point that low-field mobilities exceeding $10^7$ cm$^2$/Vs have been realized [3].

Such high mobilities open many possibilities in fundamental research as well as potentially practical applications, but are not by themselves especially useful for electronic devices such as field-effect transistors (FETs) in which high electric fields

*Negative Differential Resistance and Instabilities in 2-D Semiconductors*
Edited by N. Balkan *et al.*, Plenum Press, New York, 1993

83

heat the electrons far above the cryogenic temperatures at which such mobilities are observed. For AlGaAs/GaAs modulation-doped FETs, mobilities higher than about $3 \times 10^4$ cm$^2$/Vs do not appear to enhance the transistor performance [4]. What is far more important than low-field mobility for FET performance, is the entire velocity-field characteristic and the number of electrons available. For this reason, techniques such as Monte Carlo simulations have become very popular as a starting point for FET simulation.

Higher field velocities of the 2DEG in modulation-doped heterostructures have also been measured experimentally [5, 6]. These results will be briefly reviewed in section 2 of this paper, and, in summary, show that the velocity is higher in the modulation-doped AlGaAs/GaAs heterostructure than in heavily doped GaAs for all electric fields at least up to 10 kV/cm.

On the other hand, sheet electron density is limited when one uses a single heterointerface structure; in AlGaAs/GaAs, this limit is about $1 \times 10^{12}$ cm$^{-2}$. Thus, for applications requiring large driving currents other solutions are tempting. Recently Feng has shown that MESFETs with no heterojunction can operate as fast as MODFETs when the materials and lithography are comparable [7]. Even in such advanced MESFETs, however, the electron density should be confined to a very thin plane in order to maintain a useful aspect ratio of gate length to charge depth of about 10 to 1. Thus, whether one is studying MOSFETs, MODFETs, or advanced MESFETs, the details of electron transport in Q2D systems are what determines the device performance.

## 2. Electron Velocity in Modulation-Doped Heterostructures

This section reviews results previously published such as in Ref. [6]. Transport of electrons in doped quantum wells as described in sections 3 and 4 will be compared to these results. In a modulation-doped heterostructure, the donor atoms are placed only into the material with the higher-energy conduction band. This arrangement results in the electrons residing in the lower conduction-band material which is not doped. Mobilities in such a structure may be extremely high because of the relatively high electron density and low ionized impurity density. Especially when the donors are spaced far enough away from the 2DEG is the mobility high. At higher lattice temperatures, the mobility is limited by phonon scattering, rather than ionized-impurity scattering; thus, at higher lattice temperatures, the mobility is rather independent of the details of the structure, provided the donors are not too close to the 2DEG and the background impurity concentration in the GaAs is not too high.

Similarly, at higher electric fields, one finds that the increased electron temperature limits the mobility. Already in the warm electron regime between 40 and 400 V/cm (for low-field mobilities between $2 \times 10^4$ and $8 \times 10^4$ cm$^2$/Vs), the 77 K velocities are converging [4]. By 1000 V/cm, the 77 K velocity is fairly independent of low-field mobility when it is in the range mentioned above.

At room temperature, the mobility of the 2DEG or of electrons in undoped GaAs is relatively constant up to about 3 kV/cm; the velocity then reaches a maximum and then decreases to a lower saturation velocity. Figure 1 shows the velocity versus electric field for lightly-doped GaAs and for AlGaAs/GaAs modulation-doped structures with different Al mole fractions. These data indicate that although all three of these samples have similar low-field mobilities, the peak velocity of the two-dimensional electron gas (2DEG) is lower than that of electrons in lightly doped bulk GaAs and occurs at a somewhat lower electric field. This behavior is explained through two mechanisms. First, because the electrons are spatially confined, the density of states is step-like with zero states at the bottom of the $\Gamma$ valley conduction band. The lowest energy available for electron states coincides with the lowest $\Gamma$ valley subband with an energy of about 40 meV above the bottom of the band [8]. The lack of electron states below this subband results in the effective $\Gamma-L$ energy separation being reduced from about 0.31 eV to about 0.27 eV. Since intervalley transfer and, therefore, peak velocity followed by NDM depends on this energy difference in an exponential manner, this reduction will be important.

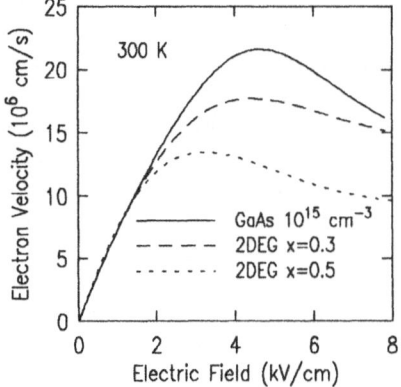

**Figure 1.** Velocity versus electric field for electrons in lightly-doped GaAs and two modulation-doped $Al_xGa_{1-x}As$/GaAs heterostructures with x = 0.30 and x = 0.50 at 300 K.

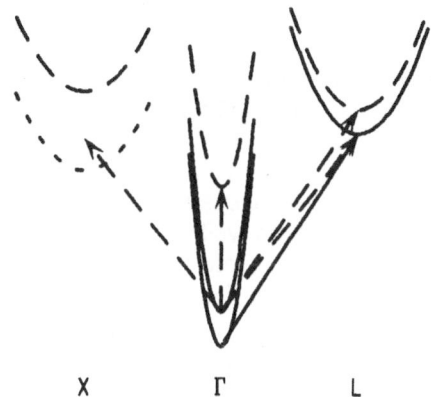

**Figure 2.** A representation of the conduction band minima in GaAs and in the $Al_xGa_{1-x}As$/GaAs heterostructure. The 2DEG has an increased scattering rate as described in the text.

The second effect contributing to the lower peak velocity of the 2DEG as compared with electrons in bulk undoped GaAs is real space transfer. In the heterostructure, besides being able to transfer from the $\Gamma$ valley into the L valleys in GaAs, the electrons may also transfer into the $\Gamma$ valley in the AlGaAs as well as into the L and X valleys in the AlGaAs. All of these valleys are characterized by higher electron effective mass and also by the addition of alloy scattering, which means that electrons scattering into these valleys will have much reduced velocities compared to those in the $\Gamma$ valley of GaAs.

These various valleys are depicted in Figure 2. The solid parabolas of Figure 2 represent the $\Gamma$ and L valleys in GaAs. The lower solid parabola at $\Gamma$ is in bulk GaAs, while the higher one is the first subband in the GaAs at the AlGaAs/GaAs heterointerface. The dashed parabolas represent the valleys in $Al_{0.30}Ga_{0.70}As$ and the dotted parabola represents the X valleys in $Al_{0.5}Ga_{0.5}As$. The solid arrow depicts the transfer of electrons in bulk GaAs from the $\Gamma$ valley into the L valleys. The dashed arrows depict the transfer of electrons in the heterostructure from the quantized $\Gamma$ valley in GaAs into the L valleys in GaAs and also into the various conduction band minima in the AlGaAs.

When considering real space transfer in these structures, there are several important issues to be understood. First is the energy separation between each valley and the Γ valley of GaAs. The density of states and the number of valleys of each type must also be known. Finally, the coupling of electronic states in the valleys must be calculated. The latter issue is probably the most difficult. Even without knowing all of these details precisely, we can qualitatively predict the effects of real space transfer in these structures. As mentioned above, even without real space transfer, the transfer of electrons within GaAs from Γ to L will be enhanced because of the reduced energy separation. Now, in addition the electrons may transfer from Γ in GaAs to L in AlGaAs. The energy separation is similar to that of Γ—L in bulk GaAs; the density of states is also similar, except that the AlGaAs is on only one side of the heterointerface. Finally, the coupling between Γ and these AlGaAs L states is probably reduced, again, in part because the AlGaAs is on but one side of the 2DEG. Therefor, although, an exact value cannot be assigned to this contribution, it is clear that any effect at all will be to add to the transfer of electrons out the GaAs Γ valley. The real space transfer from Γ to Γ has been considered within the framework of Monte Carlo calculations. This effect should be important, but with the small density of states in the Γ valley of AlGaAs, it is likely that transfer into the satellite valleys will dominate. This result is dependent, of course, on the relative coupling as well as on the density of states. Finally, from Figure 1, we see that for higher Al mole fraction AlGaAs, the velocity reaches a smaller maximum at a smaller electric field. Increased scattering of the electrons into the X valleys may account for this measured lower peak electron velocity in $Al_{0.5}Ga_{0.5}As$ when compared to $Al_{0.3}Ga_{0.7}As$.

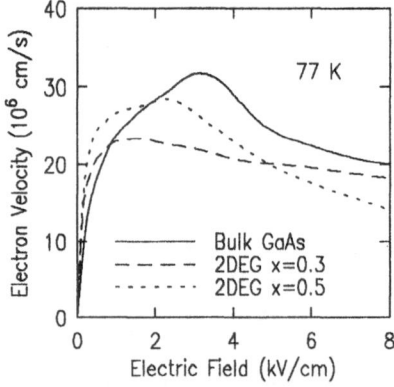

**Figure 3.** Electron velocity as a function of electric field measured at 77 K. The bulk GaAs has $n \simeq 10^{15}$ cm$^{-3}$. The two 2DEG samples are $Al_xGa_{1-x}As/GaAs$ modulation-doped heterostructures with $x = 0.3$ and $x = 0.5$.

Figure 3 depicts the velocity-field relations for the same samples at 77 K. Similar to what we found at 300 K, the selectively-doped samples at 77 K have lower peak velocities than does the lightly doped bulk GaAs sample. The two reasons given above for this behavior are still valid. In addition, when the lattice is 77 K, at about 1 kV/cm the differential mobility of the electrons in the 2DEG samples abruptly decreases, whereas this effect is not so pronounced in the bulk GaAs. Such behavior is consistent with calculations by Ridley *et al*.[9, 10] who predict a sharp decrease in the differential mobility because of an enhanced scattering of electrons with polar optical phonons. In Ridley's picture, the step-like density of states for two-dimensional electrons results in an independence of density of states with respect to energy for energies close to the optical phonon energy, $\hbar\omega_0$. Thus, there is a sudden threshold for the emission of optical phonons for two-dimensional structures. In three-dimensional structures, on the other hand, the strong energy dependence of density of states yields a more gradual turn-on of optical phonon emission. When electron-electron collisions are frequent enough to thermalize the electron energy distribution below $\hbar\omega_0$, but not frequent enough to thermalize the distribution above $\hbar\omega_0$, there will exist a sharp "knee" in the electron distribution at $\hbar\omega_0$, which has the effect of preventing the existence of high energy electrons. This effect will result in a noticeable saturation of electron velocity or even a "scattering-induced" negative differential mobility under certain conditions. The calculations of Ridley *et al* indicate that this decrease should occur at about 1 kV/cm for the AlGaAs/GaAs system which is consistent with our measurements. This agreement indicates that the Ridley picture is at least a possible explanation for this feature observed in the experimental data. The NDM occurring at higher fields is probably due to intervalley and real space transfer. The steady decrease in velocity above 2.2 kV/cm in the $x = 0.5$ sample may be in part due to electrons cooling in the AlGaAs and not transferring back to the GaAs for relatively long times. Thus, the true velocity for the $x = 0.5$ sample may be somewhat higher than indicated in Figure 3 at fields above 2.2 kV/cm.

## 3. Electron Mobility in Doped Quantum Wells

In section 2, we summarized measurements of electron velocity in the modulation-doped heterojunction system; there, the ionized-impurity scattering is kept very small by locating the impurities away from the 2DEG. In this section we describe ionized impurity scattering of electrons in a Q2D system in which the electrons and ionized impurities occupy the same region. The introduction of small concentrations of impurities into the region of the 2DEG dramatically degrades the mobility [11-13]. This degraded mobility is expected because when the added impurity concentration is small compared to the 2DEG concentration the electron screening is essentially unaffected, but the effective number of scattering centers is increased.

In the case of a bulk semiconductor, the total wavefunction of the electrons overlaps with the ionized impurities which are the same in number as the electron density. If such a uniformly doped bulk semiconductor is confined in one dimension, the overlap of the electronic wavefunction with the ionized impurities (an effective impurity concentration) will be altered because the electronic wave function will be peaked in the center, but vanishing at the edges of the resulting quantum well.

Here we describe in detail how ionized-impurity scattering in a Q2D system differs from that in bulk. Specifically we show that the ionized impurity scattering in uniformly doped AlGaAs/GaAs quantum wells is greater than in a similarly doped GaAs. This result is explained through a detailed calculation of the ionized impurity scattering rates in a Q2D system.

To compare ionized impurity scattering in three dimensions to that in two dimensions, we compare Si-doped bulk GaAs to uniformly doped GaAs quantum wells with AlGaAs barriers. All samples were grown by molecular beam epitaxy under identical conditions on (100) oriented undoped GaAs substrates. All sources were solid and a low substrate temperature of about 560° C was used. The bulk GaAs was 0.25 $\mu$m thick and doped with Si to a level of $6 \times 10^{17}$ cm$^{-3}$. The quantum wells, which were in multiple-quantum-well samples with 25 to 50 periods, were either 100 or 50 Å wide with, respectively, 34 or 17 Å wide barriers of Al$_{0.4}$Ga$_{0.6}$As. These barriers are narrow enough to ensure that the two dimensional concentration of traps in the AlGaAs is small compared to the two dimensional electron and dopant concentrations, yet are thick enough to contain the wavefunction in the GaAs quite well. All quantum wells are also doped to an average three dimensional concentration of $6 \times 10^{17}$ cm$^{-3}$. Both 100 Å wells and 50 Å wells were prepared with progressively narrower doping profiles, maintaining the same average doping concentration. The widest doping profile comprised 84% of the well width; the entire well was not doped in order to prevent D-X center formation near the interfaces. The narrow-profile limits of these sample series includes samples delta-doped [14, 15] with sheet densities of $6 \times 10^{11}$ cm$^{-2}$ in the 100 Å wells and $2.8 \times 10^{11}$ cm$^{-2}$ in the 50 Å wells. Wells 100 Å wide were also delta doped off-center similar sheet density of $6 \times 10^{11}$ cm$^{-2}$. Figure 4 shows the conduction band edge and dopant profiles for representative samples.

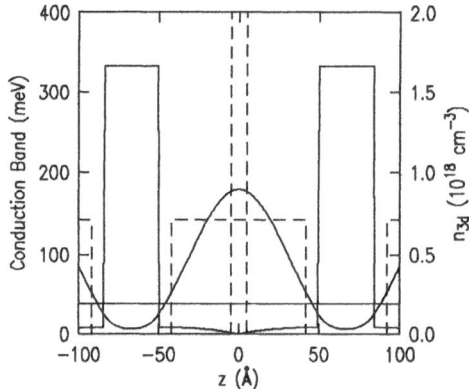

**Figure 4.** Conduction band edge, dopant location, and electron concentration for representative doped quantum wells. Here the well width is 100 Å and two different dopant profiles, both with sheet electron concentration of $6 \times 10^{11}$ cm$^{-2}$ are shown. The horizontal line is the Fermi level.

Some studies of Si $\delta$-doping diffusion in GaAs indicate that the Si atoms deposited in one atomic plane diffuse less than one atomic layer during MBE growth [16, 17]. Other studies [18-20] show evidence of much wider final Si atom profiles. Reference [19] indicates that the three-dimensional concentration of active Si in the $\delta$-doping spike is limited by the D-X center to the maximum Si concentration in bulk GaAs, which results in a maximum sheet concentration of about $7 \times 10^{12}$ cm$^{-2}$. In any case, the Si concentrations in this study are much lower than in Refs.

[16-20] and we expect negligible diffusion. Additionally, some diffusion will not affect our results because the electron wavefunction is quite flat at the well centers and therefore spreading of the doping spike in the center of the well has no effect on the scattering.

Van der Pauw-Hall measurements were made on the samples between 10 and 300 K using low electric and magnetic fields. In heavily doped semiconductors, such as the samples studied here, the Hall factor is very close to unity, regardless of scattering mechanism. The data of Figure 5 show the temperature dependence of the mobilities of three uniformly doped samples: bulk GaAs, 100 Å wells, 50 Å wells. The mobility of the bulk sample is what is typically measured at this doping level and is well understood theoretically [21]. Mobilities for the doped quantum wells are significantly lower than that observed in bulk GaAs for all temperatures. We have also examined samples with well widths of 200 and 400 Å and find that progressively wider wells lead to progressively higher mobilities, apparently asymptotically approaching that for bulk GaAs. These data are depicted in Figure 6.

Previous data [22, 23] indicate that interface roughness scattering of a 2DEG in 100 Å wide wells is quite small. These same data, on the other hand, also indicate that in 50 Å wells it can be significant. Structures with 50 Å wells and AlAs barriers have mobilities limited to about 1000 $cm^2$/Vs [22]; when the barrier alloy is $Al_xGa_{1-x}As$ with x between 0.30 and 0.35, mobilities of about $10^4$ $cm^2$/Vs were measured [23]. Electrons in 100 Å wells achieve mobilities of $10^5$ $cm^2$/Vs even with the rougher AlAs barriers [22]. From this we conclude that the interface roughness scattering in our samples with well width $\geq$ 100 Å and with $Al_{0.4}Ga_{0.6}As$ barriers is not important compared to the ionized impurity scattering, but almost certainly needs to be accounted for in the 50 Å well samples. This interface-roughness scattering in the 50-Å wells is noticeable also in Figure 6.

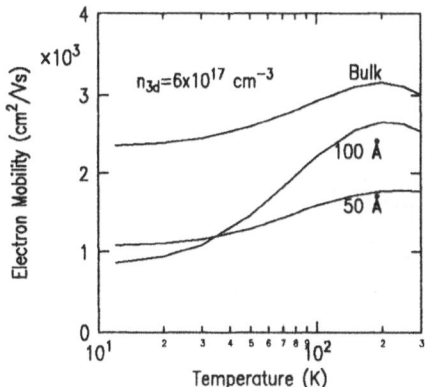

**Figure 5.** Measured Hall mobilities as functions of temperature for bulk GaAs, 100 Å wide quantum wells, and 50 Å wide quantum wells.

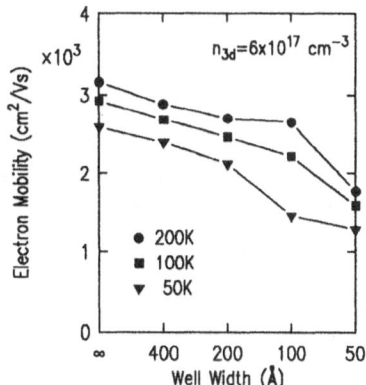

**Figure 6.** Measured Hall mobilities as functions of quantum well width for several temperatures and average doping concentrations of $6 \times 10^{17}$ $cm^{-3}$.

The confinement of the electronic wavefunction affects the ionized impurity scattering in two ways: first, it changes the way the wavefunction overlaps with the impurities; and second, it changes the electron screening term in the dielectric function. Following Ando [24], the relaxation time for electrons in one subband scattering from ionized impurities is given by

$$\frac{\hbar}{\tau_c(k)} = 2\pi \int dz\, N(z) \sum_q \left[ \frac{2\pi e^2}{q\varepsilon(q)} \right] |F(q,z)|^2 (1 - \cos\theta)\, \delta(\varepsilon_k - \varepsilon_{k-q}), \qquad [1]$$

where $N(z)$ is the doping profile, $q = 2k\sin(\theta/2)$ and $\varepsilon_k = \hbar^2 k^2/2m$ where $\mathbf{k}$ is the wave vector, and $\theta$ is the scattering angle. The static dielectric function is given by

$$q\varepsilon(q) = q + \frac{2\pi e^2}{\kappa} \frac{2m}{2\pi\hbar^2} F(q), \qquad [2]$$

where $\kappa$ is the dielectric constant and m is the electron effective mass. Here we use the $T=0$ formula for the polarization term which introduces only small errors away from $T=0$ in the case of these degenerate samples. In two dimensional systems, degenerate statistics result in no explicit $n_{2d}$ dependence of the polarization. The form factor appearing in Eq. 1, $F(q,z)$, is a measure of the overlap of the electron density with charge centers at location z and is given by

$$F(q,z) = \int dz'\, |\zeta(z')|^2 \exp(-q|z-z'|), \qquad [3]$$

where $\zeta(z)$ is the z component of the total electronic envelope wavefunction $\Psi(\mathbf{r},z) = \zeta(z)\exp(i\mathbf{k}\cdot\mathbf{r})$. The other form factor $F(q)$ which appears in Eq. 2 may be thought of as the "screening of the screening" [25] and is given by

$$F(q) = \int dz \int dz'\, |\zeta(z)|^2 |\zeta(z')|^2 \exp(-q|z-z'|); \qquad [4]$$

because it never exceeds unity, it only decreases the screening. The sum over q in Eq. 1 may be replaced by an integral over $\theta$ so that Eq. 1 is written

$$\frac{\hbar}{\tau_c(k)} = \frac{4\pi m e^4}{\hbar^2} \int_0^\pi d\theta\, (1 - \cos\theta)\, \frac{1}{(q\varepsilon(q))^2} \int dz\, |F(q,z)|^2 N(z). \qquad [5]$$

The mobility is finally given by

$$\mu(T) = \frac{e<\tau_c>}{m}. \qquad [6]$$

The temperature dependence of $\mu(T)$ enters in the averaging of $\tau_c(k)$ over k. We choose a simple wavefunction,

$$\zeta(z) = \sqrt{\frac{2}{W}}\, \cos\left(\frac{\pi z}{W}\right)$$

where W is the width of the quantum well. This wavefunction along with the one-subband approximation is justified by noting that even with such narrow barriers, especially in the 100 Å case, the wavefunction squared is close to zero at the interface and that the second subband is nearly unpopulated. The self-consistent solution of Schrödinger's and Poisson's equations used to generate Figure 4 shows that compared to the conduction band offset, the band bending with these doping levels is small. The doping profile N(z) for symmetrically doped samples is given by

$$N(z) = \begin{cases} \dfrac{1}{uW} \, n_{2d} & |z| < \dfrac{uW}{2} \\ 0 & |z| > \dfrac{uW}{2} \end{cases}$$

where u is the fraction of the well which is doped. In the case of the delta-doped wells,

$$N(z) = n_{2d} \, \delta(z - z_\delta)$$

where $z_\delta$ is the position of the delta-doping spike. It is then straightforward to evaluate the integrals of Eqs. 3 and 4 analytically and of Eqs. 5 and 6 numerically to obtain the theoretical mobilities as functions of dopant distribution, temperature, and well width. We ignore certain corrections which have been incorporated into more detailed calculations of ionized impurity scattering the bulk semiconductors [26] such as multiple scattering, as well as the effect of well-to-well screening which we expect to be small.

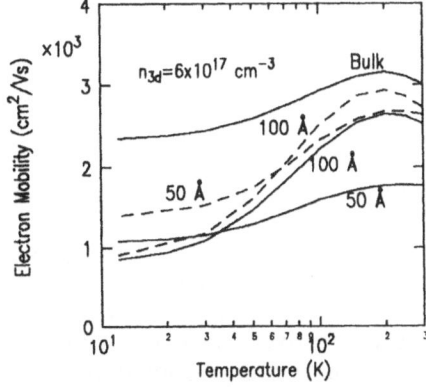

**Figure 7.** Measured Hall mobilities as functions of temperature for bulk GaAs, 100 Å wide quantum wells, and 50 Å wide quantum wells. All of these samples have average doping concentrations of $6 \times 10^{17}$ cm$^{-3}$. The solid curves are measured data; the dashed curves are adjusted as described in the text in order to make an appropriate comparison with bulk.

The data displayed in Figure 5 and by the solid curves of Figure 7 do not really provide a definitive comparison of ionized impurity scattering in two dimensions to ionized impurity scattering in three dimensions because although the bulk sample is uniformly doped, the wells are doped in the center 84% which will result in lower mobility than if they were uniformly doped. (Uniform doping over 100% of the wells was avoided to prevent D-X center formation at the interfaces.) The quantum well mobilities may be corrected, however, by using Matthiessen's rule to subtract the extra scattering rate resulting from the slightly more concentrated dopant profile. In the case of the 50 Å wells, an additional (assumed temperature-independent) scattering rate equivalent to a mobility of 6000 cm$^2$/Vs was also subtracted. The dashed curves of Figure 7 depict the measured mobilities of the 100 Å and 50 Å wells corrected as described. These data are slightly different from the raw data of the solid curves and demonstrate unambiguously that Q2D confinement results in a decrease of mobility and therefore an increase in ionized impurity scattering over a broad temperature range. This effect may be compared with an analogous enhancement of scattering processes for 2D excitons over those for 3D excitons [27].

Conceptually, one is tempted to make a connection to three-dimensional scattering by associating the three-dimensional case with a quantum well system, but with $\zeta(z)$ constant. If one makes this assumption, mobilities are, indeed, calculated to be higher than in the Q2D case described. This connection is not justified, however, because in three-dimensions the screening is different, as well as the density of states. To truly go continuously from two to three dimensions in this way, one would need to consider an infinite number of subbands in the calculation. More realistically, the dimensionality-enhanced scattering of electrons by ionized impurities results, in part, from the details of the screening and overlap with the impurities. The overlap with impurities can be thought of as an effective doping concentration, $N_{eff}$, where

$$N_{eff}(q) = \frac{1}{W} \int |F(q,z)|^2 N(z) \, dz$$

We easily see that

$$\lim_{q \to 0} N_{eff}(q) = \frac{n_{2d}}{W}$$

and that $N_{eff} \geq N_{2d}/W$ for all q. In three dimensions, $N_{2d}/W$ is replaced by $N_{3d}$, with $N_{eff} \geq N_{3d}$.

The screening term is also completely different in two dimensions than it is in three dimensions. For large q, the three dimensional scattering is smaller ($\sim q^{-2}$) than it is in two dimensions ($\sim q^{-1}$). In the opposite limit of large r, the three dimensional system is much more effectively screened, with the scattering matrix element going as $e^{-qr}$ compared to the two dimensional system in which it goes as $r^{-3}$ [28]. Additionally, we see that the most effective scattering events are those of back scattering, where $\theta = \pi$ and that back-scattering is relatively easier in a two-dimensional system, since $\Delta\theta$ enters only linearly. Another feature to notice in Figure 7 (all curves) is the positive value of $\partial\mu/\partial T$. This temperature dependence comes out of the integral of $\tau_c(k)$ over k with the ionized impurity scattering being stronger for smaller k.

From Figure 7 it is clear that uniformly doped quantum wells have a lower mobility than identically doped bulk GaAs and, of course, studies of modulation-doped quantum wells show that when the dopants are largely moved out of the quantum well, the mobility is higher. Figure 8 shows the effect of concentrating the dopants into the center of the well. In each case, the areal concentration is about 6 $\times 10^{11}$ cm$^{-2}$, but the distributions vary. When the fraction of the well which is doped reaches 0, the well is delta-doped in the center; when the fraction is 1.0, it is uniformly doped. The solid curves of Figure 8 are calculated as described above with no corrections or adjustable parameters. (Phonon scattering is not included in the calculations.) The symbols are experimental data. We see that the doping profile has a significant impact on the ionized impurity mobility and that the model described above describes the physics adequately, except at low temperatures. At higher temperatures (and especially for larger fractions of the well which is doped), phonon scattering becomes important. A structure was also prepared which was otherwise identical to the delta-doped quantum well sample except that it contains no AlGaAs barriers. This sample consists of delta-like spikes of $6 \times 10^{11}$ cm$^{-2}$ located every 100 Å; the mobility $\mu(T)$ for this sample is nearly identical to that of the bulk GaAs of identical doping for all temperatures.

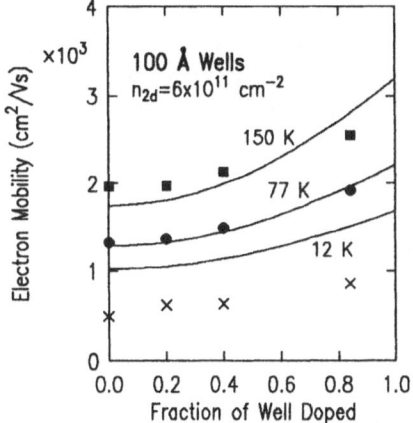

**Figure 8.** Electron mobility for 100 Å quantum wells as a function of temperature and fraction of the well which is doped. The curves are calculated as described in the text and the symbols are experimental data taken at 150 K (■), 77 K (●), and 12 K (×).

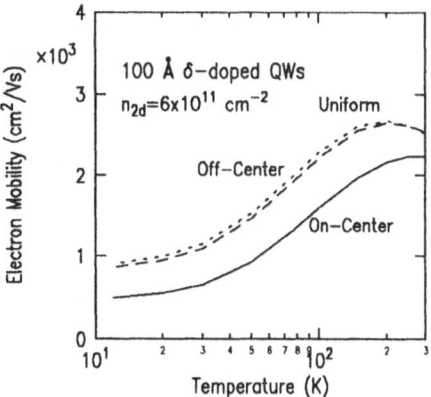

**Figure 9.** Measured Hall mobilities as functions of temperature for 100 Å wide GaAs quantum wells uniformly doped (dashed curve), δ-doped in the center (solid curve), and δ-doped midway between the center and an edge (dotted curve). The sheet density in each well is $6 \times 10^{11}$ cm$^{-2}$.

From Figure 8 we see that the lowest mobility is observed in the wells with delta-doping in the center. This effect results from all of the ionized impurities being localized where the electronic wavefunction is maximum (i.e., maximizing $N_{eff}$). By moving the delta-doping away from the center, we can increase the mobility. Figure 9 shows the mobility versus temperature for three samples. The solid curves are experimental. The one labeled "Uniform" is uniformly doped in the center 84% of the well. The lowest curve is for the center-delta-doped wells. The third curve is from a sample which is also delta-doped, but with the doping midway between the center and the edge of the well. From Figure 9, we see that the mobility is higher than when the dopants are all at the center of the well. If the dopants were moved farther from the center, the mobility would be even higher; by moving them entirely out of the wells, we are back to the modulation-doped example.

Summarizing, the electron mobility in doped quantum wells is lower than for electrons in similarly doped bulk GaAs, indicating an enhancement of the ionized impurity scattering rate of quasi-two-dimensional carriers. Our calculations of electron mobility in doped quantum wells show that this enhanced scattering rate is characteristic of the confinement in one dimension and is due to a decrease in screening, an increase in the effective overlap of electronic wavefunction with the dopant atoms, and an increase in large-angle scattering in confined systems.

### 4. Electron Velocity in Doped Quantum Wells

Transport properties of the electron gas were measured at high fields using a sinusoidally varying electric field with frequency of 35 GHz [6] to avoid the formation of charge and field domains in the samples. This technique has been shown reliable in the measurement of electron velocity in bulk GaAs [5, 29] and was also used to obtain the data in section 2. Figure 10 shows the measured electron velocity as a function of the electric field for three samples, each with a three-dimensional doping density of $6 \times 10^{17}$ cm$^{-3}$: the bulk sample of total thickness 0.25 $\mu$m, the sample with 100 Å quantum wells, and the sample with 50 Å quantum wells described in section 3.

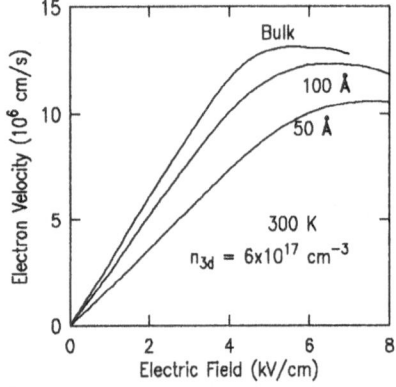

**Figure 10.** Measured velocities as functions of temperature for bulk GaAs and 100 Å and 50 Å wide GaAs quantum wells uniformly doped with Si.

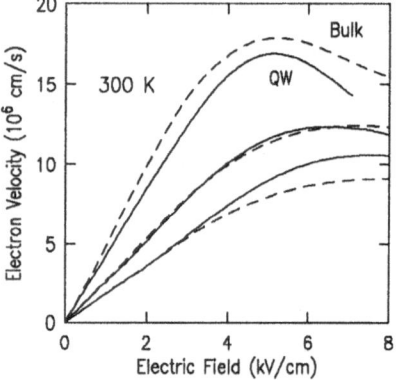

**Figure 11.** Pairs of doped GaAs samples with similar low-field mobilities. In each pair, the dashed curve is for doped bulk and the solid curve for a doped quantum well.

From Figure 10 we see that even with the same average 3-D electron density, the electron gas confined to narrower quantum wells has a lower peak velocity than in bulk GaAs. This result is qualitatively similar to lower peak velocity found in in the two-dimensional electron gas in undoped GaAs when compared to that of 3-D electrons in bulk undoped GaAs as described in section 2 and depicted in Figure 1. Comparing Figure 10 and Figure 1, however, we see that while in Figure 1, the mobility at low electric fields of the 2-D case is essentially the same as in the 3-D case, this is not true in Figure 10. In doped bulk GaAs, as the donor concentration is increased, both the peak electron velocity *and* the low field mobility decrease. The decrease of both of these values is due to an increase in ionized-impurity scattering resulting from a higher ionized impurity concentration. A fairer comparison to make regarding the data of Figure 10, then, is to compare the peak velocities as functions of low-field mobility. Consider several sample pairs, each with one doped quantum well sample and one bulk GaAs sample, and comparable low-field mobilities. (In order that each pair have comparable mobilities, the doped quantum well sample must have a lower doping level than the bulk sample.) Figure 11 shows when one measures the velocity versus electric field in these pairs, that there is no significant difference in higher field velocity. Thus, in both doped quantum wells and doped bulk GaAs, a given low-field mobility results in a given peak velocity. For example, if one compares the peak velocity of the 100 Å wide sample along with its low-field mobility of 2536 cm$^2$/Vs and of the 50 Å wide sample along with its low-field mobility of 1767 cm$^2$/Vs to a summary of peak velocity versus low-field mobility [6, 30], we see that these velocities are consistent with the velocities found in bulk GaAs. Therefore we conclude that what limits the high-field velocity in these structures is *not* enhanced transfer of electrons out of the Γ valley into the GaAs L valleys and (real space transfer) into the AlGaAs as occurs in modulation-doped heterostructures, but, rather, simply an enhancement of the ionized-impurity scattering.

From the analysis and data presented above, it is clear that uniformly doped quantum wells have a lower mobility than identically doped bulk GaAs. Figure 8 in section 3 shows the effect of concentrating the dopants into the center of the well.

Electron velocity at high electric fields was also measured for the δ-doped samples [31, 32]. Figure 12 depicts the measured electron velocity as a function of the electric field for the three samples whose mobilities were depicted in Figure 9. Although the on-center δ-doped quantum-well sample had the lowest mobility at low electric fields, the peak velocity of electrons in this sample is higher than in either the uniformly doped wells or in the off-center δ-doped wells. This higher velocity is possible because the differential mobility increases at electric fields between 0 and $\mathcal{E}_p$, the electric field at which the peak velocity occurs.

Particularly noticeable about Figure 12 is that the velocity of the lowest mobility sample — the on-center δ-doped sample — is the highest. This is because a super-linear increase of velocity with increasing electric field over some range of electric field. This effect is more clearly seen in Figure 13 which depicts the derivatives of the velocity-field curves of Figure 12. From Figure 13 we see that the differential mobility of the electrons in the on-center δ-doped quantum wells increases much more than that of the electrons in uniformly doped quantum wells or in off-center δ-doped quantum wells. We expect that the ionized impurity scattering rate of the on-center δ-doped wells would decrease relatively more than that of the uniformly doped wells simply because the ionized impurity scattering rate is so much greater at low electric fields and, therefore, has much farther to fall. From Figure 13, however, we see that the differential mobility of the electrons in the on-center δ-doped wells actually increases enough to be significantly greater than that for the electrons in the uniformly doped wells. The large increase in differential mobility in the on-center δ-doped sample may also be contrasted with the smaller increase in the off-center δ-doped sample's mobility.

These data may be explained as follows: The dramatic increase in the differential mobility of the on-center δ-doped samples is due to the heating of the electrons from the ground state (even parity) quantum-well subband into the first excited (odd parity) subband. This odd-parity state has a node at the center of the

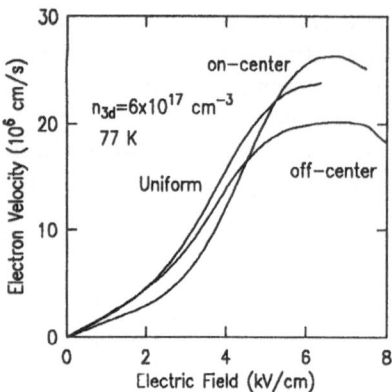

**Figure 12.** Measured Hall mobilities as functions of temperature for 100 Å wide GaAs quantum wells uniformly doped (dashed curve), δ-doped in the center (solid curve), and δ-doped midway between the center and an edge (dotted curve).

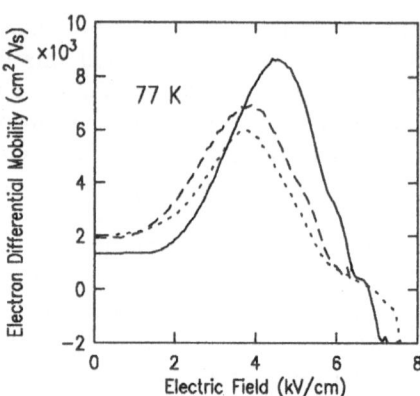

**Figure 13.** Measured differential mobilities versus electric field for 100 Å wide GaAs quantum wells uniformly doped (dashed curve), δ-doped in the center (solid curve), and δ-doped midway between the center and an edge (dotted curve).

quantum well, exactly where the ionized impurities are located. It follows, then, that heating electrons so that they begin to occupy the first excited subband should lead to higher mobility because the excited electrons experience much less ionized impurity scattering. The off-center δ-doped sample will not show this effect to the same extent because the δ-doping spike is located at one of the maxima of the first excited state instead of at the minimum.

At high electric fields, the peak velocity of quasi-two-dimensional electrons in doped quantum wells is lower than what we measure in similarly doped bulk GaAs. Detailed transport measurements indicate that this lowering of the peak velocity is not primarily due to real space transfer as is the similar effect in modulation-doped heterostructures, but, rather, is also due to the enhanced ionized impurity scattering experienced by electrons confined in one dimension. Delta-doping the quantum wells can lead to a further decrease in low-field mobility because the overlap of the electron wavefunction with the dopant atoms in increased. At high electric fields, however, the differential mobility dramatically increases, leading to a somewhat higher velocity than observed in uniformly doped quantum wells, in spite of the lower low-field mobility. This large increase in differential mobility may be the result of heating of the electrons out of the symmetric ground state into the antisymmetric first excited state.

## 5. Conclusions

This paper has described Q2D electron mobility and low and high electric fields for electrons in AlGaAs/GaAs heterojunction systems. In modulation-doped structures, the mobility is very high, limited at room temperature by phonon scattering. At elevated electric fields the velocity of these electrons reach a maximum which is smaller than that of electrons in bulk undoped GaAs. This decrease in the maximum velocity is because of enhancements in Γ-L scattering, the addition of real-space transfer of electrons to the AlGaAs, and, at 77 K, also an enhancement in polar-optical phonon scattering. All of these mechanisms follow from the quasi-two-dimensionality of the electron gas.

In the opposite extreme, when the 2DEG is immersed in an equal concentration of ionized impurities, the low field mobility is lower than what one finds for a 3D electron gas. This result also follows from the two-dimensionality of the system and will be generally observed in systems such as doped quantum wells. The higher field velocity of these structures is also lower than of 3D electrons in the same doping concentration, again because of enhanced ionized-impurity scattering.

When a quantum well is delta-doped in the center of the well, the mobility is lower than when the same 2D electron concentration is uniformly doped throughout the well because the maximum of the electronic wavefunction lies in the same region as the dopant spike. At higher electric fields, however, these center delta-doped structures have higher electron velocity; this is probably due to electrons becoming heated out of the ground state wavefunction into the first excited state and by so doing becoming spatially separated from the dopant spike.

## References

1) A.B. Fowler, F.F. Fang, W.E. Howard, and P.J. Stiles, Phys. Rev. Lett. **16**, 901-903 (1966).

2) H.L. Störmer, R. Dingle, A.C. Gossard, W. Wiegmann, and M.D. Sturge, J. Vac. Sci. Technol. **16**, 1517-1519 (1979).

3) L. Pfeiffer, K.W. West, H.L. Störmer, and K.W. Baldwin, Appl. Phys. Lett. **55**, 1888-1890 (1989).

4) W.T. Masselink, T. Henderson, J. Klem, W.F. Kopp, and H. Morkoç, IEEE Trans. Electron Devices **33**, 639-645 (1986).

5) W.T. Masselink, N. Braslau, W.I. Wang, and S.L. Wright, Appl. Phys. Lett. **51**, 1533-1535 (1987).

6) W.T. Masselink, Semicond. Sci. Technol. **4**, 503-512 (1989).

7) M. Feng, C.L. Lau, V. Eu, and C. Ito, Appl. Phys. Lett. **57**, 1233-1235 (1990).

8) F. Stern and S. Das Sarma, Phys. Rev. B **30**, 840-848 (1984).

9) F.A. Riddoch and B.K. Ridley, J. Phys. C: Solid State Phys. **16**, 6971-6982 (1983).

10) B.K. Ridley, J. Phys. C: Solid State Phys. **17**, 5357-5365 (1984).

11) R.J. Haug, K. von Klitzing, and K. Ploog, Proceedings of *19th International Conference on the Physics of Semiconductors*, Warsaw, 1988, edited by W. Zawadzki ( Inst. of Physics, Polish Academy of Sciences, Warsaw, 1988 ), pp. 307-310.

12) S. Mori and T. Ando, J. Phys. Soc. Jpn. **48**, 865-873 (1980).

13) A. Gold, J. Phys. (Paris) Colloque **C5**, 255-258 (1987).

14) C.E.C. Wood, G. Metze, J. Berry, and L.F. Eastman, J. Appl. Phys. **51**, 383-387 (1980).

15) E.F. Schubert and K. Ploog, Jap. J. Appl. Phys. Lett. **24**, L608-L610 (1985).

16) E.F. Schubert, J.B. Stark, B. Ullrich, and J.E. Cunningham, Appl. Phys. Lett. **52**, 1508-1510 (1988).

17) E.F. Schubert, J.B. Stark, T.H. Chiu, and B. Tell, Appl. Phys. Lett. **53**, 293-295 (1988).

18) A. Zrenner, F. Koch, and K. Ploog, Proceedings of *Int. Symp. GaAs and Related Compounds*, Heraklion, Greece, 1987, edited by A. Christou and H.S. Rupprecht ( Inst. of Physics Publishing Ltd., Bristol, 1988 ), pp. 171-174.

19) A. Zrenner, F. Koch, R.L. Williams, R.A. Stradling, K. Ploog, and G. Weimann, Semicond. Sci. Technol. **3**, 1203-1209 (1988).

20) J. Wagner, M. Ramsteiner, W. Stolz, M. Hauser, and K. Ploog, Appl. Phys. Lett. **55**, 978-980 (1989).

21) J.R. Meyer and F.J. Bartoli, Phys. Rev. B **36**, 5989-6000 (1987).

22) H. Sakaki, T. Noda, H. Hirakawa, M. Tanaka, and T. Matsusue, Appl. Phys. Lett. **51**, 1934-1936 (1987).

23) R. Gottinger, A. Gold, G. Abstreiter, G. Weimann, and W. Schlapp, Europhys. Lett. **6**, 183-188 (1988).

24) T. Ando, J. Phys. Soc. Jpn. **51**, 3900-3907 (1982).

25) P.J. Price, J. Vac. Sci. Technol. **19**, 599-603 (1981).

26) D. Chattopadhayay and H.J. Queisser, Rev. Mod. Phys. **53**, 745-768 (1981).

27) A. Honold, L. Schultheis, J. Kuhl, and C.W. Tu, Phys. Rev. B **40**, 6442-6445 (1989).

28) T. Ando, A.B. Fowler, and F. Stern, Rev. Mod. Phys. **54**, 437-672 (1982).

29) M.V. Fischetti, IEEE Trans. Electron Devices **38**, 634-649 (1991).

30) W.T. Masselink and T.F. Kuech, J. Electronic Materials **18**, 579-584 (1989).

31) W.T. Masselink, Appl. Phys. Lett. **59**, 694-696 (1991).

32) W.T. Masselink, Proceedings of *Int. Symp. GaAs and Related Compounds*, Seattle, USA, 1991, edited by G.B. Stringfellow ( Inst. of Physics Publishing Ltd., Bristol, 1992 ), pp. 425-430.

# APPLICATION OF A NEW MULTI-SCALE
# APPROACH TO TRANSPORT IN A GaAs/AlAs
# HETEROJUNCTION STRUCTURE

P. D. Yoder and K. Hess

Beckman Institute for Advanced Science and Technology
University of Illinois
Urbana, IL 61801

## INTRODUCTION

Electron transmission and reflection at an interface has long been recognized as a quantum-mechanical problem. With the continually decreasing scale of semiconductor devices, the presence of interfaces has an increasingly significant impact on device performance. Consequently there is a need for an accurate description of devices that includes transport over heterojunctions. The question then becomes how to simulate a small region of quantum transport which exists in an otherwise semiclassical environment. What is needed is a multi-scale approach, with the goal of describing transport far away from the band edges and in a wide range of energies.

A solution to this multi-scale problem has been provided by recent developments in the physical theory of transport in mesoscopic systems. This theory as pioneered by Rolf Landauer and Markus Büttiker [1, 2] separates the quantum transport regions from reservoirs, and treats the quantum transport problem from the viewpoint of scattering theory. One can directly apply this approach to propose a new method of simulation in which the regions of space that are large and homogeneous enough to be treated semiclassically are dealt with by semiclassical ensemble Monte Carlo; the quantum regions, i.e. regions of small size with pronounced quantum interference effects, are put on the same footage as scattering mechanisms by suitable inclusion of reflection and transmission into the ensemble Monte Carlo simulation, in complete analogy to the use of the Landauer-Büttiker formalism. Whereas Landauer and Büttiker assumed that the semiclassical regions were in thermal and diffusive equilibrium with reservoirs at temperature $T_i$ and chemical potential $\mu_i$, our approach is a natural generalization allowing for non-equilibrium transport. As a specific example, we consider here the GaAs/AlAs (1,0,0) interface system. Current-voltage characteristics for a particular diode geometry are calculated and discussed.

*Negative Differential Resistance and Instabilities in 2-D Semiconductors*
Edited by N. Balkan *et al.*, Plenum Press, New York, 1993

## THE MODEL

In device applications, electron emission into different semiconductors or insulators can occur over a wide range of energies, and a knowledge of the energy bandstructure over many electron volts is necessary to describe it quantitatively. Within the framework of effective mass theory extended by Harrison [3], our knowledge of transport over interfaces is restricted to propagation close to the band edges, and necessitates adoption of phenomenological constants to describe the coupling between states on either side of the interface. It is clear that only a general numerical approach is suitable to solve this type of problem, which naturally lends itself to a multi-scale full bandstructure Monte Carlo method. As mentioned previously, transport in the interface region is treated by scattering theory. To calculate the transmission and reflection coefficients at a perfect GaAs/AlAs interface, one can start from single electron eigenstates described by a Hamiltonian of the form:

$$\hat{H} = \frac{-\hbar^2}{2m_0}\nabla^2 + V_{GaAs}(\mathbf{r})\theta(-x) + V_{AlAs}(\mathbf{r})\theta(x) + V_{dipole}(x),$$

making the usual assumption that the two crystalline potentials join smoothly at the interface plane. Here $V_{GaAs/AlAs}(\mathbf{r})$ are the ionic potentials in the two bulk materials, assumed for simplicity to be local. $V_{dipole}(x)$ is the electrostatic potential associated with the rearrangement of bond charge at the interface. However, since the GaAs/AlAs system is nearly perfectly lattice matched, the interface dipole is expected to be negligible as confirmed by recent self-consistent local density functional calculations [4]. We therefore neglect this dipole in our investigation. Redistribution of free carrier charge is on a much larger length scale, and may be considered separately.

We necessarily rewrite this Hamiltonian in terms of a pseudo-Hamiltonian. The numerical solution of this pseudo-Hamiltonian is involved, and needs to preceed any Monte Carlo simulation. Our solution technique follows closely that of Marsh and Inkson [5], and we therefore present only the key concepts. In contrast to solutions of bulk crystal Hamiltonians, solutions to interface Hamiltonians involve evanescent states as well as propagating states, since the Bloch states of a bulk crystal are no longer complete in the presence of an interface; hence one must allow for the inclusion of states with one complex wavevector component $k_x$. It should be noted that the set of eigenstates for a bulk crystal do include these evenescent states, but they are ignored since they do not satisfy the periodic boundary conditions. Real energy eigenstates with $k_x$ complex trace out curves in complex $k_x$-space (see Figure 1). The goal is now to find all of the curves along which $E_n(k_x)$ is real in complex $k_x$-space. We do so using the local empirical pseudopotential method, with form factors given by [6] and [7] for GaAs and AlAs, respectively. Finding these curves is relatively easy, but computationally expensive. Solving for the necessary wavefunctions involves determining eigenvalues and eigenvectors of large matrices. It is easily shown that in the absence of dissipation and imperfections, crystal momentum parallel to the interface is conserved for transmission and reflection. Therefore, once all of the single-crystal propagating and evenescent eigenstates are found for a given value of $k_{\parallel}$, a system of equations expressing the continuity of the total wavefunction and its derivative is solved, using the single-crystal eigenstates as a basis. The eigenstates of the interface system are then known. With these eigenstates, one can infer the various reflection and transmission coefficients to any equal-energy state on either side of the interface

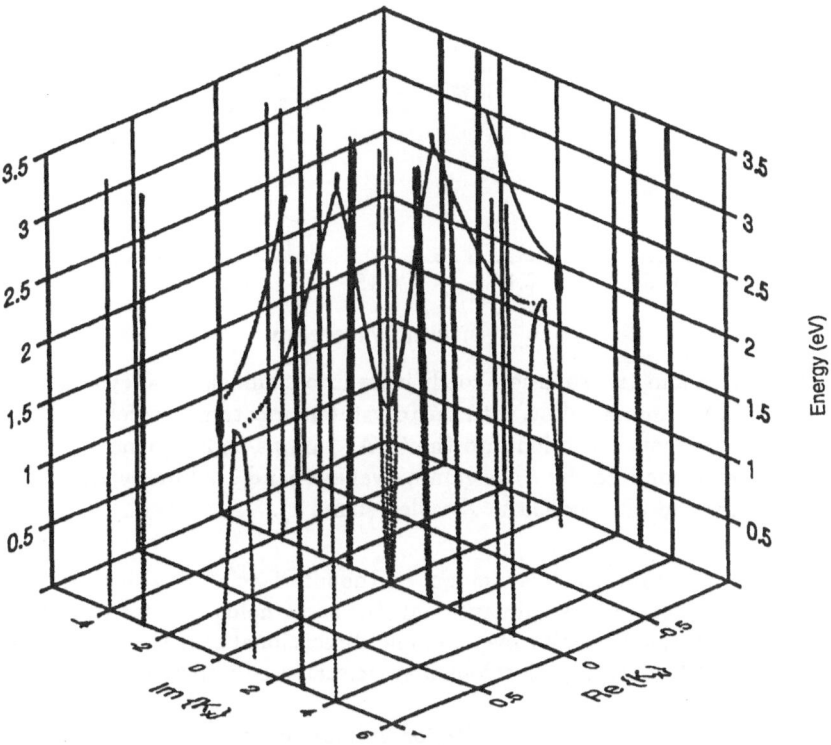

Curves of Real Energy in GaAs Complex $K_x$-Space

Fig. 1. Curves of real energy shown as functions complex $k_x$, for the case $k_{\parallel} = 0$, calculated with the local empirical pseudopotential method. For a given energy, these states represented on these curves form a basis with which to expand the total pseudowavefunction on the GaAs side of the interface with the constraint $k_{\parallel} = 0$.

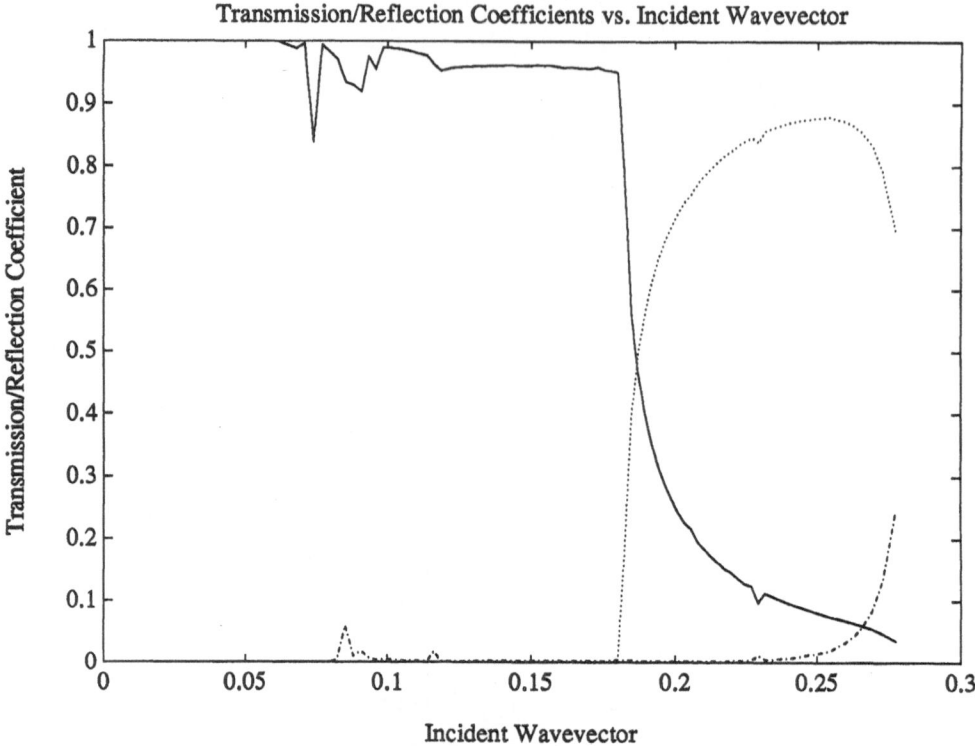

Fig. 2. Reflection and transmission coefficients into significant valleys. The incident state is in the GaAs gamma-valley, with zero crystal momentum parallel to the interface. The solid line represents reflection into the GaAs gamma-valley (band 1). The dotted line shows transmission into the AlAs gamma-valley (band 1). Finally, the dash-dotted line signifies reflection into the GaAs X-valley (band 1).

for any electron in a state which is incident on the interface. It should be noted that in making the problem truly three-dimensional (the same holds true for two dimensions) one incurs the additional penalty over the one-dimensional case (besides the coupling with evanescent states [8]) that electrons may be scattered into several states on either side of the interface.

Figures 2 and 3 show typical results for various reflection and transmission coefficients of the GaAs/AlAs interface system, with an incident electron in the GaAs Γ-valley. One significant feature is that there exist regions of energy in which the transmission probability is quite low, in spite of the fact that the electron propagates above the conduction band edge of all regions of the heterolayers. This has to do with the dissimilarity between the wavefunctions of the incident and transmitted electronic states. Figure 4 is a contour plot of the magnitude squared of the Bloch function of a GaAs gamma-valley electron in the interface plane, while figure 5 is a similar plot corresponding to an X-point electron in AlAs. Notice that the pseudowavefunctions have corresponding peaks close to lattice points, but differ substantially in the interstitial regions. As described previously, our study of charge transport at interfaces proceeds then by dividing real space into three distinct regions. On either side of an interface, charge carriers are assumed to behave semiclassically in a half-space which begins a distance away from the interface. In the bounded region between the two half-spaces,

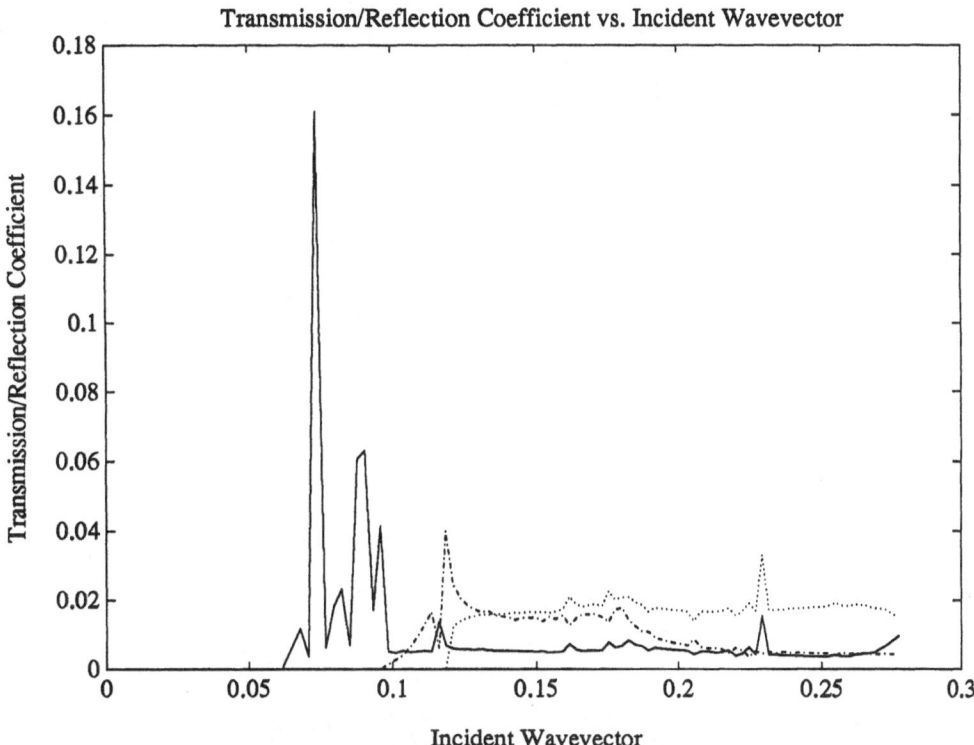

Fig. 3. Reflection and transmission coefficients into less significant valleys. The incident state is in the GaAs gamma-valley, with zero crystal momentum parallel to the interface. The solid line shows transmission into the AlAs X-valley (band 1). The dash-dotted line signifies transmission into the AlAs X-valley of band 2. Also, the dotted line shows transmission into the GaAs X-valley in band 2.

GaAs Γ-Point Wavefunction in Interface Plane

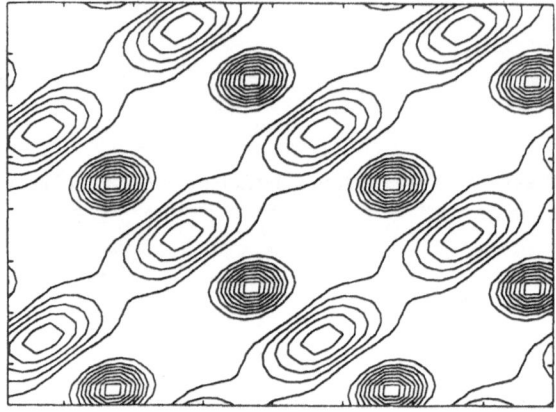

Fig. 4. Contour plot of the magnitude squared of a gamma-valley GaAs electronic wavefunction in the interface plane. The peaks in pseudocharge density correspond to the regions of the basis atoms.

AlAs X-Point Wavefunction in Interface Plane

Fig. 5. Contour plot of the magnitude squared of an X-valley AlAs electronic wavefunction in the interface plane. The peaks in pseudocharge density correspond to the regions of the basis atoms.

charge carriers are assumed to behave purely quantum mechanically, depending critically on the form of the potential in this region. The width of the quantum transport region can not, of course, be precisely defined except perhaps if a certain doping profile favors classical (contacts) and quantum regions in a natural way. A typical choice for the width would be the inelastic mean free path of the electrons. We find that within the limitations of the model, this choice does not significantly influence the results.

The transmission and reflection coefficients, calculated on a mesh in $(k_\parallel, E)$-space (Figures 2 and 3), can then be used in our multi-scale Monte Carlo approach as described above. This will result in a zero-order approach for interface transport. First-order corrections involving various types of scattering in the mesoscopic interface region and the influence of external fields on the transmission and reflection coefficients are currently under investigation, but not considered here. Our Monte Carlo simulator features transport in three full pseudopotential conduction bands for both GaAs and AlAs half-spaces. We consider scattering by all significant phonon processes, as well as ionized impurity scattering with third-body exclusion [9]. Impact ionization is treated by the method of Keldysh [10]. This technique may be seen as a unified approach, because the same pseudopotentials which are used to generate the bandstructure for the semiclassical transport are also used to calculate the quantum transport parameters at the interface.


## RESULTS AND DISCUSSION

We have simulated transport in a $0.3\mu m$ GaAs/AlAs $n^+n/nn^+$ heterojuction structure, for a range of applied potentials $-5\mathbf{V} < V_{ap} < 5\mathbf{V}$ at a lattice temperature of 300K. The doping level is $N_d = 1.0e + 17cm^{-3}$ on both sides of the heterojunction, except for the contacts which are doped at $N_d = 1.0e + 18cm^{-3}$. The simulation procedure described in the preceeding section was used to generate the J-V curve shown in figure 6.

For negative biases, there emerges a thin sheet of charge on the AlAs side of the interface. To explain this, we examine the heterojunction band lineup with valley offsets, as shown in figure 7. We use a conduction band edge discontinuity of 0.19eV, in close accord with the value 0.20eV of Beltram et. al. [11]. Also, valley offsets are deduced from bulk pseudopotential analysis, consistent with our work. Because the X-valleys of AlAs are lowest in energy, we expect them to be the most heavily populated. As is well known, two fundamental principles of transport at an interface are that, in the absence of first order scattering mechanisms, energy is conserved along with the parallel component of crystal momentum. We notice then, from figure 7 that there is no energy-conserving final state in the GaAs for an initial state near the bottom of either of the AlAs X-valleys with their major axes parallel to the interface which also conserves parallel crystal momentum. This is because the X-valleys in GaAs have their minima approximately .18 eV higher than those in the AlAs. Electrons in such incident states remain in a cycle of reflection from and incidence onto the interface – as the electric field and the channel-blocking effect of the heterojunction work competitively – until it is broken by interaction with phonons, ions, or other electrons. Similar channel-blocking effects occur for band 2 X-valley AlAs electrons. Only low-energy incident states in the AlAs X-valley which has its major axis in the direction perpendicular to the interface may be transmitted into the energy-conserving GaAs Γ-valley. But the coupling from this AlAs X-valley to the GaAs Γ valley is weak, and even most of these incident electrons are initially reflected back into the AlAs. This may be interpreted as a

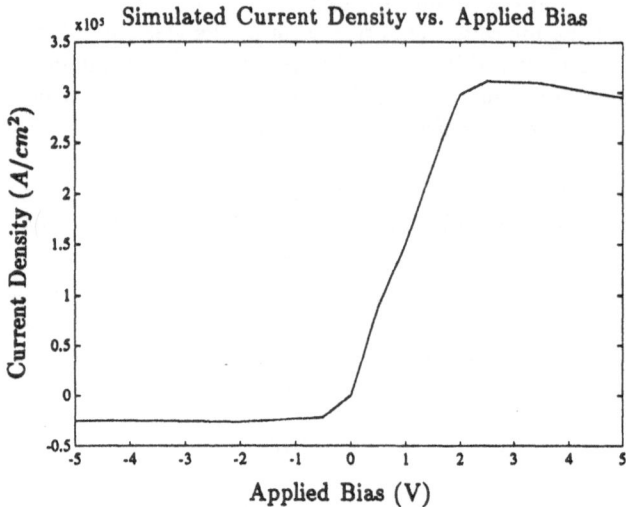

**Fig. 6.** Voltages are defined with respect to the GaAs contact. Both contacts are considered to be ohmic. The somewhat counterintuitive result that the current is lowest when the bias is negative is explained in the text.

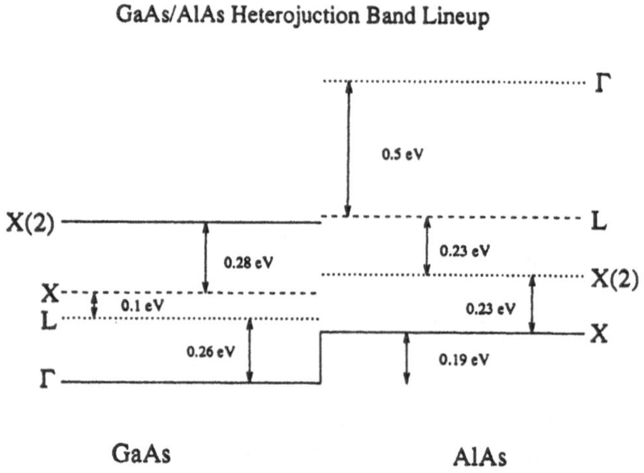

### GaAs/AlAs Heterojuction Band Lineup

**Fig. 7.** GaAs/AlAs heterojunction band lineup and valley separations calculated with empirical pseudopotentials, and used in the simulation. Positions of some subsidiary valleys show the known differences from experiment. Parentheses indicate the second conduction band. Notice the relatively wide energy separation of corresponding symmetry points. This has significant consequences for transport.

virtual channel-blocking effect, due to the dissimilarity of the incident and transmitted electronic wavefunctions – an effect which is pervasive throughout the bandstructure whenever a significant change in perpendicular crystal momentum is required for the transmission.

This accumulation of sheet charge on the AlAs side of the interface for negative biases tends to screen the applied potential, with the result that most of the voltage drop appears across the GaAs. (This qualitative effect has also been predicted for the direct heterojunction $Ga_{0.75}Al_{0.25}As/GaAs$ [12].) Hence, the current is limited by the low-field mobility of AlAs, explaining the relatively low reverse bias current. For positive biases, a similar though smaller accumulation of sheet charge on the GaAs side of the interface also screens the applied potential. Conversely, most of the potential drop appears across the AlAs, and the current is limited largely by just the GaAs transport.

Both macroscopic and microscopic effects can reduce the magnitude of the sheet charge. It is certain that if the electric field is high enough and if the distance from the contact to the interface is long enough for the ensemble to gain a significant amount of energy, the distribution at the interface will be considerably spread out in energy, thereby reducing the probability of reflection for a given particle. Obviously, the net effect of first order scattering processes in the quantum transport region (assuming any reasonable distribution function) is to facilitate the transmission, thereby tending to reduce the magnitude of the sheet charge. Nonetheless, the emergence of a sheet charge is a "zero order" phenomenon, and will persist in spite of these first order effects.

## CONCLUSIONS

We have presented a numerical approach to electronic transport in the presence of an interface, which involves ensemble Monte Carlo simulation of transport in semiconductor half-spaces with quantum transport in a nanostructure interface subsystem. We believe that this multi-scale approach represents a major step towards what can be accomplished short of solving the full quantum mechanical interface transport problem including dissipation. This technique is also ideally suited for application to device simulation; for with the current device designs, quantum coherence will only be essential in small fractions of the devices, which lends itself directly to the multi-scale approach.

## ACKNOWLEDGMENTS

This research was supported by the Army Research Office. Computer time donated by Cray Research is gratefully acknowledged.

## References

[1] R. Landauer, IBM J. Res. Develop. **1**, 223 (1957).

[2] M. Büttiker, Phys. Rev. B **35**, 4123 (1988), 12724 (1988), IBM J. Res. Develop. **32**, 306 (1988).

[3] W. A. Harrison, Phys. Rev. **123**, 85 (1961).

[4] C. G. Van de Walle, and R. M. Martin, Phys. Rev. B **35**, 8154 (1986).

[5] A. C. Marsh and J. C. Inkson, J. Phys. C. **17**, 6561 (1984).

[6] M. L. Cohen and T. K. Bergstresser, Phys. Rev. **141**, 789 (1966).

[7] J. B. Xia, Phys. Rev. B **41**, 3117 (1990).

[8] A. A. Grinberg and S. Luryi, Phys. Rev. B **39**, 7466 (1989).

[9] B. K. Ridley, in *Quantum Processes in Semiconductors*, 2nd ed., Clrendon Press, Oxford (1988).

[10] J. Y. Tang, Ph.D. Thesis, University of Illinois (1983).

[11] F. Beltram, F. Capasso, J. F. Walker and R. J. Malik, Appl. Phys. Lett. **53**, 376 (1988).

[12] A. Al-Omar and J. P. Krusius, Solid State Electronics **31**, 329 (1988).

# NEGATIVE DIFFERENTIAL RESISTANCE, INSTABILITIES, AND CURRENT FILAMENTATION IN GaAs/Al$_x$Ga$_{1-x}$As HETEROJUNCTIONS

J.H. Wolter, J.E.M. Haverkort, P. Hendriks and E.A.E. Zwaal

Eindhoven University of Technology, Department of Physics, P.O.B. 513
5600 MB Eindhoven, The Netherlands

## ABSTRACT

In this paper we discuss the transport properties of 2-dimensional electron gases in modulation doped heterostructures subjected to high electric fields. We report on the observation of negative differential resistance and oscillatory behaviour of the current for electric fields above 1 kV/cm. We develop a model in which the understanding of these phenomena is provided by the ohmic contacts to the 2-dimensional electron gas. The key phenomenon is that at a high electric field, well below the threshold field for real space transfer across the interface between the GaAs and the Al$_x$Ga$_{1-x}$As, injection of electrons from the contacts into the Al$_x$Ga$_{1-x}$As layer opens a conductive channel in the Al$_x$Ga$_{1-x}$As parallel to the 2-dimensional electron gas in the GaAs layer.

We will also present evidence that avalanche ionization in the Al$_x$Ga$_{1-x}$As layer plays an important role at these electric fields. This avalanche ionization leads to current filamentation in the Al$_x$Ga$_{1-x}$As layer. In this way the parallel conduction of a modulation doped heterostructure becomes highly spatially inhomogeneous. We studied this behaviour by means of a novel technique which we developed for this purpose: the technique of time resolved optical beam induced current. By means of this technique we were able to image the current filamentation for various experimental conditions.

## INTRODUCTION

In bulk semiconductors, current nonlinearities and instabilities induced by avalanche ionization of donors are well-known phenomena[1,2,3,4,5]. Not only shallow[1,2,3,4] but also deep impurities[5,6] can be involved in the ionization process. During avalanche ionization, the current is a highly nonlinear function of the applied electric field: current jumps with hysteresis-effects are observed in the current-voltage characteristics, corresponding to an oscillatory or chaotic temporal behaviour of the conductivity. The frequency range in which instabilities have been observed is from a few Hz up to

*Negative Differential Resistance and Instabilities in 2-D Semiconductors*
Edited by N. Balkan *et al.*, Plenum Press, New York, 1993

109

several MHz under varying experimental conditions[1,2,3,7,8]. The complex temporal behaviour in the avalanche ionization regime is often related to the formation of highly conductive current filaments. Current filaments in bulk GaAs have first been visualized by Mayer et al. using a scanning electron microscope[2]. In these current filaments a very high mobility was reported[2], which was attributed to a reduced phonon scattering.

Not only in bulk materials but also in GaAs/Al$_x$Ga$_{1-x}$As heterostructures current instabilities have been observed[9,10,11,12]. The observed phenomena are only partly understood. Most explanations of the instabilities in heterostructures are based on the properties of the two-dimensional electron gas (2DEG), including phenomena like real space transfer across the GaAs/Al$_x$Ga$_{1-x}$As barrier. We assume that, at high electric fields, electrons can be injected out of the ohmic source contact directly into the Al$_x$Ga$_{1-x}$As conduction band minimum which is located only 50 meV above the Fermi level[9]. In a previous paper[9] we reported on current collapse in GaAs/Al$_x$Ga$_{1-x}$As heterostructures. This phenomenon was explained by this electron injection into the Al$_x$Ga$_{1-x}$As layer. Electron injection is usually observed as a collapse of the current, since the electrons in the Al$_x$Ga$_{1-x}$As layer have a much lower mobility than in the 2DEG and are easily trapped by DX-centers.

In this paper we discuss the transport properties of two-dimensional electron gases in modulation doped heterostructures subjected to high electric fields. After a review of the phenomena mentioned above, together with the model which provides the understanding of these phenomena, we will focus on new experiments on avalanche ionization induced current jumps which occur a critical time delay after the electron injections into the Al$_x$Ga$_{1-x}$As layer has started. We will show that this avalanche ionization leads to current filamentation in the Al$_x$Ga$_{1-x}$As layer. In this way the parallel conduction of a modulation doped heterostructure becomes highly spatially inhomogeneous.

## THE CONCEPT OF ELECTRIC FIELD INDUCED PARALLEL CONDUCTION

In order to perform experiments on the electrical transport of hot electrons in the 2DEG of an Al$_x$Ga$_{1-x}$As/GaAs heterostructure, one needs low ohmic, non-rectifying, and homogeneous ohmic contacts to the 2DEG. Although the well-known technology based on AuGe/Ni provides excellent ohmic contacts to bulk devices (MESFETs) it often fails for the case of modulation doped heterostructures. The physical reason is that the band bending between the Al$_x$Ga$_{1-x}$As and the GaAs, providing the confining potential for the 2DEG, is strongly changed in the direct vicinity of such a contact. This leads to an unpredictable behaviour of the transport properties of the contacted 2DEG. While for low electric fields applied to the contacts these problems have been solved, mostly by trial and error, it is still unclear how for high electric fields the contacts affect the transport properties of the whole configuration, i.e. of the 2DEG plus the contacts. Such high electric fields of typically 1 kV/cm do not just occur under exotic circumstances. Such fields are normal in HEMTs with small gate lengths of typically 1 $\mu$m.

In principle, the doped Al$_x$Ga$_{1-x}$As-layer provides a second conducting channel parallel to the 2DEG. Dependent on doping concentration, layer thickness and aluminum concentration this layer can be fully depleted. In this case no parallel conduction through the Al$_x$Ga$_{1-x}$As layer is expected. Below we present evidence that this simple argument is not true in the case of a sample with contacts. Due to the fact that the contacts run through the Al$_x$Ga$_{1-x}$As layer, electron injection from the contacts

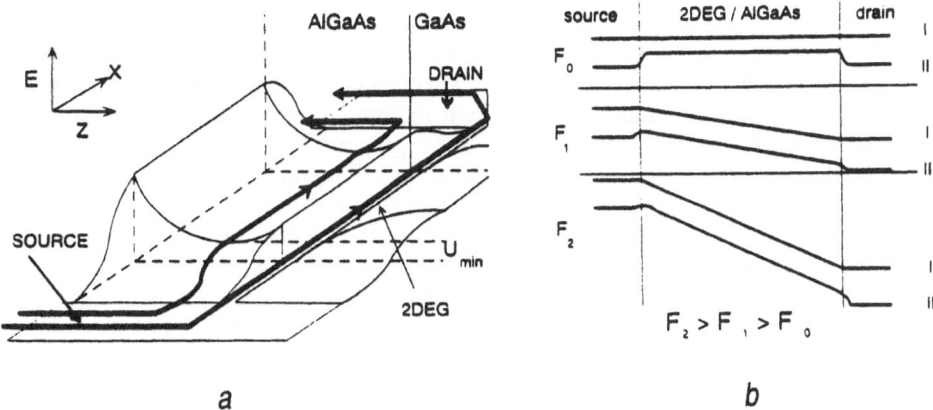

*a*                                                        *b*

**Figure 1a.** Three-dimensional picture of the band edge of a GaAs/Al$_x$Ga$_{1-x}$As heterostructure connected to ohmic contacts. Indicated are paths I and II as used in b). $U_{min}$ is the minimum of the conduction band of the Al$_x$Ga$_{1-x}$As.

**Figure 1b.** The energy as a function of position through the 2DEG (curves I) and the Al$_x$Ga$_{1-x}$As (curves II) for three electric fields. Indicated are the case of zero field ($F_o$), the warm-electron case ($F_1$), and the hot-electron case ($F_2$).

into the Al$_x$Ga$_{1-x}$As layer takes place at high electric fields, leading to all kind of spurious effects such as current instabilities, current switching and even oscillatory behaviour.

Fig. 1 shows a schematic representation of the conduction band in a GaAs/Al$_x$Ga$_{1-x}$As heterostructure, together with ohmic contacts and possible current paths through the 2DEG and the Al$_x$Ga$_{1-x}$As layer. Obviously the contact regions run through both the 2DEG and the Al$_x$Ga$_{1-x}$As layer. This is due to the fact that the 2DEG is buried about 100 nm below the free surface of the sample and one has to alloy through the top layers, mostly a GaAs cap-layer, a Si-doped Al$_x$Ga$_{1-x}$As layer, and eventually an undoped Al$_x$Ga$_{1-x}$As spacer-layer. Since in a properly designed heterostructure the conduction band of the Al$_x$Ga$_{1-x}$As layer lies sufficiently far above the Fermi level, only conduction through the 2DEG occurs in the situation illustrated in Fig. 1a. However, under the influence of an applied electric field, this situation changes dramatically. In Fig. 1b we illustrate the effect of an electric field and track the energy of an electron for two different paths: the first path starts in contact 1 (source), runs through the 2DEG and ends in contact 2 (drain). This is indicated in Fig. 1b by curve I. The second path starts in the source as well, but runs through the minimum of the Al$_x$Ga$_{1-x}$As conduction band. It also ends in the drain. (curve II in Fig. 1b).

In the case of zero electric field ($F_o$), as shown in the upper two curves of Fig. 1b, electrons flow only through the 2DEG, since they cannot gain enough energy to overcome the potential barrier and to move into the Al$_x$Ga$_{1-x}$As. This situation remains almost the same, when we raise the field to $F_1$. Only a few electrons have enough energy to tunnel through the potential barrier and to move from the source contact into the Al$_x$Ga$_{1-x}$As. In the case of a high electric field ($F_2$), however, the situation becomes totally different. Now, the barrier is sufficiently pulled down to allow the electrons to move into the Al$_x$Ga$_{1-x}$As. This is the case, when electric field induced parallel conduction in the Al$_x$Ga$_{1-x}$As layer can occur. The height of the barrier between the contact and the Al$_x$Ga$_{1-x}$As layer depends strongly on the material parameters.

In the $Al_xGa_{1-x}As$ layer the scattering rate for the electrons is high due to the usually high concentration of the silicon donors, typically $10^{18}$ cm$^{-3}$. These donors can also act as traps for the injected electrons. Even though the electron mobility in the $Al_xGa_{1-x}As$ is low, and thus the main contribution to the total conduction of the heterostructure originates from the 2DEG, the properties of the 2DEG are strongly affected by the extra electrons in the $Al_xGa_{1-x}As$. The total charge density in the $Al_xGa_{1-x}As$, which is normally positive due to the ionized silicon donors, decreases, when extra electrons are injected from the contact into the $Al_xGa_{1-x}As$ layer. This immediately modifies the band structure at the $Al_xGa_{1-x}As$-GaAs interface. Remember, that the band bending is governed in a self-consistent way by the Poisson equation, the Schrödinger equation and the demand for charge neutrality: each extra electron in the $Al_xGa_{1-x}As$ layer removes one electron from the 2DEG. Thus the electron density in the 2DEG, in particular in the regions close to the injecting source contact, is changed by the electrons injected into the $Al_xGa_{1-x}As$. In this way the properties of the $Al_xGa_{1-x}As$ influence the behaviour of the 2DEG substantially. Note, that no real space transfer of electrons across the $Al_xGa_{1-x}As$ barrier is required for changing the electron density in the 2DEG.

As we calculated with the self-consistent algorithm taking into account a depletion at the sample surface[9], the minimum of the conduction band ($U_{min}$) reduces linearly with increasing electron density of the 2DEG. When the barrier thus becomes lower the probability of the transfer of an electron from a contact to the $Al_xGa_{1-x}As$ increases exponentially. This decrease of the contact barrier can be accomplished by illumination of the sample.

The physical model described above explains how current instabilities and current collapse can occur in modulation-doped heterostructures at high electric fields. However, as we will show later, also other observed types of time-dependent current instabilities may be observed. For example, it is experimentally found that, after the current has collapsed, it suddenly increases again. We will argue that this effect is due to avalanche ionization of donor impurities in the $Al_xGa_{1-x}As$ layer. Avalanche breakdown in GaAs bulk material and other semiconductors is a well-known process[13-18]. Our experiments indicate that this process which we will describe below also occurs in the $Al_xGa_{1-x}As$ layer of the heterostructure under certain circumstances.

## IMPACT IONIZATION IN A GaAs/$Al_xGa_{1-x}As$ HETEROSTRUCTURE

In the following, we assume that the applied electric field is high enough to allow thermionic emission of the electrons from the source contact into the conduction band minimum of the $Al_xGa_{1-x}As$ layer, which is located between the surface depletion layer and the thin depletion layer near the GaAs/$Al_xGa_{1-x}As$ interface[9]. We subsequently consider the time dependence of the trapping-detrapping processes in the $Al_xGa_{1-x}As$ layer. Trapping of electrons by $DX$ centers is described with a trapping coefficient $T$ and impact ionization of trapped electrons out of these $DX$ centers with an impact ionization coefficient $X$, which is a function of the electric field[19,20,21,22]. Monte Carlo calculations[22] of the latter process have shown that at electric fields of several kV/cm, a significant fraction of the electrons in the $Al_xGa_{1-x}As$ layer has sufficient energy (> 80 meV) to ionize a $DX$-center. As a result, the impact ionization rate has been calculated[22] to sharply increase above 2 kV/cm. A diagram showing the processes involved, is shown in Fig. 2. In this diagram, $I_{in}$ and $I_{out}$ are the number of electrons per unit time flowing into and out of a segment of the $Al_xGa_{1-x}As$ layer, respectively,

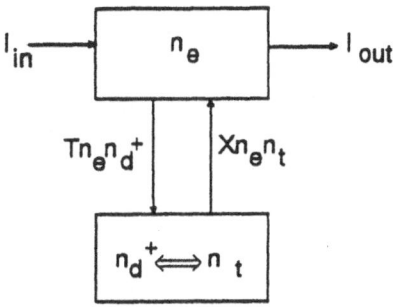

**Figure 2.** Schematic representation showing the processes taking place in the Al$_x$Ga$_{1-x}$As layer. $I_{in}$ and $I_{out}$ are the electron currents flowing into and out of a segment of the Al$_x$Ga$_{1-x}$As layer, $T n_e n_d^+$ is the trapping rate, and $X n_e n_t$ is the ionization rate.

$T n_e n_d^+$ is the trapping rate, and $X n_e n_t$ is the impact ionization rate. We take into account only the deep $DX$ level of the Si-donor, since the shallow donor level lies above the Fermi level, and therefore does not enter the generation-recombination statistics. The corresponding rate equation of the Al$_x$Ga$_{1-x}$As conduction electrons is

$$\frac{dn_e}{dt} = I_{in} + X n_e n_t - T n_e n_d^+ - I_{out} , \tag{1}$$

where $n_e$ is the sheet electron density in the Al$_x$Ga$_{1-x}$As layer, and $n_t$ and $n_d^+$ are the densities of occupied and ionized donors in the Al$_x$Ga$_{1-x}$As layer, respectively. We now first derive an expression for the time dependence of the injection and trapping rate in the Al$_x$Ga$_{1-x}$As layer. When an electrical pulse is applied, initially the number of ionized donors $n_d^+(0)$ is large and thus the trapping rate is much larger than the impact ionization rate. During the injection into the Al$_x$Ga$_{1-x}$As, the ionized donors are gradually occupied due to trapping. The ionized donor density as a function of time is given by

$$n_d^+(t) = n_d^+(0) \exp(-T n_e t) . \tag{2}$$

In the heterostructure, the neutralization of donors due to electron injection into the Al$_x$Ga$_{1-x}$As corresponds to a current collapse in the 2DEG with a time constant equal to $1/T n_e$. When the number of neutral donors increases, the impact ionization rate also increases with respect to the trapping rate. When the electric field is high enough, the impact ionization rate eventually will exceed the trapping rate at a critical time $t_c$ during the pulse. At this time, avalanche ionization is initiated and a rapid current increase or current jump is observed. Due to the avalanche ionization, the carrier density $n_e$ increases until a new steady state of higher conductivity has been reached. The time dependence of the density of ionized donors, the electron density in the Al$_x$Ga$_{1-x}$As layer, and the electron density in the 2DEG are depicted schematically in Fig. 3. The

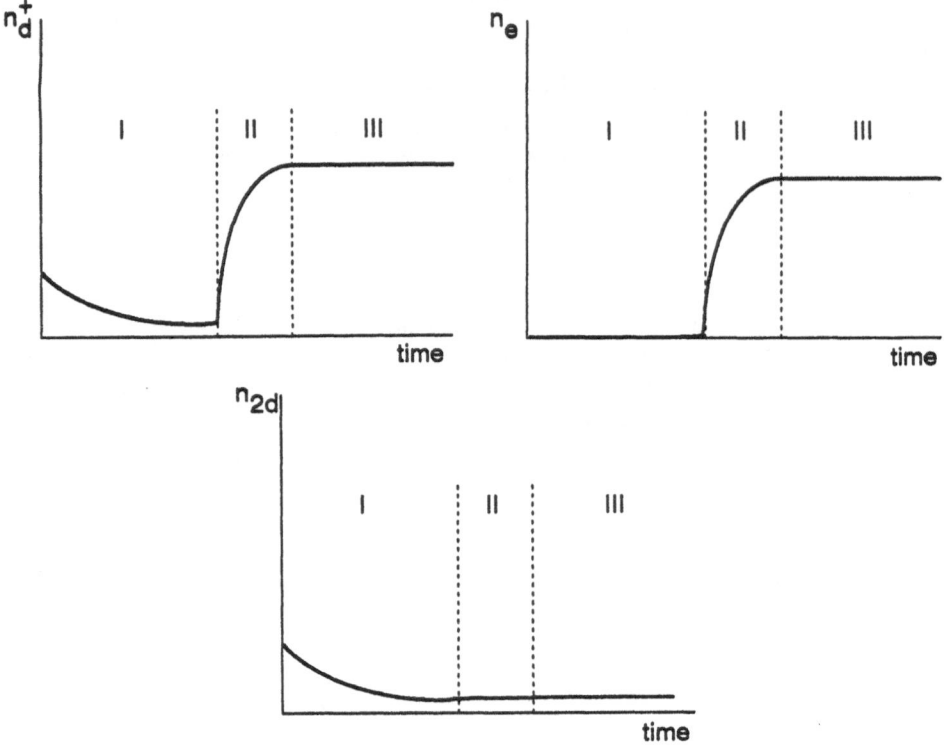

**Figure 3.** The density of ionized donors ($n_d^+$), the electron density in the Al$_x$Ga$_{1-x}$As layer ($n_e$), and the electron density in the 2DEG ($n_{2d}$) as a function of time. I indicates the trapping regime, II the avalanche regime, and III the steady state.

critical time at which the avalanche is initiated is

$$t_c = \frac{1}{Tn_e} \ln \frac{X+T}{X} \frac{n_d^+(0)}{n_d} , \qquad (3)$$

where $n_d$ is the total donor density in the Al$_x$Ga$_{1-x}$As layer. In the expression for $t_c$ the factors $1/Tn_e$ and $n_d^+(0)/n_d$ primarily depend on temperature and illumination, whereas $(X+T)/X$ is mainly determined by the electric field. Inspection of Eq. (3) tells us that increasing the electron density by illumination or heating, or increasing the impact ionization coefficient $X$ through the electric field shifts the avalanche induced current jump to an earlier time in the pulse.

We now continue with the dependence of the avalanche ionization process on the electric field and derive an expression for the steady state current under avalanche conditions. This analysis is based on the theoretical work by Schöll[19], who calculated the carrier concentration in bulk semiconductors under avalanche conditions. We adapted this theory to the case of avalanche ionization in the Al$_x$Ga$_{1-x}$As layer of a two-dimensional GaAs/Al$_x$Ga$_{1-x}$As heterostructure. In the steady state, the trapping rate is equal to the ionization rate

$$X n_e n_t = T n_e n_d^+ . \qquad (4)$$

We find two stable steady-state solutions

$$n_e = 0 \qquad\qquad for\ X < X_c$$

and

$$n_e = \frac{X\,(n_d - n_{2d}) - T\,(n_{2d})}{X + T}, \qquad\qquad for\ X > X_c, \qquad (5)$$

where $n_{2d}$ is the electron density in the 2DEG and $X_c$ is the critical value of the avalanche ionization coefficient, corresponding to the threshold electric field for avalanche ionization. Using Eqs.(5), $X_c$ can be expressed as

$$X_c = \frac{T\,(n_{2d})}{n_e + n_t}, \qquad\qquad (6)$$

showing that avalanche ionization occurs at lower electric fields when the trapping coefficient is smaller or when more electrons or occupied donors are available in the $Al_xGa_{1-x}As$ layer. The calculated dependence of the carrier density on the impact ionization coefficient is depicted schematically in the Fig. 4 by the solid line[19]. Here we clearly observe that the carrier density shows a sharp rise at the critical value $X_c$. Until now, we assumed that optical and thermal ionization of the $DX$ centers plays a negli-

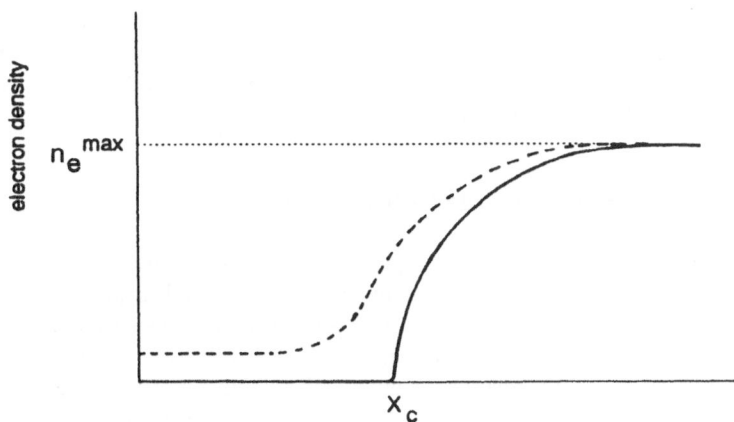

**Figure 4.** Steady state carrier concentration as a function of the ionization coefficient $X$ [Ref. 19]. The solid line represents the carrier concentration for the case that thermal and optical ionization is neglected, the dashed line includes a weak thermal and optical ionization, in addition to impact ionization.

gible role compared to impact ionization. A non-zero value of the thermal and optical generation coefficient destroys the sharp phase transition by smoothing out the discontinuity of $dn_e/dX$ at the critical field. This is shown by the dashed line[19]. For large $X$ the carrier density saturates at a value $n_e = n_d - n_{2d}$, corresponding to a level where all donors in the $Al_xGa_{1-x}As$ layer are ionized.

## EXPERIMENTAL DETAILS

In our experiments we used three types of samples. Samples I are modulation-doped $GaAs/Al_{0.38}Ga_{0.62}As$ heterostructures grown by MBE on a semi-insulating GaAs substrate. Subsequently are grown a 1 $\mu$m GaAs buffer layer, an undoped 6 nm $Al_xGa_{1-x}As$ spacer layer, a 50 nm Si-doped $Al_xGa_{1-x}As$ layer and a 20 nm undoped GaAs cap-layer. AuGe/Ni ohmic contacts were made by evaporation of 175 nm of eutectic AuGe and 36 nm of Ni, and by alloying the contacts in a $H_2$ ambient for 3 minutes. The contacts were arranged in a transmission line geometry of 100 $\mu$m wide and a contact spacing of 60, 30, 15 and 5 $\mu$m, enabling us to determine the contact resistances. The contact resistances, measured with these transmission lines, were about 0.3 $\Omega$ x mm. Samples II have a 1 $\mu$m GaAs buffer layer, 1.7 nm $Al_{0.25}Ga_{0.75}As$ spacer-layer, a 40 nm $1.5 \times 10^{18}$ cm$^{-2}$ Si-doped $Al_{0.25}Ga_{0.75}As$, and a 20 nm undoped GaAs cap-layer, with AuGe/Ni contacts, at 600 $\mu$m spacing. The difference in Al content of the $Al_xGa_{1-x}As$ layer influences the incorporation of the Si donor atoms and results in a different behaviour in high electric fields. Samples III have on a semi-insulating substrate a 50 nm GaAs buffer layer and 25 periods of a 5 nm GaAs/AlAs superlattice followed by a 55 nm GaAs layer, a 10 nm undoped $Al_{0.33}Ga_{0.67}As$ spacer-layer, a 35 nm Si-doped $Al_{0.33}Ga_{0.67}As$-layer and a 17 nm GaAs cap-layer. A rectangular mesa-etched structure was defined with a width of 200 $\mu$m and a contact spacing of 60 $\mu$m. The contact resistances are about 0.6 $\Omega$ mm at 10 K and 10 $\Omega$ mm at room temperature. The low-field resistivity of the structure is 20 $\Omega$ at 4 K (illuminated) and about 2-20 k$\Omega$ at 300 K, dependent on illumination.

To prevent lattice heating we performed pulsed experiments. Both, the applied voltage and the current were measured across the 50 $\Omega$ terminator of a sampler. The pulse length was always smaller than 3 $\mu$s with a repetition rate below 1000 Hz, insuring that no lattice heating occurs.

## EXPERIMENTAL RESULTS AND DISCUSSION

The experiments on samples I and II presented here were carried out at 77 K. For electric fields below typically 800 V/cm samples I show the well-known decrease of the mobility[9] as a function of electric field. There are no indications of electron injection into the $Al_xGa_{1-x}As$ layer. When we apply higher electric fields, we observe time-dependent features in the current. In samples II the current drops sharply (Fig. 5a) directly after the voltage pulse has been switched on. The corresponding resistance is shown in Fig. 5b.

Measurements of the geometrical magneto-resistance ($\Delta R/R$ vs. B) at different times within the pulse confirmed that the electron mobility remains unchanged during the pulse. As a consequence, Fig. 5 indicates that the number of electrons in the 2DEG is reduced during the pulse. We note that in this hot electron regime the mobility is independent of the electron density. In samples I we also observe a collapse of the cur-

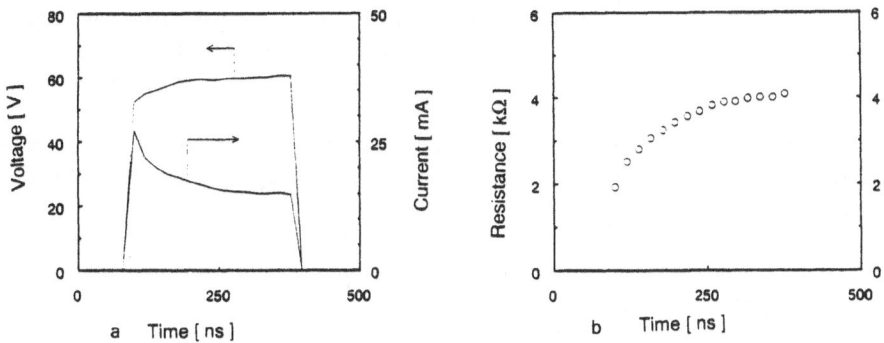

**Figure 5a.** Current and voltage pulse as a function of time. One notices that the current drops during the first 200 ns, while the voltage is still increasing.

**Figure 5b.** The resistance as a function of time as deduced from a). The resistance drastically increases during the first 200 ns and then remains almost constant.

rent. The most striking observation, however, is the occurrence of oscillations in the current, when we apply electric fields between 800 V/cm to 1200 V/cm (Fig. 6).

As indicated above, we ascribe the observed time-dependent phenomena to the injection of hot electrons from the source contact into the $Al_xGa_{1-x}As$ layer. Here the electrons are trapped by silicon impurities. In this way the electron density in the 2DEG changes immediately in the vicinity of the source contact. Since in these experiments we apply repetitive pulses with a repetition rate of 40 Hz, the experiments show that the detrapping time must be smaller than 25 ms. This detrapping time is expected to depend strongly on how the Si-doped $Al_xGa_{1-x}As$ has been grown; it depends on the doping level, the aluminum concentration, and the incorporation of the silicon in the lattice as shallow or as deep donors. We attribute the oscillations observed in samples I to the successive population and depopulation of the $Al_xGa_{1-x}As$ layer with electrons injected from the source contact. The population time, which is equivalent to the time

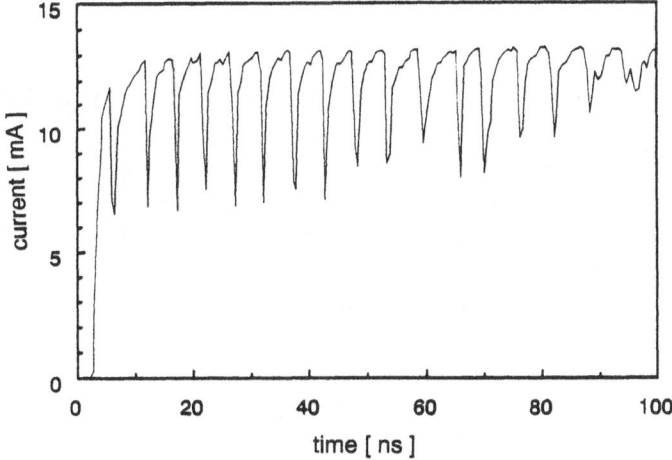

**Figure 6.** Current as a function of time for an electric field of 1 kV/cm. The current oscillates with frequency of about 200 MHz.

necessary to let the current collapse, is determined by the time the electrons need to overcome the potential barrier and to settle in the $Al_xGa_{1-x}As$. On the other hand, the depopulation time is attributed to the detrapping and consecutive release of the electrons to the drain contact. Both processes are highly non-linear.

To understand the oscillatory behaviour we suggest that the electrons trapped by the Si donors in the $Al_xGa_{1-x}As$ layer can be released by the applied electric field. From experiments in bulk GaAs[13-18] it is well-known that this process causes current fluctuations. Such a system can show steady-state, oscillatory, and even chaotic behaviour due to the non-linear process of detrapping. We propose that this process occurs in the $Al_xGa_{1-x}As$ layer as well.

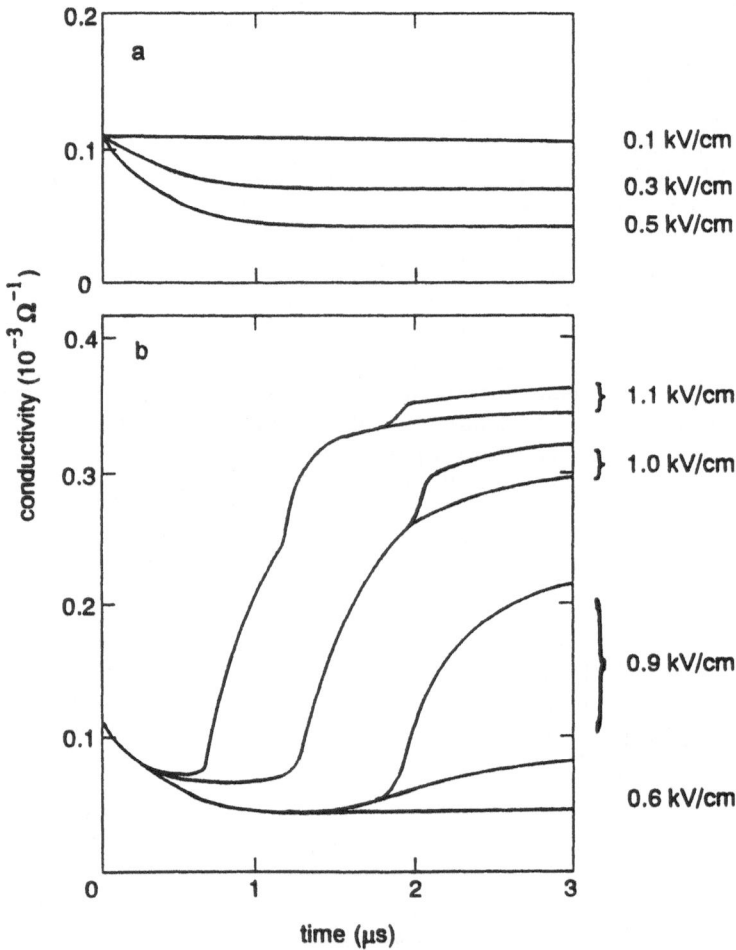

**Figure 7.** Conductivity as a function of time for various electric fields. The traces have been measured at room temperature without illumination. At 0.9 kV/cm, 1.0 kV/cm, and 1.1 kV/cm, one, two, and three current jumps are observed respectively. At these three fields the current switches between the low and the upper conductivity traces.

In Fig. 7 we present experimental evidence for sample III that also avalanche ionization in the $Al_xGa_{1-x}As$ layer can occur. Here we show the time resolved conductivities as a function of electric field when the sample is not illuminated. We see that at an electric field of 0.1 kV/cm the current is independent of time. Here, the sample conductivity corresponds to the low-field ohmic conductivity. At 0.3 kV/cm and 0.5 kV/cm we observe a current collapse as a result of carrier injection into the $Al_xGa_{1-x}As$ layer. At electric fields between 0.6 and 1.1 kV/cm, we observe the development of up to three successive current jumps at several well-defined values of the electric fields due to avalanche ionization of $DX$-centers in the $Al_xGa_{1-x}As$ layer.

We studied in detail the behaviour of current collapse and avalanche ionization for the case that the sample is illuminated with white light. We find that the current collapse becomes more pronounced under illumination. This is consistent with the physical concept described above. Illumination decreases the potential barrier between the contact and the $Al_xGa_{1-x}As$ layer and thus enhances the electron injection and

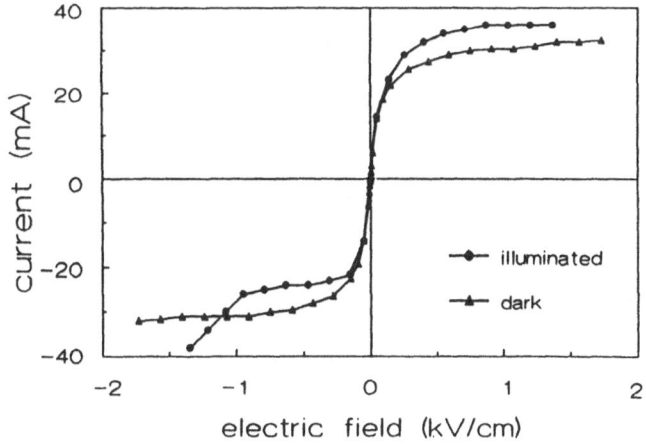

**Figure 8.** Current-field characteristics at 10 K in the dark and with illumination.

therefore the current collapse. Illumination, on the other hand, is found to reduce the contribution of the avalanche ionization to the total current. This is also consistent with the concept described, since less non-ionized donors are present in this case.

Note, that from these experiments we cannot conclude that all electrons generated by avalanche processes in the $Al_xGa_{1-x}As$ layer show up only in this layer. When an electron has moved through the $Al_xGa_{1-x}As$ layer and has reached a drain contact, another electron is ejected from the source contact. This electron, however, can be injected into the $Al_xGa_{1-x}As$ layer or into the 2DEG.

We now turn to experiments on samples III at a temperature of 10 K. The pulsed current field measurements do not show any time-dependent behaviour such as injection of electrons and avalanche ionization as long as the samples are in dark. The current field characteristics show a normal current saturation at fields up to 2 kV/cm (Fig.8).

From these measurements we conclude that at low temperature without illumination the electrons do not have enough energy to overcome the barrier between the source contact and the $Al_xGa_{1-x}As$ layer. These measurements in the dark indicate that the contact resistances to the 2DEG are perfectly ohmic at low electric fields. When we switch on the weak homogeneous illumination of the room lights, we see that the current field curve, represented by the solid line, is not symmetrical anymore. The positive branch shows a 2D behaviour with a saturation current higher than the one observed in the dark. The reason for this is that illumination ionizes additional donors in the $Al_xGa_{1-x}As$ layer and increases the electron density in the 2DEG because of the persistent photo conductivity (PPC) effect[23]. All current pulses observed in this regime are time independent, thus indicating that no electron injection into the $Al_xGa_{1-x}As$ layer takes place. In the negative branch of the I-V curve, however, current collapse is observed for electric fields between 0.3 and 1.0 kV/cm. This leads to a smaller steady-state current as shown in Fig. 8. At higher electric fields avalanche shaped current pulses are observed. It is our understanding that in this case we have two conducting channels again: first the saturated current through the 2DEG and secondly the parallel conducting $Al_xGa_{1-x}As$ layer, showing nearly ohmic behaviour after the steady-state has been reached. The asymmetry under illumination shows that the contacts for source and drain are different, i.e. the barriers from the contacts to the $Al_xGa_{1-x}As$ layer are not equal. In fact, the experiments show that the potential barrier is lower at the contact, to which the voltage is applied, than at the ground contact. Electron injection then occurs when the applied voltage is negative.

Once avalanche ionization has started the electron transport becomes highly spatially inhomogeneous. This leads to current filaments which we will study below in detail by means of optically beam induced currents. First of all, however, we estimate the electron mobility (at 10 K) in the $Al_xGa_{1-x}As$ layer in the avalanche regime. From the 2D saturation current in the injection regime between 0.3 and 1.0 kV/cm, a 2D

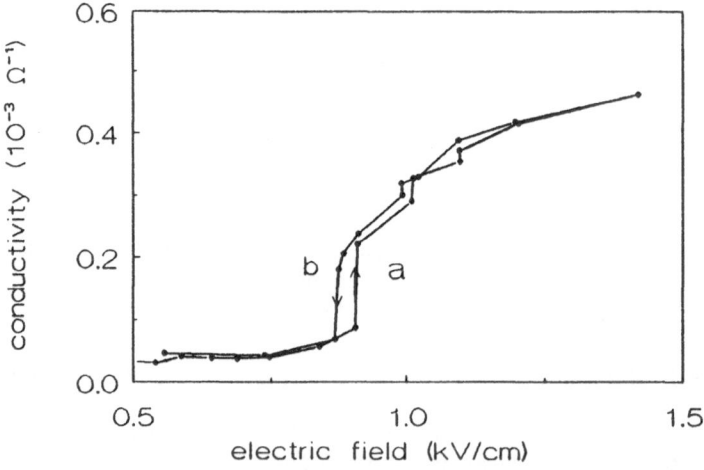

**Figure 9.** Observation of hysteresis effects for the onset of the current jumps measured at increasing (a) and decreasing (b) electric field. The plotted steady state conductivities are measured 2.5 $\mu$s after the beginning of the voltage pulse at room temperature.

electron density of $4 \times 10^{11}$ cm$^{-2}$ can be determined assuming a saturation velocity of $2 \times 10^7$ cm/s. At high electric fields, all donors in the Al$_x$Ga$_{1-x}$As layer are ionized, resulting in a total electron concentration of roughly $8 \times 10^{11}$ cm$^{-2}$ and thus an Al$_x$Ga$_{1-x}$As electron density of about $4 \times 10^{11}$ cm$^{-2}$. Since we determine an Al$_x$Ga$_{1-x}$As conductivity of $1.4 \times 10^{-3}$ $\Omega^{-1}$ from Fig. 8, this results in a low-temperature Al$_x$Ga$_{1-x}$As mobility of $2 \times 10^4$ cm$^2$/Vs in the filament. Our observed low-temperature filament mobility is exceptionally high as compared to the low-field mobility in bulk $n$-Al$_x$Ga$_{1-x}$As at low temperatures, which is lower than 100 cm$^2$/Vs for comparable donor densities[24]. At room temperature, a similar calculation yields a filament mobility of 4000 cm$^2$/Vs from the data in Fig. 9. The latter value is in accordance with the mobility limit for optical phonon scattering in Al$_{0.3}$Ga$_{0.7}$As at 300 K[25], but is also much larger than the bulk $n$-Al$_x$Ga$_{1-x}$As ($n = 10^{18}$ cm$^{-3}$) room temperature values of 250 and 500 cm$^2$/Vs which have been reported[26,27]. Comparably high mobilities under avalanche conditions have also been reported for bulk $n$-GaAs[2] at low temperatures. The high mobility in the latter case has been attributed to a reduced phonon scattering since all optical phonons emitted by the hot electrons leave the narrow filament and therefore do not contribute to the scattering. Since Monte Carlo calculations[22] indicate that the electron drift mobility in Al$_x$Ga$_{1-x}$As increases with an increasing electron density we suggest that the high mobility in the filaments might be caused by an enhanced screening of the ionized impurity scattering due to this high electron density inside the current filaments. It should be noted that the combined effects of screening and the absence of hot optical phonons inside the current filament yield a calculated[22] mobility of $5 \times 10^3$ cm$^2$/Vs, proving that a filament mobility can indeed be higher than a bulk Al$_x$Ga$_{1-x}$As mobility.

We now turn to the experiments which we carried out in order to image in the current filaments discussed above.


## IMAGING OF CURRENT FILAMENTS

We developed an imaging technique, using a scanning laser microscope and pulsed electric fields. This technique enables us to study the high electric field induced current patterns in the Al$_x$Ga$_{1-x}$As layer of a GaAs/Al$_x$Ga$_{1-x}$As heterostructure. We measured the laser beam induced current as a function of the position of illumination at several selected times during the pulse.

## EXPERIMENTAL DETAILS ABOUT TROBIC

From the modulation-grown structure a rectangular sample was photolithographically defined with dimensions of 200 $\mu$m x 200 $\mu$m. The contact resistances to the 2DEG are about 1.5 $\Omega$ mm at 15 K, and were determined in a transmission line geometry. Conventional optical beam induced current (OBIC) images, which reveal the homogeneity of the sample through the lateral photo effect[28], show that the 2DEG is perfectly homogeneous when no electric field is applied and that the contacts to the 2DEG are homogeneous as well.

The time-resolved optical beam induced current (TROBIC) images[29] under high electric field conditions were obtained as follows. We performed pulsed current-voltage measurements in a 50 $\Omega$ coaxial circuit. Both the applied voltage and the current were measured by means of a 12 GHz sampling oscilloscope. The 3 $\mu$s electric pulses were applied at a low repetition rate of 1000 Hz to avoid lattice heating. Using the sample-

and-hold mode of the oscilloscope, we measured the current at a selected time in the pulse and applied this signal to the analog input of the Zeiss scanning laser microscope. The sample was scanned by the focused He-Ne laser beam of the laser microscope. The optical beam induced changes of the sample current were measured as a function of the position of illumination, and were stored in the 512 x 512 frame store memory of the scanning laser microscope. A large value of the optical beam induced current signal is obtained at those positions, where the optically excited carriers induce the largest conductivity increase in the sample. This corresponds to a bright region in the current image. The sample was illuminated with a laser power of about 3 nW, using a Zeiss long working-distance objective with a spot size of 2 $\mu$m. To obtain a maximum spatial resolution, we scanned the laser beam very slowly at a rate of 800 $\mu$m/s, resulting in a digitizing error of less than 1 $\mu$m. The sample was mounted in an optical flow cryostat to allow measurements at low temperatures. All TROBIC images were measured at 15 K.

## EXPERIMENTAL RESULTS OF TROBIC

The current patterns observed in the modulation doped GaAs/Al$_x$Ga$_{1-x}$As heterostructure at 15 K were found to be closely related to the time-dependent current behaviour. When the sample is not illuminated, three different regimes in the time-dependent behaviour of the current can be distinguished. Current pulses, corresponding to these three regimes, are shown in Fig. 10. (i) At electric fields lower than 1 kV/cm

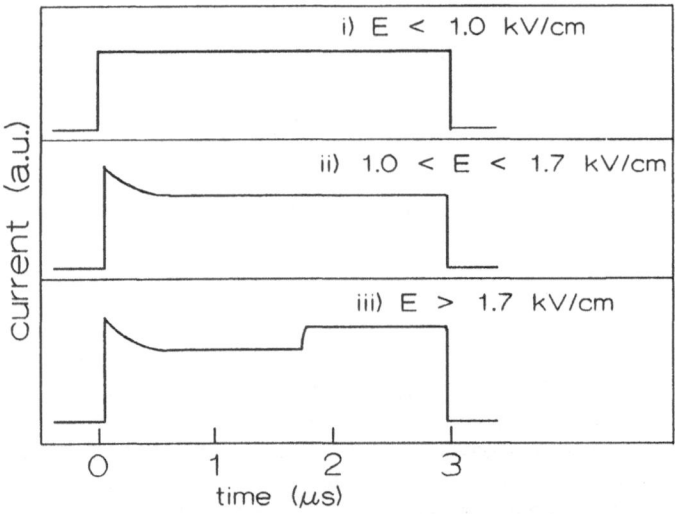

**Figure 10.** Current pulses at 15 K, observed in the three different regimes of the electric field. (i) At electric fields lower than 1.0 kV/cm the current is independent of time. (ii) Current collapse at fields between 1.0 and 1.7 kV/cm. (iii) At electric fields higher than 1.7 kV/cm impact ionization of Si donors in the Al$_x$Ga$_{1-x}$As layer gives rise to a current jump during the pulse.

we observe a current which is constant during the pulse, corresponding to a purely two-dimensional behaviour. (ii) At electric fields between 1.0 and 1.7 kV/cm we observe current collapse in the heterostructure. This is explained by the injection of electrons into the $Al_xGa_{1-x}As$ layer. (iii) At electric fields higher than 1.7 kV/cm an abrupt current increase is observed during the pulse. In this avalanche regime, electrons are injected into the $Al_xGa_{1-x}As$ and induce there impact ionization of Si donors.

**Figure 11.** TROBIC image at an electric field lower than 1.0 kV/cm. The electric field is applied in vertical direction. The bottom side of the picure is the negative contact.

At electric fields below 1.0 kV/cm [regime (i)] the TROBIC images are homogeneous. An example is given in Fig. 11. In this image a relatively high laser power of 4 $\mu$W was used. We see that the sample is brighter than the background, since illumination increases the current through the sample. If we take a current image in this regime, no beam induced current is detected and again a homogeneous current image is obtained. The TROBIC images presented hereafter all correspond to electric fields in regime (ii).

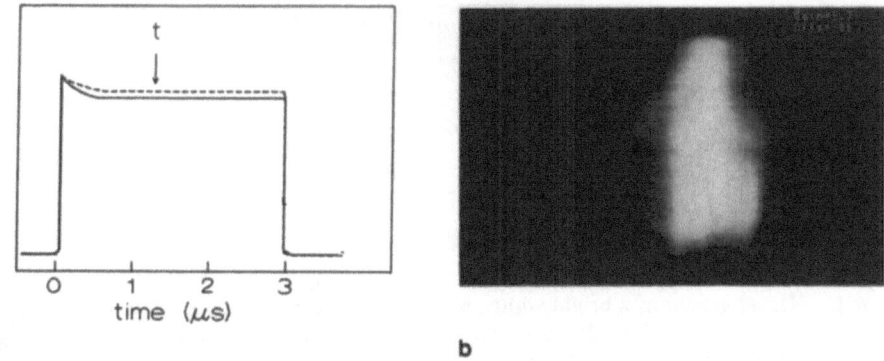

**Figure 12.** Current image of the heterostructure at 15 K and an electric field of 1.4 kV/cm taken 1.2 $\mu$s after the beginning of the pulse.

Let us now consider the TROBIC images which we get in regime (ii). We show an image at an electric field of 1.4 kV/cm. The sample is dark (Fig. 12). The electric field is applied in vertical direction and the bottom side of the picture is the injecting source contact. We observe that the TROBIC image is strongly inhomogeneous and shows a bright channel, which corresponds to a large value of the optical beam induced current signal. We ascribe this local increase of the TROBIC signal inside the channel to the formation of an optical beam induced filament. At an electric field below the threshold for spontaneous avalanche breakdown, a light-induced increase of the free carrier density locally enhances the impact ionization rate and thus initiates the formation of a current filament. This is most likely to occur in a region where a high density of injected and trapped electrons is present. We therefore suggest that the bright region is our image corresponds to a channel where the electrons are injected out of the source contact into the $Al_xGa_{1-x}As$ layer.

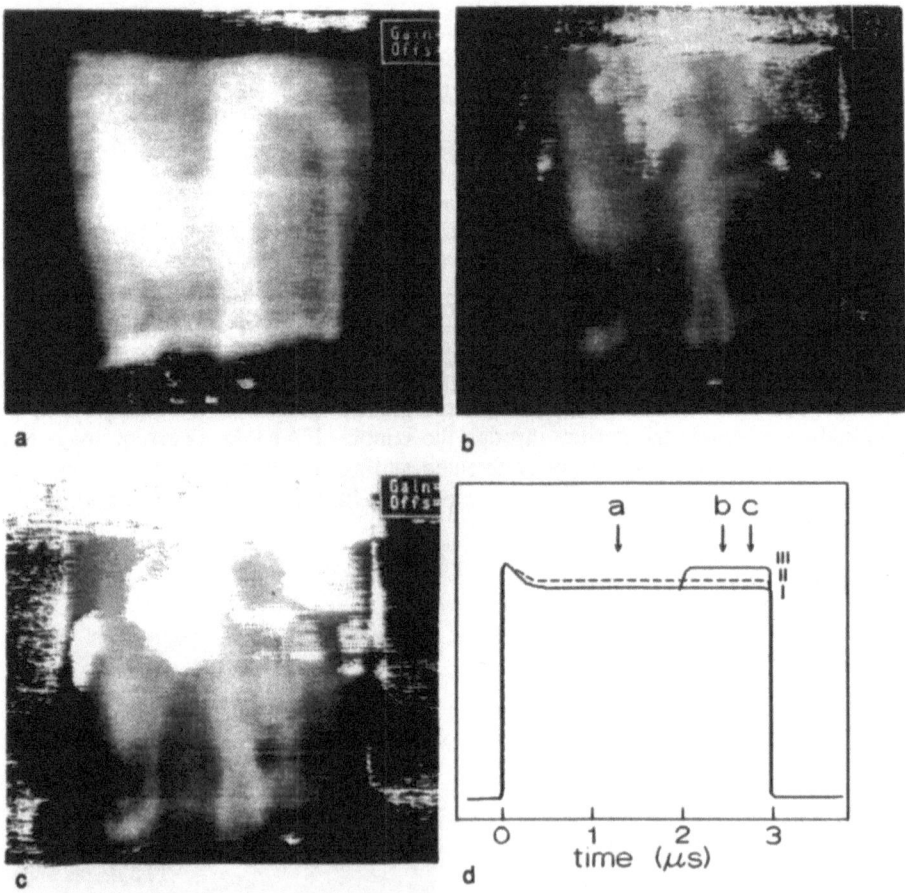

**Figure 13.** Development of a bright white spot near the drain contact. Current images at 1.6 kV/cm at three different times during the pulse, as indicated by the arrows. Pulse I is measured when the laser spot is positioned outside the channels, pulse II corresponds to illumination in the current channels, and pulse III corresponds to the bright spot close to the positive contact.

We now present some TROBIC images corresponding to experimental conditions where we found a current jump during the pulse. In the images presented below we will see that the current in this regime shows an unexpected sensitivity to illumination near the drain contact. In Fig. 13 we show the time dependence of the current filaments at 1.6 kV/cm. In the images(b) and (c), a bright spot appears close to the positive contact. This bright spot corresponds to the switching of the sample current from pulse I to pulse III [see Fig. 13(d)], which means that spontaneous avalanche ionization is initiated, ionizing all donors in the sample. We thus conclude that illumination in the positive contact region lowers the threshold voltage for avalanche ionization from 1.7 to 1.6 kV/cm.

The reason for this sensitivity of the positive contact region is not exactly clear at present. A possible explanation is the presence of a Schottky barrier between the $Al_xGa_{1-x}As$ layer and the ohmic contact. The electrons injected out of the source contact are trapped and detrapped while moving in the direction of the positive contact. When the field is higher than the threshold field, the electrons arriving at the positive contact cross the barrier and avalanche ionization takes place. At a lower value of the electric field the electrons do not have enough energy to cross the barrier and accumulate in traps near the positive contact. Illumination in this positive contact region can lower the Schottky barrier and thus decrease the threshold field for spontaneous avalanche breakdown. However, a complete model of this process is not available at present and additional experiments have to be performed to investigate this phenomenon.

## CONCLUSIONS

In conclusion we have shown that a number of effects like negative differential resistance, oscillatory behaviour of the current and instabilities in the current-voltage characteristics can be consistently described by means of a model which is based on the injection of electrons from the contacts into the modulation doped $Al_xGa_{1-x}As$ layer and subsequent avalanche ionization in this layer. We have presented experimental evidence that current filamentation in this $Al_xGa_{1-x}As$ layer takes places which introduces a highly spatially inhomogeneous current pattern in the structure.

## REFERENCES

1.  A. Brandl, W. Kröninger, W. Prettl, and G. Obermaier, *Phys. Rev. Lett.* **64**, 212, (1990).
2.  K.M. Mayer, J. Parisi, and H.P. Huebener, *Z. Phys. B-Cond. Matt.* **71**, 171 (1988).
3.  R. Obermaier, W. Böhm, W. Prettl, and P. Dirnhofer, *Phys. Lett.* **105A**, 149 (1984).
4.  K. Aoki, T. Kobayashi, and K. Yamamoto, *J. de Physique* **C7**, 51 (1981).
5.  K. Piragas, Yu. Pozhela, A. Tamashyavichyus, and Yu. Ul'bikas, *Sov. Phys. Semicond.* **21**, 335 (1987).
6.  G.A. Maracas, W. Porod, D.A. Johnson, D.K. Ferry, and H. Goronkin, *Physica* **134B**, 276 (1985).
7.  G.N. Maracas, W. Porod, D.A. Johnson, D.K. Ferry, and H. Goronkin, *Physica* **134B**, 276 (1985).
8.  R.P. Huebener, J. Peinke, and J. Parisi, *Appl. Phys.* **A48**, 107 (1989).
9.  P. Hendriks, E.A.E. Zwaal, J.G.A. Dubois, F.A.P. Blom, and J.H. Wolter, *J. Appl. Phys.* **69**, 302 (1991). See also:
    P. Hendriks, E.A.E. Zwaal, J.E.M. Haverkort, and J.H. Wolter, *SPIE Vol. 1362*

*Physical Concepts of Materials for Novel Optoelectronic Device Applications II: Device Physics and Applications* 217 (1990).

10.  P. Hendriks, A.A.M. Staring, R.G. van Welzenis, J.H. Wolter, W. Prost, K. Heime, W. Schlapp, and G. Weimann, *Appl. Phys. Lett.* **54**, 2688 (1989).

11.  N. Balkan and B.K. Ridley, *Semicond. Sci. Technol.* **3**, 507 (1988).

12.  N. Balkan, B.K. Ridley, and J.S. Roberts, *Superl. and Microstruct.* **5**, 539 (1989).

13.  A. Brandl, W. Kröninger, W. Prettl, and G. Obermaier, *Phys. Rev. Lett.* **64**, 212, (1990).

14.  K.M. Mayer, J. Parisi, and R.P. Huebener, *Z. Phys. B - Cond. Matt.* **71**, 171, (1988)

15.  R. Obermaier, W. Böhm, W. Prettl, and P. Dirnhofer, Phys. Lett. **105A**, 149 (1984).

16.  K. Aoki, T. Kobayashi, and K. Yamamoto, *J. de Physique* **C7**, 51 (1981).

17.  G.A. Maracas, D.A. Johnson, and H. Goronkin, *Appl. Phys. Lett.* **46**, 305 (1985).

18.  K. Piragas, Yu. Pozhela, A. Tamashyavichyus, and Yu. Ul'bikas, *Sov. Phys. Semicond.* **21**, 335 (1987).

19.  E. Schöll, *Nonequilibrium Phase Transitions in Semiconductors* (Springer Verlag, Berlin) 40-48 (1987).

20.  V.A. Kuz'min, N.N. Kryukova, and A.S. Kyuregyan, *Sov. Phys. Semicond.* **9**, 1136 (1976).

21.  D.J. Robbins, *Phys. Stat. Sol.* (b) **98**, 11 (1980).

22.  E.A.E. Zwaal, Ph.D. Thesis, Eindhoven University of Technology (1991).

23.  S.T. Pantelides, *Deep Centers in Semiconductors*, Gordon and Beach Science Publishers, London (1986).

24.  P.F. Fontein, P. Hendriks, J.H. Wolter, R. Peat, D.E. Williams and J.P. Andre, *J. Appl. Phys.* **64**, 3085 (1988).

25.  L. Reggiani, in *Hot Electron Transport in Semiconductors*, Topics in applied physics, **58**, ed. by L. Reggiani (Springer Verlag, Berlin, Heidelberg, New York, Tokyo) 66 (1985).

26.  S.T. Pantelides, *Deep Centers in Semiconductors* (Gordon and Beach Science Publishers, London) (1986).

27.  D.M. Collins, D.E. Mars, B. Fischer, and C. Kocot, *J. Appl. Phys.* **54**, 857 (1983).

28.  P.F. Fontein, P. Hendriks, J.H. Wolter, R. Peat, D.E. Williams, and J.P. André, *J. Appl. Phys.* **64**, 3085 (1988).

29.  E.A.E. Zwaal, M.J.M. Vermeulen, P. Hendriks, J.E.M. Haverkort, and J.H. Wolter, *J. Appl. Phys.* **71**, 3330 (1992).

# HOT ELECTRON INSTABILITIES IN QWs:
# ACOUSTOELECTRIC EFFECT AND
# TWO-STREAM PLASMA INSTABILITY

Rita Gupta, N. Balkan and B. K. Ridley

Department of Physics
University of Essex
Colchester CO4 3SQ
United Kingdom

## I  INTRODUCTION

In this paper we discuss two types of instabilities in two-dimensional semiconductor structures when the electron drift velocity, in an applied electric field, is in excess of the velocity of the acoustic modes of the system: acoustic phonons and acoustic plasmons, the latter being a feature of coupled carrier-carrier plasmas. Under suitable conditions, the piezoelectric coupling of electrons with acoustic phonons can result in an amplification of the modes, present in thermal equilibrium or deliberately injected from a transducer. This growth in acoustic flux leads to electrical non-linearities which result in the formation of acoustoelectric domains accompanied with current oscillations. These acoustoelectric domains, formed due to the bunching of electrons in the troughs of the acoustic wave, travel with the velocity of sound, and, therefore, the period of current oscillations is just the time needed for the acoustic wave to travel from one end of the crystal to the other.

Amplification of acoustic phonons, and the current instabilities associated with it, have been predicted and observed in bulk[1,2] as well as two-dimensional(2D)[3,4] semiconductors. About three decades ago, Pines and Schrieffer[5] derived conditions for the existence of plasma wave instabilities in the electron-hole plasma in bulk semiconductor; any search for these instabilities has yielded negative results so far. In low-dimensional structures conditions for the amplification of the optic plasmons have been obtained by Krasheninnikov

*Negative Differential Resistance and Instabilities in 2-D Semiconductors*
Edited by N. Balkan *et al.*, Plenum Press, New York, 1993

127

and Chaplik[6] for a coupled 2D electron-hole plasma, while, Gupta and Ridley[7] considered the possibility of inducing growth instability in the low-frequency acoustic plasma mode in two-dimensional degenerate systems. Also, instability in one component plasma has been considered by Kempa[8], and the growth of bulk and surface plasmons in type II superlattices has been explored by Hawrylak and Quinn[9]. Once again, no definite experimental evidence of two-dimensional plasma wave instability has been reported as yet.

In the following Section we present the theory for acoustoelectric effect in quantum well structures, and, also, the results of our studies on single and multiple GaAs/GaAlAs quantum well structures which clearly exhibit current instabilities in an applied electric field, probably associated with the acoustoelectric effect. In Section III we outline the theory and conditions for obtaining two-stream plasma instabilities and discuss some of the reasons why they have not yet been observed. Some concluding remarks are presented in Section IV.

## II ACOUSTOELECTRIC EFFECT

### II.1    Theory [3]

In a piezo-electric crystal, stress T, strain S, field E and the electric displacement D are interrelated:

$$T_{ij} = C_{ijk\ell} S_{k\ell} - e_{ijk} E_k , \qquad (1)$$

$$D_i = e_{ijk} S_{jk} + \varepsilon_{ij} E_j \qquad (2)$$

where c and e are the elastic and piezo-electric co-efficients, $\varepsilon$ is the static permittivity, and, summation over repeated indices is implied. The equation of motion for a piezoelectric-active mode in GaAs (propagating in the x direction) is

$$d \frac{\partial^2 u}{\partial t^2} = c \frac{\partial^2 u}{\partial x^2} - e_{14} \frac{\partial E_x}{\partial x} \qquad (3)$$

where d is the density and $e = e_{14}$ is the sole non-zero piezoelectric tenser component. Equations (1) - (3) lead to the identification of the acoustic waves that incorporate longitudinal electric fields and, can, therefore, bunch free carriers or be affected by them. From Eqs. (1) - (3), one obtains

$$\frac{\partial^2}{\partial t^2} (D_x - \varepsilon E_x) = v_s^2 \frac{\partial^2 D_x}{\partial x^2} - v_p^2 \frac{\partial^2}{\partial x^2} (\varepsilon E_x) , \qquad (4)$$

where

$$K^2 = e_{14}^2/c\varepsilon \quad ; \quad v_p^2 = v_s^2\,(1 + K^2) \quad ; \quad v_s^2 = c/d. \tag{5}$$

Coupling of the acoustic wave to the carriers is effected through the space-charge wave equation (once again, propagating in the x direction):

$$\frac{\partial \rho}{\partial t} = -\sigma\frac{\partial E_x}{\partial x} - v_d\frac{\partial \rho}{\partial x} + \mathcal{D}\frac{\partial^2 \rho}{\partial x^2}, \tag{6}$$

where, $\rho$ and $\sigma$ are the space charge density and the surface charge density, respectively, $v_d$ is the drift velocity of the carriers and $\mathcal{D}$ is the diffusion constant. The space charge density $\rho$ and the electric displacement D, in a layered system, are related as

$$D_r = -\frac{i}{2}\int \mathbf{k}\,\frac{\rho_k}{k}\exp(i\mathbf{k.r})\exp(-k|z|)dk$$

$$D_z = \pm\frac{1}{2}\int \rho_k\,\exp(i\mathbf{k.r})\exp(-k|z|)dk \tag{7}$$

where $\mathbf{r}$ and z are the spatial coordinates parallel and perpendicular to the interface. Assuming a time dependence of the form $\exp(-i\omega t)\exp(-\gamma t)$ and linearizing the space-charge equation, one obtains

$$\left[(\omega - i\gamma)^2 - v_d^2 k^2\right]D_k - \left[(\omega - i\gamma)^2 - v_p^2 k^2\right]\varepsilon E_k = 0 \tag{8}$$

$$\left[\omega - i\gamma - kv_d + ik^2\mathcal{D}\right]D_k + ik(\sigma_k/2\varepsilon)\varepsilon E_k = 0 \tag{9}$$

where the subscript k denotes the Fourier component. In general, equations (8) and (9) describe two oppositely propagating acoustic waves and one space-charge wave. The acoustic wave solution for the frequency $\omega_A$ and the amplification factor $\alpha_A$ ($=-2\gamma/v_s$) is

$$\omega_A^2 = k^2\,v_s^2\left(1 + K^2\,\frac{(1+kL)kL + \Gamma^2 L^2}{(1+kL)^2 + \Gamma^2 L^2}\right), \tag{10}$$

$$\alpha_A = \frac{K^2\Gamma kL}{(1+kL)^2 + \Gamma^2 L^2}, \tag{11}$$

where $K^2$ is the appropriate electromechanical coupling constant, k is the wavevector of the acoustic wave and L is the Debye screening length. In Eq. (1), $\Gamma = (v_d - v_s)/\mathcal{D}$, where $v_s$ is the sound velocity. Equations (10) and (11) hold for a non-degenerate semiconductor and for the case $kl \ll 1$, l being the electron mean free path.

Amplification of acoustic phonons is obtained when their coefficient for acoustoelectric growth, $\alpha_A$, exceeds the coefficient for lattice attenuation for these waves, $\alpha_L$. For bulk GaAs the lattice attenuation coefficient $\alpha_L$ is[10]

$$\alpha_L = 1.33 \; f^{1.8} \; cm^{-1} \, , \tag{12}$$

where $f$ is the frequency of the waves (in units of THz). Equation (12) is also assumed to represent the lattice attenuation of acoustic waves in quantum well structures considered here. The net amplification of the acoustic phonon is obtained as

$$\alpha = \alpha_A - \alpha_L \tag{13}$$

Equation (3) results in a cutoff value for electron concentration, $n_c$, so that for $n > n_c$, the lattice attenuation suppresses the acoustoelectric amplification. A value of $n_c = 3 \; 10^{11} \; cm^{-2}$ is obtained for GaAs quantum well structures.

## II.2    Experimental Results

The GaAs/AlGaAs samples studied at Essex were grown by MBE and MOCVD techniques, and the details of the samples are shown in table 1. The samples were frabricated into simple or Hall bar shapes long enough to accommodate the acoustic wave. Ohmic contacts were made by In or Au-Ge diffusion. The crystallographic orientation of all the samples was <110> within 5°. Pulsed electric fields a few microseconds wide, and with a

**Table 1.** Sample Parameters (GaAs/Ga$_{1-x}$Al$_x$As)

| Sample | %AL | No.of layers | Barrier | $n_H/cm^{-2}$ | $\mu_H/cm^2 \, V^{-1} sec^{-1}$ | Cladding layer thickness |
|--------|-----|--------------|---------|---------------|----------------------------------|--------------------------|
| MB662  | 30  | 10 x 50A°    | 84A°    | T = 60K  $1.9 \; 10^{11}$ | At T = 60K  $3.2 \; 10^4$ | 1.25 μm, top  1.5 μm,bottom |
| CPM279 | 30  | 50 * 60A°    | 100A°   | T = 4.2K  $9.10^{10}$ | T = 4.2K  3000 | 0.1 μm, top  0.9 μm,bottom |
| OC82   | 30  | 100 * 75A°   | 100A°   | T = 77K  $9.7 \; 10^{11}$ | T = 77K  4300 | No top cladding |
| MV281  | 45  | SQW 40A°     | –       | T = 77K  $8.9 \; 10^{11}$ | T = 77K  $4.3 \; 10^4$ | 600A°, top  0.5 μm,bottom |
| MV280  | 45  | DQW 40A°     | 100A°   | T = 77K  $1.14 \; 10^{12}$ | T = 77K  $3.7 \; 10^4$ | 600A°, top  0.5 μm,bottom |

duty cycle of less than 1%, were applied along the layers and the current waveforms were observed on a GHz oscilloscope and plotted by using a box-car and x-t recorder assembly. Photoluminescence (PL) and electroluminescence (EL), collected while the pulse was on, were dispersed and dedected by using the gating techniques.

Figure 1 shows the typical damped oscillation of the current pulse in multiple quantum well (MQW) sample at lattice temperature T=59K. The oscillations persisted up to T~200K, but the amplitude decreased with increasing temperature, indicative of the increased lattice attenuation. The period of oscillation, however, remained independent of temperature and corresponded to a transit velocity of ~3.6 $10^5$ cm s$^{-1}$, close to the velocity of transversely polarised acoustic waves propagating in the <110> direction. The threshold field for the oscillations corresponded to an electron drift velocity ~$10^6$ cm s$^{-1}$. The amplitude of the oscillations dropped and their period increased as the field F was increased above 2 kV/cm, and eventually the effect was destroyed. At this point the ohmic conductivity of the sample was seen to drop to half its initial value. This could be the result of the NDR effect associated with real-space transfer and capture of carriers in these layered structures.

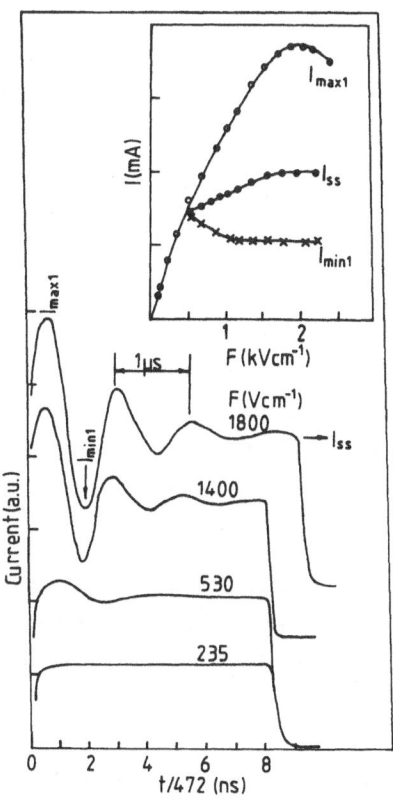

Figure 1.   Current pulse shapes at below and above the oscillation threshold field
for a MQW sample. The inset shows the field dependence of the maxima
$I_{max}$, minimum, $I_{min}$ and steady state, $I_{ss}$ current at T=59K.

The samples were illuminated by the 676nm line of a cw Kr laser, and an enhancement in the amplitude of the current oscillations was observed, together with a drop in the photoluminescence by 40% with respect to zero applied field value. Also, the oscillations in the PL signal were in phase with the current. Measurements of the electron temperature, using the hot electron PL technique, showed a marked cooling of electrons with time (figure 2), accompanied with a slight shift of the excitonic peak to lower energy.

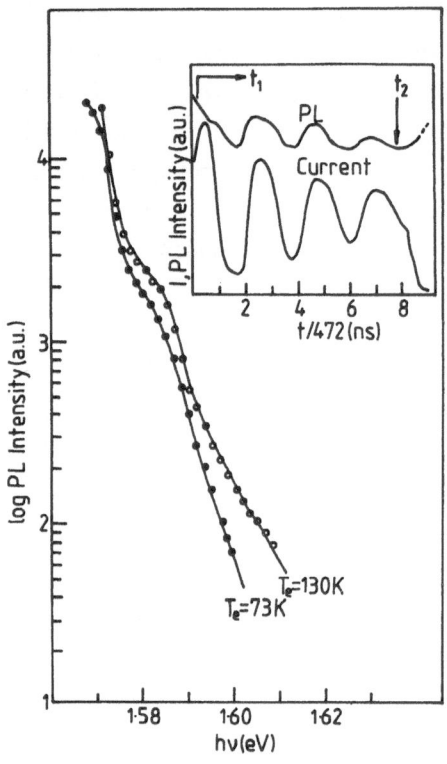

Figure 2.   The high energy tail of the PL spectra at 59K and F=1600 V/cm
(for MB662-2 when the luminescence is gated and collected at t=$t_1$
and t=$t_2$ as indicated in the insert.

All the MQW samples exhibited damped oscillations, while, continuous oscillations were observed in the single quantum well (SQW) structures; in the latter case, the period of oscillations was seen to decrease with the increasing applied field (figure 3) upto F~560 V/cm after which it remained constant upto F~750 V/cm, and above this field the period increased slightly. The double quantum well (DQW) structures showed two distinct regions of (continuous) current oscillations: one between 150-350 V/cm where the oscillations were unstable in time, and, the second region (at higher fields) where the oscillations were more stable. In this region, the initially fast oscillations rapidly slowed down and then stabilized with a constant period. The increase in the period was accompanied by an increase in the ohmic resistance and a shift of the threshold field to higher values, the effect being similar to that observed in SQW samples.

Figure 4 shows the single-probe measurements of the voltage distribution in a DQW sample. The side-arms of the Hall bar were used as probing points, and only steady-state voltage distribution was measured due to the long time constants associated with the side-arm resistances. The figure shows that at electric fields below the oscillation threshold, the field is uniformly distributed along the sample. However, at fields above the threshold, the field

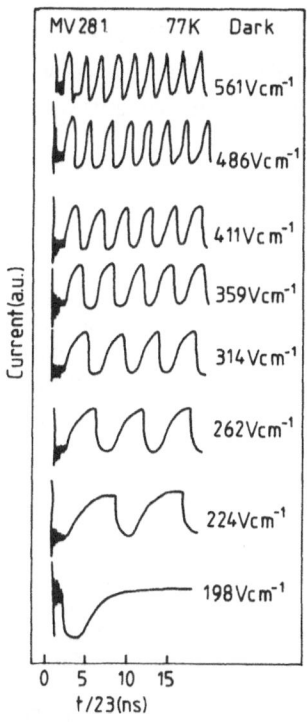

Figure 3.  Current pulse shapes at fields above the oscillation threshold in a 1 mm
SQW sample at T=77K.

distribution becomes highly nonuniform and a high-field domain is developed in the middle of the sample.  As the applied voltage is increased further, this high-field domain spreads and covers most of the sample in the anode region, and, the field in the cathode region is seen to fall below the threshold field necessary to sustain the oscillations. The uniformity of the field inside the sample was restored when the oscillations had decayed.  This result is a clear demonstration of the association of the current oscillations with the domain formation in the samples studied here.

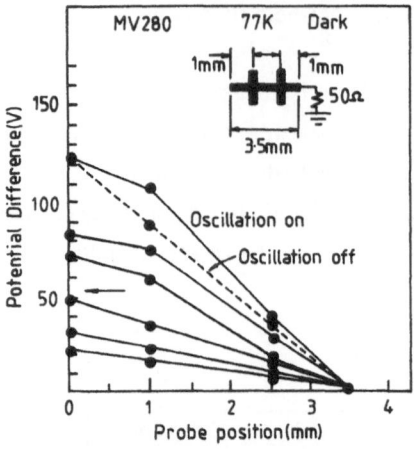

Figure 4.  Steady state voltage distribution along a 3.5 mm Hall bar shaped DQW sample
the arrow indicates the oscillation threshold voltage.

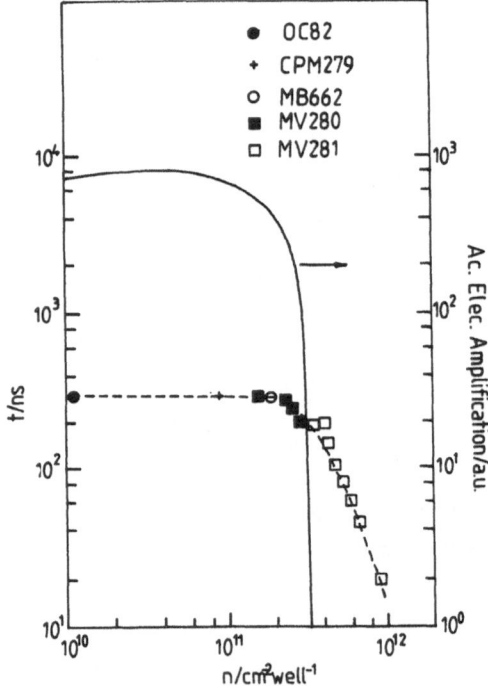

Figure 5.  Oscillation period (normalised to 1 mm sample length, for the samples
investigated, is shown against the carrier concentration per well.  The
solid line shows the calculated acoustoelectric amplification vs carrier
concentration, using sample parameters of MB662.

In figure 5 we have plotted the amplification coefficient for maximum growth condition ($k^2L^2=1+\Gamma^2L^2$) as a function of sheet carrier concentration. In the calculations, we used the experimental parameters for the sample MB662, at the oscillation threshold at T=60K. The value $K^2=0.0036$ (for bulk GaAs) was used for the coupling constant. The observed oscillation periods, for all the samples studied, are also shown in the figure. Here it was assumed that, in nominally undoped MQW samples with thick cladding layers, most of the carriers were in one well, an assumption supported by the Shubnikov-de Haas together with Hall measurements. The theory is in remarkably good agreement with the experiment.

## III   TWO-STREAM PLASMA INSTABILITY

The collisionless coupled Boltzmann equation has been used to study the collective excitations of a separated two-dimensional two-component in presence of external drift velocities. Consider two charge sheets of charges $q_1$ and $q_2$, masses $m_1$ and $m_2$, and separated by a distance d. In the self-consistent linear approximation, the dispersion relation for plasmons in the coupled system is obtained by solving the following equation for $\omega$ [7],

$$1 + \chi_1 + \chi_2 + \chi_1 \chi_2 \, p \; = \; 0, \tag{14}$$

where

$$p \; = \; 1 - e^{-2kd} \;\; , \tag{15}$$

$$\chi_i \; = \; \frac{Q_i}{k} \; g \left[ \frac{\omega - \mathbf{k} . \mathbf{v}_{Di}}{k v_{Di}} \right] , \tag{16}$$

and $Q_i$ is the Thomas-Fermi wavevector

$$Q_i \; = \; \frac{2 m_i e^2}{\varepsilon \hbar^2} \;\; . \tag{17}$$

In the above equation, k is the wavevector of the plasma mode, $v_{Di}$ and $v_{Fi}$ are the drift velocity and the Fermi velocity, respectively, of the $i^{th}$ component, and, $\varepsilon$ is the dielectric constant. The function g is defined as

$$g(\lambda_i) \; = \; \begin{cases} 1 + i\lambda_i \,/\, (1-\lambda_i^2)^{1/2}, & \lambda_i^2 < 1 \\ 1 - \lambda_i \,/\, (\lambda_i^2-1)^{1/2}, & \lambda_i^2 > 1 \end{cases} \tag{18}$$

Equations (4)-(8) yield two solutions for $\omega$. The low-frequency mode $\omega_-$ corresponds to the acoustic plasmon, and, it is found that the high-frequency optic mode becomes disallowed for very large separations (d $\rightarrow \infty$) and k > $Q_2/2$ (where $v_{F2} > v_{F1}$).

The plasma waves are undamped in the high-frequency region, $\omega$-$\mathbf{k}\cdot\mathbf{v}_{Di} > kv_{Fi}$ (i=1,2), since both $\chi_1$ and $\chi_2$ are real in this case. However, the value of $\chi_2$ is complex in the low-frequency region, $\omega$-$\mathbf{k}\cdot\mathbf{v}_{D1} > kv_{F1}$ and $\omega$-$\mathbf{k}\cdot\mathbf{v}_{D2} < kv_{F2}$ so that Eq. (4) admits to complex solutions for $\omega$, $\omega = \omega_k + i\gamma_k$. The amplification factor $\alpha_p$ associated with the acoustic plasma wave is

$$\alpha_p = 2\gamma_k/\omega_k , \tag{19}$$

where

$$\gamma_k = \frac{m_1}{m_2} \frac{(\lambda_1^2 - 1)^{1/2}}{(1 - \lambda_2^2)^{1/2}} \frac{(\omega_k^0 + \mathbf{k}\cdot\mathbf{v}_{D1} - \mathbf{k}\cdot\mathbf{v}_{D2})}{(1 + Q_2p/k)^2} e^{-kd} \tag{20}$$

and $v_s$ (= $\omega_k/k$) is the phase velocity of the mode. In Eq. (20), $\lambda_i = (\omega_k - \mathbf{k}\cdot\mathbf{v}_{Di})/kv_{Fi}$ (i=1,2) and $\omega_k^0$ is the frequency of the plasma mode in the absence of any external drift velocities. Eq. (20) has been obtained in the approximation $\gamma_k/\omega_k \ll 1$. In the same approximation, the frequency of the mode is obtained as

$$\omega_k = \frac{Q_1(1+Q_2p/k) + k(1+Q_2/k)}{\left\{ (k+Q_2)[k+Q_2 + 2Q_1(1+Q_2p/k)] \right\}^{1/2}} v_{F1} k + \mathbf{k}\cdot\mathbf{v}_{D1} . \tag{21}$$

From Eq. (20), one observes that external drift velocities can induce two-stream growth instability in the acoustic plasma wave for

$$\mathbf{k}\cdot\mathbf{v}_{D2} > \omega_k^0 + \mathbf{k}\cdot\mathbf{v}_{D1} . \tag{22}$$

In figure 6 we show the k-variation of the power gain coefficient $\alpha_p$ for the case of electron-electron (e-e) plasma and the hole-electron (h-e) plasma for intersheet separations of d= 50A and 100A. Values of the other parameters used in the calculations are listed in table 2. It is observed from the figure that the drift velocities of $v_{D2} = 1.3 \times 10^7$ cm/s for the e-e system, and $v_{D2} = 10^7$ cm/s for the h-e system, are just sufficient to induce growth instability for k greater than some wavevector, say, $k_c$. This instability, or, indeed, the damping for k < $k_c$, is greater for the h-e two-stream plasma. The effect of separation between the two charge sheets is to suppress the growth instability, the effect, once again, being larger for the electron-hole plasma. It is also noticed that a low value of $n_1$, the carrier density of charge sheet with lower Fermi energy, results in a low value of the threshold drift velocity, needed to induce plasma growth instability. The onset of the instability is not critically dependent on charge density $n_2$. However, a low value of $n_2$ results in a higher growth rate.

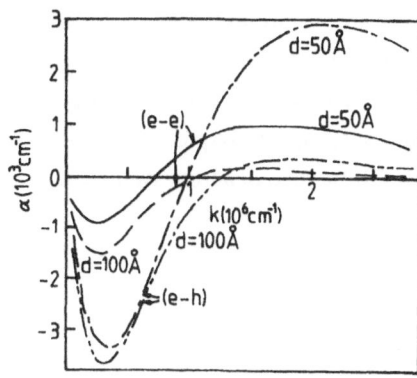

Figure 6. Gain $\alpha = 2\gamma_k/v_k$ associated with the acoustic plasma wave in the
*e-e* plasma and *e-h* plasma for the parameters listed in Table 2.

**Table 2.** Various parameters used for the calculation of phase velocity and gain factor for the low-frequency plasma mode

|  | $n_1 (10^{12}\ cm^{-2})$ | $n_2 (10^{12}\ cm^{-2})$ | $v_{D1} (10^7\ cm\ s^{-1})$ | $v_{D2} (10^7\ cm\ s^{-1})$ |
|---|---|---|---|---|
| electron-electron plasma | 0.05 | 1 | 0.13 | 1.3 |
| hole-electron plasma | 0.5 | 1 | 0.0 | 1 |

electron mass $m_e = 0.068 m_o$, hole mass $m_h = 0.5\ m_o$.

The requirement that drift velocities be in the region of $10^7$ cm/s, in order to achieve plasma instability, implies that the electrons will be hot. It is unlikely that the hot-electron effects will be strong enough to affect the two-stream coupling significantly. However, it has been shown that hot-phonon effects become important at these large electron concentrations and the high fields required to reach the threshold drift velocities of $10^7$ cm/s. It has also been demonstrated that these non-equilibrium phonons are non-drifting, and, are, therefore, instrumental in suppressing electron drift velocities at high fields. [11] This is demonstrated in Fig. 7, which plots electron drift velocity vs. field for three samples with different carrier concentration. The electron drift velocity is seen to saturate at a lower value as the carrier concentration, and, therefore, the effect of non-drifting hot phonons increases. The loss of phonon drift is a result of the elastic scattering of these modes from the interface roughness (IFR) and well-width fluctuations,[12,13] which are inevitably present in semiconductor heterostructures considered here. An estimate of IFR in some GaAs/AlGaAs samples[13] has yielded the IFR width $\Delta$ = 2-4 monolayers, and, the corresponding momentum relaxation rates for phonons are large. Therefore, it may become possible to induce two-stream plasma instability once quantum well structures with smooth interfaces become available.

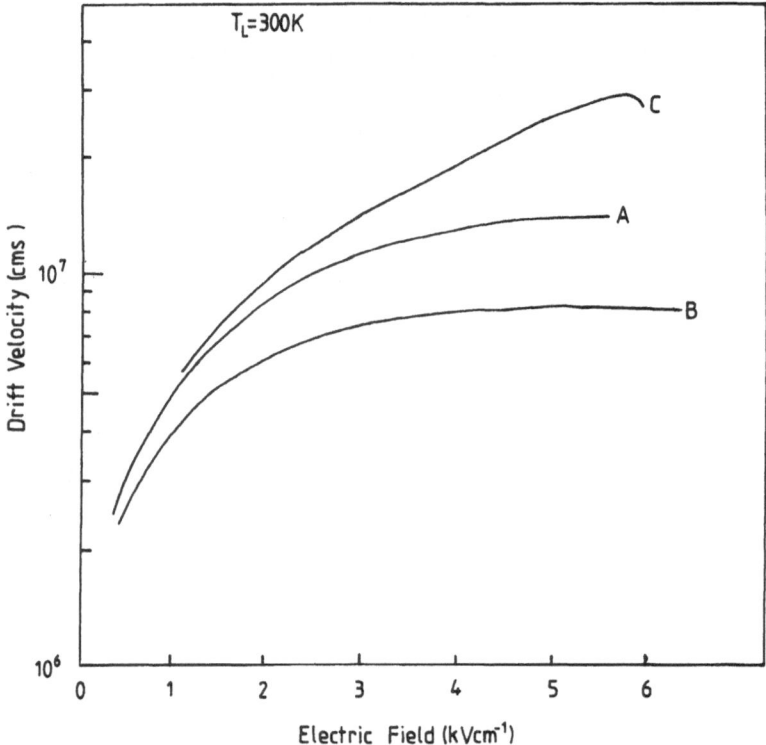

Fig. 7. Electron drift velocity as a function of field. for samlpe A (n=1.45 $10^{18}$ cm$^{-3}$), samlpe B (n=2.18 $10^{18}$ cm$^{-3}$) and sample C (n=2 $10^{16}$ cm$^{-3}$).

## IV    CONCLUSIONS

We have presented a study of current instabilities associated with the acoustoelectric effect in GaAs/GaAlAs single and multiple quantum well structures. The possibility of two-stream plasma growth instability, associated with acoustic plasmon, was also considered, and, it is hoped that one would be able to observe this instability in future, as the quality of quantum well structures improves.

## V  ACKNOWLEDGEMENT

We thank the Office of Naval Research (RG,BKR) and SERC (BKR,NB) for the support of this work.

## VI  REFERENCES

1. D. L. White, *J. Appl. Phys.* **33**, 2547 (1962).

2. R. S. Sussman and B. K. Ridley, *J. Phys. C: Solid State Phys.* **5**, 199 (1972).

3. B. K. Ridley, *Semicond. Sci. Technol.* **3**,542 (1988).

4. N. Balkan and B. K. Ridley, *Semicond. Sci. Technol.* **3**, 507 (1988); N. Balkan, B. K. Ridley, *Superlattices and Microstructures* **5**, 539 (1989).

5. D Pines and J. R. Schrieffer, *Phys. Rev.* **124**, 1387 (1961).

6. M. V. Krasheninnikov and A. V. Chaplik, *Zh. Ekop. Teor. Fiz.* **79**, 555 (1980) [*Sov. Phys. JETP* .**52**,279 (1980)].

7. R. Gupta and B. K. Ridley, *Phys. Rev. B* **39**, 6208 (1989).

8. K. Kempa, *Proc SPIE* **792**,320 (1978).

9. P. Hawrylak and J. J. Quinn, *Appl. Phys. Lett.* **49**, 280 (1986).

10. R. S. Susman, "Acoustoelectric effect in GaAs", Thesis. University of Essex (1973).

11. R. Gupta, N. Balkan and B. K. Ridley, *Phys. Rev. B*, 15 September 1992.

12. R. Gupta and B. K. Ridley, *Proc. SPIE* **1362**,792 (1991).

13. R. Gupta and B. K. Ridley,submitted to Phys. Rev. B.

# HYBRID OPTICAL PHONONS IN LOWER DIMENSIONAL SYSTEMS AND THEIR INTERACTION WITH HOT ELECTRONS

N.C. Constantinou and B.K. Ridley

Department of Physics
University of Essex
Colchester C04 3SQ
England

## INTRODUCTION

Any continuum model which is proposed in order to describe the confined optical modes of a GaAs/AlAs multi-quantum well system, must be critically compared to microscopic lattice dynamical calculations[1-3]. Two contrasting continuum models have been employed to date in order to describe electron-optical phonon interactions. The first of these is the so-called dielectric continuum (DC) model[e.g.4] which treats the confined modes within the Einstein approximation. This model demands vanishing potentials at the interfaces, and thus automatically obeys electromagnetic boundary conditions. On the other hand, it flouts the mechanical boundary conditions. In contrast, the hydrodynamic (HD) model[5,6], has the virtue of including bulk LO and TO mode dispersion, and obeying mechanical boundary conditions. Unfortunately, it thus has the difficulty of allowing for a discontinuity in the potentials.

Huang and Zhu[2] have looked critically at the DC model in the light of results from their simple microscopic calculations. Although they were critical of the DC model, a reconciliation can be achieved provided the macroscopic modes are correctly identified. A major conclusion then being that the DC model may safely be employed in calculations of the electron-phonon interaction, at least in the GaAs/AlAs system. More recently, Rucker et al[7] have arrived at much the same conclusions. Nevertheless, the DC model is unable to predict many of the important features associated with the confined optical vibrations. In particular, LO/interface optical mode hybridization and the associated mode conversions[8].

Here, a continuum model is outlined which has just the desired features that might be asked for in such a theory. As such it is easily applicable to transport calculations, where in the hot electron regime the dominant relaxation mechanism is via the interaction with these polar optical modes.

*Negative Differential Resistance and Instabilities in 2-D Semiconductors*
Edited by N. Balkan *et al.*, Plenum Press, New York, 1993

141

## DISPERSIVE BORN-HUANG MODEL

In what follows, it is assumed that the material is elastically isotropic, and that retardation effects are negligible. In the long wavelength regime, the relative ionic displacement field (RIDF) $u$ satisfies the following mechanical equations of motion[5]

$$-\omega^2 u = b_{11} u + b_{12} E - \alpha^2 \nabla(\nabla.u) - \beta^2 \nabla^2 u \qquad (1)$$

$$P = b_{12} u + b_{22} E \qquad (2)$$

where $E$ is the electric field, and $P$ the polarization field. The b coefficients are well known[9], and will not be presented here. The phenomelogical parameters $\alpha$ and $\beta$ are chosen in order to correctly reproduce the zone-centre LO and TO dispersions. In the bulk, equations (1) and (2) together with Maxwell's equations may be solved if the RIDF is decomposed into longitudinal and transverse components viz:

$$u = u_L + u_T \quad (\nabla \times u_L = 0, \nabla. u_T = 0) \qquad (3)$$

and $u_L$ and $u_T$ satisfy

$$\nabla^2 u_L + [(\omega_L{}^2 - \omega^2/v_L{}^2] u_L = 0 \qquad (4)$$

$$\nabla^2 u_T + [(\omega_T{}^2 - \omega^2)/v_T{}^2] u_T = 0 \qquad (5)$$

where $v_L{}^2 = (\alpha^2 + \beta^2)$, and $v_T = \beta$. The TO component has a vanishing electric field whilst the LO component does, of course have an associated field, but is also characterised by a vanishing electric displacement field (EDF), $D$. In the presence of interfaces, additional excitations are allowed which do have electric fields in the non-retarded regime, in contrast to the TO modes, together with a non-vanishing EDF in contrast to the LO modes. These are the well known interface optical modes[10]. It is a simple exercise to demonstrate that their RIDF, $u_I$ satisfies

$$\nabla^2 u_I = 0 \qquad (6)$$

which just states that the electric field associated with $u_I$ obeys Laplace's equation.

The field equations (4)-(6) may readily be solved for any geometry given the appropriate boundary conditions. For modes with frequencies within the GaAs reststrahl band AlAs is to a very good approximation mechanically rigid. Hence, the mechanical boundary condition is simply the vanishing of the RIDF. Together with electromagnetic boundary conditions, a triple hybrid scheme is required in order for all the boundary conditions to be satisfied[11]

$$u = u_L + u_T + u_I \qquad (7)$$

The TO component of the hybrid has zero electric field and the fields associated with the other components are given by

$$E_L = -\rho_0 r u_L , \quad D_L = 0 ; E_I = -\rho_0 s u_I , \quad D_I = \varepsilon_0 \varepsilon E_I \qquad (8)$$

$\rho_0 = e^*/\varepsilon_0 V_0$ , with $e^*$ the effective charge, $V_0$ the volume of the unit cell, $\varepsilon_0$ the permittivity of free space and $\varepsilon$ the dielectric function. In equation (8), s is given by

$$s = (\omega^2 - \omega_T^2)/(\omega_L^2 - \omega_T^2) \qquad (9)$$

with $\omega_{L(T)}$ the zone-centre LO(TO) frequency ($\omega_L = 5.5 \times 10^{13} s^{-1} = 1.09 \omega_T$ for GaAs), and r is a frequency dependent factor close to unity in the GaAs reststrahl band.

A less stringent mechanical boundary condition requiring that the component of the RIDF normal to the boundary vanish is found to be an excellent approximation for modes with frequencies close to the GaAs LO mode. This is tantamount to ignoring shear, or, equivalently $v_T \to 0$. It only requires LO-Interface mode hybrids in order for these boundary conditions to be satisfied. This approximation has been compared favourably with the more involved triple hybrid scheme, and will be used in what follows. It is close in spirit to the recent work of Zianni et al[12] although these authors only dealt with an isolated quantum well.

## SUPERLATTICE OPTICAL HYBRID PHONONS

In this section, the theory outlined in the previous section is applied to describe the optical hybrids of an GaAs/AlAs superlattice. The growth direction is taken to be along the z-axis, with the GaAs layer having width a and characterized by a bulk dielectric function

$$\varepsilon_1 = \varepsilon_\infty(\omega^2 - \omega_L^2)/(\omega^2 - \omega_T^2) \qquad (10)$$

and the AlAs barriers of width b and dielectric constant $\varepsilon_2$. As AlAs is mechanically rigid, u=0 in the barriers, although there may of course be electric fields due the interface mode component of the hybrid. It is a straightforward matter to write down both the RIDF in the GaAs layer which satisfy the wave equations (4)-(6) viz:

$$u_L = A\exp(ik_L z) + B\exp(-ik_L z) \qquad 0 < z < a \qquad (11)$$

$$u_I = C\exp(k_x z) + D\exp(-k_x z) \qquad 0 < z < b \qquad (12)$$

In the barrier region

$$E_I = F\exp(k_x z) + G\exp(-k_x z) \qquad a < z < b \qquad (13)$$

The common factor $\exp[i(k_x x - \omega t)]$, where $k_x$ is the in-plane wavevector is assumed. It is straightforward to apply the boundary conditions at z=a and z=b. Bloch's theorem then relates equivalent points in neighbouring unit cells via a phase exp(iQL) where L(=a+b) is the superlattice period , and the Bloch wavevector in the growth direction Q is restricted to $0 \leq QL \leq \pi$. From the solvability condition, the following simple dispersion relation is obtained for the superlattice optical hybrids

$$\kappa\sinh(k_x b)\{s + \frac{1}{2}\kappa\sinh(k_x a)\sin(k_L a) - s\cosh(k_x a)\cos(k_L a)\}$$
$$+\varepsilon' s\kappa\sinh(k_x a)\{\cos(QL) - \cosh(k_x a)\cos(k_L a)\} + \varepsilon' s^2 \sin(k_L a)D = 0 \tag{14}$$

where $\varepsilon' = \varepsilon_1/\varepsilon_2$ and $\kappa = k_x/k_L$. The confinement wavevector $k_L$ satisfies

$$k_L^2 = (\omega_L^2 - \omega^2 - v_L^2 k_x^2)/v_L^2 \tag{15}$$

and D is simply given by

$$D = \cos(QL) - \cosh(k_x a)\cosh(k_x b) - \frac{1}{2}(\varepsilon' + \varepsilon'^{-1})\sinh(k_x a)\sinh(k_x b) \tag{16}$$

It is recognised that the condition D=0 is just the familiar dispersion relation for 'pure' superlattice interface optical phonons[13]. In fact, equation (15) may be regarded as a generalisation of the above to describe optical hybrids. In the limit $v_L \to 0$ and $\omega \neq \omega_L$ (14) collapses to just this condition. On the other hand, if $\omega = \omega_L$, $v_L \to 0$ then the superlattice terms (those terms involving Q and b) disappear in (14). If further $k_x a \ll 1$(but non-zero) then it reduces to a couple of simple relations for the confinement wavevector of the even and odd symmetry hybrids

$$k_L a = 2n\pi \qquad n=1,2,3... \qquad \text{(even)} \tag{17}$$

$$\tan(k_L a/2) = k_L a/2 \qquad \text{(odd)} \tag{18}$$

where the solutions to equation (18) are $k_L a/\pi$ = 2.86,4.92,6.94 etc. These are just the well known Huang and Zhu[2] results.

Figure 1(a) illustrates the superlattice hybrid mode frequencies which are calculated from equation (14) as a function of in-plane wavevector for QL = $\pi$. Note, in particular, the conversion of LO1 to LO3 at a finite in-plane wavevector due to the hybridization of LO1 with the odd symmetry interface mode. This behaviour is best illustrated in figure 1(b) where the mode frequency is plotted against propagation angle $\theta = \tan^{-1}(k_x/Q)$ for vanishing in-plane wavevector ($k_x a$=0.15). This is often the preferred presentation employed in the literature[1,8]. The important point to notice is the conversion of LO1 to LO3 and that for mode propagation **strictly** normal to the superlattice axis ($\theta = \pi/2$ ), LO1 has vanished altogether with LO2 the highest energy mode. This regime is the one most often encountered in the electron-phonon interactions in  parallel transport .It is encouraging that such a simple dispersion relation can predict the highly non-trivial behaviour of the superlattice optical modes that are obtained from the complex lattice dynamical calculations as soon as one has propagation off-axis.

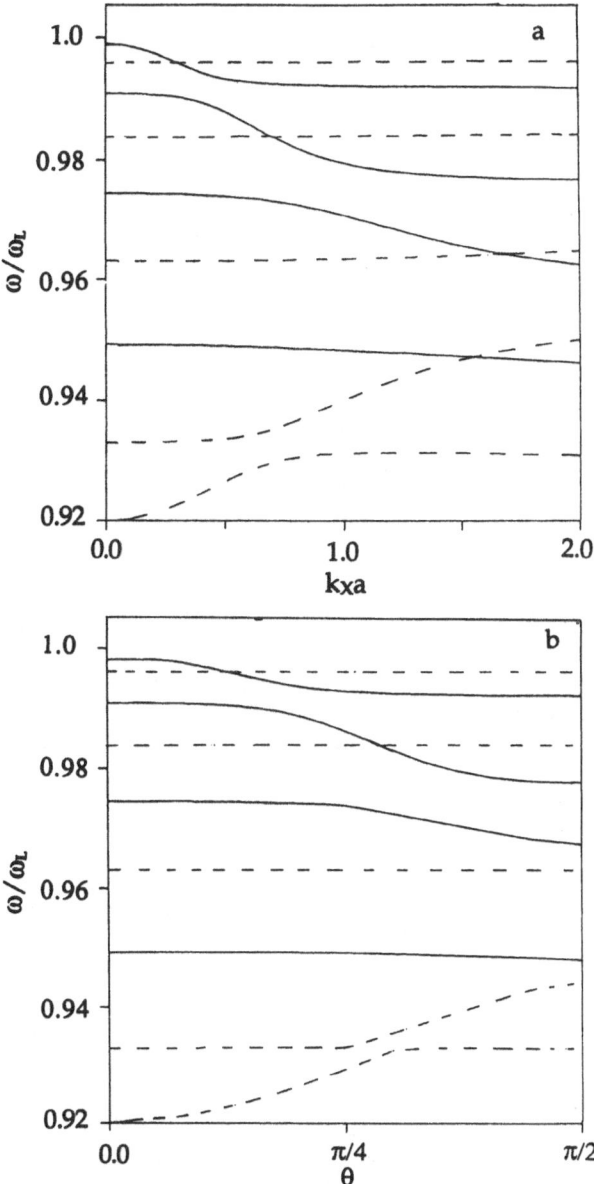

**Figure 1** Hybrid dispersion for a=50A and b=100A for (a) QL=π and (b) as a fuction of propagation angle θ. The dashed curves are the even modes and the full curves the odd modes.

Mode patterns are displayed in figure 2 for the highest energy odd mode of figure 1 for typical wavevectors. For intrasubband scattering, it is of course the even symmetry modes that participate and figure 3(a) shows $u_z$ for the highest frequency even mode of figure (1), LO2 with its associated parallel electric field,$E_x$, shown in 3(b). Note that both the LO and interface contributions are separately discontinuous at the interfaces, but the **total** $E_x$ is of course continuous. More importantly though, the parallel electric field has **just** the form predicted by the simple DC model which explains its success in calculating intrasubband scattering rates[7].

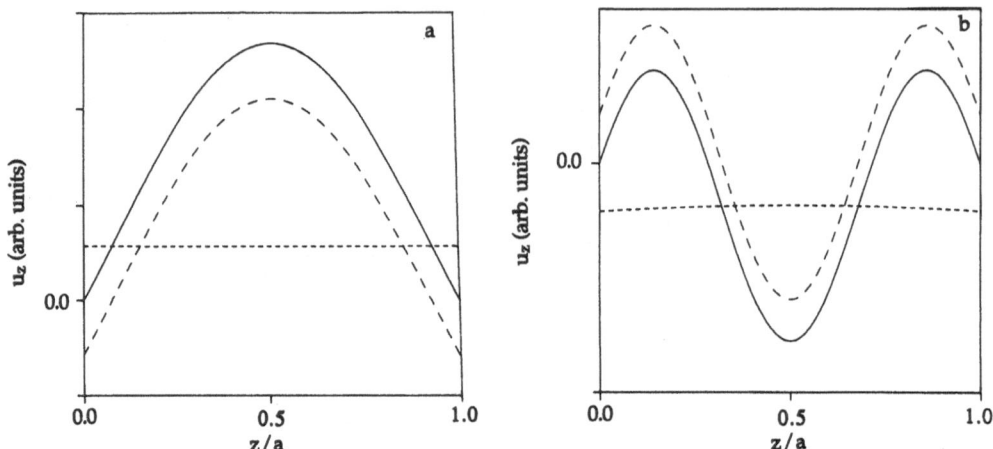

**Figure 2** The RIDF ($u_z$) of the highest energy odd mode of figure 1(a) for (a) $k_x a=0.1$ and (b) $k_x a=1.0$. The long dashes correspond to the LO component, and the small dashes to the interface component, the full curve is the sum of the two.

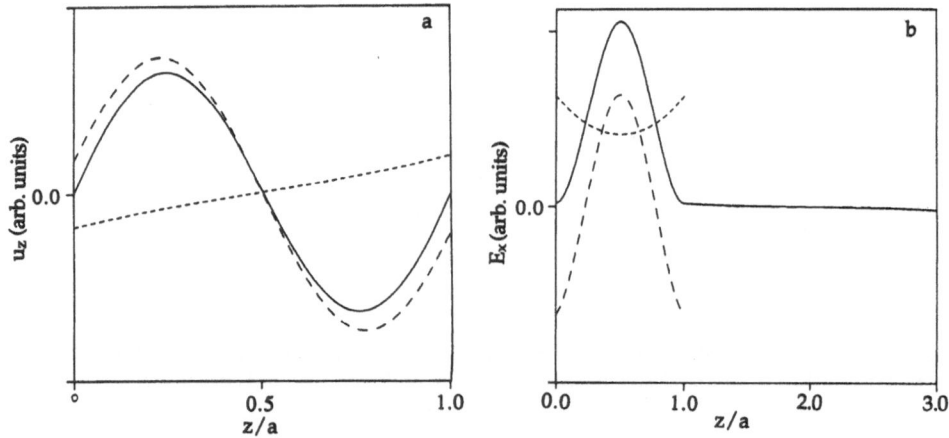

**Figure 3** (a) The RIDF ($u_z$) and (b) the in-plane electric field ($E_x$) for the highest energy even mode of figure 1(a).

# INTERACTION OF ELECTRONS WITH OPTICAL HYBRIDS

In this section the interactions of electrons with optical hybrids is briefly discussed. A full treatment would require a quantum theoretical approach which is beyond the scope of this work. Here a simple quantisation procedure is utilised for brevity. Further, the complications of a superlattice are dispensed with and a single well is considered. The single well results are obtained from the superlattice results in the limit $b \rightarrow \infty$. The dispersion relations for the odd and even modes are obtainable from equation (14) in this limit and are simply[11,12]

$$\cot(k_L a/2) = \kappa p_1 \text{ (odd)} \qquad \tan(k_L a/2) = -\kappa p_2 \text{ (even)} \qquad (19)$$

where

$$p_1 = \{s[\varepsilon' + \tanh(k_x a/2)]\}^{-1} \text{ and } p_2 = \{s[\varepsilon' + \coth(k_x a/2)]\}^{-1} \qquad (20)$$

The displacement fields within the well also simplify considerably and are expressible as

$$\mathbf{u}_L = A_+\{\cos(k_L z), 0, i\kappa^{-1}\sin(k_L z)\} \qquad \text{(even)} \qquad (21)$$

$$\mathbf{u}_L = A_-\{\sin(k_L z), 0, -i\kappa^{-1}\cos(k_L z)\} \qquad \text{(odd)} \qquad (22)$$

$$\mathbf{u}_I = A_1\{\cosh(k_x z), 0, i\sinh(k_x z)\} \qquad \text{(even)} \qquad (23)$$

$$\mathbf{u}_I = A_2\{\sinh(k_x z), 0, -i\cosh(k_x z)\} \qquad \text{(odd)} \qquad (24)$$

$$A_1 = -A_+p_2\cos(k_L a/2)/\cosh(k_x a/2) \qquad (25)$$

$$A_2 = -A_-p_1\sin(k_L a/2)/\cosh(k_x a/2) \qquad (26)$$

The origin of the z-axis is set at the mid-point of the well for convenience. The fields are given by equation (8), and the $A_\pm$ coefficients are mode amplitudes . The electric fields in the barrier regions are simply proportional to $\exp(-k_x|z|)$.

The normalisation of the mode amplitudes is formally achieved via quantum theoretical methods. Here an informal approach is adopted in which the mode energy is equated to that of a simple harmonic oscillator. The energy U associated with a particular hybrid is a sum of a mechanical and electromagnetic terms viz:

$$U = (M\omega^2/2V_0)\sigma \int_{-a/2}^{a/2} \mathbf{u}^*.\mathbf{u} \, dz + \frac{1}{2}\varepsilon_0\varepsilon\sigma \int_{-\infty}^{\infty} \mathbf{E}_I^*.\mathbf{E}_I dz \qquad (27)$$

where $\sigma$ is the surface area and M the reduced mass. The electrical energy is solely due to the optical interface mode component of the hybrid since $\mathbf{D}_L$ is identically zero for the LO component. This electromagnetic energy term is, in any case very small compared to the mechanical contribution and is ignored here for convenience (typically the mechanical contribution is an order of magnitude greater). Thus introducing a harmonic oscillator coordinate for the even (+) and odd (-) hybrids, $\chi_\pm$ , the equality in the energies leads to

$$\chi_\pm^2 = (\sigma/V_o) \int\limits_{-a/2}^{a/2} u_\pm^* . u_\pm \, dz \qquad (28)$$

Equation (26) defines the amplitudes $A_\pm$ which are expressible in terms of the harmonic oscillator coordinate as

$$A_\pm^2 = k_x^2 \, (2/N)\Delta_\pm^{-2}\chi_\pm^2 \qquad (29)$$

where N is the number of unit cells in the well and

$$\Delta_\pm^2 = k_x^2 + k_L^2 + a^{-1}\{\sin(k_L a)(k_L - k_x\kappa) - 4k_x f_\pm\} \qquad (30)$$

$$f_+ = p_2\cos^2(k_L a/2)\{2 - p_2\coth(k_x a/2)\} \qquad (31)$$

$$f_- = p_1\sin^2(k_L a/2)\{2 - p_1\tanh(k_x a/2)\} \qquad (32)$$

The above straightforward determination of the amplitudes allows for a calculation of the electron-hybrid optical phonon interaction rate and hence with a comparison with Rucker et al[7].

In what follows, the electrons are confined by infinite walls to the quantum well, their effective mass wavefunctions are well known and not repeated here. The interaction Hamiltonian between the electrons and the optical hybrid is given by

$$H_{int} = -e\phi_L + (e/m^*)A_I . p \qquad (33)$$

where $\phi_L$ is the scalar potential associated with the LO component of the hybrid $(E_L = \nabla\phi_L)$, $A_I$ is the vector potential for the interface optical component $(E_I = i\omega A_I)$ and $p$ the momentum of the electron. This distinction between a vector potential for the hybrid and simply taking an $e\phi$ coupling stems from the modes being transverse solutions to Maxwell's equations. In what follows this distinction is unimportant[15]. It is assumed that only emission processes are relevant and that the scattering is confined to the lowest subband, hence from symmetry considerations, only the even order optical hybrids participate. Further, to facilitate analytic results it is assumed that the electron has just sufficient energy to emit an optical mode. The more general situation in which the electron has an arbitrary in-plane kinetic energy requires a numerical evaluation.

The scattering rate, $\Gamma$, is obtainable from the golden rule

$$\Gamma = \frac{2\pi}{\hbar} \sum \int |<f|H_{int}|i>|^2 \, \delta(E_i - E_f) \, dN_f \qquad (34)$$

where the integration is over all final states, and i and f refer to the initial and final states respectively. The integrations involved in evaluating the rate are straightforward. The total rate is obtained by summing over all the allowed even modes consistent with the dispersion relation. The magnitude of these contributions reduce with mode number n as $n^{-2}$. If further it is assumed that $k_x a < 1$ (narrow wells), which in any case is the regime of interest in most applications, the rate becomes

$$\Gamma = \frac{1}{2}\Gamma_0 \left(\frac{\hbar\omega}{E_1}\right)^{1/2} \left\{ \frac{5}{4} + \frac{\pi^2}{6} \right\} \tag{35}$$

$$\Gamma_0 = \frac{e^2}{4\pi\epsilon_0\hbar} \left(\frac{2m^*\omega}{\hbar}\right)^{1/2} \left(\frac{1}{\epsilon_\infty} - \frac{1}{\epsilon_0}\right) \tag{36}$$

where $\Gamma_0$ is the typical bulk scattering rate ($\approx 8.7\times10^{12}\text{s}^{-1}$ for GaAs), and $E_1 = \pi^2\hbar^2/2m^*a^2$ the subband energy. This rate is plotted in figure 4 as a function of well width.

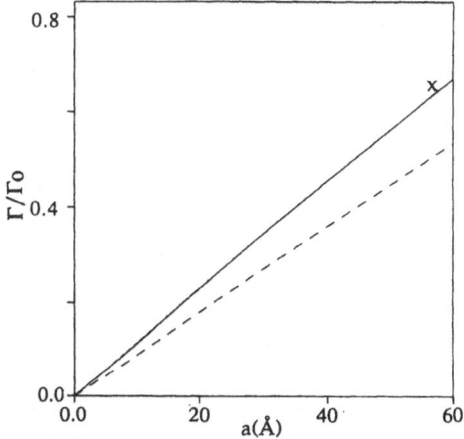

**Figure 4** The scattering rate as a function of well width. The cross corresponds to the calculation of Rucker et al. The dashed curve illustrates the contribution of LO2.

At a width of 56Å, the rate obtained by Rucker and co-workers using an ab initio treatment of the optical vibrations is marked by an X. This coincidence may be slightly fortuitous as these workers took into account the finite penetration of the wavefunction into the barrier regions, which will tend to reduce the rate, and a finite temperature of 300K which will increase the rate due to the thermal factor. Nevertheless if these two effects do more or less cancel one another, then this simple theory gives an excellent account of itself. It must be said that the DC model also agrees with the calculations **provided** that the contributions of interface optical modes are also included.

The calculations in this section are by no means complete, and it is envisaged that a fully quantum theoretical approach taking into account superlattice effects on both the phonons and electrons will be undertaken in the future. One further important point needs to be mentioned. In this calculation no account was taken of the contribution to the rate of the interface optical vibrations in the AlAs reststrahl band. These are known from theoretical[7,16,17] and experimental work[18] to dominate the interaction at small well widths. The AlAs bulk LO dispersion is very small[3] and hence hybridization of the modes in the barrier is less severe. The modes can then be treated via conventional theory[16,17].

## CONCLUSIONS

The main result of this work is the dispersion relation given by equation (14). This analytic result illustrates in a straightforward manner the main conclusions of the complex lattice dynamical calculations. In particular the conversion of LO1 to LO3 at finite in-plane wavevector and the disappearance of LO1 for **strictly** in-plane propagation. The highest frequency even mode, LO2 is seen to have parallel electric fields which are close in form to those predicted by the DC model. It is these modes which are allowed to participate in intrasubband events, which accounts for the success of the DC model. Further, equation (14) reduces in the limit of zero bulk mode dispersion to two important cases, namely, that for pure superlattice interface vibrations when $\omega \neq \omega_L$, and to the Huang and Zhu modes when $\omega = \omega_L$. This theory should serve as a useful compliment to the microscopic calculations which, after all, are computationally very intensive. As a continuum theory it can be applied to any geometry and extended to other materials provided the relevant information is known, in this case the zone centre bulk LO dispersion. But of most importance is its application to transport calculations.

The electron-hybrid optical phonon intrasubband interaction was determined in a simple manner for a thin single well and shown lead to results consistent with microscopic approaches. The hybridization of the modes effectively leads to a weakening of the interaction in very narrow wells, ignoring for the moment the AlAs-like interface mode discussed briefly above. This leads to the important possibility of engineering structures in order to reduce the important electron- polar phonon interactions. A very interesting proposal has recently been put forward by Stroscio et al[19] in which the barrier is replaced by a metal. The requirement that the tangential component of the electric field also vanish does drastically restrict the number of modes and reduce the electron-phonon interaction. What is really required, as is apparent from from our calculation is a barrier material which is effectively **rigid** at the frequencies within the GaAs restrahl band, and which has its own resonances well away from typical in-plane carrier energies, AlAs satisfies the first criterion but not the second. A metal of course, satisfies to a good approximation both of these conditions. These and other matters are left for future investigations.

## ACKNOWLEDGMENTS

We would like to thank the SERC for financial support and we are grateful to M.Babiker, O.Al-Dossary, S.R.P.Smith and M.Chamberlain for useful discussions.

# REFERENCES

1. S-F.Ren, H.Chu and Y-C.Chang, Anisotropy of optical phonons and interface modes in GaAs-AlAs superlattices, Phys.Rev.**B37** 8899 (1988).

2. K.Huang and B.Zhu, Dielectric continuum model and Frohlich interaction in superlattices, Phys.Rev.**B38** 13377 (1988).

3. E.Molinari, S.Baroni, P.Giannozi, and S.de Girondi, Theory of phonons in GaAs/AlAs superlattices, in 20th International Conference on the Physics of Semiconductors, E.M.Anastassakis and J.D.Joannopolous eds., World Scientific, Singapore (1990).

4. J.K.Jain and S.Das Sarma, Role of discrete slab phonons in carrier relaxation in semiconductor quantum wells, Phys.Rev.Lett. **62** 2305 (1989).

5. M.Babiker, Longitudinal polar phonons in semiconductor quantum wells, J.Phys.**C19** 683 (1986).

6. B.K.Ridley, Electron scattering by confined LO polar phonons in a quantum well, Phys.Rev.**B39** 5282 (1989).

7. H.Rucker, E.Molinari and P.Lugli, Microscopic calculation of the electron-phonon interaction in quantum wells, Phys.Rev.**B45** 6747 (1992).

8. H.Gerecke and F.Bechstedt, Influence of bulk-phonon-branch dispersion on displacement patterns and the intermixing of interface and confined optical phonons in superlattices, Phys.Rev.**B43** 7053 (1991).

9. M.Born and K.Huang,"Dynamical Theory of Crystal Lattices",Clarendon Press, Oxford.

10. R.Fuchs and K.L.Kliewer, Optical modes of vibration in an ionic crystal slab, Phys.Rev.**A140** 2076 (1965).

11. B.K.Ridley, Continuum theory of optical phonon hybrids and their interaction with electrons in a quantum well, Proc. SPIE **1675** 492, Somerset,New Jersey (1992).

12. X.Zianni. P.N.Butcher and I.Dharssi, Macroscopic behaviour of longitudinal phonons in a AlAs/GaAs/AlAs quantum well, J.Phys.Condens. Matter **4** L77 (1992).

13. A.K.Arora, A.K.Ramdas,M.R.Melloch and N.Otsuka, Interface vibrational Raman lines in GaAs/$Al_xGa_{1-x}$As superlattices,Phys.Rev.**B36** (1987).

14. N.C.Constantinou, O. Al-Dossary and B.K.Ridley, A simple dispersion relation for superlattice optical phonons, submitted to J.Phys.Condens.Matter (1992).

15. K.Ridley, The electron-Hybridon interaction in a quantum well, submitted to Phys.Rev**B** (1992).

16. N.Mori and T.Ando, Electron-optical-phonon interaction in single and double heterostructures,Phys.Rev.**B40** 6175 (1989).

17. O.Al-Dossary,M.Babiker and N.C.Constantinou, Fuchs-Kliewer interface polaritons and their interactions with electrons in GaAs/AlAs double heterostructures, Semicond. Sci.Technol. **7** B91 (1992).

18. K.T.Tsen, K.R.Wald, T.Ruf, P.Y.Yu and H.Markoc, Electron-optical-phonon interactions in ultrathin GaAs/AlAs multi quantum wells, Phys.Rev.Lett. **67** 2557 (1991).

19. M.A.Stroscio, K.W.Kim, G.I.Iafrate, M.Dutta and H.Grubin, Dramatic reduction in the longitudinal-optical phonon emission rate in polar semiconductor quantum wires, Philosophical Magazine Letters **65** 173 (1992).

# NEGATIVE DIFFERENTIAL RESISTANCE IN
# SUPERLATTICE AND HETEROJUNCTION
# CHANNEL CONDUCTION DEVICES

Steven W. Kirchoefer

Naval Research Laboratory
Washington, DC 20375-5320

## INTRODUCTION

The application of epitaxial III-V compound crystal growth techniques for negative differential resistance (NDR) high speed electronic devices has long been a subject of considerable interest. The earliest work employing crystals grown by molecular beam epitaxy (MBE) consisted primarily of devices with current flow perpendicular to the growth interfaces.[1] Interest in such devices remains high, and experimental successes have served as benchmarks of the improvement in state-of-the-art MBE growth.[2,3] However, development of devices with current flow parallel to the growth interfaces has been somewhat slower. With the notable exception of various types of field effect transistor (FET) structures,[4-6] the innovative use of MBE-based superlattices and heterostructures for planar devices remains at a very early stage of development.

A number of device concepts involving current flow parallel to the growth interfaces have been suggested. One of the most promising of these ideas was first proposed by Hess,[7,8] and involved the use of a real-space transfer effect for electrons in heterojunctions to produce an NDR device. The conceptual extension of real-space transfer involving bulk heterolayers into the regime of quantum-well heterostructures was first proposed by Price.[9] Phonon induced NDR in superlattice structures has been proposed by Ridley,[10] but has not been experimentally observed. The experimental observation of NDR in a bi-level superlattice quantum-well heterostructure[11] suggests that an extension of the real-space transfer effect into the quantum-well regime could be made.

The quantum state transfer device was the first structure employing real space effects of quantum states to realize a functional microwave source. Attempts to utilize this structure were partially successful.[11,12] One of the principle problems in utilizing this effect was the inability to demonstrate unambiguously that the observed negative differential resistance (NDR) originated from quantum state interactions. Many devices of various geometries were fabricated in attempts to isolate the effect, and hence show it to be different from NDR commonly observed in other channel structures.

The following article reviews the current state of research on quantum state transfer devices. A large number of heterostructure combinations and device geometries have been investigated in the effort to understand the underlying conduction effects. Several measurement techniques have been employed. These include dc conductivity,[11] low-frequency ac conductivity,[13] microwave measurements,[12] and photoconductivity studies. The results of this work have revealed certain insights into the operating states and conductivity properties of these devices.

*Negative Differential Resistance and Instabilities in 2-D Semiconductors*
Edited by N. Balkan *et al.*, Plenum Press, New York, 1993

# DISCUSSION

NDR in channel conduction devices can originate from a number of effects. Either the intrinsic properties of the semiconductor material or the device geometry effects interacting with the intrinsic material properties can give rise to NDR. The effects of primary interest here are those which are derived from conduction properties of the heterostructure alone and are not dependent upon the device geometry. These effects merit attention since they allow the device geometry design to be dictated by high-frequency engineering considerations. It is hoped that circuits can be developed using epitaxially grown NDR layers that will avoid the parasitic effects that limit frequency response in conventional devices.

The basis of the early real-space transfer and quantum-state transfer devices was the analog these devices posed to the Gunn effect. The essence of the Gunn effect lies in the existence of the low-energy, high-mobility $\Gamma$ minimum and the high-energy, low-mobility X minima. When sufficiently heated, electrons will make transitions from $\Gamma$ to X, resulting in the familiar domain formation and propagation properties of the Gunn effect. As originally proposed by Hess[7], one can build a heterostructure utilizing appropriate materials (in this case GaAs and AlGaAs) and produce a real-space analog to the Gunn effect, whereby electrons in the low-energy, high-mobility GaAs can become heated by an externally applied electric field and make transitions to the high-energy, low-mobility AlGaAs. Since these transitions occur in real space and not k-space, one may well expect to be free from domain formation consequences of the Gunn effect.

A similar effect can be achieved using quantum wells. The requisite real-space confinement to different regions of the structure is achieved by using the inherent localization properties of confined particles in dissimilar wells separated by thin barriers. Consider the case of two sets of dissimilar sets of wells, a type A set and a type B set, which have been grown in an alternating sequence and are separated by 80 Å barriers. If both sets of wells are identical, electrons in the confined particle states have equal probability of occupying either set. However, any differences in either alloy composition or well width between A and B can result in a splitting of the Kronig-Penney miniband into two minibands. This results in wavefunction localization to either A or B. Thus, electrons in different energy states occupy different physical layers. These electrons will exhibit conduction properties that are consistent with the composition and dimensions of the layers (set of wells) to which they are confined. This effect can be used to engineer an NDR device.[11]

The electric field at which state transitions would be expected can be estimated. Assuming an electron mean free path of 500 Å at 300K and an energy separation of 50 meV between the ground states of A and B, electrons will attain sufficient energy for the quantum transition at an electric field of ~$10^4$ volts/cm. These electrons will have a minimum distribution in energy of kT at zero field. The actual energy distribution width can be expected to be larger at high fields where the electron temperature will exceed the lattice temperature. This could results in the onset of the electron transitions occurring at lower applied electric fields.

It should be pointed out that the electric field intensities cited in the preceding argument are well in excess of the peak of the velocity-field characteristic for bulk GaAs, although arguments have been presented supporting the theory that intervalley scattering in two dimensions requires higher fields.[14,15] Similar electric field thresholds are predicted for real-space transfer in bulk layers[7] and for phonon-induced NDR.[10] Therefore, it is difficult to experimentally verify the source of NDR in any particular sample.

Another source of NDR is also possible in channel conduction devices. When semiconducting materials of differing composition (alloy and/or doping) form a junction, transverse (perpendicular to the grown layers) built-in electric fields and accompanying depletion regions are present. These depletions are modified by channel currents. The effect is similar to that responsible for channel pinch-off in a FET. Any mechanism responsible for NDR in perpendicular conduction devices may produce NDR in a channel conduction device through this interaction. The movement of depletions through quantum wells has been shown to result in NDR in other devices,[16] and could be present here. It is important to realize that all channel conduction devices will, at sufficiently high bias levels, experience depletion effects similar to pinch-off and thus give rise to pinch-off related NDR.[14,15] Explanations for deviations from linear resistance should always consider these possibilities.

## EXPERIMENT

In order to design a conduction channel device whose properties are dominated by a heterostructure-based mechanism, it is necessary to exercise some care. If doping levels and layer thicknesses are not chosen to correctly place depletion layers, the channel can be dominated by depletion effects. Should such depletion effects occur, the high-field region associated with pinch-off may cause stationary Gunn domains resulting in NDR which might be mistaken as originating from the heterostructure. Many of the devices presented here have been designed with $10^{17}$ Si-doped AlGaAs cladding layers. Although inclusion of these cladding layers was shown to be unnecessary for NDR,[12] they were important to prevent depletion of the superlattice. In any actual high frequency application, the cladding layers would not be included in the design, and the superlattice thickness would be increased so as to accommodate the native substrate and surface depletions. It was found in the early quantum-state transfer devices that parasitic effects from the cladding layers limited the NDR frequency response to ~1 MHz.[12]

### Device Design

An additional limitation to high frequency performance, posed by present device construction, is the parasitic resistance due to the contacts. In the present devices, contact is made to a $10^{18}$ cm$^{-3}$ Si-doped 0.25 μm GaAs layer at the surface. Injection into the structure through the top cladding layers to the channel is expected to dominate device properties at low applied voltages. In order to separate contact-related parasitics from channel conduction phenomena, a four-terminal device design has been utilized for device characterization. This configuration is used in a manner similar to the standard four-point probe method of measuring conductivities. Current is injected at the outer contacts located at the ends of a 40 μm long by 20 μm wide bar. Contacts separated by 20 μm are placed 10 μm from each injection contact and are used to sense the voltage distribution along the channel. The four-terminal geometry used in most of these measurements is shown in Fig. 1. Situations where NDR is occurring in a specific place in the channel can be detected with this geometry. In most of these cases, the greater part of the bias drop is observed near the a) contact. This is because the a) contact is the positive electrode in the label notation used here, and so the depletion effects and resulting channel pinch-off occur at this end of the mesa. Using this device construction, one can measure channel conductance as opposed to contact-related or Gunn domain-related effects.

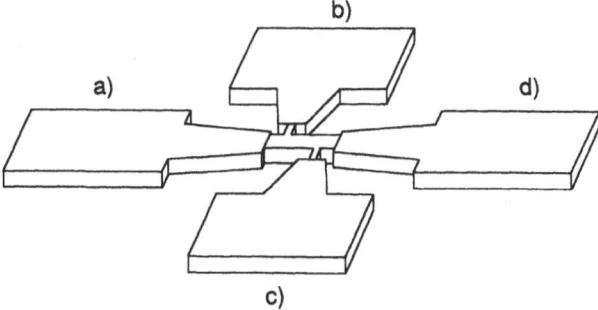

**Figure 1.** Four-terminal device structure. Current is injected through terminals a) and d), and voltages are sensed with terminals b) and c).

A typical heterostructure used in these experiments is grown by MBE. The epilayers from substrate to surface typically consist of a 0.5 μm buffer layer of undoped GaAs. This is followed by a 0.5 μm Si-doped $Al_{0.3}Ga_{0.7}As$ cladding layer. A 1 μm superlattice is then grown. The superlattice is composed of a repeated unit cell structure with two well types (differing in alloy composition and/or thickness) separated by 80 Å $Al_{0.3}Ga_{0.7}As$ barriers. Another $Al_{0.3}Ga_{0.7}As$ cladding layer, identical to the first one, is then grown. This is followed by a cap layer of $10^{18}$ cm$^{-3}$ Si-doped GaAs for ohmic contact purposes. A number of variations from this basic structure have been investigated to better understand the NDR mechanism and to improve its observed frequency range.

The devices are fabricated using standard photolithographic techniques. Initially, windows are opened in photoresist and a GaAs selective etch ($H_2O_2:NH_3OH$) is employed to remove the heavily doped GaAs contact layer over the regions between current injectors and sense electrodes. Photoresist liftoff is used to define thermally evaporated ohmic contacts (250 Å Sn - 750 Å Au), which are alloyed at 300 C for 15 seconds. A third photoresist step is used to pattern the bar structure. A 5:1:1 $H_2O:H_2O_2:H_2SO_4$ etch is used to complete the mesa fabrication.

## Low-Frequency Measurements

Low-frequency ac and dc conductivity measurements are made using lock-in amplifier techniques. A schematic of the setup is shown in Figure 2. The output of the internal signal generator of a PAR 5210 lock-in amplifier with a 600 Ω internal impedance is placed across a 1 Ω fixed resistance. This resistor is connected to the low voltage side of a Keithley 601 voltage source. This source configuration provides dc voltages ranging from -100 to +100 volts, with a small-signal voltage of variable amplitude and frequency impressed on the dc. For most measurements, a signal amplitude of 2 mV RMS at a frequency of 1 KHz is used. The output of the sources is connected to the series combination of the device under test and a 100 Ω current-sensing resistor. The ac voltage across the device under test is measured by a PAR 5208 lock-in amplifier with the TTL synchronization signal from the 5210 supplying the reference signal for the 5208 lock-in. A 2.7 μs delay exists in the TTL signal, and is corrected by the data acquisition software. The 5210 monitors the ac voltage across the

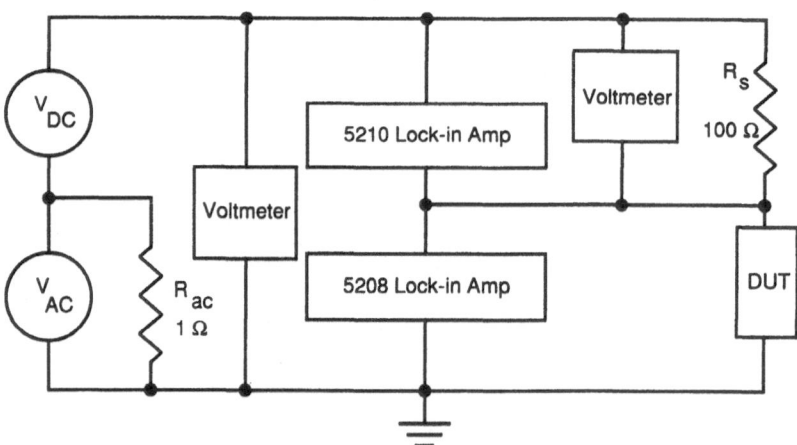

**Figure 2.** Circuit diagram for experimental setup. $R_s$ is the current-sensing resistor. $R_{ac}$ is the dc isolation resistor for the ac supply. $V_{ac}$ is the internal oscillator source of the 5208 lock-in amplifier. The box labeled DUT is the device under test.

current-sensing resistor. A pair of HP3478A multimeters are used to measure dc voltage levels. The entire apparatus is controlled by a personal computer through an IEEE-488 bus system. Calibration experiments have shown this setup to be accurate to within ±2° phase over a frequency range from 5 Hz to 50 KHz.

## Photoconductivity Measurements

Photoconductivity measurements are performed with a similar setup to that described above. The experimental method used here makes use of a lock-in amplifier to isolate the small photoconductive component of the device current. The device under investigation is biased at the operating point of interest by means of a dc power supply connected in series with a 100 ohm current sensing resistor. A PAR 5210 lock-in amplifier monitors the voltage across the sensing resistor. The device is illuminated with light coming from a monochromatic source consisting of a tungsten-filament lamp and a Spex 6810 double grating spectrometer. The output of the monochromator is chopped at about ~300 Hz. A synchronization signal from the chopper is fed to the reference channel of the lock-in, which is connected across the current sensing resistor. This permits the component of device current produced by the photons to be detected by the lock-in amplifier. The wavelength of the monochromatic light is swept, and data is collected on the relative intensity of the induced photocurrent as a function of photon wavelength. This spectrum exhibits peaks at the location of light-hole-to-conduction-band and heavy-hole-to-conduction-band transitions in the quantum wells within the heterostructure. Since these transitions have well defined and understood energy shifts with applied field, it is possible to ascertain the transverse electric fields experienced by these quantum wells. Once the internal electric fields are known, the supporting charge distributions and potential band bending can be calculated. With this information, one can look for critical quantum state alignments and charge distributions that might produce enhanced tunneling or other energy storage and conduction effects. The known current vs. voltage characteristics of the device under test can be correlated with this information.

## RESULTS

A large variety of device designs and heterostructure configurations have been tested. Four different basic measurement techniques were used. These are dc conductivity, low-frequency ac conductivity, microwave testing, and photoconductivity measurements. The microwave testing, while technologically important, has contributed little to the basic understanding of the conduction properties of these devices, and so will not be covered in depth here.

## DC Conductivity Measurements

The dc conductivity experimental results for all devices displaying NDR are summarized in Table 1. These are typical results, representing a large number of tested devices of each type. Anomalies originating from detectable fabrication origins such as bad ohmic contacts have been discarded. The electric field thresholds of the devices measured here are the minimum possible values. They are obtained by dividing the applied bias voltage at the maxima of the current vs. voltage characteristic (I-V) by the length of the conduction channel. If the applied bias is not dropped uniformly over the length of the channel, then the electric field threshold calculated by this simple method will be too low. Given this constraint, it is apparent that the field thresholds for all the NDR devices tested here are similar. The measured peak to valley ratios are also displayed. In many cases, the maximum possible peak to valley ratio is unknown, since the devices failed during testing and the current minimum was never measured. In these cases the lowest current measured before device failure is considered to be the valley current. The best NDR devices to date have been fabricated from samples utilizing bi-level superlattices. The only samples to show NDR at microwave frequencies have been those grown without cladding layers.[12]

**Table 1.** Summary of results. Superlattice unit cell is designated by layers $L_1$ to $L_4$ grown from substrate to surface, respectively. This unit cell is repeated as necessary to achieve the total designated superlattice thickness ($L_z$).

| Sample | Substrate Cladding Layer | Superlattice unit cell $L_1$ | $L_2$ | $L_3$ | $L_4$ | total $L_z$ | Surface Cladding Layer | Ohmic Contact Layer | Electric Field V/cm | Peak to Valley Ratio |
|---|---|---|---|---|---|---|---|---|---|---|
| A) | 0.5 μm x=0.36 $10^{17}$ Si | 80 Å x=0 $10^{15}$ Si | 80 Å x=0.3 $10^{15}$ Si | 80 Å x=0.06 $10^{15}$ Si | 80 Å x=0.3 $10^{15}$ Si | 1 μm | 0.5 μm x=0.36 $10^{17}$ Si | 0.5 μm x=0 $10^{18}$ Si | $6.2 \times 10^3$ | 1.318 |
| B) | " | " | " | 160 Å x=0.06 $10^{15}$ Si | " | " | " | " | $7.8 \times 10^3$ | 1.058 |
| C) | 0.1 μm x=0.36 $10^{17}$ Si | " | " | 80 Å x=0.06 $10^{15}$ Si | " | " | 0.1 μm x=0.36 $10^{17}$ Si | " | $3.3 \times 10^3$ | 1.039 |
| D) | 0.5 μm x=0.36 $10^{17}$ Si | bulk GaAs $10^{18}$ Si | | | | .25 μm | 0.5 μm x=0.36 $10^{17}$ Si | 0.25 μm x=0 $10^{18}$ Si | $6.0 \times 10^3$ | 1.014 |
| E) | " | bulk GaAs $10^{15}$ Si | | | | " | " | " | $3.1 \times 10^3$ | 1.167 |
| F) | " | 80 Å x=0 $10^{15}$ Si | 80 Å x=0.3 $10^{15}$ Si | 45 Å x=0 $10^{15}$ Si | 80 Å x=0.3 $10^{15}$ Si | 1 μm | " | " | $3.0 \times 10^3$ | 1.045 |
| G) | " | " | 80 Å x=0.3 $10^{17}$ Si | " | 80 Å x=0.3 $10^{17}$ Si | " | " | " | $3.5 \times 10^3$ | 1.236 |
| H) | 0.5 μm x=0.3 $10^{17}$ Si | " | 80 Å x=0.3 $10^{15}$ Si | 50 Å x=0 $10^{15}$ Si | 80 Å x=0.3 $10^{15}$ Si | " | 0.5 μm x=0.3 $10^{17}$ Si | " | $2.4 \times 10^3$ | 1.021 |
| I) | " | " | 80 Å x=0.3 $10^{17}$ Si | " | 80 Å x=0.3 $10^{17}$ Si | " | " | " | $5.5 \times 10^3$ | 1.077 |
| J) | none | " | " | " | " | 2 μm | none | " | $5.3 \times 10^3$ | 1.023 |
| K) | " | " | 80 Å x=0.3 $10^{15}$ Si | 50 Å x=0 $10^{17}$ Si | 80 Å x=0.3 $10^{15}$ Si | " | " | " | $7.8 \times 10^3$ | 1.191 |
| L) | " | " | 80 Å x=0.3 $10^{15}$ Si | none | none | " | " | " | $6.8 \times 10^3$ | 1.019 |

Column headers for "Grown Layers (x = mole fraction Al, $Al_xGa_{1-x}As$)": Substrate Cladding Layer; Superlattice unit cell ($L_1$, $L_2$, $L_3$, $L_4$); total $L_z$; Surface Cladding Layer; Ohmic Contact Layer. Results: Electric Field V/cm; Peak to Valley Ratio.

Initial results for the bi-level superlattice, as discussed above, were previously reported[11] and are summarized in Table 1 as Samples A) and B). The two-terminal I-V curves for Sample A) and Sample B) are shown in Fig. 3. These I-V curves were obtained using a pulsed measurement technique. This was intended to reduce the affects of heating which occur when conventional curve tracing is used. Sample A) showed the highest peak to valley ratio of any of the devices shown in Table 1. This may be attributed to the pulsed measurement scheme since later samples were measured with a transistor curve tracer or at dc. Biasing of the devices under steady state conditions at fields much above the threshold often led to a failure mechanism that is discussed below. For this reason the valley minimum was not reached for many of the devices.

Sample B), although exhibiting NDR, had a less pronounced NDR than Sample A). Two factors may contribute to this relative difference between Sample A) and Sample B).

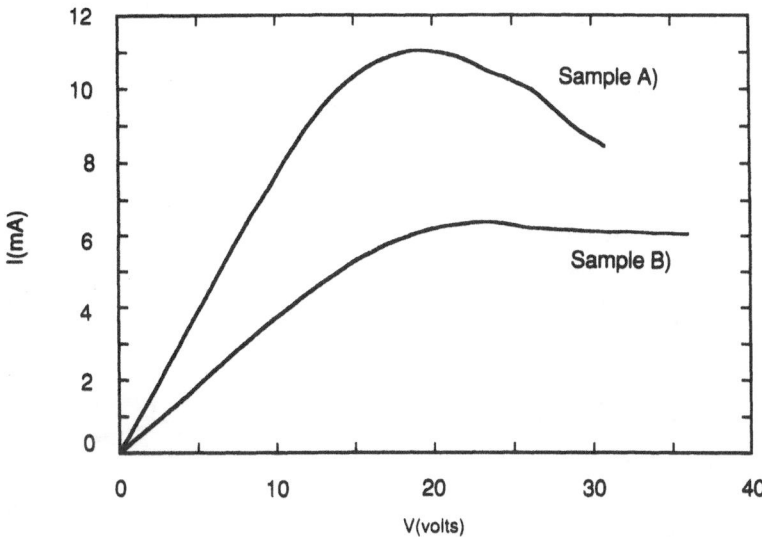

**Figure 3.** Current vs. voltage characteristics for two-terminal devices fabricated from Sample A) and Sample B).

The conductance of the high-energy set of wells in Sample B) is approximately twice that of the conductance of the high-energy set of wells for Sample A). Since the conductance of the low-energy set of wells is the same for both samples, the differential conductance between the low-energy set of wells and the high-energy set of wells is greater for Sample A). This gives rise to a larger value of NDR. In addition, the energy separation between the ground state of the low-energy well and the high-energy well is smaller for Sample B) than for Sample A). The sharpness of the transition is expected to be proportional to kT and inversely dependent on the energy separation of the ground states. Thus, at the same temperature, it would be expected that Sample A) would have a sharper transition than Sample B).

The existence of NDR in channel conduction devices is not confined to bi-level superlattice structures but can also be observed in bulk heterojunction devices.[7,8] The I-V curve of Fig. 4 shows the NDR resulting from the real-space transfer of electrons from the 0.25 μm GaAs layer into the $Al_{0.36}Ga_{0.64}As$ cladding layers of Sample E). In contrast, there was very little NDR shown by Sample D) which had a very similar heterostructure. The primary difference between Samples D) and E) was the doping of the GaAs layer. The doping of the GaAs layer in Sample D) was chosen such that the depletion regions would expand into the AlGaAs layers with increasing applied bias. The doping of the GaAs layer of Sample E) was less than the doping of the cladding layers, resulting in the depletion layer extending into the GaAs channel as bias is increased. For this reason, a pinch-off region can be expected to form in the GaAs channel yielding a high-field stationary-Gunn-domain region which improves the efficiency of the real-space transfer mechanism.[4] This Gunn domain results in the non-uniform distribution of bias drop along the channel as observed in Fig. 4. Since a pinch-off region is not readily formed in the GaAs channel of Sample D), the real-space transfer efficiency is expected to be less, resulting in a less pronounced NDR.

NDR at microwave and millimeter-wave frequencies can be utilized to make both oscillators and amplifiers. A straightforward method of determining the frequency response of the NDR is to configure the device as a reflection amplifier. The amplitude and phase of the reflected signal as referenced to the incident wave characterizes the frequency response of the NDR. Samples such as A) and B) exhibited no NDR above about 1 MHz. Parasitics associated with the AlGaAs cladding layers were believed responsible for the frequency limitation. This was confirmed when reflection amplification was observed up to 2.5 GHz from a bi-level superlattice without the AlGaAs cladding layers.[12] The I-V characteristic of the device, fabricated from Sample K), which produced amplification is shown in Fig. 5. Since there are no cladding layers, the NDR cannot result from real-space transfer of

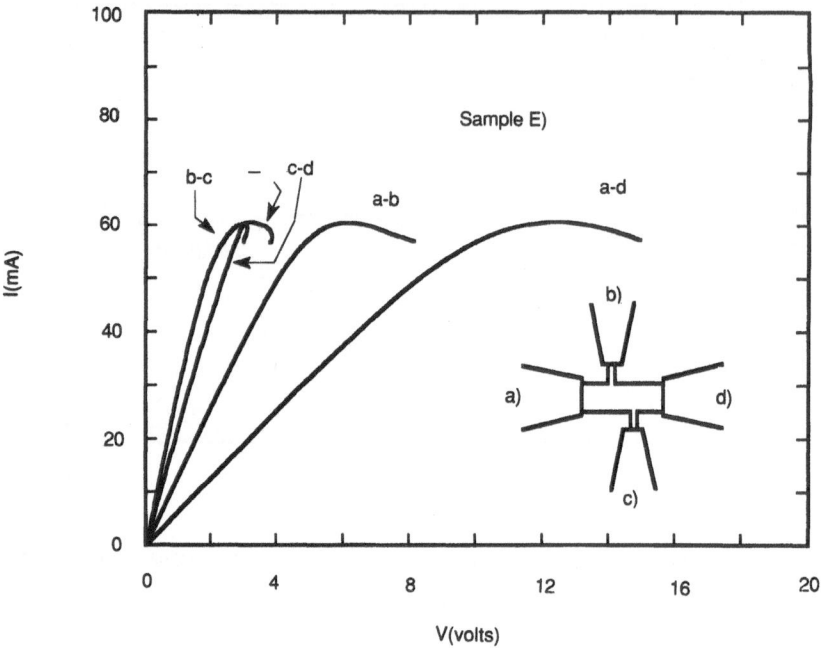

**Figure 4.** Current vs. voltage characteristic for a four-terminal device fabricated from Sample E). Each curve represents the current from a) to d) as a function of voltage between the terminals shown in the inset diagram.

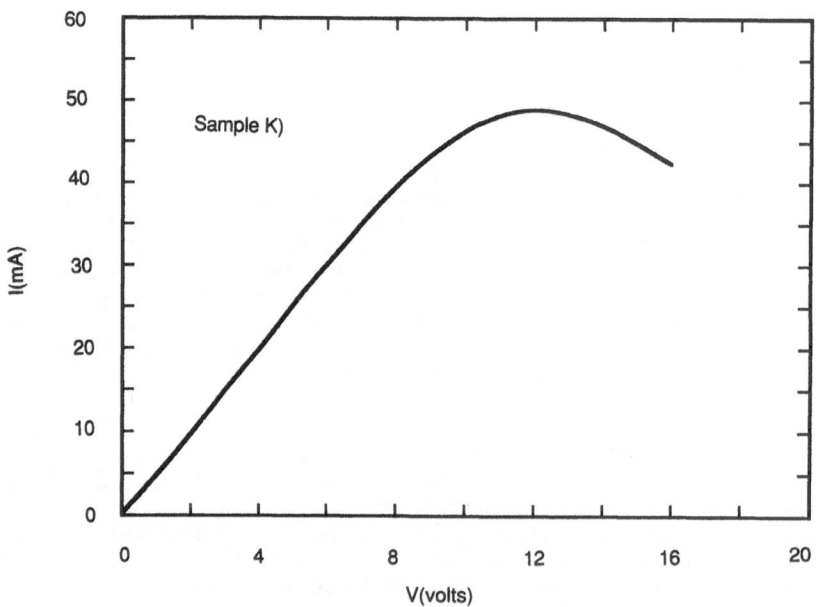

**Figure 5.** Current vs. voltage characteristic for a two-terminal device fabricated from Sample K).

electrons from the superlattice region to bulk AlGaAs layers as may be possible with Samples A) and B). The device whose I-V curve is shown in Fig. 3 had a different active channel geometry than the devices which yielded Fig. 5, accounting for the different biasing conditions at which the onset of NDR occurs. With this difference in mind, it should be noted that the I-V curve shapes are remarkably similar for Samples A) and K).

An example of a typical four-terminal I-V characteristic is shown in Fig. 6. The four curves shown represent the I-V as measured from each of the terminals represented in the inset figure. The relative scaling of the measured voltage as a function of current for each terminal is consistent with the geometry of the structure, and allowing for contact resistance, indicates that the applied bias is dropped uniformly along the channel. It is apparent from these data that this NDR is not a localized effect. Fig. 7 shows similar data for a device in which the potential distribution is non-uniform. Both figures represent devices fabricated from Sample L) of Table 1. It appears that observations of uniformly distributed NDR in a particular heterostructure can not be used to preclude the possibility of non-uniformly distributed NDR in other devices fabricated from the same sample. This non-uniformly distributed NDR could result from defects in the MBE crystal or could arise from a small difference in the contacts or geometry.

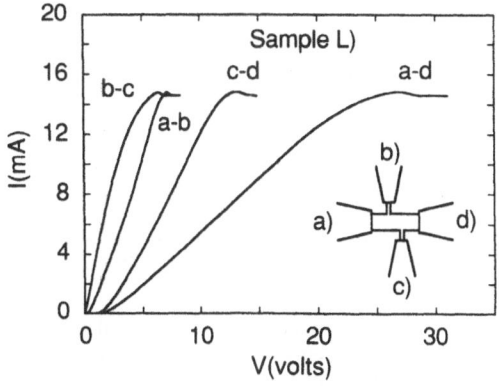

**Figure 6.** Current vs. voltage characteristic for a four-terminal device fabricated from Sample L), showing uniformly distributed NDR along the channel. Each curve represents the current from a) to d) as a function of voltage between the terminals shown in the inset diagram.

The failure mechanism observed in these devices deserves separate comment. Many of the devices exhibit instabilities in their I-V characteristics when biased into the NDR regime. This behavior has been correlated with permanent damage occurring to the channel regions of these devices. This degradation always occurs only when the devices are biased in the NDR region. Observations of this damage have been made in a scanning electron microscope (SEM) as the device was subjected to an applied bias. With the device under bias, dark lines form which extend back toward the cathode. What appear to be gallium droplets form concurrently, and move along the surface of the channel toward the anode. The timing of the breakdown event is too rapid to be measured with the SEM. After one of these events occurs, the device continues to function, only at lower currents for any given voltage bias level. These discrete events occur repeatedly with continued increasing bias. After considerable damage has been done to the channel, the device fails catastrophically with the contact metal migrating and shorting the cathode to the anode. The observation of this same failure mechanism in a large number of samples only when biased in the NDR region suggests that it is directly related to the NDR conduction mechanism.

### Low-Frequency Conductivity Measurements

Examination of these devices with the low frequency ac lock-in amplifier techniques has produced new evidence of the contribution of the quantum state transfer effect to the conductivity.[13] Typical current vs. voltage characteristics and numerically integrated conductance vs. voltage data taken at 300 K are shown in Fig. 8 for Sample A) of Table 1. The most striking single feature of this data is the fact that the curves do not lie directly over each other. This is due in some part to experimental error, the cumulative effect of which is compounded by the numerical integration. Since the measured conductance is significantly smaller in magnitude compared to $dI_{dc}/dV$ calculated from measured current vs. voltage characteristics for all devices tested when biased in the NDR regime, this implies that the observed NDR at dc may be caused to some extent by bias point shifts and not by low-frequency small signal effects.

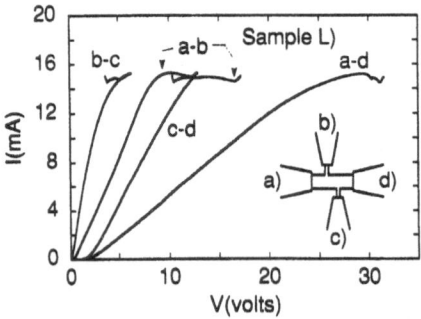

**Figure 7.** Current vs. voltage characteristic for a second four-terminal device fabricated from Sample L), showing non-uniformly distributed NDR. Each curve represents the current from a) to d) as a function of voltage between the terminals shown in the inset diagram.

Real and imaginary admittance data are shown in Figure 9 for Sample A) of Table 1. This device exhibits a substantial imaginary admittance component. This initially occurs at ~8 volts, which is the point of the real admittance (conductance) maxima. It is important to recognize that this particular device does not actually exhibit a negative conductance at 1 KHz, although many other devices do. Other than the fact that the ac conductance is of consistently greater magnitude than the dc dI/dV, these curves do not exhibit any apparent differences. This device shows a significantly increasing imaginary admittance component that reaches a maximum at the same bias point as the minimum in real admittance at ~22 volts. The imaginary component then extinguishes with increasing bias as the device is pressed beyond the region of dc NDR.

Considerable insight into the quantum state transfer related NDR has been gained through characterization of the imaginary admittance component observed in bi-level superlattice structures. Interactions between the measurement equipment and an active (i. e. not a simple resistive) NDR device can be complicated, and careful calibration of the measurement

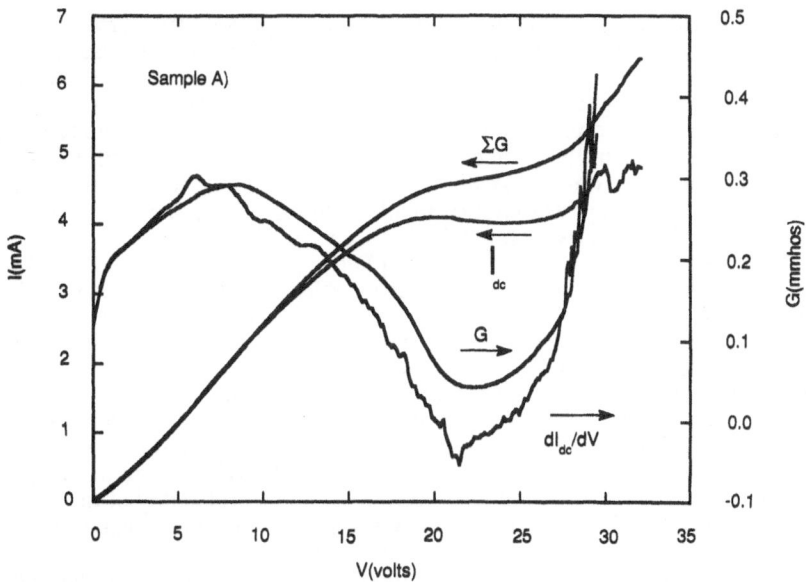

**Figure 8.** Combined plot of current ($I_{dc}$) and integrated conductance ($\Sigma G$) vs. voltage and conductance (G) and numerical differentiated current ($dI_{dc}/dV$) vs. voltage for the device described in the text. The arrows indicate the relevant Y-axis scales.

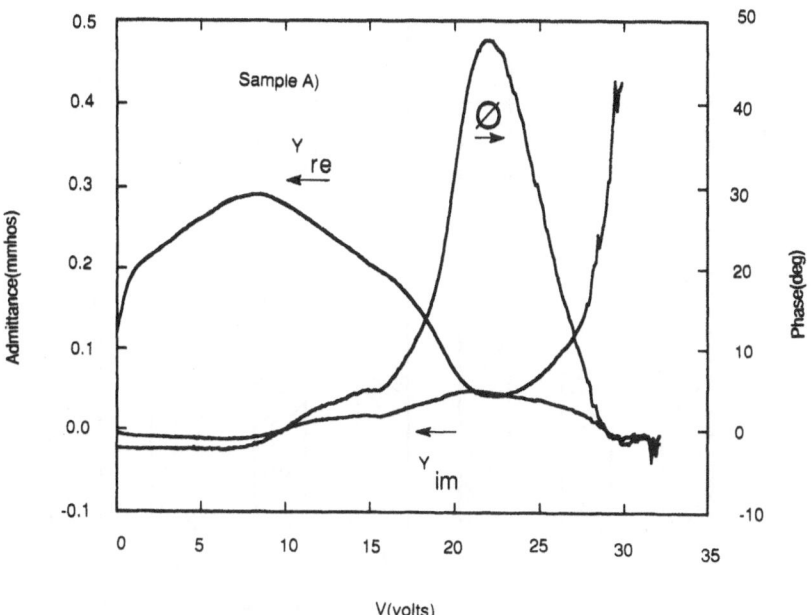

**Figure 9.** Combined plot of real ($Y_{re}$) and imaginary ($Y_{im}$) admittance vs. voltage and the phase angle between real and imaginary admittance vs. voltage for the same device as displayed in Fig. 8. The arrows indicate the relevant Y-axis scales.

apparatus is necessary. The imaginary component of admittance measured here is essentially capacitive in nature and exhibits a peak value of 0.05 mmhos. An analysis of the imaginary component as a simple capacitance is instructive. With the excitation frequency of 1 KHz, the effective capacitance is $G_{im}/\omega = 8$ nF. This is an extremely large value for an electronic device of these physical dimensions. The simplest way to understand this is to consider the case of a parallel plate capacitor with a GaAs - like dielectric. The area of the active channel in these devices is 800 $\mu m^2$. Since one might conceive of the unlikely situation that the contact pads might be involved in the effect, one can include their areas. This results in the effective area for the plates of 20000 $\mu m^2$. Using the relation $C = \varepsilon_r\varepsilon_0A/d$, one obtains only 3 Å for the plate separation. Since the smallest physical dimension likely to contribute to this effect is the 80Å size of the $Al_{0.3}Ga_{0.7}As$ barriers, it is obvious that no single interface is responsible for the observed effect. If the number of interfaces is increased to that contained in the superlattice of the device, it becomes possible to hypothesize the mechanism responsible for $G_{im}$.

The following model is proposed to explain this data. This structure contains two interleaved superlattices, with ground-state minibands centered on energies $E_1$ and $E_2$, as shown in Fig. 10a). It is fairly easy to determine that an electron in either miniband is strongly localized to its miniband of origin, and has a relatively high probability of tunneling into quantum wells of the same miniband compared to the probability of tunneling into wells of the interleaved miniband. When the device is under sufficient bias to populate $E_2$, an effective charge distribution is formed which can be thought of as separate conducting regions connected in an interleaved fashion by tunneling. The presence of these two spatially separated electron gases allows the applied ac to induce dipole imbalances of charge between adjacent dissimilar quantum wells. The resulting effect is similar to that present in an interleaved parallel plate capacitor, as shown in Figure 10b). The occupied quantum states act as the plates of the structure, with the connection between plates occurring via tunneling. Having a low probability of electron occupation from either miniband, the barriers act as the

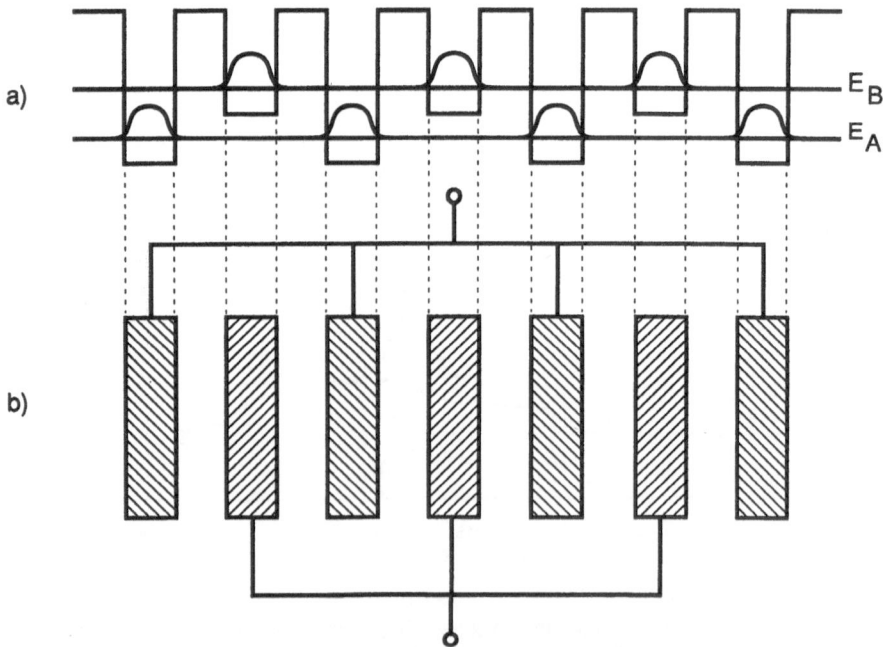

**Figure 10.** Part a) represents the conduction band of the bi-level superlattice with the superimposed positions and probability densities of the ground-state minibands of the two interleaved well types A and B. Part b) is a schematic of the manner in which the capacitances of the individual barriers are linked together to form a parallel combination between the conduction electrons in miniband A and miniband B.

dielectric. However, it is inaccurate to call this effect a capacitance in the conventional sense of a discrete device. It is more accurately viewed as energy storage in the induced transverse electric fields by the collected electrons in the quantum wells.

The data presented here is of a device at 300 K. If the device is cooled to 77 K, the imaginary admittance component vanishes, although the NDR in real admittance persists. Since the proposed model attributes the imaginary admittance component to the coupling of similar-type wells, this observation suggests that the coupling has been disturbed when the device is cooled. This should not happen if the mechanism of coupling was simple non-thermal quantum tunneling. Since this structure has relatively thick 80 Å $Al_{0.3}Ga_{0.6}As$ barriers, one should not expect to observe any direct effect of superlattice miniband-like coupling. However, the above data are strong evidence that coupling of similar quantum wells into minibands does occur by some temperature-dependent mechanism in these structures at 300 K, and is responsible for the observed anomalous imaginary admittance.

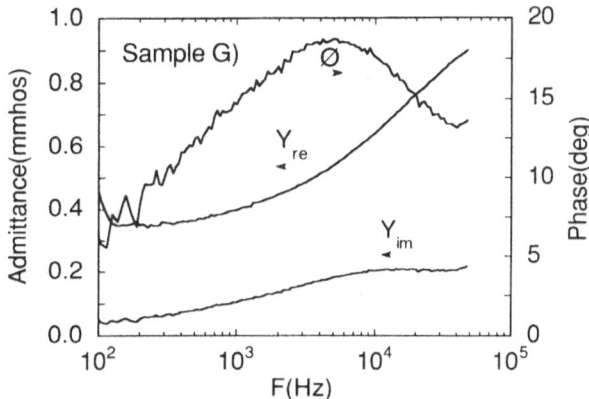

**Figure 11.** Combined plot of real ($Y_{re}$) and imaginary ($Y_{im}$) admittance vs. frequency and the phase angle between real and imaginary admittance vs. frequency for a device fabricated from Sample G) of Table 1, biased in the regime of NDR. The arrows indicate the relevant Y-axis scales.

Measurements have been taken in which devices are held at a constant bias within the regime of NDR, and the frequency is swept from 100 Hz to 50 KHz. Fig. 11 shows data collected for a device fabricated from Sample G) of Table 1. The device was biased at 20 volts, which is in the midrange of the NDR. This data shows the dependence of imaginary admittance on frequency is essentially a power law function over most of the range of measurement, as one might expect for a capacitive-like element.

It should be emphasized that a significant number of superlattice structures have been tested, including single-well-type superlattices and multi-well-type superlattices comprising a variety of parameter combinations. Only structures with two interleaved superlattices of dissimilar well types exhibit this effect. Superlattices of a single well type have shown no consistent imaginary admittance properties, even in the few cases where NDR were observed. These experimental results give considerable credence to the present hypothesis that the imaginary admittance and the observed NDR in bi-level superlattices come from the interaction of the spatially-dependent wavefunctions of the two ground-state minibands.

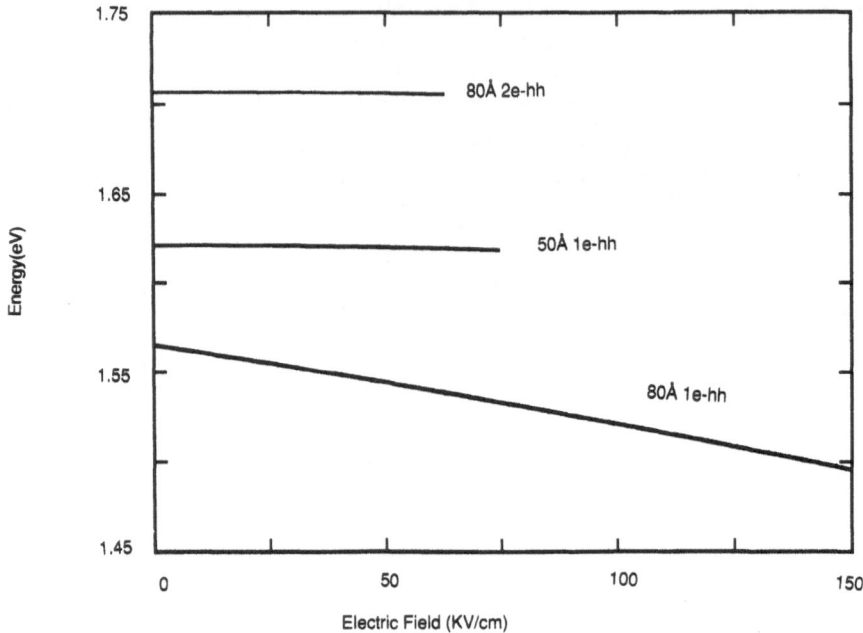

**Figure 12.** Calculated data plot of the energy shifts of the quantum states as a function of transverse electric field. The ground state electron-to-heavy-hole transitions for the 80Å and 50Å wells and the first excited state electron-to-heavy-hole transition are displayed. These state energies are calculated from zero electric field to the electric field value at which carrier confinement is lost to tunneling out of the well.

## Photoconductivity Measurements

Photoconductivity measurements of superlattice quantum state transfer devices have been useful in measuring more of the conductivity properties of these structures. Since the mesa structure of these devices allows easy illumination of the active area, existing devices can be used for examination with the photoconductivity technique. Photoconductivity data exhibit strong exciton peaks, which can be used as markers to energy levels within the interior of the bandstructure.[17-20] This allows one to follow the changes in energy levels with applied bias on the device, and to correlate these changes with induced transverse electric fields and their supporting charge distributions.

The transverse electric field can be determined by the shift in energy of the photoconductivity peaks with applied bias. The skewing of band edge with the transverse field results in a lowering of the energy of the transitions. This effect is greatest for those states nearest to the band edge. The amount of energy shift can be calculated numerically, and the result for 80Å and 50Å GaAs wells confined by 80Å $Al_{0.3}Ga_{0.7}As$ barriers is shown in Figure 12. The presence of transverse electric fields is consistent with the presence of charge carrier accumulations or depletions in adjacent layers of the superlattice. One can expect to see large relative shifts in the photoconductivity peaks when large transverse electric fields exist. Such a situation occurs when the channel becomes pinched off or when stationary Gunn domains form.[13,14]

Measurements were made on these devices at 4 K and 300 K on Sample K) of Table 1. Fig. 13 shows the photoconductivity response of the device under constant voltage bias varying over the operating range of the device. The relative amplitude of the signal between separate bias sweeps, while accurately represented here, is not always repeatable and can therefore only be used for qualitative interpretation. The ground state heavy- and light-hole to conduction-band exciton peaks are plainly evident in the data. The relative energy spacings between the peaks are nearly constant from one bias point to the next, indicating that no large

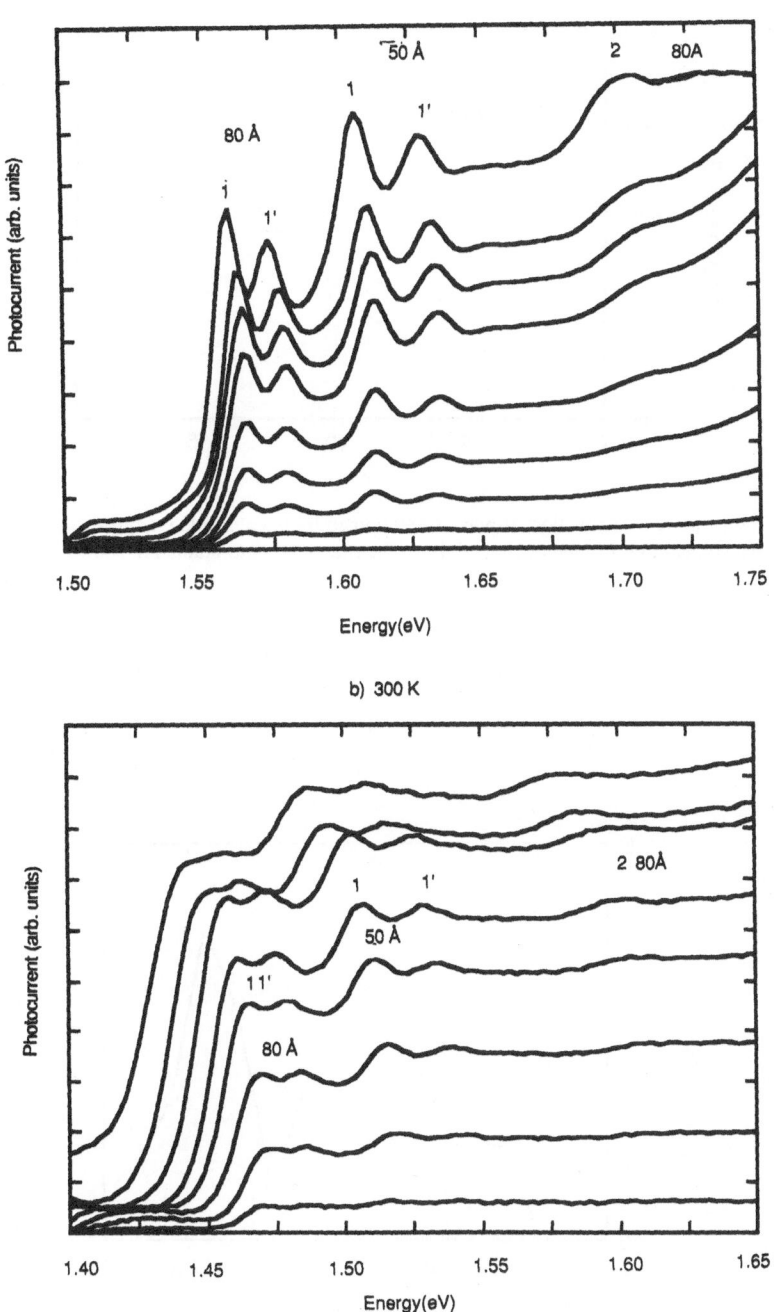

**Figure 13.** Photocurrent signal as a function of incident photon energy for temperatures of 4 K and 300 K for Sample K). The signal intensity increases monotonically with applied bias for the data shown. The device bias levels are, from lowest vertical amplitude to highest, 6, 8, 10, 12, 14, 18, 22, and 28 volts for the 4 K data and 4 to 18 volts in 2-volt increments for the 300 K data. The exciton peaks are labeled with unprimed numbers for electron-to-heavy-hole transitions and primed numbers for electron-to-light-hole transitions.

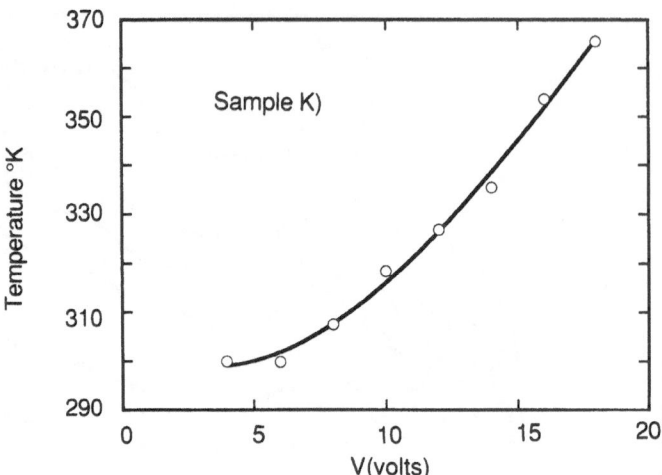

**Figure 14.** Temperature as a function of bias as determined by the absolute energy shifts of the quantum state transitions as measured by photoconductivity. The measured data points are shown as circles on the plot.

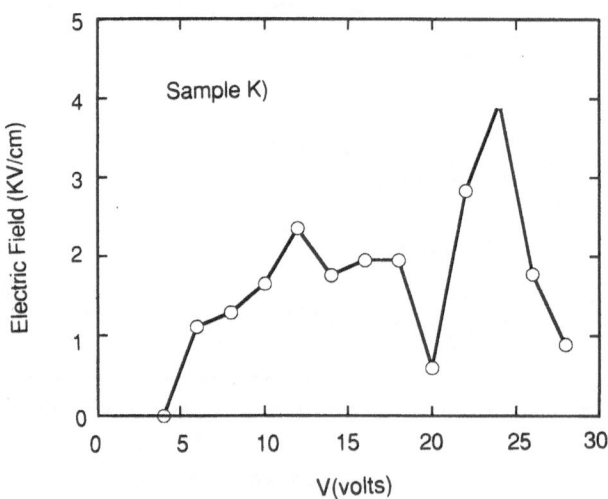

**Figure 15.** Transverse electric field as a function of applied bias as measured from the relative energy shifts between the ground-state-80Å-well and ground-state-50Å-well transitions. The displayed result is determined from the 4 K data.

electric fields are being induced perpendicular to the grown layers. The exciton peaks shift uniformly to lower energies as the bias voltage is increased. This shift can be related directly to device heating. As can be seen in the 4 K data of Fig. 13a), when the device temperature is held nearly constant in a helium cryostat, there is comparatively little peak shift to lower energy as contrasted to the 300 K data of Fig. 13b). The well-known relationship of GaAs energy gap to lattice temperature can be used to calculate the lattice temperature of the superlattice region of the device from this data. This is shown in Fig. 14. These temperatures are higher than initially suspected, but are not unreasonable given the thermal conductivity of GaAs and geometry of the mesa device employed here, and may play a role in the failure mechanism associated with the NDR mentioned earlier. The data verifies that the power dissipation must be occurring in the mesa in order for this large of a temperature gradient to be supported. Heat generation in other regions would be more readily conducted away and would not result in the high temperatures that the restrictive geometry of the mesa supports.

Neglecting the small common-mode energy shifts due to the slight presence of heating effects in the 4K data, the relative positions of the 80Å ground-state-electron to ground-state-heavy-hole and 50Å ground-state-electron to ground-state-heavy-hole peaks can be analyzed more closely. The divergence in energy of these two peaks with applied bias can be superimposed on the curves for Fig. 13a) to project a perpendicular electric field vs. applied bias data plot, as give in Fig. 15. The large scatter in the data demonstrates the difficulty in determining the precise position of the exciton peaks of Fig. 13a), due to the small size of the energy shift measured. The differential resistance for this device begins decreasing toward negative values at 15 volts. This is likely to have contributed to the scatter in the data at higher voltages. At bias levels less than 5 volts, injection across the heterolayers under the ohmic contacts has not been established, and so little current is injected and little field is measured in the channel of the device. The absence of a large transverse electric field is significant, because it demonstrates the absence of interface- or depletion-related effects contributing to the observed negative differential resistance.

## CONCLUSION

Various MBE-grown GaAs-AlGaAs heterostructures, discrete device designs, and measurement techniques have been used to study the phenomena of NDR in parallel conduction quantum state transfer devices. The large number of structures examined here permits several broadly-stated conclusions to be asserted. The existence of a spatially-distributed NDR mechanism in these devices has been verified. This mechanism was identified in the original work as quantum-state transfer. Since similar, less pronounced spatially-distributed NDR has been observed in a few devices composed of superlattice structures other than bi-level superlattices, the possibility of a completely different NDR mechanism is apparent. All channel conduction devices with demonstrated spatially-distributed NDR likewise have similar phenomenology in dc IV. Low frequency admittance measurements reveal differences not apparent in dc measurements. Low-frequency ac measurements have demonstrated energy storage effects that are unique to bi-level superlattice structures. This effect manifests itself through the presence of imaginary admittance that is far in excess of values supported by conventional capacitance-based models. The presence of an imaginary admittance in the same bias region as the NDR suggests that both effects originate from the same physical source, and both have been postulated to be directly related to carrier populations in different quantum states of the bi-level superlattice structure. Photoconductivity measurements have been used to investigate these structures. These measurements have shown the absence of large transverse electric fields when the devices are under high bias conditions. This data verifies the absence of any carrier depletion or Gunn-effect related NDR mechanisms and is consistent with the model of NDR via transfer of the electrons between quantum states.

## ACKNOWLEDGMENTS

The author wishes to thank Dr. J. M. Pond for many helpful comments and suggestions. The technical efforts of N. F. Gardner and X. C. Liu are also appreciated. This work was supported by the Office of Naval Research.

# REFERENCES

1.  R. Tsu and L. Esaki, Tunneling in a finite superlattice, Appl. Phys. Lett. 22:562 (1973).
2.  T. C. L. G. Sollner, P. E. Tannenwald, D. D. Peck, and W. D. Goodhue, Quantum well oscillators, Appl. Phys. Lett. 45:1319 (1984).
3.  H. Morkoç, J. Chen, U. K. Reddy, T. Henderson, and S. Luryi, Observation of a negative differential resistance due to tunneling through a single barrier into a quantum well, Appl. Phys. Lett. 49:70 (1986).
4.  A. Kastalsky and S. Luryi, Novel real-space hot-electron transfer devices, Electron Device Lett. EDL-4:334 (1983).
5.  H. Albrecht, Observation of negative differential resistance in $In_{0.53}Ga_{0.47}As/InP:Fe$ JFETS's, Electron. Lett. 20:930 (1984).
6.  A. Kastalsky, J. H. Abeles, R. Bhat, W. K. Chan, and M. A. Koza, High-frwquency amplification and generation in charge injection devices, Appl. Phys. Lett. 48:71 (1986).
7.  K. Hess, H. Morkoç, H. Shichijo, and B. G. Streetman, Negative differential resistance through real-space electron transfer, Appl. Phys. Lett. 35:469 (1979).
8.  P. D. Coleman, J. Freeman, H. Morkoç, K. Hess, B. G. Streetman, and M. Keever, Demonstration of a new oscillator based on real-space transfer in heterojunctions, Appl. Phys. Lett. 40:493 (1982).
9.  P. J. Price, Mesostructure electronics, IEEE Trans. Electron Devices 28:911 (1981).
10. B. K. Ridley, The electron-phonon interaction in quasi-two dimensional semiconductor quantum well structures, J. Phys. C. 15:5899 (1982).
11. S. W. Kirchoefer, R. Magno, and J. Comas, Negative differential resistance at 300K in a superlattice quantum state transfer device, Appl. Phys. Lett. 44:1054 (1984).
12. J. M. Pond, S. W. Kirchoefer, and E. J. Cukauskas, Microwave amplification to 2.5 GHz in a quantum state transfer device, Appl. Phys. Lett. 47:1175 (1985).
13. S. W. Kirchoefer, Occupation of quantum states determined by energy storage in superlattice quantum state transfer devices, Appl. Phys. Lett. 57:1143 (1990).
14. K. Yamaguchi, S. Asai, and H. Koder, Two-dimensional numerical analysis of stability criteria of GaAs FET's, IEEE Trans. Electron Devices ED-23:1283 (1976).
15. T. A. Fjeldly and J. S. Johannessen, Negative differential resistance in GaAs MESFET's, Electron. Lett. 19:649 (1983).
16. S. W. Kirchoefer and H. S. Newman, Superlattice doping interfaces, Superlattices and Microstructures 4:87 (1988).
17. K. Fujiwara, K. Kawashima, and K. Kobayashi, Electro-optical bistability in strained $In_xGa_{1-x}As/Al_{0.15}Ga_{0.85}As$ multiple quantum wells, Appl. Phys. Lett. 57:2234 (1990).
18. S. H. Park, J. F. Morhange, A. D. Jeffery, R. A. Morgan, A. Chavez-Pirson, H. M. Gibbs, S. W. Koch, and N. Peyghambarian, Measurements of room-temperature band-gap-resonant optical nonlinearities of GaAs/AlGaAs multiple quantum wells and bulk GaAs, Appl. Phys. Lett. 52:1201 (1988).
19. J. Feldmann, E. Gobel, and K. Ploog, Ultrafast optical nonlinearities of type II $Al_xGa_{1-x}As/AlAs$ multiple quantum wells, Appl. Phys. Lett. 57:1520 (1990).
20. E. S. Snow, S. W. Kirchoefer, and O. J. Glembocki, Spectrally-Dependent Photocurrent Measurements in n+-i-n+ Heterostructure Devices, Appl. Phys. Lett. 54:2023 (1989).

# ELECTRONIC TRANSPORT IN A LATERALLY
# PATTERNED RESONANT STRUCTURE

M. Cemal Yalabık

Department of Physics
Bilkent University
06533 Ankara, Turkey

## INTRODUCTION

Non-linearities in electron transport through a mesoscopic structure due to quantum effects have generated considerable scientific interest[1], and some of these structures may indeed have potential technological applications. The resonant tunneling diode is one such structure. In this device, one obtains strong nonlinearities in the current-potential difference relationship due to the change in the energy of the incident electron states with respect to a resonant state.

In this study, the current-potential difference relationship in a different type of geometry is considered: This geometry still has nonlinear characteristics, not due to the existence of a barrier confinement, but due to the lateral patterning of the electron flow path. The geometry of the structure is shown in Figure 1. This system will be analyzed with the assumption that the electrons confined to a plane (parallel to the plane of the figure) are restricted from leaving the patterned area by an infinite well potential. First, the results for the case of non-interacting electrons will be reported, which can be presented in scale independent form. Then the results with the inclusion of a self consistent potential, for a structure patterned on a GaAs substrate with specific dimensions will be reported.

## TRANSPORT CALCULATION

Since potential is constant inside the pattern, and infinitely large outside it, the analysis of the current carrying states of the associated single particle Hamiltonian is straightforward: The wave-function in all regions is a superposition of separated functions of the form $\Psi_n(x,y) = \psi_n(x)\phi_n(y)$ with

$$\psi_n(x) = e^{\pm \imath k^{(n)} x} \quad \text{and} \quad \phi_n(y) = \sqrt{\frac{2}{w}} \sin \frac{n\pi}{w} y \tag{1}$$

*Negative Differential Resistance and Instabilities in 2-D Semiconductors*
Edited by N. Balkan *et al.*, Plenum Press, New York, 1993

171

where $w$ is the appropriate width of the channel in region I, II, or III in Figure 1. The linear combinations of these functions and their $x$-derivatives at the boundaries of these regions are then matched. The energy $E$ of the corresponding state is related to the wavenumber $k$ through the relation $E = \hbar^2/(2m^*w^2)[n^2 + (kw)^2]$, with $\hbar$ the Plank constant and $m^*$ the effective mass. Quantization in the $y$ direction leads to overlapping bands of energy, with which we associate the index $n$. (The results reported in this study were obtained with the inclusion of 6 bands in regions I and III, and 12 bands in region II.) It was assumed that the confinement in the direction perpendicular to the plane of the pattern is sufficiently large so that quantization in this direction leads to bands well separated in energy.

We will assume that for relatively small potential differences across the structure, the influx of electrons from both sides can be represented by the appropriate Fermi distributions, with Fermi energies differing by an amount corresponding to the potential difference. The current in one direction through the structure can then be calculated[2] by summing over the current carried by each state, labelled by a band and a wavenumber $k$:

$$I_{\text{I}\rightarrow\text{III}} = \sum_n \frac{e\hbar}{\pi m^*} \int_{k_{\text{minI}}^{(n)}}^{\infty} dk \; k \; T^{(n)}(k) F_{\text{I}}^{(n)}(k) \tag{2}$$

where $n$ is the band index, $k_{\text{minI}}$ is the minimum value of the wavenumber of an electron incident from region I that has available to it a corresponding state with same energy in region III (a limit that may be imposed by the existence of a conduction band edge at the contacts to the structure), $e$ is the electronic charge, $T(k)$ is the transmission coefficient of electrons with incident wave number $k$, and $F(k)$ is the Fermi distribution. The superscript $(n)$ indicates that the associated quantity depends on the band. The total current is then the difference of two quantities $I_{\text{I}\rightarrow\text{III}} - I_{\text{III}\rightarrow\text{I}}$ as computed by Eqn. 2 for transport in two directions.

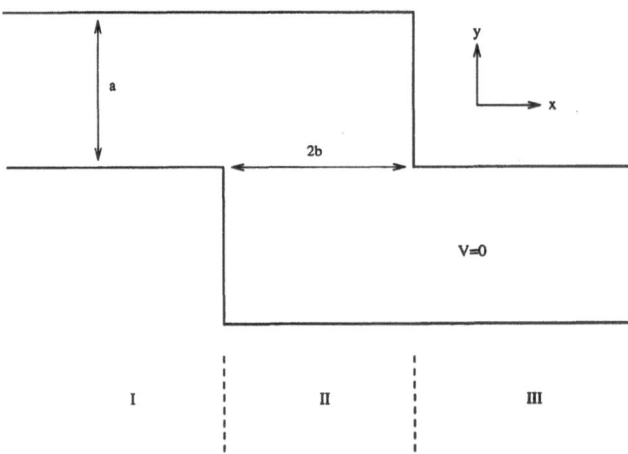

**Figure 1.** The geometry of the structure discussed in this work. The potential outside the kink shaped region is infinitely large.

Figure 2 shows the transmission coefficient $T^{(1)}(k)$ as a function of $ka$ for a ratio of the structure size parameters $b/a = 0.6$. There are strong resonance effects, due to the bouncing of the electron wave-function inside region II. Because of the richness of the

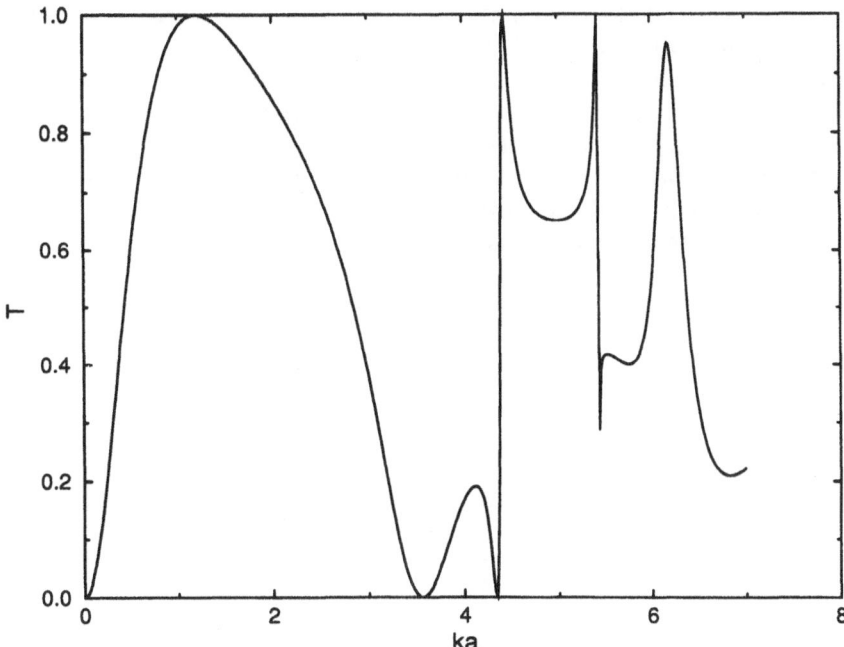

**Figure 2.** The total transmission coefficient for electrons incident from the first band as a function of the scaled wave-number.

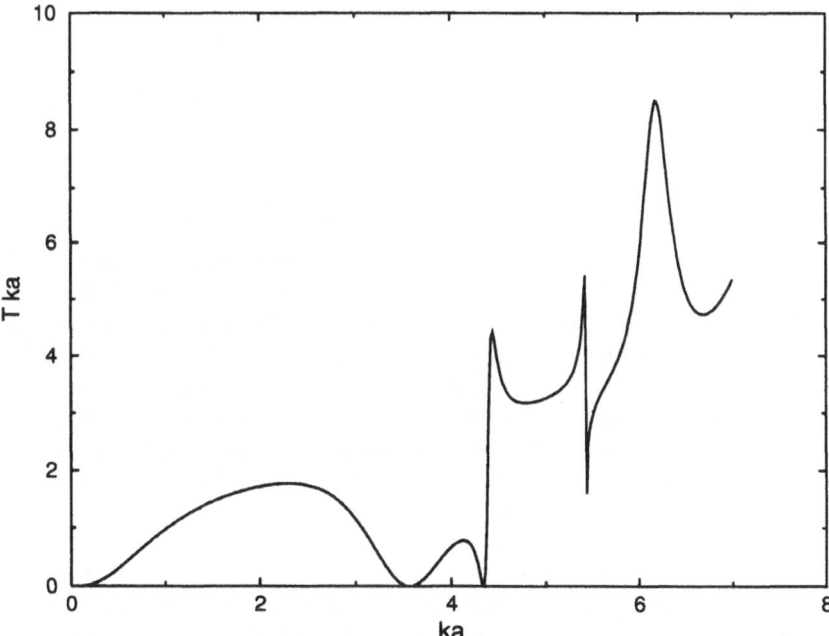

**Figure 3.** The wavenumber times the sum of the transmission coefficients from all bands as a function of the scaled wave-number. The wavenumbers in all of the band have been referred to the one in the lowest energy band.

resonance structure, and the complications introduced by the existence of bands, the utility of this geometry is not trivially evident. Note that the presence of a gap in $T(k)$ could result in a decrease in the current as the wavenumbers of the incident electrons take values within this gap. This may then lead to a negative differential resistance. Figure 3 shows that a plot of $\sum_n T^{(n)}(k)k$, does have a gap for values of $b/a \approx 0.6$ . (In Figure 3, the wave-number $k^{(n)}$ in the $n$'th band has been referred to the wave-number with the same energy in band one by the relation $k^2 = k^{(n)^2} - 2m^*\Delta E/\hbar^2$ , where $\Delta E$ is the energy difference of the two bands.)

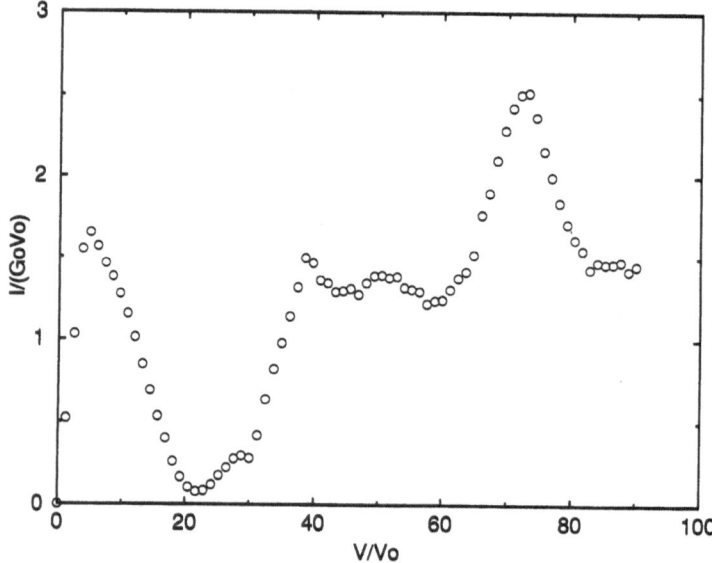

**Figure 4.** The current in the structure as a function of the potential difference. $G_0 = 2e^2/h$ is the universal conductance.

If the potential due to the presence of charges in the structure is neglected, for a Fermi energy of $E_F = 10eV_0$ the current as a function of the potential difference is as shown in Figure 4. The electron distributions on both sides are taken to correspond to a zero temperature Fermi distribution, with allowed values of the scaled energy between 0 and 10 (in units of $eV_0 \equiv \hbar^2/2a^2m^*$) relative to the lowest energy current carrying state in the structure. A pronounced negative differential resistance effect is apparent in the figure. Although the higher potential difference portion of this figure will not probably be relevant to an actual implementation, it was included to demonstrate the related charging effects in this region.

The practical implementation of this structure necessitates the fabrication of geometric feature sizes comparable to the quantum mechanical wavelength of the carriers. For the value of $a = 10$nm and an effective mass of 0.067 times the bare mass of the electron (corresponding to GaAs), the potential difference scale shown in Figure 4 is set by the value of $V_0 = 5.7$mV. (The qualitative form of the dependence of current on the potential difference does not change appreciably for a Fermi distribution corresponding to a temperature of 300K.) It is of course not obvious that the above model contains a sufficient amount of the physics of an actual device constructed on GaAs so that the current-potential difference dependence shown in Figure 4 can be assumed to be realistic. A number of factors that would affect the performance of the actual structure would be the Coulomb interaction between the electrons, the shape of the potential profile corresponding to the walls, and the actual shape of the geometry that ends up on the wafer. A rough estimate of the effects of the Coulomb interaction will now be made, which probably is the most significant factor that needs to be considered.

The change in the potential inside the channel due to the existence of electrons can be taken into consideration using a self consistent approximation. This complicates the problem considerably, and the results of such a computation in progress will be reported separately. One can, however, make an estimate of the change in the potential by using the charge densities found in the non-interacting model.

At this point, one has to take into account that this structure can have a number of bound states, in fact for the dimensions leading to Figure 4, there is just one such state, with the possibility of trapping two electrons. Note that due to the relatively large feature sizes in this structure relative to typical resonant tunneling devices, the charging effects would be expected to be smaller. In fact, with an electron trapped in the structure with the given dimensions, the change in the potential can be estimated by assuming that the electrons are confined to a disk of radius $R$. This value is then

$$\Delta V = \frac{e^2}{2\pi \epsilon R}. \tag{3}$$

Substituting the permittivity $\epsilon = 12.5\epsilon_0$ for GaAs and an estimated value of $R = 10$nm, one obtains $\Delta V = .023$V $\approx 4V_0$. In comparison to the potential scale of Figure 4, this is an appreciable amount of potential difference.

## SELF CONSISTENT CALCULATION OF CHARGING

In order to include, admittedly in a somewhat crude fashion, the effects of the Coulomb interaction, it will be assumed that the potential is fairly constant in a region which includes the I/II and II/III boundaries of the structure. (A computation of the potential distribution corresponding to charge densities under various types of conditions obtained with the above model indicates that the potential profiles are fairly smooth, and quite flat in the central region.) It will further be assumed that electrons incident from both sides will be transmitted without reflection to the central region if

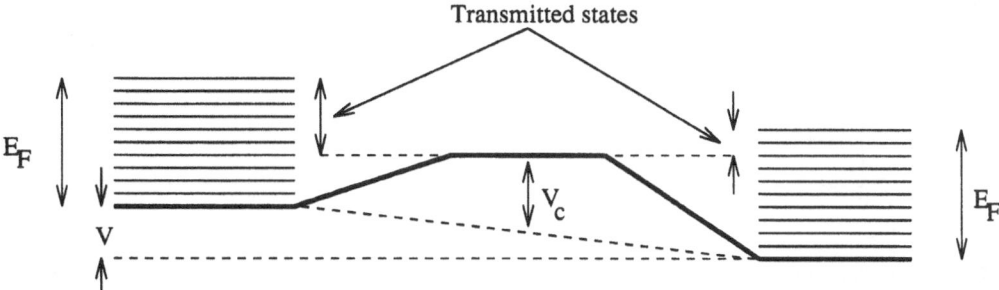

**Figure 5.** The model used to include a self consistent potential. The boundaries between the regions I/II and II/III are assumed to be in the central flat part. $V_c$ is the potential of this part, determined from the charge in region II.

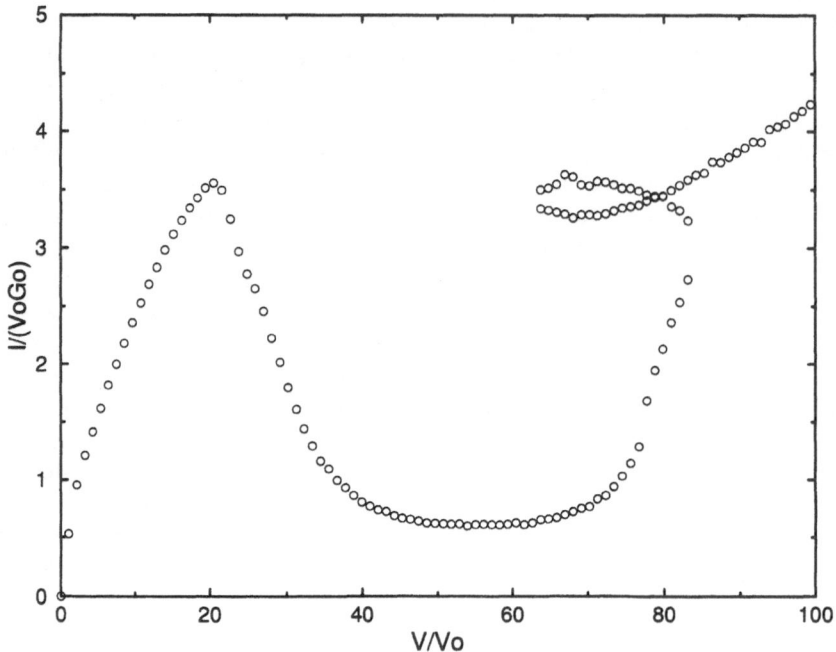

**Figure 6.** The current in the structure as a function of the potential difference.

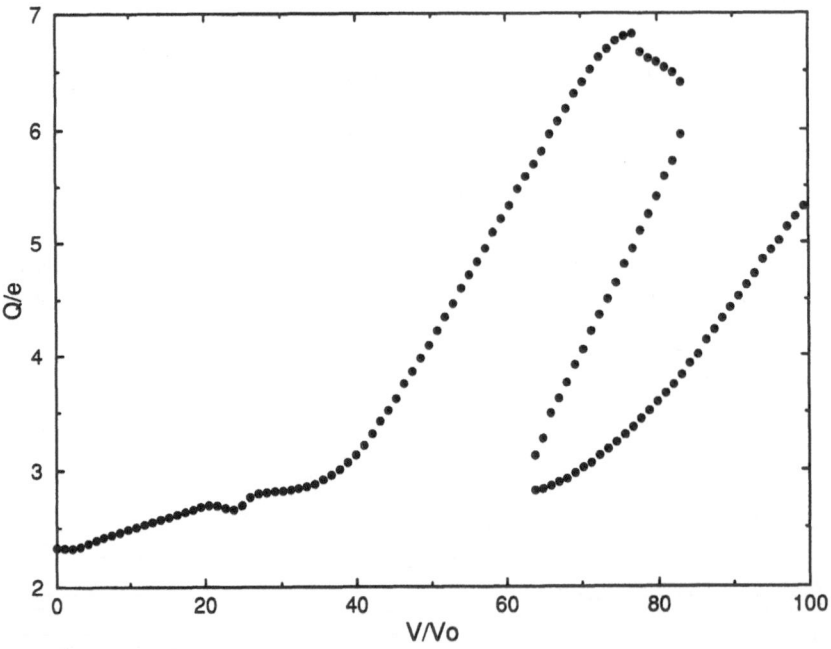

**Figure 7.** The charge in the central region of the structure as a function of the potential difference.

they have sufficient energy to get there, otherwise they will be totally reflected. Similarly, an electron leaving the central region will be assumed to reach the corresponding contact if there are available to it states with same energy. The potential of the central region is then determined self consistently, based on the charge present in region II. The new model is demonstrated in Figure 5. Although this is a somewhat crude modeling of the Coulomb interaction, it would still be expected to give a good qualitative description of the effects produced by it.

Figure 6 shows the dependence of the current on the potential difference. Note that there is more than one solution to the self consistency condition at higher energies, resulting in a hysteresis. (Only the two extremal solutions out of the three indicated on the figure is expected to be stable.) This region corresponds to a resonance of the structure (that can also be identified in Figure 1), and the phenomena is in principle the same as those seen in resonant tunneling devices. Note however that there is no such effect corresponding to the negative resistance region at lower energies, since this cannot be associated with a resonance of the same type. In fact, the charge in the structure shows no anomaly at these potential differences as evidenced in Figure 7.

## CONCLUSIONS

An analysis of transport in a laterally patterned structure indicates that negative differential resistance and (with the inclusion of self consistent treatment of charging effects) hysteresis can be observed in such geometries. It was also shown that negative differential resistance may appear due to gaps in the transmission at low electron energies, with very little charging effect. This results in qualitatively different characteristics when charging effects are included in the analysis. The author acknowledges many fruitful discussions with Prof. V. A. Kochelap.

## REFERENCES

1. For example, see Y. Takagaki and D.K. Ferry, Double quantum point contacts in series, *Phys. Rev.* B45:13494 (1992) and the references therein.
2. L. Landauer, Spatial variations of currents and fields due to localized scatterers in metallic conduction, *IBM J. of Res. and Dev.* 1:223 (1957).

# TRAVELLING DOMAINS IN MODULATION-DOPED
# GaAs/AlGaAs HETEROSTRUCTURES

R. Döttling and E. Schöll

Institut für Theoretische Physik
Technische Universität Berlin
Hardenbergstr.36
D-1000 Berlin 12
Germany

## ABSTRACT

We study charge transport parallel to the layers of a modulation doped $GaAs/Al_xGa_{1-x}As$ heterostructure far from thermodynamic equilibrium. An applied electric field leads to the heating of the high-mobility electrons in the $GaAs$ channel and to real space transfer of hot electrons into the low-mobility $Al_xGa_{1-x}As$ layer. The transport processes perpendicular and parallel to the layers are modelled by a system of nonlinear partial differential equations for the carrier densities, the potential barrier between the two layers, and the electric field. In the monostable regime the existence of stable solitary waves of the electric field and the carrier distribution is found using methods of singular perturbation theory. These correspond to travelling field domains and pulses of real-space transferred electrons.

## INTRODUCTION

We consider parallel transport in a modulation-doped $GaAs/Al_xGa_{1-x}As$ heterostructure, as schematically shown in Fig.1. The $Al_xGa_{1-x}As$ layer is heavily n-doped with donor density $N_D$. Between the conduction bands in the two layers there appears a band edge discontinuity $\Delta E_c$ as sketched in Fig.1b.

At low bias $U_0$ the electrons reside in the $GaAs$ channel, where they are separated from their parent donors and thus experience strongly reduced impurity scattering. Therefore the mobility $\mu_1$ of the electrons in the $GaAs$ will be high. A small electric field $\mathcal{E}_\parallel$ ($< 1\,kV/cm$) parallel to the layer interface will result in a current essentially due to the electrons in the $GaAs$ layer, because the carrier

*Negative Differential Resistance and Instabilities in 2-D Semiconductors*
Edited by N. Balkan *et al.*, Plenum Press, New York, 1993

179

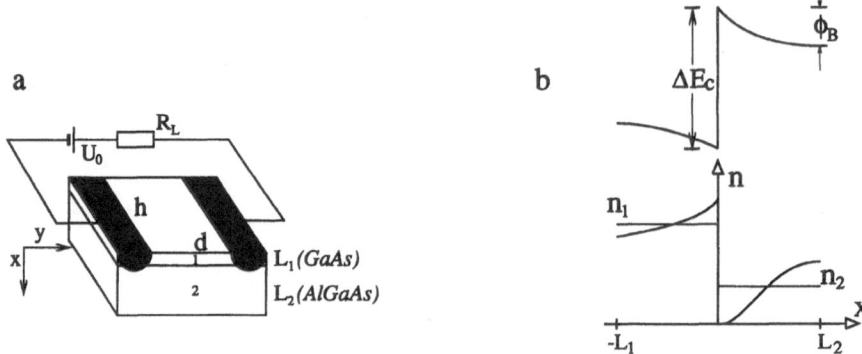

**Figure 1.** (a) Schematic sample and circuit configuration of a modulation doped $GaAs/Al_xGa_{1-x}As$ heterostructure with heterolayer widths $L_1$ and $L_2$, respectively, and lateral dimensions $h$, $d$. (b) Energy-band diagram (top) and carrier density (bottom) versus the perpendicular coordinate $x$ of the heterolayer (schematic).

density and mobility are much larger in the $GaAs$ channel than in the $Al_xGa_{1-x}As$ layer. A high electric field ($\approx 2\,kV/cm$) induces carrier heating. If the electrons gain enough kinetic energy, thermionic emission across the barrier into the $Al_xGa_{1-x}As$ layer is possible, where their mobility, $\mu_2$, is much lower due to strongly enhanced impurity scattering. The low-mobility carriers in the $Al_xGa_{1-x}As$ layer are not heated significantly since the power input $e\mu_2\mathcal{E}_{\parallel}^2$ is small. This real-space transfer of hot electrons from the high-mobility to the low-mobility layer causes an N-shaped current-voltage characteristic with a regime of negative differential conductivity (NNDC)[1–8] in anology with the intervalley transfer in the Gunn-Hilsum effect. However, the physical mechanism is different, because the electrons are transferred in real space rather than in momentum-space, and the mobility in the two states is different because of different scattering times rather than because of different effective masses. The switching speed is much higher due to the fast thermionic emission time ($\approx 10^{-12}s$). Also the static and dynamic characteristics can be easily controlled over a wide range by varying the material parameters such as the layer widths, the doping concentration and the aluminium fraction (which determines the band-edge discontinuity $\Delta E_c$).

Spontaneous current oscillations of $2 - 200\,MHz$ have been experimentally observed under ac-driven[9] and dc-driven[10,11] conditions. A physical mechanism of a real-space transfer oscillator which gives hysteretic switching transitions between oscillatory and stationary states[12] and periodic or chaotic self-generated oscillations at much higher frequencies ($20 - 100\,GHz$) has recently been investigated.[6,13,14] This analytic approach is based on the coupled nonlinear dynamics of real space electron transfer, the dielectric relaxation of the electric field and of the space charge in the $Al_xGa_{1-x}As$ layer, which controls the interface potential barrier $\Phi_B$. Static domains have also been considered.[15]

The aim of this paper is to study the bifurcation of spatio-temporal structures occurring in the case of NNDC due to real-space transfer of hot electrons in a semiconductor heterostructure. The separation of the time scale of dielectric relaxation from the much smaller time-constant of electron-transfer out of the $GaAs$ channel leads to an excitable wave-supporting medium. This means that small perturbations are rapidly damped out, but suprathreshold disturbances trigger an abrupt response. Thereby stable solitary waves with a constant speed can be excited. These solutions are transformed at their periphery into the stable uniform state of the system. For the mathematical treatment of the travelling wave solutions it is important to note that they correspond to well-defined trajectories in the phase space of dynamical variables in a moving coordinate frame. In the phase space a solitary pulse corresponds to a homoclinic orbit (saddle-to-saddle loop), and a periodic pulse train is represented by a limit cycle. The abruptness of the response can be exploited by the methods of singular perturbation theory to obtain a mathematical description of the wave propagation.

In our simple analytical model we describe the internal state of the semiconductor by a set of local variables, whose time evolution is given by nonlinear transport equations coupled to Maxwell's equations.

## MODEL

On a mesoscopic level the semiconductor heterostructure is described by the following variables: the density of the carriers $n(\vec{x}, t)$, the mean carrier momentum $\vec{p}(\vec{x}, t)$ and the mean carrier energy $E(\vec{x}, t)$.[16,17] These variables obey the equation of continuity

$$\frac{\partial n}{\partial t} + \vec{\nabla} \cdot (n\vec{v}) = 0 \tag{1}$$

and the transport equations

$$\frac{\partial \vec{p}}{\partial t} + (\vec{v} \cdot \vec{\nabla})\vec{p} + \frac{1}{n} \vec{\nabla} (nk_B T_e) + e\vec{\mathcal{E}} = -\frac{\vec{p}}{\tau_m} \tag{2}$$

$$\frac{\partial E}{\partial t} + (\vec{v} \cdot \vec{\nabla})E + \frac{1}{n} \vec{\nabla} (nk_B T_e \vec{v}) + e\vec{v} \cdot \vec{\mathcal{E}} = -\frac{(E - E_L)}{\tau_E} \tag{3}$$

Here $\tau_m, \tau_E$ are the momentum and the energy relaxation time, respectively, and $E_L$ is the energy of the carriers at thermal equilibrium. If the band structure is non-degenerate and isotropic parabolic with an effective mass $m^*$, the mean momentum and energy are related to the mean group velocity $\vec{v}$ and the electron temperature $T_e$ by $\vec{p} = m^*\vec{v}$ and $E = (1/2)m^*\vec{v}^2 + (3/2)k_B T_e$.

The transport equations are coupled to the local electric field via Maxwell's inhomogeneous equations

$$\varepsilon_0 \varepsilon_s \vec{\nabla} \cdot \vec{\mathcal{E}} = \rho \tag{4}$$

with the local charge density $\rho = e(N_D - n)$ ($N_D$ is the donor density and $\varepsilon_0, \varepsilon_s$ are the absolute and relative permittivity, respectively) and

$$\vec{\nabla} \times \vec{H} = \varepsilon_0 \varepsilon_s \frac{\partial \vec{\mathcal{E}}}{\partial t} + \vec{j} = \vec{j}_{tot}(\vec{x}, t) \tag{5}$$

with conduction current density $\vec{j} = -en\vec{v}$ and magnetic field $\vec{H}$.

By forming the divergence $\vec{\nabla} \cdot (\vec{\nabla} \times \vec{H})$ it follows that the total current density $\vec{j}_{tot}(\vec{x}, t)$ including displacement current and conduction current is divergence-free. This is a form of Ampère's law which estabilishes that for transport problems with one spatial coordinate, the sum of displacement current and the conduction current at any point of space is to be equal to a constant $\vec{j}_{tot}(t)$.

Typically, momentum relaxation occurs faster than other processes, so that $\vec{p}$ can be eliminated adiabatically from (2) setting $\partial \vec{p}/\partial t + (\vec{v} \cdot \vec{\nabla})\vec{p} = 0$. Introducing the mobility $\mu = e\tau_m/m^*$ and the diffusion coefficient $D$ via Einstein's relation $eD = \mu k_B T_e$, the current density is given by the drift-diffusion equation:

$$\vec{j} = en\mu\vec{\mathcal{E}} + eD\vec{\nabla}n \quad . \tag{6}$$

In the following we derive a reduced set of nonlinear partial differential equations for the carrier densities $n_1, n_2$ in the two layers, the dielectric relaxation of the applied lateral field $\mathcal{E}_\parallel$ and the potential barrier $\Phi_B$ as a function of the lateral coordinate $y$ (the direction parallel to the layer interface) and the time $t$. This can be achieved by integrating the state variables over the space coordinate $x$ in transverse direction and neglecting spatial dependence in the $z$ direction. The spatially averaged carrier density in the $GaAs$ layer $n_1(y,t) \equiv \int_{-L_1}^{0} n(x,y,t)\,dx/L_1$ as a function of time and the lateral coordinate $y$ is governed by the averaged continuity equation

$$\frac{\partial n_1}{\partial t} = \frac{1}{eL_1}(J_{1\to2} - J_{2\to1}) + \mu_1\frac{\partial}{\partial y}(n_1\mathcal{E}_\parallel) + D_1\frac{\partial^2 n_1}{\partial y^2} \tag{7}$$

where we have assumed that the carrier concentration and the parallel electric field are smoothly varying functions in the direction perpendicular to the layer interface. This is correct only within a finite distance from the heterojunction barrier which is comparable to the mean free distance of collisionless flight of the electrons. Within this finite length scale it is possible to express the drift current in (6) approximately as the product of the carrier concentration $n_1$ in the $GaAs$ channel and the electric field $\mathcal{E}_\parallel$ parallel to the semiconductor layers. Away from the semiconductor barrier the average velocity of the carriers in the $x$ direction is much smaller due to the enhanced collisions with phonons and the barrier and the assumption of constant carrier concentration in transverse direction breaks down.[18]

To simplify matters we have further assumed the mobility $\mu_1$ and the diffusion coefficient $D_1$ in the $GaAs$ to be independent of the other state variables. In (7)

$$J_{1\to2} \equiv -en_1\sqrt{\frac{E_1}{3\pi m_1^*}}\exp\left(-\frac{3\Delta E_c}{2E_1}\right) \tag{8}$$

$$J_{2\to1} \equiv -en_2\sqrt{\frac{E_2}{3\pi m_2^*}}\exp\left(-\frac{3\Phi_B}{2E_2}\right) \tag{9}$$

are the thermionic-emission current densities ($m_i^*$ are the effective masses) given by Bethe's theory,[18] and $E_i = (3/2)k_B T_i$ ($i = 1,2$) are the mean carrier energies given by the carrier temperatures $T_i$. The thermionic emission theory corresponds to the following physical picture. Electrons in the $GaAs$ with energy less than $\Delta E_c$ cannot propagate into the adjacent $Al_xGa_{1-x}As$ layer; all electrons with higher energy are emitted across the barrier without collisions. This assumption is correct only within a certain regime that is of the order of the mean free paths of electrons. If the $GaAs/Al_xGa_{1-x}As$ layers are wider, the thermionic emission current represents the current only close to the interface, and perpendicular diffusion (transverse dissipation of carriers in the $Al_xGa_{1-x}As$ layer) will play a major role. Intervalley transfer has been shown[19] to be negligible compared to real-space transfer at electric fields $\leq 4kV/cm$. This means that the transport processes are dominated by the $\Gamma$-valley, if real-space transfer occurs in the low-field range. At higher electric fields ($\approx 8kV/cm$) intervalley scattering will become more important. Nevertheless in Monte-Carlo simulations reasonable agreement is found between calculations which include the effects of $k$-space transfer in the $GaAs$ channel and the surrounding $Al_xGa_{1-x}As$ layer, and those which neglect transfer into higher valleys.[19] Quantum effects like the quantum-transmission coefficient or tunneling through the barrier are also disregarded, cf discussion in Ref.[20]. Size quantization effects, which arise if the layer widths are smaller than 100 Å, are also neglected, since the current-voltage characteristic is not essentially affected by the quasi-two dimensional subbands below the barrier, except that the critical field for the onset of real-space transfer is shifted to higher values.[3]

In conclusion, the thermionic emission theory is a simplified approach for the interaction of the semiconductor layers at high electric fields. Nevertheless, we expect that our model correctly describes the qualitative features of the complex transport phenomena.

To proceed in deriving a reduced set of local variables we average Poisson's equation together with the boundary conditions for the perpendicular electric field $\mathcal{E}_\perp(-L_1, y, t) = \mathcal{E}_\perp(L_2, y, t) \equiv 0$. This yields

$$\varepsilon_0 \varepsilon_s \frac{\partial \mathcal{E}_\|}{\partial y} = \frac{e}{L_1 + L_2} ((N_D - n_2)L_2 - n_1 L_1) \tag{10}$$

for the spatially averaged electric field $\mathcal{E}_\|(y, t)$ parallel to the layer interface, with $n_2(y, t) \equiv \int_0^{L_2} n(x, y, t) \, dx / L_2$.

The energy transfer between the heterolayers is described by the energy-balance equations containing Joule heating, convective, diffusive, and electron-pressure induced heat flow, and energy loss due to polar-optical-phonon scattering.[6] Adiabatic elimination of the mean energy and neglecting spatial derivatives yields in a first order approximation

$$E_1 \approx E_L + \tau_{E_1} e \mu_1 \mathcal{E}_\|^2 \ , \quad E_2 \approx E_L + \tau_{E_2} e \mu_2 \mathcal{E}_\|^2 \ , \tag{11}$$

with the thermal equilibrium energy $E_L = (3/2)k_B T_L$ and the energy relaxation time $\tau_{E_1}$ in the GaAs-channel and $\tau_{E_2}$ in the $Al_x Ga_{1-x} As$ layer. This simplified approach is sufficient to gain physical insight into the mechanism of real-space induced spatio-temporal instabilities.

The dielectric relaxation of the parallel electric field is given by

$$\varepsilon_0 \varepsilon_s \frac{\partial \mathcal{E}_\|}{\partial t} = j_\| - \frac{1}{L_1 + L_2} \left( (e n_1 \mu_1 L_1 + e n_2 \mu_2 L_2) \mathcal{E}_\| + e D_1 L_1 \frac{\partial n_1}{\partial y} + e D_2 L_2 \frac{\partial n_2}{\partial y} \right) \ . \tag{12}$$

This equation can be obtained by averaging the parallel component of eq. (5). In (12) $j_\|(t) \equiv \int_{-L_1}^{L_2} j_{tot\|}(\vec{x}, t) \, dx / (L_1 + L_2)$ is a first integral equal to the current density flowing through the external circuit (the displacement current in a metallic wire is negligible) and is determined by the particular external circuit used in the experiment. Under current bias $j_\|$ is a fixed constant (control parameter). For a purely resistive circuit $j_\|(t)$ is a function of time determined by Kirchhoff's law such that the applied bias voltage $U_0 = \mathcal{E}_0 d$ is a constant: $j_\|(t) = \sigma_L (\mathcal{E}_0 - \int_0^d \mathcal{E}_\|(y, t) \, dy / d)$ where $d$ is the sample length, $\sigma_L \equiv d/(A R_L)$ is connected to the load resistance $R_L$, and $A$ is the cross section of the current flow.

As shown in Ref.[13] the dynamics of the interface potential barrier $\Phi_B(y, t) \equiv -e \int_0^{L_2} \mathcal{E}_\perp(x, y, t) \, dx$ is governed by the space-charge dynamics in the $Al_x Ga_{1-x} As$ layer and the resulting internal electric field $\mathcal{E}_\perp(x, y, t)$. Integrating the perpendicular component of (5) and using Poisson's equation (10) yields

$$\frac{\partial \Phi_B(y, t)}{\partial t} = -\frac{e \mu_2 N_D}{\varepsilon_0 \varepsilon_s} \Phi_B + \frac{e \mu_2}{2} \left( \frac{e}{\varepsilon_0 \varepsilon_s} (N_D - n_2)L_2 - L_2 \frac{\partial \mathcal{E}_\|}{\partial y} \right)^2 + \mu_2 \Phi_B \frac{\partial \mathcal{E}_\|}{\partial y} \tag{13}$$

183

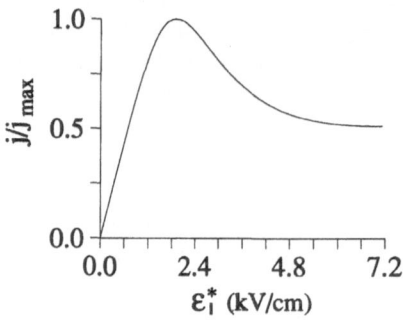

**Figure 2.** Static current density-field characteristic as a function of the static electric field $\mathcal{E}_{\parallel}^{*}$.

where we have neglected the diffusive contributions in transverse directions, and the integral $\int_0^{L_2} j_{tot_\perp}(x, y, t)\,dx$, which is appropriate if $L_2$ is less or comparable to the mean free path of the electrons. The steady uniform state (denoted by an asterisk) is given by $\Phi_B^* = e^2 L_1^2 n_1^{*2}/(2\varepsilon_0\varepsilon_s N_D) = e^2 N_D L_w^2/(2\varepsilon_0\varepsilon_s)$ which corresponds to the depletion approximation in the $Al_xGa_{1-x}As$ layer within the effective depletion width $L_w = n_1^* L_1/N_D$.

The equations (7), (10), (12) and (13) represent a simplified, but analytically tractable description of the complex transport phenomena between the semiconductor layers at high electric fields. These equations have to be solved on a segment $0 < y < d$, with suitable boundary conditions that depend on the nature of the metal-semiconductor contact. Solving the equations plus initial and boundary conditions and current bias constitutes a mathematical problem which is called the *direct* problem.[21]

## TRAVELLING DOMAINS

Due to the great number of parameters appearing in the equations it is more convenient to write the dynamical equations in a non-dimensional form. We introduce the dimensionless variables $\tilde{n}_i \equiv n_i/N_D$, $\tilde{\mathcal{E}}_{\parallel} \equiv \mathcal{E}_{\parallel}(\mu_1/v_{ds})$, $\tilde{j}_{\parallel} \equiv j_{\parallel}/(eN_D v_{ds})$, $\tilde{E}_i \equiv E_i/\Delta E_c$, $\tilde{\Phi}_B \equiv \Phi_B/\Delta E_c$, $\tilde{t} \equiv t/\tau_r$, $\tilde{y} \equiv y/L_D$ and the dimensionless parameters $l \equiv L_1/L_2$, $\gamma \equiv \tau_0/\tau_r$, $\lambda^2 \equiv \mu_2/\mu_1$, $\beta \equiv \sqrt{m_1^*/m_2^*}$, $\kappa \equiv e^2 L_1^2 N_D/(2\varepsilon_0\varepsilon_s\Delta E_c)$, $\sigma \equiv (h(L_1 + L_2)R_L e\mu_1 N_D/d)^{-1}$, $\delta \equiv \lambda\tau_r v_{ds}/L_D$ where $v_{ds}$ is the drift saturation velocity, $L_D \equiv \sqrt{\varepsilon_0\varepsilon_s k_B T_L/(e^2 N_D)}$ is the Debye length, $\tau_r \equiv \varepsilon_0\varepsilon_s/(e\mu_2 N_D)$ is a dielectric relaxation time and $\tau_0 \equiv (3\pi m_1^*/\Delta E_c)^{1/2} L_1 \exp(3/2)$ is an effective thermionic emission time.[6] Omitting the tilda we get

$$\frac{\partial n_1}{\partial t} = \frac{1}{\gamma}(\beta g_2(\mathcal{E}_{\parallel}, \Phi_B)n_2 - g_1(\mathcal{E}_{\parallel})n_1) + \delta\frac{\partial}{\partial y}(n_1\mathcal{E}_{\parallel}) + \frac{\partial^2 n_1}{\partial y^2} \tag{14}$$

$$\frac{\partial \Phi_B}{\partial t} = -\Phi_B + \frac{\kappa}{l^2}\left(\delta\lambda^2\frac{\partial \mathcal{E}_{\parallel}}{\partial y} + n_2 - 1\right)^2 + \delta\lambda^2\,\Phi_B\frac{\partial \mathcal{E}_{\parallel}}{\partial y} \tag{15}$$

$$\lambda^2\frac{\partial \mathcal{E}_{\parallel}}{\partial t} = j_{\parallel}(t) - \frac{l}{1+l}\left(n_1 + \lambda^2\frac{(1-ln_1)}{l}\right)\mathcal{E}_{\parallel} + \delta\lambda^4\,\mathcal{E}_{\parallel}\frac{\partial \mathcal{E}_{\parallel}}{\partial y} \tag{16}$$

$$- \frac{l}{1+l}\frac{1-\lambda^2}{\delta}\frac{\partial n_1}{\partial y} + \lambda^4\frac{\partial^2 \mathcal{E}_{\parallel}}{\partial y^2}$$

a                                          b

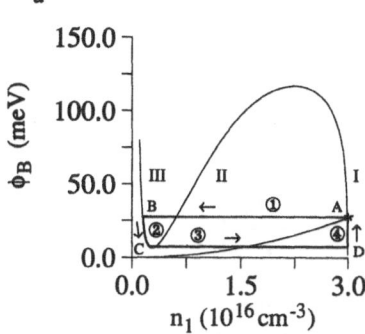
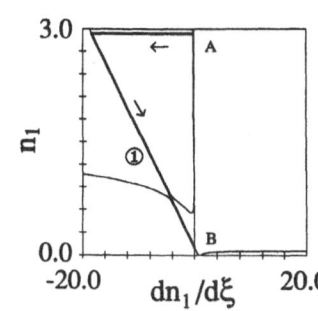

**Figure 3.** (a) Phase plane of potential barrier $\Phi_B$ vs carrier concentration $n_1$. The intersection of the null-isoclines yield one stable uniform steady state (A). For small $\gamma$ the system becomes an excitable medium capable of pulse propagation according to the homoclinic trajectory (1-2-3-4) with the trailing edge corresponding to (1) and the leading edge correspondig to (3). (b) Phase plane of carrier concentration $n_1$ vs $dn_1/d\xi$ ($n_1$ in units of $N_D$; $dn_1/d\xi$ in units of $N_D/L_D$). The heteroclinic orbit for constant $\Phi_B$ is denoted by (1).

$$n_2 = 1 - l n_1 - \delta\lambda^2(1+l)\frac{\partial \mathcal{E}_{\|}}{\partial y} \tag{17}$$

with $g_1(\mathcal{E}_{\|}) \equiv \sqrt{E_1}\exp{(3(1-1/E_1)/2)}$ and $g_2(\mathcal{E}_{\|},\Phi_B) \equiv \sqrt{E_2}\exp{(3(1-\Phi_B/E_2)/2)}$. Using values appropriate for a $GaAs/Al_xGa_{1-x}As$ heterostructure, i.e. $N_D = 10^{16}cm^{-3}$, $\mu_1 = 10^4 cm^2/Vs$, $\mu_2 = 50 cm^2/Vs$, $L_1 = 200$ Å, $L_2 = 600$ Å, we get $\gamma \equiv \tau_0/\tau_r = 0.025$ and $\lambda^2 \equiv \mu_2/\mu_1 = 0.005$.

Our analysis will now be based on the time-scale separation of the dielectric relaxation time compared to the smaller time-constant of electron-transfer out of the $GaAs$ channel, and the large separation of mobilities in the semiconductor layers.

The homogeneous stationary states of the system are obtained by setting the spatial and temporal derivatives in (14)-(17) equal to zero. The resulting static current-density-field characteristic is shown in Fig.2. Real-space transfer of hot electrons leads to an N-shaped characteristic with a regime of negative differential conductivity.[12] In the following we discuss the formation of stable wave solutions of the dimensionless system (14)-(17) along the infinitely extended $y$-direction under current bias. We assume that the dimensionless sample length satifies $d/L_D \gg 1$ and therefore our solutions are expected to represent reasonable approximations of the full problem far from the boundaries. This means that we neglect the influence of the metal-semiconductor contacts at $y = 0$ and $y = d$.

In this paper we will focus on the creation of stable *domains* in the monostationary regime of the current-densitiy-field characteristic, where there is only one stable stationary state under current bias. The formation of stable *kink*-shaped solutions (*fronts*) will be discussed elsewhere.

The spatial profile and velocity of the propagating domains in the $GaAs$ channel will be calculated in the following by using methods of singular perturbation theory.[22] We assume the existence of physical (bounded) propagating solutions and consider only such cases for which the profile of the propagating structure is time independent in a reference frame moving at constant velocity with the profile. Introducing the relative coordinate $\xi = y + tv/\gamma$ into (14)-(17), we obtain a system of ordinary differential equations. The full analysis of (14)-(17) is very complicated even in the limit of well separated time scales $\gamma \ll 1$ and length scales $\lambda^2 \ll 1$. In zeroth order of singular perturbation theory we find:

$$v \frac{dn_1}{d\xi} = \beta g_2(\mathcal{E}_\parallel, \Phi_B)(1 - l n_1) - g_1(\mathcal{E}_\parallel) n_1 \qquad (18)$$

$$v \frac{\Phi_B}{d\xi} = \gamma(-\Phi_B + \kappa n_1^2) \qquad (19)$$

and the electric field is enslaved by $n_1$ and $dn_1/d\xi$:

$$\mathcal{E}_\parallel = \left( \frac{1+l}{l} j_\parallel - \frac{1}{\delta} \frac{dn_1}{d\xi} \right) \left[ n_1 + \lambda^2 \frac{1 - l n_1}{l} \right]^{-1} \qquad (20)$$

(Note the term of the order $\lambda^2$ in (20) must be taken into account in order to obtain in this lowest order the same uniform state as from the full system.) Since $\gamma \ll 1$, we have $|dn_1/d\xi| \gg |d\Phi_B/d\xi|$ except in a neighborhood of the null-isocline:[23]

$$\Phi_B(n_1) = E_2 \left( 1 - \frac{2}{3} \ln \frac{g_1(\mathcal{E}_\parallel) n_1}{\beta \sqrt{E_2}(1 - l n_1)} \right) . \qquad (21)$$

Thus the solution $\Phi_B(\xi) = const$ approximates the flow away from the null-isocline increasingly well as $\gamma \to 0$. Near the null-isocline, and in particular when $|dn_1/d\xi|$ is of the order $\gamma$ both solution components are comparable and hence after entering this boundary layer the trajectories turn and follow *slowly* the null-isocline.

In the monostable regime small disturbances are rapidly damped out, but perturbations with a sufficiently large amplitude can trigger a single pulse or a periodic wave train of pulses. The trajectory in the phase space is then a homoclinic or a periodic orbit. From the $(\Phi_B, n_1)$ phase plane shown in Fig.3a we see that as the leading edge of a pulse passes a given point, the carrier concentration jumps *fast* from state (A) to point (B) on the branch III of the null-isocline at approximately constant $\Phi_B$ in the limit $\gamma \to 0$. For a certain value of the wave velocity $v$ there is a trajectory joining (A) in the limit $\xi \to -\infty$ and (B) in the limit $\xi \to +\infty$, for constant $\Phi_B$. In Fig.3b this orbit (denoted by (1))is shown in the $(n_1, dn_1/d\xi)$ phase plane. The wake of the leading edge is on the longer time scale and corresponds to the trajectory (2) along the branch III. At the point (C) the phase point jumps along (3) to the branch I at essentially constant $\Phi_B$ to point (D). Finally the pulse profile is completed by the long time scale evolution (4) along the branch I to the stable steady state. At (C) an *up-jump* wave develops travelling with the same speed as the *down-jump* wave[22,24] (1). In the phase space this is a homoclinic trajectory and this solitary pulse corresponds to a packet of real-space transferred electrons moving with a constant speed. A global bifurcation of periodic orbits is connected with this nontransverse homoclinic orbit[23] in the phase space. These orbits correspond to wave trains of pulses and can be created by local periodic stimulations. This means that before the system has fully recovered after a suprathreshold excitation and completed a closed loop in the phase space a new pulse is triggered.

In order to assess the quality of our lowest order approximation, the numerical solution of the full coupled partial differential equations (14)-(17) with periodic boundary conditions is shown in Fig.4. In Fig.4a. a large enough initial perturbation indeed develops into a propagating depletion pulse of real-space transferred electrons in the $GaAs$-channel. In the $Al_x Ga_{1-x} As$ the emitted carriers form an accumulation pulse moving with the same velocity as the depletion pulse in the $GaAs$. This can be explained by the following physical picture: A large enough local fluctuation of the electric field leads to real-space transfer of hot electrons into the $Al_x Ga_{1-x} As$

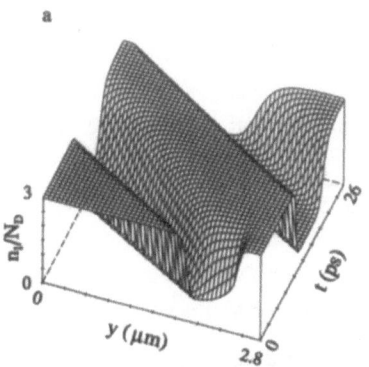
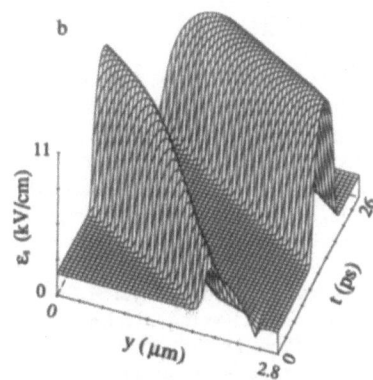

**Figure 4.** Numerical solution of the spatio-temporal dynamics (14)-(17) for conditions as in Fig. 3a and periodic boundary conditions. The initial suprathreshold perturbation develops into a propagating pulse. (a) Carrier density distribution in the *GaAs* layer. (b) Corresponding high field domain of the lateral electric field.

layer. Because of the different mobilities in the semiconductor layers, the carrier fluctuation in the *GaAs*-channel moves fast with the electric field and is readily damped out due to lateral diffusion, while the emitted carrier fluctuation in the $Al_xGa_{1-x}As$ layer remains almost unchanged. The local increase of the carrier density in the $Al_xGa_{1-x}As$ diminishes the positive space charge controlling the band bending. The potential barrier $\Phi_B$ decreases therefore with some delay due to the finite dielectric relaxation time. This leads to an increased backward thermionic emission current, which decreases the carrier density and increases the space-charge and the potential barrier $\Phi_B$ in the $Al_xGa_{1-x}As$. This local process of real-space transfer of hot electrons in the $Al_xGa_{1-x}As$ layer and backward emission into the *GaAs* channel can lead to a stable disturbance of carriers which cycle between the two semiconductor layers and move with constant velocity from the cathode to the anode.

In Fig.4b. the corresponding high field domain is shown under current bias for periodic boundary conditions.

## CONCLUSION

The spatio-temporal self-organization due to real-space transfer of hot electrons was investigated on the basis of a system of partial differential equations in one spatial dimension.

Due to the existence of well separated time and length scales in the system (the characteristic time of electron transfer out of the *GaAs* channel is much smaller than the dielectric relaxation time, and the mobility of the carriers in the *GaAs* channel is much higher than in the $Al_xGa_{1-x}As$ layer), it was possible to use methods of singular perturbation theory, and compare the results with the data obtained by direct numerical investigation. Our model predicts stable travelling wave solutions on the infinite line, i.e. in an infinitely extended semiconductor under current bias. In the monostable regime of the static current-density-field characteristic a solitary pulse of the carrier concentration corresponding to moving „packets" of real-space transferred electrons and a domain solution of the parallel electric field are found. This propagating domain solution may serve as an explanation for the current oscillations observed at low electric fields in semiconductor heterostructures.[8]

# REFERENCES

1. Z.S. Gribnikov, Sov. Phys. Semicond. 6 1204 (1973)
2. F. Pacha, F. Paschke, Electron. Commun. 32, 235 (1978)
3. K. Hess, H. Morkoc, H. Shichijo, and B.G. Streetman, Appl. Phys. Lett. 35, 469 (1979)
4. H. Shichijo, K. Hess, and B.G. Streetman, Sol. State Electron. 23, 817 (1980)
5. R. Sakamoto, K. Akai, and M. Inoue, IEEE Trans. Electron Devices 36, 2344 (1989)
6. K. Aoki, K. Yamamoto, N. Mugibayashi, and E. Schöll, Sol. State Electron. 32, 1149 (1989)
7. A. Kastalsky, M. Milshtein, L.G. Shantharama, J. Harbison, and L. Florez,
    Sol. State Electronics 32, 1841 (1989)
8. B.K. Ridley, Rep. Progr. Phys. 54, 169 (1991)
9. P.D. Coleman, J. Freeman, H. Morkoc, K. Hess, B.G. Streetman, and M. Keever,
    Appl. Phys. Lett. 40, 493 (1982)
10. A.J. Vickers, A. Straw, and J.S. Roberts, Semicond. Sci. Technol. 4, 743 (1989)
11. P. Hendriks, E.A.E. Zwaal, J.G.A. Dubois, F.A.P. Blom, and J.H. Wolter,
    J. Appl. Phys. 69, 302 (1991)
12. R. Döttling and E. Schöll, Phys. Rev. B 45, 1935 (1992)
13. E. Schöll and K. Aoki, Appl. Phys. Lett. 58, 1277 (1991)
14. E. Schöll and K. Aoki, Proc. 20th Int. Conf. Physics of Semiconductors, ed. E.H.
    Anastassakis and J.D. Joannopoulos (World Scientific Singapore 1990) p. 1125
15. S. Luryi and M.R. Pinto, Phys. Rev. Lett. 67, 2351 (1991)
16. E. Schöll, in: Handbook on Semiconductors Vol. I, 2nd. ed, Ch. 8, ed. P.T. Landsberg
    (North Holland, Amsterdam 1992)
17. E. Schöll, Current Instabilities in Semiconductors: Mechanism and Self-Organized
    Structures, in: Nonlinear Dynamics in Solids,
    ed. H. Thomas (Springer, Berlin 1992)
18. K. Hess, Advanced Theory of Semiconductor Devices (Prentice Hall, New Jersey 1988)
19. K.F. Brennan and D.H. Park, J. Appl. Phys. 65, 1156 (1989)
20. T.H. Glisson, J.R. Hauser, M.A. Littlejohn, K. Hess, B.G. Streetman and H. Shichijo,
    J. Appl. Phys. 51, 5445 (1980)
21. L.L. Bonilla and S.W. Teitsworth, Physica D 50, 545 (1991)
22. J.J. Tyson and J.P. Keener, Physica D 32, 327 (1988)
23. J. Guckenheimer and P. Holmes, Nonlinear Oscillations, Dynamical Systems, and
    Bifurcations of Vector Fields, Applied Mathematical Sciences 42 (Springer, New York 1983)
24. P. Ortoleva and J. Ross, J. Chem. Phys. 63, 3398 (1975)

# NEGATIVE DIFFERENTIAL RESISTANCE AND DOMAIN FORMATION IN SEMICONDUCTOR SUPERLATTICES

H.T. Grahn*

Max-Planck-Institut für Festkörperforschung,
Heisenbergstr.1, W-7000 Stuttgart 80, Germany

**ABSTRACT.** We review the occurrence of negative differential resistance (NDR) and its consequence for electric field domain formation in semiconductor superlattices. Transport and optical experiments are presented to determine the origin of NDR and the conditions for domain formation. The strongly nonlinear relation between drift velocity and field strength in conjunction with a large carrier density, which can be introduced either by doping or by optical excitation, leads to the formation of electric field domains in superlattices.

## INTRODUCTION

Negative differential resistance (NDR) or conductance (NDC) are a consequence of negative differential velocity (NDV). The latter is caused in some III-V semiconductors (e.g. GaAs, InGaAs or InP) by the Ridley-Watkins-Hilsum (RWH) mechanism[1] or better known as intervalley transfer. At large electric fields carriers are transferred from the $\Gamma$-minimum to the satellite minima (X- or L-minima), where the effective masses are larger. This leads to a lower drift velocity at high electric fields resulting in negative differential drift velocity. The I-V characteristic can then exhibit NDR or NDC. If the system is biased in the NDV region, a high field domain can form which propagates at the speed of the carriers through the material.[2] The effect of NDV in bulk semiconductors was first observed by Gunn.[3]

Semiconductor superlattices should also exhibit NDV according to the proposal by Esaki and Tsu.[4] The formation of a miniband leads to a field dependence of the drift velocity comparable to one of the RWH mechanism. The origin in this case is the negative effective mass for miniband states beyond the inflection point. Since the field dependence of the drift velocity is not substantially different from the one of the RWH mechanism, the physics of domain formation in the case of miniband transport should be comparable to the one of the Gunn effect. However, semiconductor superlattices can exhibit more than one region of NDV due to resonant tunneling between different subbands.[5,6] The drift velocity can display maxima at electric field strengths corresponding to the conduction subband spacings $F_{ij} = (E_{Ci} - E_{Cj})/(ed)$, where $E_{Ci}$ are the conduction subband energies and $d$ is the superlattice period. The formation of electric

*Negative Differential Resistance and Instabilities in 2-D Semiconductors*
Edited by N. Balkan *et al.*, Plenum Press, New York, 1993

189

field domains originating from the NDV due to resonant tunneling is the subject of this paper.

After the proposal of NDV in superlattices by Esaki and Tsu in 1970 the first experimental observation of NDC was reported in 1974 by Esaki and Chang.[7] The conductance showed oscillation with a period comparable to the spacing between the first and second electronic subband. This fact already demonstrated that the observed NDC was not due to miniband transport but rather due to resonant tunneling. A tentative explanation was given pointing out the formation of electric field domains in semiconductor superlattices. It took another 10 years before the topic was investigated further by a number of groups measuring I-V characteristics and conductance in GaAs-Al$_x$Ga$_{1-x}$As,[8-13] In$_{1-x}$Ga$_x$As-In$_{1-x}$Al$_x$As,[14,15] and In$_{1-x}$Ga$_x$As-InP superlattices.[16,17] Most experiments have periodic structures with a frequency directly related to the conduction subband spacing in common. This correlation gave strong evidence for domain formation by resonant tunneling.

A breakthrough in the understanding of electric field domain formation was accomplished when in addition to transport experiments optical techniques were employed on the same samples.[18-20] The generation of electric field domains by optical excitation was investigated. Furthermore, using the quantum-confined Stark effect photoluminescence (PL) experiments were used to directly demonstrate the existence of two field regions with definite field strengths. Increasing the applied field only led to a reduction in the spatial extent of the low field domain and, at the same time, to an increase of the high field domain. In this article we will review the transport and optical measurements that have been performed to achieve a detailed understanding of electric field domain formation in semiconductor superlattices. A simple theoretical model containing the important ingredients for domain formation will also be presented.


## NDR IN SUPERLATTICES

There are two possible origins for NDR in semiconductor superlattices. The first mechanism occurs for a superlattice with a wide miniband of width $\Delta$ as first proposed by Esaki and Tsu.[4] The theoretical drift velocity $v_d$ has the form

$$v_d(F) = v_0 \frac{F/F^*}{1 + (F/F^*)^2} \ , \tag{1}$$

where $v_0 = \Delta d/(2\hbar)$, $F^* = \hbar/(ed\tau)$, and $\tau$ is a scattering time. The dependence of the drift velocity on $F$ is very similar to the one of the RWH mechanism. Therefore, it is not expected that this type of NDV leads to a very different behavior than in the Gunn effect. A typical I-V characteristic of this type of NDV is shown in Fig.1, the so-called voltage-controlled negative differential resistance. The I-V characteristic in Fig.1 displays one maximum and one minimum. The voltage range between the maximum and minimum is the NDR region as indicated in Fig.1. When the system is biased in the NDR voltage range, a propagating high field domain can form. Due to NDR an instability in the electron concentration can grow exponentially, while outside the NDR region it will always decay. In the NDR region the field distribution can also change leading to a propagating high field domain. The dynamical aspect of high field domain formation in a system with one NDR region is discussed in Ref. 21.

The second mechanism is resonant tunneling.[5,6] In this case the drift velocity has the following form

$$v_d(F) = v_b(F) + \sum_{i=1}^{m} v_i \, \delta(F - F_i) \ , \tag{2}$$

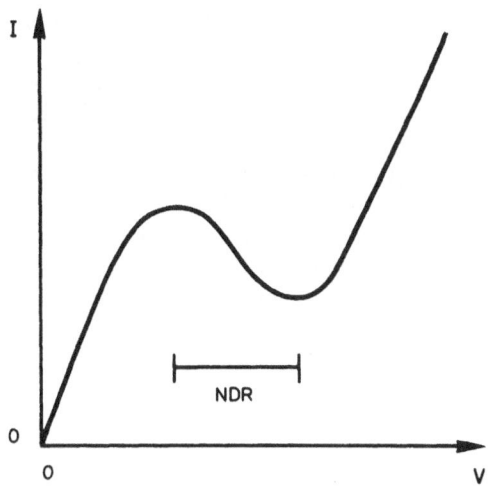

**Figure 1.** Schematic I-V characteristics of voltage-controlled NDR. The voltage range marked with a bar corresponds to the NDR region.

where $v_b(F)$ is a smoothly varying function due to non-resonant tunneling and $F_i = F_{i1} = (E_{Ci} - E_{C1})/(ed)$ are the electric field strengths of the tunneling resonances. A system which does not exhibit NDV would only have the first term of the drift velocity in Eq.2. The second term leads to as many maxima in the drift velocity as there are subbands in the system. At the same time for $m$ subbands there are $m$ regions of NDV. In reality the tunneling resonances are broadened by scattering mechanisms, but this does not change the number of NDV regions. For a homogeneous field distribution together with a small carrier density these tunneling resonances should be observable in the I-V characteristics. A typical curve for a superlattice with 40 periods is shown in Fig.2. There are four strong resonances and two weaker ones ($V_{ap} = -6.6$ and $-7.7$ V). The strong resonances correspond to resonant tunneling from the first electronic sub-

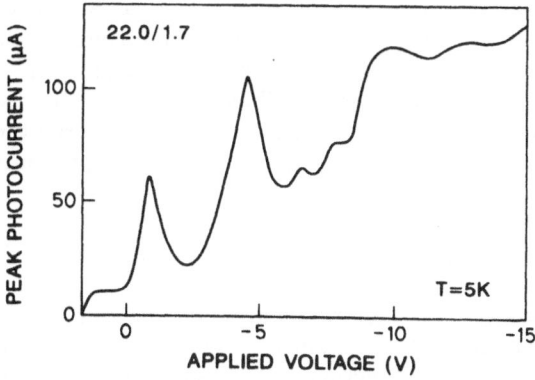

**Figure 2.** Peak photocurrent versus voltage in a 40 period GaAs-AlAs superlattice with 22.0 nm GaAs and 1.7 nm AlAs.

191

band into the first, second, third, and fourth subband, leading to four regions of NDR. The two weaker resonances originate from phonon assisted tunneling and also exhibit NDR. The smoothly varying background current is also observed and is ascribed to non-resonant tunneling. As can be seen in Fig.2 the NDR of resonant tunneling is voltage-controlled.

With regard to miniband transport this sample has a theoretical miniband width of 0.3 meV, which can be regarded as negligible. This sample therefore represents a coupled multi-quantum well system. At the resonance field $F_2$ carriers are injected from the first subband into the second subband in the adjacent well. From the second subband the carriers can either scatter down to the first subband again or tunnel non-resonantly

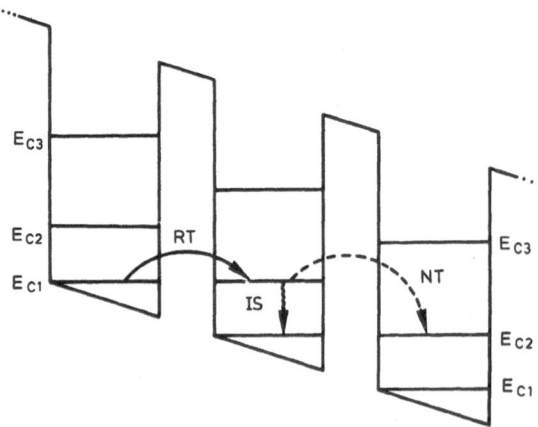

**Figure 3.** Schematic diagram of resonant tunneling (RT) and the subsequent processes such as intersubband scattering (IS) and non-resonant tunneling (NT) including a scattering process.

into the next well. This is illustrated in Fig.3 for part of a coupled multi-quantum well system. In the samples that are discussed in this paper the non-resonant tunneling channel is strongly suppressed because AlAs barriers are used. For a subband spacing above the optical phonon energy in GaAs (36 meV) the intersubband scattering time is below 1 psec.[22] The tunneling process in the discussed samples is much slower than this intersubband scattering time so that all injected carriers relax down to the first subband. In this paper we will focus on the direct tunneling resonances and consider for simplicity only two electronic subbands with energies $E_{C1}$ and $E_{C2}$. Furthermore, all presented samples have a negligible miniband width, but represent coupled multi-quantum wells. Before discussing the experimental results a brief theoretical treatment of electric field domains generated by resonant tunneling is presented.

## THEORETICAL TREATMENT OF ELECTRIC FIELD DOMAINS

In this section we will show that resonant tunneling can lead to the formation of stable electric field domains. This is a consequence of the interaction of the nonlinear behavior of the drift velocity as a function of the electric field with a large carrier density. While a negligible carrier density leads to a homogeneously distributed electric field over the region of the superlattice as shown in Fig.4(a), doping a superlattice results in a field gradient as demonstrated in Fig.4(b). A doped superlattice with domain formation will have a field distribution as shown in Fig.4(c). This follows from current conservation under steady state conditions

$$j = e\, n\, v_d = const. \, , \tag{3}$$

where $j$ is the current density and $n$ the carrier density, in conjunction with Gauss law $dF/dz = en/(\epsilon_0\epsilon)$. The result is an equation which relates the gradient of the electric field $F$ with the drift velocity

$$\frac{dF}{dz}\, v_d(F) = const. = \frac{j}{\epsilon_0\,\epsilon} \tag{4}$$

with $\epsilon$ denoting the dielectric constant. The spatial direction of the superlattice is $z$. Eq.(4) contains the basic ingredients for domain formation. If the drift velocity exhibits peaks, the solution of Eq.(4) leads to a large field strength change for a small drift velocity and a small field strength gradient for large $v_d$. Assuming a very simple field dependence of the drift velocity for two subbands as given in Eq.2, the field distribution for a constant applied voltage has the following form

$$F(z) = F_1\, \Theta(z)\Theta(z_1 - z) + F_2\, \Theta(z - z_1)\Theta(L - z) \, , \tag{5}$$

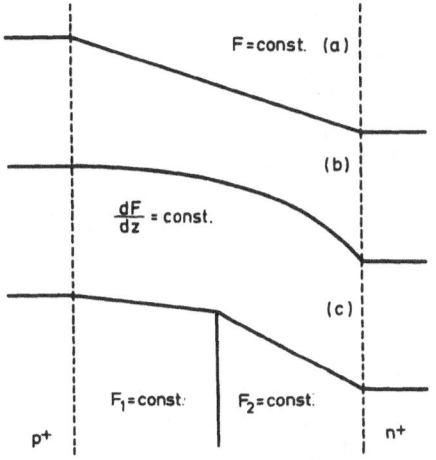

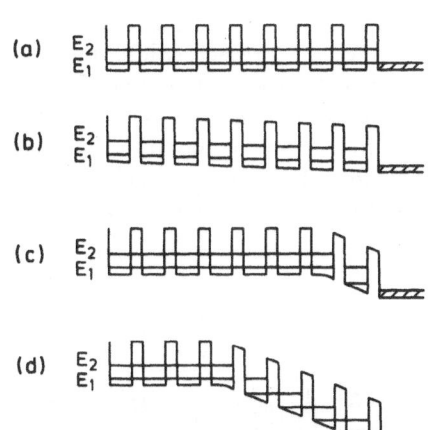

**Figure 4.** Conduction band edge distribution in the intrinsic region of a $p^+$-i-$n^+$ structure for a negligible (a), a homogeneously distributed (b), and electric field domain forming (c) carrier density. The domain boundary in (c) is marked by a vertical line.

**Figure 5.** Schematic band diagram of a doped superlattice for different applied voltages. (a) $V_{ap} = 0$, (b) $V_{ap} < (E_2 - E_1)/e$, (c) $V_{ap} = (E_2 - E_1)/e$, and (d) $V_{ap} = 4(E_2 - E_1)/e$.

where $\Theta$ is the unit step function, $L = Nd$ the total thickness of the superlattice, $N$ the number of periods, and $z_1 = L(1 - F_{ap}/F_2)$. $F_{ap}$ is the applied electric field and $z_1$ is the location of the domain boundary. The field distribution of Eq.(5) is schematically shown in Fig.4(c). Two regions with different electric field strengths are separated by a boundary. In Fig.5 the conduction band edge distribution of a superlattice exhibiting domain formation is shown for different applied voltages. The movement of the domain boundary through the superlattice as the applied voltage is increased will lead to $N$ jumps in the current-voltage characteristic. Since the field strengths are not arbitrary, but well determined by the subband spacing, the origin of domain formation in superlattices is very different than in the Gunn effect. Furthermore, recent theoretical publications[23,24] do not adequately describe the domain formation in superlattices, since these theories do not take into account the second subband. In our analysis the existence of two maxima in the drift velocity is essential to account for the experimentally observed facts, which are presented in the next section.

## ELECTRIC FIELD DOMAINS IN SUPERLATTICES

In this section we will discuss the investigation of the domain boundary by transport experiments and of the domain field strengths by photoluminescence spectroscopy. The generation of electric field domains by doping and photoexcitation will be compared.

### Transport Experiments

The I-V characteristic in Fig.2 was measured with a time-of-flight technique using a very small photoexcited carrier density. The sample is nominally undoped so that the applied electric field can be assumed to be homogeneously distributed. The field distribution can change drastically if a large carrier density is introduced. This can be done either by doping or by intense photoexcitation. In this section we we will discuss transport experiments performed on doped samples. In this case only electrons (n-doping) are present, while under photoexcitation electrons and holes are created. The current-voltage characteristics of an undoped ($n^+$-i-$n^+$) and a doped ($n^+$-n-$n^+$) superlattice with 40 periods of 9.0 nm GaAs and 4.0 nm AlAs with a doping level of $3 \times 10^{17}$ cm$^{-3}$ are compared in Fig.6. While the undoped system does not show any fine structure in the displayed field range, the doped structure exhibits 36 jumps in close correlation to the number of periods. The average voltage spacing between two jumps

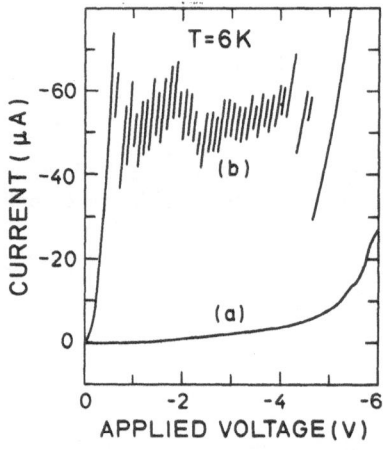

**Figure 6.** Dark current of a $n^+$-i-$n^+$ (a) and a $n^+$-n-$n^+$ (b) structure with 40 periods of 9.0 nm GaAs and 4.0 nm AlAs vs. applied voltage.

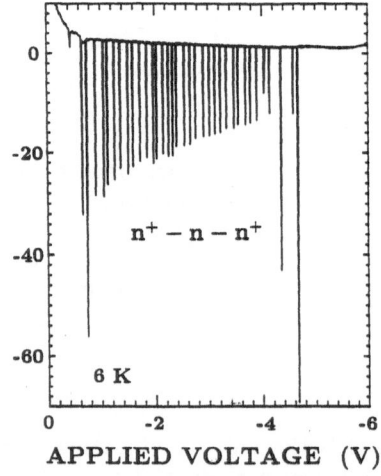

**Figure 7.** Numerical derivative of the dark current shown in Fig.6 for the $n^+$-n-$n^+$ structure vs. applied voltage.

is 127 mV, which is in good agreement with estimated subband spacing $E_{C2} - E_{C1}$ of 135 meV. A similar behavior was observed in a sample with the same well and barrier thickness, but a larger doping level of $1 \times 10^{18}$ cm$^{-3}$. Since the average current in the field range of the discontinuities is rather constant, the current must be limited by non-resonant tunneling at the domain boundary. In this case the tunneling current at the domain boundary can be written as[25]

$$I \;=\; I_0 \; exp(-\frac{2\,d_B}{\hbar}\; \sqrt{2\,m_B^*\,(V_B^C \;-\; E_{C1}))} \;,$$
(6)

where $m_B^*$ is the effective mass of the barrier material, $d_B$ the thickness of the barrier, $V_B^C$ the barrier height in the conduction band, and $E_{C1}$ the binding energy of the lowest subband. Increasing the applied voltage between two discontinuities leads to a decrease of the barrier height of $V_B^C(0) - V_B^C(F) = eFd_B$. The additional electric field at the domain boundary is given by $F = (V_{ap} - nF_2d)/d$, where $n$ is the number of periods in the high field domain. The derivative of the logarithm of the current with respect to the applied voltage can be approximately written as

$$\frac{dln(I/I_0)}{dV} \;=\; \frac{e\,d_B^2}{\hbar\,d}\; \sqrt{\frac{2\,m_B^*}{V_B^C(0)\;-\;E_{C1}}} \;.$$
(7)

The approximation is valid for GaAs-AlAs superlattices with subband energies far below the barrier height as in our samples. Taking the parameters of our samples for a single AlAs barrier at the domain boundary we calculate a value of 2.5 V$^{-1}$. In Fig.7 the numerical derivative of the I-V characteristic of the n$^+$-n-n$^+$ structure as shown in Fig.6 is plotted versus the applied voltage. The experimental values for $dln(I/I_0)/dV$ vary between 2.9 and 1.3 V$^{-1}$ in good agreement with the theoretical value. For the sample with the higher doping level the experimental values lie between 2 and 0.5 V$^{-1}$. Since the derivative in Eq.(7) depends quadratically on the barrier thickness, its value for a double barrier would be much too large.

The I-V characteristic of the doped samples exhibits bistabilities. In Fig.8 I-V characteristics of the n$^+$-n-n$^+$ structure are shown for two different sweep directions. While the number of discontinuities is about the same, the average current of the domain plateau depends on the sweep direction. Since the change of the field at the domain bound-

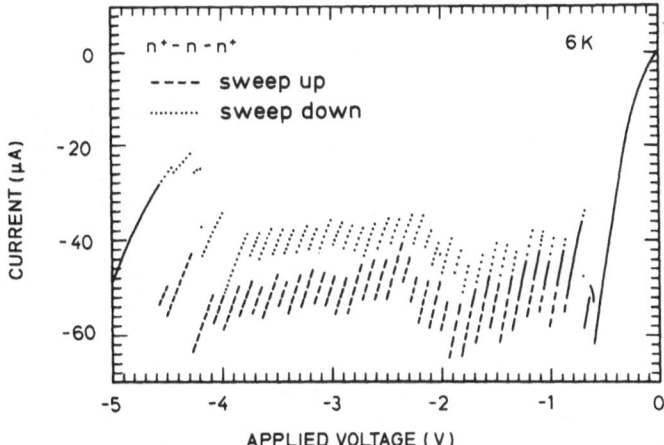

**Figure 8.** Dark current vs. applied voltage for the n$^+$-n-n$^+$ structure of Fig.6. Sweep up corresponds to a sweep from 0 to −7 V, while sweep down went from −7 to 0 V.

ary is due to a charging effect, the current varies with the sweep direction resulting in a hysteresis effect. However, the slope, i.e. $dln(I/I_0)/dV$, does not depend on the sweep direction.

## Optical Excitation and Photoluminescence Spectroscopy

Before discussing the PL experiments it is important to look at the I-V characteristic under illumination. If we use a very small light intensity on the doped sample, the I-V characteristic does not change. A different method of creating electric field domains in superlattices is using an undoped $p^+$-i-$n^+$ structure in reverse bias under strong illumination (647 nm). An I-V characteristic for a GaAs-AlAs superlattice with 40 periods of 9.0 nm GaAs and 4.0 nm AlAs is shown in Fig.9. We have subtracted the built-in voltage of 1.5 V typical for $p^+$-i-$n^+$ diodes to compare this structure directly with the I-V characteristic of the illuminated $n^+$-n-$n^+$ structure. The excitation density was comparable in both cases. It is apparent from Fig.9 that illumination of the $p^+$-i-$n^+$ structure does not result in an I-V characteristics as regular as for the doped superlat-

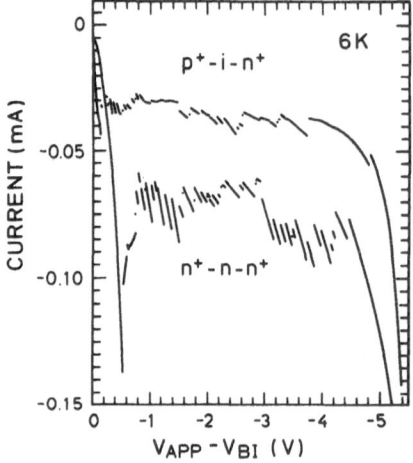

**Figure 9.** Total current (dark plus photo) of a $p^+$-i-$n^+$ and $n^+$-n-$n^+$ structure with 9.0 nm GaAs and 4.0 nm AlAs vs. applied voltage. The built-in voltage $V_{bi}$ is 1.5 V for the $p^+$-i-$n^+$ and and 0 V for the $n^+$-n-$n^+$ structure.

tice. Furthermore, even for the doped structure strong illumination leads to a partial destruction of the regular structures observed in the dark. This effect is due to the simultaneous presence of electrons and holes when the samples are illuminated. In the case of the illuminated $n^+$-n-$n^+$ structure the current also contains a contribution of the injected electrons from the contacts.

We can test the field distribution which is shown in Figs.4 and 5 by photoluminescence spectroscopy. If two different, but well-defined field strengths are present in the sample, two PL lines are expected, since a larger electric field leads to a red shift of the PL line with respect to the lower field strength. This is a consequence of the quantum-confined Stark effect.[26] When the applied voltage is increased, the position of the two PL lines should be fixed and only the intensities should change. This is exactly what is observed, when the $n^+$-n-$n^+$ structure is illuminated with very little light such that the I-V characteristic does not change. In Fig.10 PL spectra of the $n^+$-n-$n^+$ structure are shown between $-1$ and $-5$ V. The high energy line, which corresponds to the low

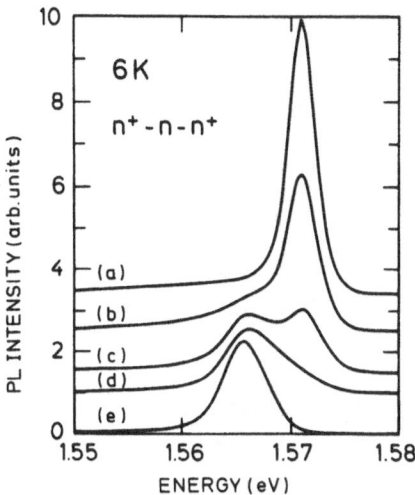

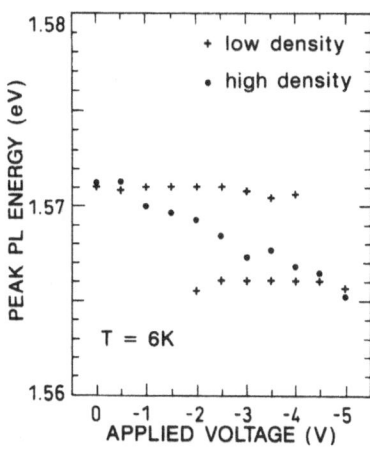

**Figure 10.** Photoluminescence spectra of the $n^+$-n-$n^+$ structure of Fig.6 at different applied voltages. (a) $-1$ V, (b) $-2$ V, (c) $-3$ V, (d) $-4$ V, and (e) $-5$ V.

**Figure 11.** Photoluminescence peak energies in the $n^+$-n-$n^+$ structure vs. applied voltage. The low density points correspond to the spectra in Fig.10, the high density spectra correspond to the conditions in Fig.9.

field domain, decreases in intensity, while the low energy line of the high field domain increases in intensity with increasing electric field. The voltage dependence of the peak energy of the PL lines is shown in Fig.11. The low density results are obtained under the same conditions as the PL spectra in Fig.10. For excitation densities corresponding to the I-V characteristic of the $n^+$-n-$n^+$ structure in Fig.9 the two PL lines can not be resolved anymore and only a single, but wide PL line is observed. Its voltage dependence is also shown in Fig.11 labelled as high density. The peak energy of this line shifts linearly with voltage, but this could also be caused by two wide lines which strongly overlap.

PL experiments have also been performed on the $p^+$-i-$n^+$ and $n^+$-i-$n^+$ structures under *strong* illumination with similar intensities as the ones in Fig.9 leading to comparable results. The PL experiments directly show the existence of two regions with well-defined field strengths. While the transport experiments give information about the domain boundary, the optical experiments reveal information about the regions of constant field strengths.

The picture that has been presented so far contained a lot of simplifications to emphasize the main features of electric field domains. We have investigated a number of additional samples to answer questions about the spatial distribution of the domains,[27] the coexistence of more than two domains at higher field strengths, and the carrier density dependence of domain formation in $p^+$-i-$n^+$ structures.[19] In Fig.12 PL spectra of a GaAs-AlAs superlattice with 50 periods of 14.4 nm GaAs and 3.4 nm AlAs in a $p^+$-i-$n^+$ configuration are shown. Due to the larger well width more subbands are below the barrier height. Therefore, tunneling resonances injecting into second, third, and fourth subband can be observed. Fig.12 displays PL spectra obtained in a field range where $F_2$ and $F_3$ coexist. Due to the larger well width and field strength the Stark shift is increased and the PL lines are clearly separated. The field dependence is

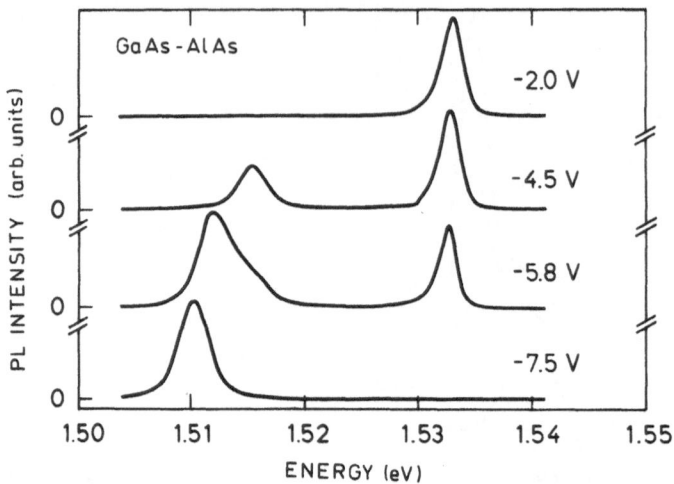

**Figure 12.** Photoluminescence spectra of a $p^+$-i-$n^+$ structure with 50 periods of 14.4 nm GaAs and 3.4 nm AlAs at different applied voltages as indicated in the figure.

comparable to the $n^+$-n-$n^+$ structure, although there is a little shift of the PL of the high field domain at lower fields. The PL spectroscopy results are summarized in Fig.13, where the PL peak energies are plotted versus the applied voltage. A total of five different domains are identified corresponding to field strengths of $F_{11}$, $F_{21}$, $F_{31}$, $F_{42}$, and $F_{41}$. There are two interesting features in Fig.13. First, in the field range of $-10$ to $-12$ V there are three coexisting domains. It has been shown theoretically that under certain conditions the field will break up into more than two domains. Second, the field strength of the domain with the peak energy in the middle ($F_{42}$) originates from a resonance between the *second* and fourth subband. As a comparison the dashed line

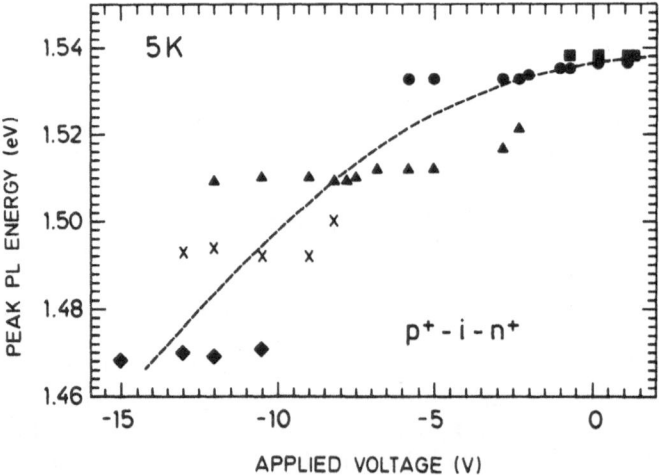

**Figure 13.** Photoluminescence peak energies in the $p^+$-i-$n^+$ structure of Fig.12 vs. applied voltage. The dashed line represents the Stark shift of the PL line observed under weak illumination.

shows the experimentally observed Stark shift measured at low excitation densities on the same sample by Tarucha and Ploog.[28] At certain field strengths the whole super-lattice is resonantly coupled, i.e. the field is homogeneously distributed. In Fig.13 the intersection of the dashed line with the data points occurs at these resonance field strengths $F_i$.

Using a $p^+$-i-$n^+$ structure instead of a $n^+$-n-$n^+$ structure has the disadvantage that electrons and holes are simultaneously present. However, to study the carrier density dependence of domain formation the $p^+$-i-$n^+$ structure exhibits a great advantage over the $n^+$-n-$n^+$ structure, since only one sample is needed instead of many samples with different doping concentrations. We recorded the PL spectra of the $p^+$-i-$n^+$ structure of Fig.12 for different laser intensities. Some selected spectra taken at $-5$ V are shown in Fig.14. The photocurrent at $-8$ V is used as a measure for the light intensity or car-

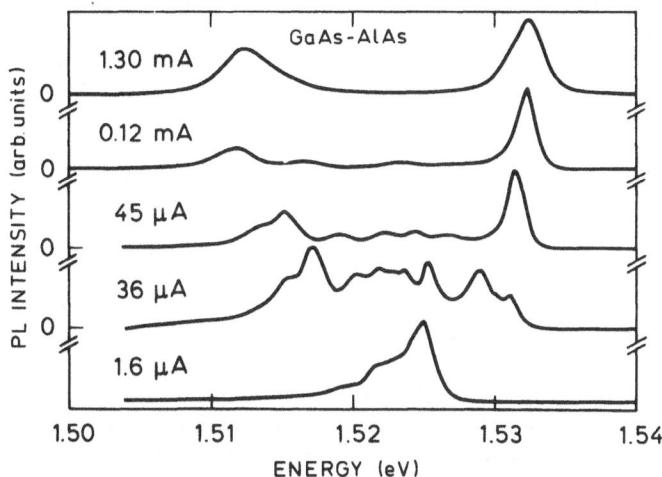

**Figure 14.** Photoluminescence spectra of the $p^+$-i-$n^+$ structure of Fig.12 at $-5$ V for different excitation intensities. The photocurrent at $-8$ V is used as a measure of the laser intensity (1 mA corresponds to about 50 W cm$^{-2}$).

rier density. At small intensities (1.6 $\mu$A) only one PL line is visible corresponding to a homogeneous field distribution. At large intensities (1.3 mA) we again observe two PL lines, one at lower energies corresponding to the high field domain and one at higher energies corresponding to the low field domain. At intermediate intensities (e.g. 36 $\mu$A) a very broad PL line with a lot of fine structure is observed. The results are summarized in Fig.15, where the PL peak energies are plotted as a function of the photocurrent, i.e. carrier density, on a semilogarithmic scale. At low and high densities we observe the already discussed scenario. At approximately 50 $\mu$A the PL spectra show a very rich structure. This region can be identified as the threshold region for domain formation. The field distribution in this region is rather complicated and may not even be stable, i.e. the fine structure of the PL spectra in this region might not be exactly reproducible. Stable domains are probably established only at a photocurrent of more than 1 mA.

Recently, we have investigated the behavior of electric field domains under the application of a magnetic field parallel and perpendicular to the superlattice direction. In the perpendicular configuration the tunneling process at the domain boundary can be studied.[29,30] The single barrier at the domain boundary can be viewed as a tunneling barrier between two strongly confined 2-dimensional systems.

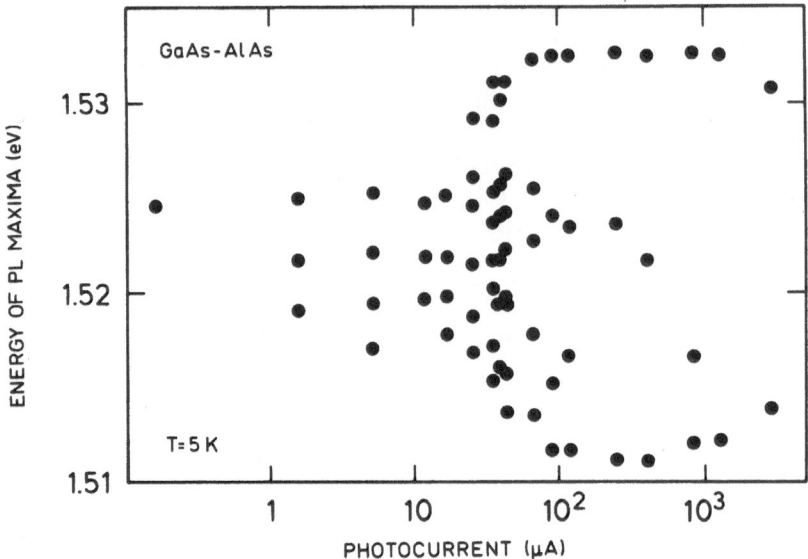

**Figure 15.** Photoluminescence peak energies in the p$^+$-i-n$^+$ structure of Fig.12 vs. the photocurrent measured at $-8$ V. The experimental conditions are identical to the ones in Fig.14.

## CONCLUSIONS

Semiconductor superlattices exhibit negative differential resistance for two reasons, miniband transport and resonant tunneling between different subbands. The second type of NDR leads together with a large carrier density to the formation of stable electric field domains. Combining optical with transport experiments has led to detailed understanding of electric field domain formation in semiconductor superlattices. The transport experiments demonstrate that the current is limited by non-resonant tunneling at the domain boundary. The photoluminescence experiments clearly show the existence of two regions with well-defined electric field strengths, which are determined by the subband spacing. The origin of these electric field domains is the non-linear field dependence of the drift velocity due to resonant tunneling with at least two maxima in combination with a large carrier density. The existence of one region of negative differential velocity alone as for the Gunn effect or miniband transport cannot explain the field strengths of the domains.

## ACKNOWLEDGMENTS

The author would like to thank A. Fischer and K. Ploog for sample growth, and W. Müller, H. Schneider, and K. von Klitzing for their collaboration. This work was supported in part by the Bundesminister für Forschung und Technologie.

---

* Present address: Paul-Drude-Institut für Festkörperelektronik, Hausvogteiplatz 5-7, O-1086 Berlin, Germany.

# REFERENCES

1. B.K. Ridley and T.B. Watkins, The Possibilty of Negative Resistance Effects in Semiconductors, *Proc. Phys. Soc.* 78:293 (1961).
2. S.M. Sze, "Physics of Semiconductor Devices", $2^{nd}$ edition, Wiley, New York (1981), p.641.
3. J.B. Gunn, Microwave Oscillations of Current in III-V Semiconductors, *Solid. State Comm.* 1:88 (1963).
4. L. Esaki and R. Tsu, Superlattice and Negative Differential Conductivity in Semiconductors, *IBM J. Res. Develop.* 14:61 (1970).
5. F. Capasso, K. Mohammed, and A.Y. Cho, Sequential Resonant Tunneling through a Multiquantum Well Superlattice, *Appl. Phys. Lett.* 48:478 (1986).
6. H. Schneider, H.T. Grahn, and K. v. Klitzing, Tunneling Resonances and Miniband Conduction in Superlattices, *Surf. Sci.* 228:362 (1990).
7. L. Esaki and L.L. Chang, New Transport Phenomena in a Semiconductor "Superlattice", *Phys. Rev. Lett.* 33:495 (1974).
8. T. Furuta, K. Hirakawa, J. Yoshino, and H. Sakaki, Splitting of Photoluminescence Spectra and Negative Differential Resistance Caused by the Electric Field Induced Resonant Coupling of Quantized Levels in GaAs-AlGaAs Multi-Quantum Well Structures, *Jap. J. Appl. Phys.* 25:L151 (1986).
9. K.K. Choi, B.F. Levine, R.J. Malik, J. Walker, and C.G. Bethea, Periodic Negative Conductance by Sequential Resonant Tunneling through an Expanding High-Field Superlattice Domain, *Phys. Rev.* B35:4172 (1987).
10. K.K. Choi, B.F. Levine, C.G. Bethea, J. Walker, and R.J. Malik, Multiple Quantum Well 10 $\mu$m GaAs-Al$_x$Ga$_{1-x}$As Infrared Detector with Improved Responsivity, *Appl. Phys. Lett.* 50:1814 (1987).
11. H.S. Newman and S.W. Kirchoefer, Electronic Properties of Symmetric and Asymmetric Quantum-Well Electron Barrier Diodes, *J. Appl. Phys.* 62:706 (1987).
12. M. Helm, P. England, E. Colas, F. DeRosa, and S.J. Allen, Jr., Intersubband Emission from Semiconductor Superlattices Excited by Resonant Tunneling, *Phys. Rev. Lett.* 63:74 (1989).
13. E.S. Snow, S.W. Kirchoefer, and O.J. Glembocki, Spectrally Dependent Phototcurrent Measurements in $n^+$-n-$n^+$ Heterostructure Devices, *Appl. Phys. Lett.* 54:2023 (1989).
14. Y. Kawamura, K. Wakita, H. Asahi, and K. Kurumada, Observation of Room Temperature Current Oscillations in InGaAs/InAlAs MQW Pin Diodes, *Jap. J. Appl. Phys.* 25:L928 (1986).
15. Y. Kawamura, K. Wakita, and K. Oe, Current Oscillations Related to N=3 Subband Levels up to Room Temperature in InGaAs-InAlAs MQW Diodes, *Jap. J. Appl. Phys.* 26:L1603 (1987).
16. T.H.H. Vuong, D.C. Tsui, and W.T. Tsang, High-Field Transport in an InGaAs-InP Superlattice Grown by Chemical Beam Epitaxy, *Appl. Phys. Lett.* 52:981 (1988).
17. R.E. Cavicchi, D.V. Lang, D. Gershoni, A.M. Sergent, H. Temkin, and M.B. Panish, Sequential Screening Layers in a Photoexcited In$_{1-x}$Ga$_x$As/InP Superlattice, *Phys. Rev.* B38:13474 (1988).
18. H.T. Grahn, H. Schneider, and K. von Klitzing, Optical Detection of High-Field Domains in GaAs/AlAs Superlattices, *Appl. Phys. Lett.* 54:1757 (1989).
19. H.T. Grahn, H. Schneider, and K. von Klitzing, Optical Studies of Electric Field Domains in GaAs-Al$_x$Ga$_{1-x}$As Superlattices, *Phys. Rev.* B41:2890 (1990).
20. M. Helm, J.E. Golub, and E. Colas, Electroluminescence and High-Field Domains in GaAs/AlGaAs Superlattices, *Appl. Phys. Lett.* 56:1356 (1990).
21. M. Shur, "Physics of Semiconductor Devices", Prentice Hall, Englewood Cliffs (1990), p.546.
22. H.T. Grahn, H. Schneider, W.W. Rühle, K. von Klitzing, and K. Ploog, Nonthermal Occupation of Higher Subbands in Semiconductor Superlattices via Sequential Resonant Tunneling, *Phys. Rev. Lett.* 64:2426 (1990).
23. B. Laikhtman, Current-Voltage Instabilities in Superlattices, *Phys. Rev.* B44:11260 (1991).
24. D.V. Averin, A.N. Korotkov, and K.K. Likharev, Quantization of Static Domains in Slim Superlattices, *in*: "Quantum Well and Superlattice Physics IV", G.H. Döhler and E.S. Koteles, eds., SPIE Proc. Vol. 1675, Bellingham (1992), in press.

25. C.B. Duke, Theory of Metal-Barrier-Metal Tunneling, *in*: "Tunneling Phenomena in Solids", E. Burstein and S. Lundqvist, eds., Plenum Press, New York (1969), p.31.
26. D.A.B. Miller, D.S. Chemla, T.C. Damen, A.C. Gossard, W. Wiegmann, T.H. Wood, and C.A. Burrus, Band-Edge Electroabsorption in Quantum Well Structures: The Quantum-Confined Stark Effect, *Phys. Rev. Lett.* 53:2173 (1984).
27. H.T. Grahn, H. Schneider, and K. von Klitzing, Spatial Distribution of High-Field Domains in GaAs-AlAs Superlattices, *Surf. Sci.* 228:84 (1990).
28. S. Tarucha and K. Ploog, Sequential Resonant Tunneling Characteristics of AlAs/GaAs Multi-Quantum-Well Structures, *Phys. Rev.* B38:4198 (1988).
29. K.K. Choi, B.F. Levine, N. Jarosik, J. Walker, and R. Malik, Anisotropic Magnetotransport in Weakly Coupled GaAs-Al$_x$Ga$_{1-x}$AsMultiple Quantum Wells, *Phys. Rev.* B38:12362 (1988).
30. H.T. Grahn, R.J. Haug, W. Müller, and K. Ploog, Electric-Field Domains in Semiconductor Superlattices: A Novel System for Tunneling between 2D-Systems, *Phys. Rev. Lett.* 67:1618 (1991).

# HIGH FREQUENCY DC INDUCED OSCILLATIONS IN 2D

A.J.Vickers, E.S-M.Tsui, and A.Straw

Department of Physics
University of Essex
Colchester CO4 3SQ
United Kingdom

## I. INTRODUCTION

In recent years DC induced current oscillations have been observed in many different types of 2d structures. In this paper we present results on GaAs/AlGaAs MQW structures that exhibit this behaviour at low temperatures (<260K) and low applied electric fields (<1KV/cm). The frequency of the oscillations depends to some extent on the applied electric field but was generally approximately 15MHz. The magnitude of the electric field at which the oscillations occur and the frequency of the oscillations mean that the well known mechanisms of DC induced current oscillations such as the Gunn Effect and the acousto-electric effect cannot be used in this instance. There are three current models[1,3,4] proposed to explain oscillations in 2d systems. Two of these models involve the interaction of carriers in the quantum well and the barrier. The third involves only carriers in the quantum well

Model 1 (Hendriks et al[1]) is a descriptive one and is used to explain a number of types of observed behaviour, including current collapse and oscillations (200MHz occuring at 77K with an applied electric field of approximately 1KV/cm). The oscillations are explained as being due to the successive population and depopulation of the AlGaAs layer with electrons injected from the source contact. The carriers trapped by the Si donors in the AlGaAs can be released by the applied electric field. Experiments in bulk GaAs[2] have shown that this process can lead to current oscillations and even chaotic behaviour.

Model 2 (Schöll and Aoki[3]) is a full theoretical treatment of the mechanism of real space transfer between the 2d channel and the barrier. In this case the predicted oscillations occur at fields of the order of 2KV/cm and have frequencies in the range 20-60GHz.

*Negative Differential Resistance and Instabilities in 2-D Semiconductors*,
Edited by N. Balkan *et al.*, Plenum Press, New York, 1993

203

Model 3 (Ridley[4,5]) is a full theoretical treatment of a hot electron percolating system. Electrons are modelled as travelling along a percolation path. They gain energy from the applied electric field, become hot and transfer out of the percolation path into localised states. The model predicts that the behaviour will only occur at low temperatures as at high temparatures the regeneration rate of the captured electrons might be similar to the capture rate.

Model 2 appears to require applied fields slightly higher than was applied in the case of our experiments. However it is possible that the field applied to the samples is not dropped uniformly across the sample creating regions of high electric field and low electric field. The high electric field regions could then be high enough to allow real space transfer between the 2d channel and the barrier. The dimensions of the samples which exhibit this phenomena are 1mm in length making standard techniques to evaluate the potential profile across the sample, such as point probing, impossible. In this paper we have probed the voltage profile in the samples using the technique of electro-optic probing which is only dimensionally limited by the limitation on focussing a laser beam which is considerably less than 1mm. We present results for probes taken at room temperature which shows some very interesting features. We also discuss the possibility of using the probe to study the field profile under the oscillation conditions so as to ascertain whether the oscillations are characterized by travelling high field domains as in the case of the Gunn Effect.


## II. THE ELECTRO-OPTIC PROBE

The Pockels effect is the basis of all electro-optic probing and sampling. It only occurs in crystals which have no centre of inversion symmetry. These are otherwise known as non-centrosymmetric crystals and one of the most technologically important of these is Gallium Arsenide.

When an electric field is applied across such a crystal birefringence is induced. This comes about because the electric field causes an alteration in the refractive indices of the crystal in the three crystallographic directions. Therefore the group velocity of the two components of polarisation of a light beam propagating through the crystal are changed. This means that the polarisation components of the beam move at different speeds in the crystal and therefore the transmitted beam has a different phase to the incident beam.

This change in phase may be understood by first describing the refractive index by means of the index ellipsoid. This is a tool used to describe the optical properties of an isortopic medium under the influence of an applied electric field, $E_z$. The index ellipsoid relates the refractive indices of the crystal, along the three principal axes, to $E_z$. The principal axes are defined by the crystallographic directions x, y, and z, and the direction of the applied electric field is chosen to correspond to one of these axes, i.e. the z direction in this case. Therefore the index ellipsoid may be defined as a solid revolution about $E_z$ and is a function of the medium and the applied field only.

The values of the refractive indices in the directions perpendicular to $E_z$, i.e. $n_x$ and $n_y$, are obtained by considering the intersection of the plane perpendicular to $E_z$ and the index

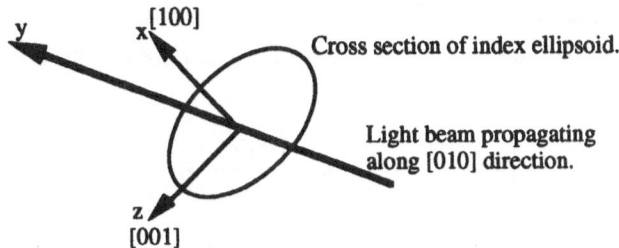

Figure 1.    Pictorial representation of the intersection
             of the index ellipsoid of a light beam propagating
             along the <010> axis of a crystal, and the plane
             perpendicular to the direction of propagation.

ellipsoid. This yields an ellipse whose semi-major and -minor axes give the values for the indices of refraction $n_x$ and $n_y$

Mathematically the index ellipsoid is described by:

$$\left(\frac{x^2+y^2+z^2}{n_0^2}\right) 2r_{kl}\left( zyE_x+xzE_y+xyE_z \right) = 1 \qquad \text{Eq. 2.1}$$

In this equation x, y and z are the principal axes of the crystal, $n_0$ is the unperturbed refractive index and $\mathbf{E} = (E_x, E_y, E_z)$ is the electric field present in the crystal. Also $r_{kl}$ is the electro-optic tensor of the crystal, which is defined by the equation describing the polarisation of the light beam propagating through the crystal:

$$P = \varepsilon_0(\chi_1 E + \chi_2 E^2 + \chi_3 E^3 + \dots\dots) \qquad \text{Eq. 2.2}$$

In this equation $\varepsilon_0$ is the permittivity of free space. $\chi_1$ is the linear susceptibility of the crystal, $\chi_2$ and $\chi_3$ are the quadratic and cubic susceptibilities known as the Pockels and Kerr coefficients, where $\chi_2$ is proportional to $r_{kl}$. For GaAs the only non-zero componet of this tensor is $r_{41}$. The second and third order terms in this equation are only finite for large elctric fileds , as $\chi_1 \gg \chi_2 \gg \chi_3$. Hence for low fields the polarisation of the beam is directly proportional to $E_z$.

The perturbation of the refractive index due to the electric field $E_z$ is described by the equation:

$$\Delta\left(1/n_0^2\right) = r_{41}E_z \cdot \qquad \text{Eq. 2.3}$$

The values of the perturbed refractive index in the x and z directions, as functions of the unperturbed refractive index and the applied field, are described by the relationships:

$$n_x' = n_0 + \left(\frac{n_0^3 r_{41}}{2}\right) E_z \quad \text{and} \quad n_y' = n_0 - \left(\frac{n_0^3 r_{41}}{2}\right) E_z \qquad \text{Eq. 2.4}$$

The effect of this perturbation in $n_x$ and $n_y$, on the beam of light propagating through the crystal may be seen from the following analysis. For the purposes of this analysis, and the experiment described in this work, the light is assumed to propagate parallel to the electric field, $E_z$, i.e. the longitudinal mode of electro-optic probing.

It can be shown, intuitively and from equations 2.3 and 2.4, that when the applied electric field is zero there is no perturbation of the refractive index, i.e. $n_x' = n_y' = n_0$ and the index ellipsoid is spherical. In other words the light propagates with the same velocity in all directions. Therefore when a circularly polarised beam of light passes through the crystal, the polarisation components remain unchanged throughout the crystal and the light emerges circularly polarised. However when an electric field is applied in the direction of propagation the refractive indices in the plane perpendicular to the field are no longer equal, as shown by equations 2.3 and 2.4. Hence the components of polarisation no longer propagate with the same velocity. As a result they are shifted with respect to one another and the light emerges elliptically polarised. The magnitude of this shift, or retardation, is directly proportional to the shift in the refractive index. This in turn is directly proportional to the magnitude of the electric field applied.

The magnitude of the retardation, $\Gamma$, is given by the difference between the phases of the polarisation components, i.e.:

$$\Gamma = \Phi_x' - \Phi_y' \qquad \text{Eq. 2.5}$$

where $\Phi_x'$ and $\Phi_y'$ are the perturbed polarisation components in the crystallographic direction, x and y.

The phases of $\Phi_x'$ and $\Phi_y'$ are proportional to the refractive indices and the length of crystal, $L_z$, through which the beam passes, as shown in equation 2.6:

$$\Phi = \frac{2\pi L_z}{\lambda_0} n' \qquad \text{Eq. 2.6}$$

This allows the retardation to be described in terms of the perturbed refractive indices:

$$\Gamma = \frac{2\pi L_z}{\lambda_0} \left(n_x' - n_y'\right) \qquad \text{Eq. 2.7}$$

206

where $\lambda_0$ is the free space wavelength of the propagating light. This retardation may be related to the applied electric field using equation 2.4:

$$\Gamma = \frac{2\pi L_z}{\lambda_0} n_0^3 r_{41} E_y \qquad\qquad \text{Eq. 2.8}$$

However this may be simplified by defining the half wave voltage, $V\pi$, as the voltage required to shift the polarisation of the beam by $\pi$ radians where:

$$V_\pi = \frac{\lambda_0}{2n_0^3 r_{41}} \qquad\qquad \text{Eq. 2.9}$$

Therefore the shift in polarisation of the beam may be written as :

$$\Gamma = \frac{\pi L_z E_z}{V_\pi} \qquad\qquad \text{Eq. 2.10}$$

or :
$$\Gamma = \frac{\pi V_z}{V_\pi} \qquad\qquad \text{Eq. 2.11}$$

where $V_z$ is the voltage applied to produce the field $E_z$.

This mathematically shows that the retardation of the polarisation of a light beam, propagating through a crystal which has an electric field applied to it in the direction of propagation, is directly proportional to the magnitude of the electric field. Hence, by measuring the shift in polarisation of a beam of light passing through such a crystal, the applied voltage may be found. This forms the basis of electro-optic voltage probing (EOVP) and may be used, in any non-centrosymmetric crystal which has a homogeneous unperturbed refractive index, to measure d.c. electrical signals. The same effect may be used to measure a.c. signals but for this procedure sampling techniques are required because of the transient nature of the field[6].

## III. EXPERIMENTAL DETAILS

### III.1 Sample details

All the samples studied were MOCVD grown single GaAs/Al$_{0.45}$Ga$_{0.55}$As quantum wells of width 90Å with a well carrier concentration of $10^{18}$ cm$^{-3}$ and AlGaAs doping of $6\times10^{16}$ cm$^{-3}$ at room temperature. Hall mobilities of 47 544cm$^2$V$^{-1}$s$^{-1}$ and 7330cm$^2$V$^{-1}$s$^{-1}$ at 77K and 298K respectively were measured. The samples were made into 1mm long and

100μm wide Hall bars, along the <110>, <100>, and <010> directions. Electrical connections were made through Au/Ge contacts.

## III.2 Current Measurements

Electrical measurements were made by applying a stable pulsed voltage across the two end contacts of the sample, supplied by a Velonex 570 pulse generator set at 4kHz repetition rate with a pulse width of 350ns. The resulting current was observed by monitoring the voltage across a series resistor (50Ω). The resultant signal was observed on an oscilloscope and ultimately stored on an OPUS PC microcomputer through the use of an A/D converter. The temperature dependence of the current oscillations was recorded between 3K and 300K by means of a pumped bath cryostat. Time dependence studies of the oscillation frequencies were investigated and will be discussed in the experimental results section.

## III.3 Electro-optic Measurements

For the purposes of the EOVP experiments the sample used was polished on the substrate side and a thin gold coating was applied to provide a reference for the 2DEG potential. In EOVP the 2DEG potential is detected by sending a circular polarized probe beam through the GaAs substrate of the sample along the <100> direction where it experiences a phase retardation that is directly proportional to the voltage difference between the 2DEG and the sample back surface. This is due to a birefringence in the xy plane (parallel to the QW) induced by the field component in the z or <100> direction. By placing the sample between crossed polarizers voltage information in the form of phase retardations is converted into light intensities which are then detected by conventional optoelectronics with the aid of a lock-in amplifier. The reference for the lock-in amplifier is provided by pulsing the voltage supply to the device in the same way as described for the current oscillation experiments.

## IV. EXPERIMENTAL RESULTS

### IV.1 Current Measurements

Initial assessment of all the samples was made by studying the low-field current-voltage characteristics at room temperature in order to check the ohmic quality of the contacts.

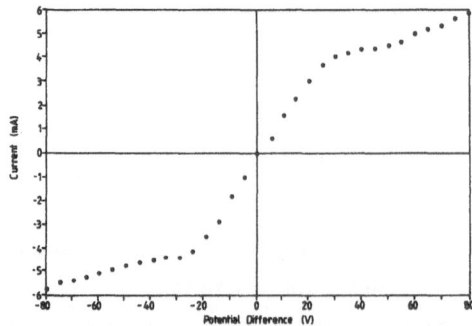

Figure 2.    Current-voltage characteristics of sample
CPM451 <100> measured at 295K.

All the samples studied showed good linear behaviour at low electric fields. At high electric fields, at room temperature, all the structures exhibited a non-linear behaviour and no current oscillations were observed (figure 2)

The point at which the ohmic behaviour broke down was not consistent, ranging from 300 to 650V/cm. When the temperature of the sample was reduced to typically below 160K, current oscillations were observed (figure 3) over the field range 300-1000V/cm, which showed no dependence on the details of the external circuit. At fields below 300V/cm the current showed no time dependence. The electric field dependence of the oscillation frequency is shown in figure 4, where apart from random fluctuations a clear turn on field can be observed and little field dependence of the frequency above threshold. There was also no dependence of the oscillation frequency with pulse width.

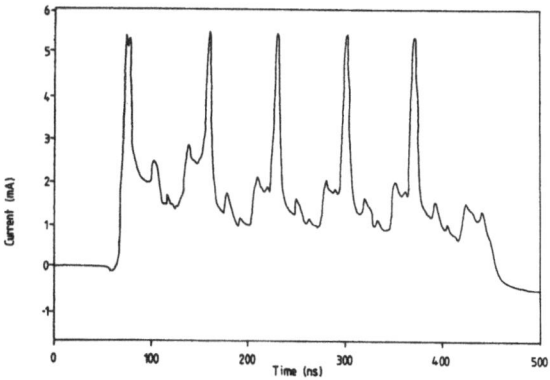

Figure 3.    Current oscillations observed at T=74K
             when the applied field was 800V cm$^{-1}$
             across sample CPM451 <110>.

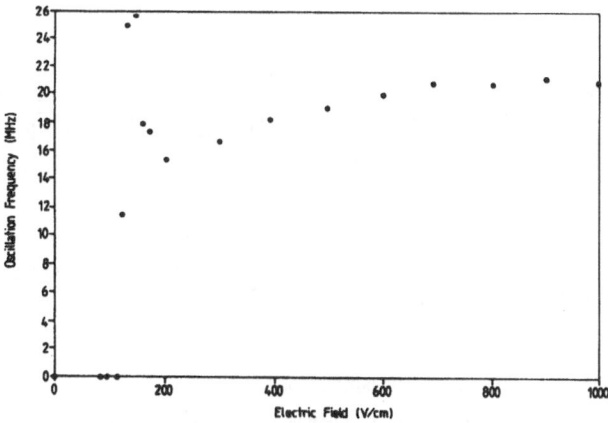

Figure 4.    Magnitude of the oscillation frequency as
             a function of applied electric field at
             T=74K for CPM451 <010>.

Figure 5 shows a very interesting feature. At high electric fields the oscillation frequency is seen to fall to approximately half its value at low electric fields. This we believe is an example of period doubling, which has been observed by other workers in bulk material and is indicative of chaotic behaviour.

## IV.2 EOVP

All EOVP measurements were carried out between zero and 50V biases applied along the QW layer. To ensure that the profile shapes were accurate we took account of the transmission variations in the sample by normalising the electro-optic signal, $V_{eo}$, by the average DC light level, $V_o$, of the transmission beam.

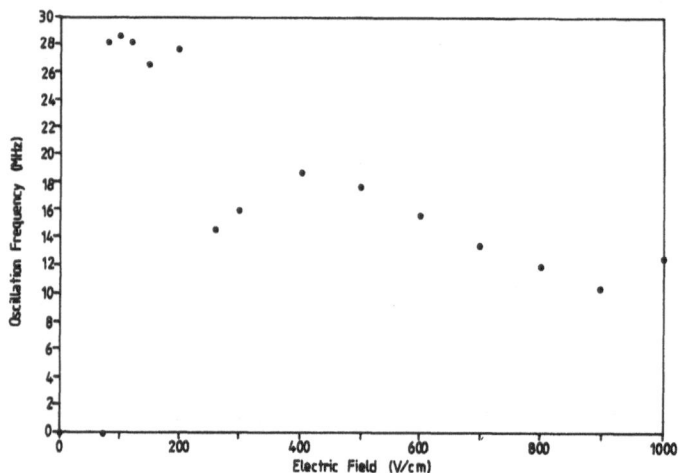

Figure 5.   Period doubling observed in the <010> direction. T=74K

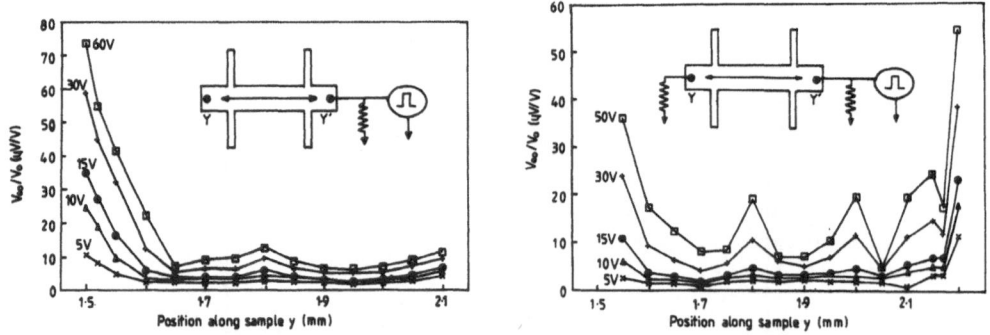

Figure 6.   Potential distribution along the length of the sample
at different applied voltages under (a) open-circuit
and (b) closed-circuit conditions. ( In each inset the
line YY' indicates the path along which the voltage
scans were made.)

210

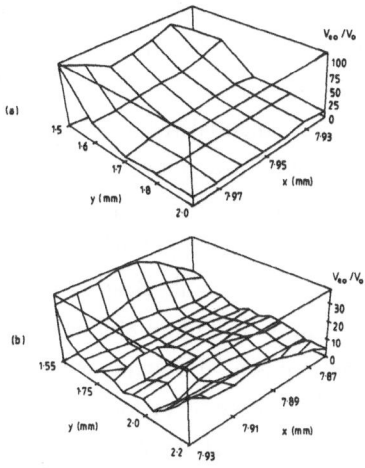

Figure 7.    Potential profiles at 50V under (a)open-circuit
and (b) closed-circuit conditions.


We first profiled the 2DEG potential distribution along the sample when no current
was flowing. The voltage profile shown in Fig 6a reveals a very unusual   potential
distribution. The dominant feature is a large potential hill situated at Y near one end of the
sample (se inset of Fig 6a). Two much smaller features can also be seen: one at Y and the
other half way between the terminals. These non-uniformities tend to become more
prominant as the bias is raised. In Fig 7a we have a 2D map of the potential distribution at the
highest field available. The variation of the 2DEG potential at Y with applied voltage, $V_a$, is
given in Figure 8. the initial rise of the potential is linear and can be extrapolated back to the
origin. The curve starts to become sublinear at an applied voltage of approximately 15V
(150V/cm), which is about half the oscillation threshold value. This appears to be
accompanied by a slight spatial enlargement of the potential hill. At position y=1.6mm,
which is near the foot of the hill, the increase of the potential with applied bias has actually
become linear.

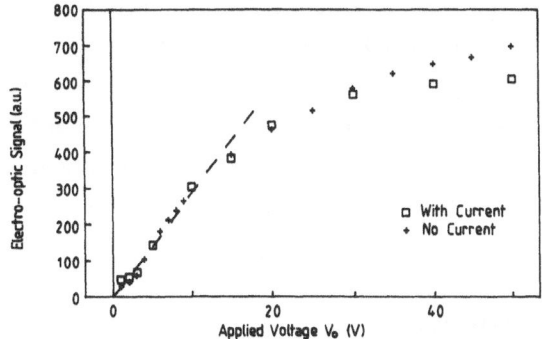

Figure 8.    Electro-optic signal as a function of applied voltage
under open-cicuit and closed-circuit conditions.

When current was allowed to flow through the sample, we found more structure to the potential profile. This can be seen in Figure 6b and 7b. Apart from the increased complexity, a small decrease in the potential hill at Y and a significant rise rise in potential at Y are noted. Now, if we compare the potential versus bias curves taken at Y for both open-circuit and closed-circuit cases (Figure 8) we notice they begin to depart from one another when the applied voltage is approximately 30V, and the curve of the latter case saturates in a more pronounced way. Under closed-circuit conditions one would expect to see a potential gradient along the length of the sample. This is not clearly visible from the potential profile of Figure 6b. However close examination of Figure 8b does indicate that a route from one contact to the other along which the potential is continually falling is possible. This would imply that the current route from one contact to the other is not along a straight line joining the two contacts. This observation supports model 3 in which it is suggested that the instabilities arise from hot electron percolation within the GaAs layer.

## V. CONCLUSION

The results presented here show the presence of high frequency DC induced current oscillations in GaAs/AlGaAs 2DEG structures. The field dependence and magnitude of the frequency observed rules out mechanisms such as the Gunn Effect and the acousto-electric effect. The results of the EOVP indicate that at room temperature the electric field in the sample is not uniform. It is possible that this feature is retained at low temperatures, and is a feature of the oscillation mechanism. It would of course be very satisfying to be able to confirm which of the three models presented in the introduction was the most likely explanation of the oscillations. However it is possible to use the EOVP results to defend all three models. The presence of a large voltage drop near one contact could be related to the the effect of injecting carriers into the AlGaAs from the contact or it could be indicative of the presence of a large electric field in the GaAs well near the contact making possible the transfer from the well to the barrier. It is also possible that the voltage profile (Figure 6b) is indicative of the type of profile required to produce a percolating system. At low temperatures it may be that potential shallows will develop that could provide localised trapping centres, just what would be required to support model 3.

The outcome of this work is therefore that there appear to be potential variations across the sample which could be linked to the oscillation mechanism. It is not possible to use the EOVP results to absolutely say which of the three models presented is the explanation for the oscillations. Further measurements at low temperature may assist by revealing potential shallows. It would also be revealing to attempt a measurement of the field profile under oscillating conditions. This would determine whether or not travelling domains were present

## VI. ACKNOWLEDGEMENT

We thank the SERC for the support of this work.

# REFERENCES

1. P.Hendriks, E.A.E. Zwaal, J.E.M. Haverkort, and J.H. Wolter, *SPIE Proc.* **1362**, 217 (1990).

2. A.Brandl, W.Kröninger,W. Prettl, and G. Obermaier, *Phys. Rev. Lett.* **64**, 212 (1990)

3. E. Schöll, and K. Aoki, *Appl. Phys. Lett.* **58**,1277 (1991).

4. B.K.Ridley, *Solid State Electronics.* **33**, 859 (1990).

5. B.K.Ridley, this book (1992).

6. B.H.Kolner, *SPIE Proc.* **795**,310 (1987).

# HOT ELECTRON INDUCED IMPACT IONIZATION AND LIGHT EMISSION IN GaAs BASED MESFETs, HEMTs, PM-HEMTs AND HBTs

C. Canali (°), C. Tedesco (*), E. Zanoni (*), M. Manfredi (+) and A. Paccagnella(*)

(°) Facolta' di Ingegneria, Universita' di Modena, Via Campi 213/b, 41100 Modena Italy
(*) Dipartimento di Elettronica ed Informatica, Universita' di Padova, Via Gradenigo 6/A, 35131 Padova, Italy
(+) Dipartimento di Fisica, Universita' di Parma, Viale delle Scienze, 43100 Parma, Italy

## I INTRODUCTION

As device dimensions become smaller, the increase of electric field in the active regions induces hot carrier effects such as impact ionization and light emission phenomena. Owing to their importance on advance device performances hot carrier effects have been investigated intensively both in Si and GaAs based devices. In fact the occurrence of impact ionization gives rise to multiplication and breakdown phenomena which limit the power handling capability of devices. As a consequence the study of impact ionization is very important to design device structure.

On the other hand a proper understanding and interpretation of light emitted from devices, operated at high electric fields and in the impact ionization regime, may provide a deeper understanding of the physics of hot carriers. Furthermore in heterostructure devices the experimental investigation of impact ionization induced currents and of light emission can provide insights into the vertical distribution of electrons and holes and of impact ionization events when devices are operated at high electric fields.

However the breakdown phenomena, in particular in GaAs based FET devices, and light emission mechanisms have not been fully understood.

In particular, unfortunately, there is no agreement as to the dominant mechanism for the emitted light. In general there are two important types of photoemission mechanisms in semiconductors: radiative recombination involving both carrier types, and radiative recombination which involves only one type of carrier. The former will be referred to as conduction-band to valence-band (c-v) radiation, and the latter as either conduction to conduction-band (c-c) or valence to valence-band (v-v) radiation, Fig. 1 [1]. These

*Negative Differential Resistance and Instabilities in 2-D Semiconductors.*
Edited by N. Balkan *et al.*, Plenum Press, New York, 1993

215

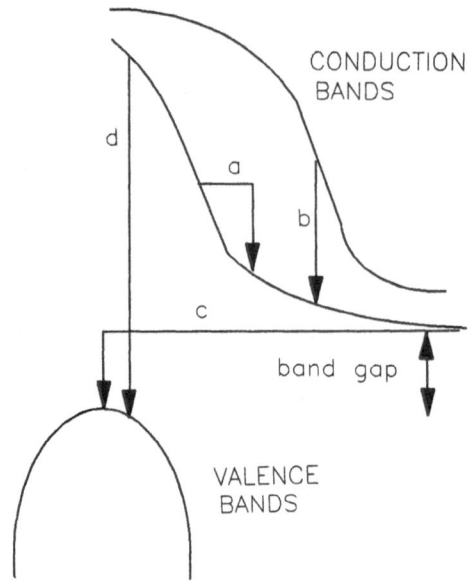

Fig. 1. Distinction of various luminescence mechanisms in a
realistic band structure: (a) indirect c-c, (b) direct c-c, (c)
indirect c-v, (d) direct c-v

processes can be further broken down into either direct transitions, in which a single photon provides both energy and momentum conservation, or indirect transitions, in which a photon provides the energy and an auxiliary interaction (phonon or impurity scattering) provides the momentum. The ionized impurity assisted indirect c-c (or v-v) emissions are conventionally referred as bremsstrahlung radiation.

Since the carriers involved in impact ionization and light emission processes are typically hot carriers, realistic band structure should be considered and sophisticated transport simulators, such as Monte Carlo, employed, [2,3].

Despite their importance few experimental results have been reported in technical literature on the correlation between impact ionization phenomena and light emission. This work presents experimental studies on impact ionization and light emission phenomena in GaAs based devices operated at high electric fields. Studied devices were: Metal Semiconductor Field Effect Transistors (MESFETs), High Electron Mobility Transistors (HEMTs), PsedoMorphic HEMTs (PM-HEMTs) and Heterostructure Bipolar Transistors (HBTs).

The main purposes of this work are:
i) to study impact ionization phenomena and to evaluate impact ionization coefficients by measuring the excess gate current (in FET devices) or base current (in bipolar devices);
ii) to analyze intensity and energy distribution of light emission in the 1.1-3.1 eV range;
iii) to correlate light emission with impact ionization phenomena in order to find out the dominant emission mechanism of photons with energy greater than Eg.

The following is an outline of the topics covered here. Section II presents samples and the experimental apparatus for electrical and optical characterization; Sections II, IV, V and VI provide a detailed description of examined samples and a survey of impact ionization and light emission in GaAs based MESFETs, HEMTs, PM-HEMTs and AlGaAs/GaAs HBTs respectively, and Section VI gives the Conclusions.

## II SAMPLES AND EXPERIMENTAL APPARATUS

Experimental results reported in this work refer to both commercially available and to laboratory GaAs-based MESFETs, HEMTs, PM-HEMTS and HBTs, see Tab. I. A more detailed description of each sample is reported in the following sections.

Electrical characterization has been performed by means of Hewlett-Packard HP 4142 and 4145 Semiconductor Parameter Analyzer. In particular impact ionization phenomena in pre-avalanche conditions have been characterized by measuring gate and substrate current in FET devices and base current changes in HBTs respectively. To avoid self-heating effects measurements have been performed also by pulsing the Vds or $V_{CB}$ bias with a duty cycle as low as 0.2 % using the HP 4142 pulse mode.

Optical measurements were performed using the experimental setup shown in Fig. 2, [4]. Devices were mounted on a temperature-controlled holder and the light emitted was collected by an optical fiber, 6 mm in diameter, placed at about 1 mm from the device. Spectral analysis of the emitted photons was performed using an OCLI 4000-1200 grating narrow band monochromator and a set of photomultipliers, cooled to improve the signal-to-noise ratio. Three different photomultipliers have been used: (a) an EMI 9684AM with a S1 response type (1.13-5.9 eV energy band), (b) an extended S20 EMI 9816QAM photomultiplier (1.4-6.2 eV energy band) and (c) a conventional S20 EMI 9816QB

TABLE 1. Devices and suppliers used in the present work

| DEVICES | SUPPLIER | LABORATORY | COMMERCIAL |
|---|---|---|---|
| MESFET | ALENIA | * | |
| | MITSUBISHI | | MGF1403 |
| HEMT (AlGaAs/GaAs) | TOSHIBA | | S8901/2 |
| | NEC | | NE20200 |
| PM-HEMT (AlGaAs/InGaAs) | TRW | * | |
| HBT (AlGaAs/GaAs) | BELL LABS | * | |

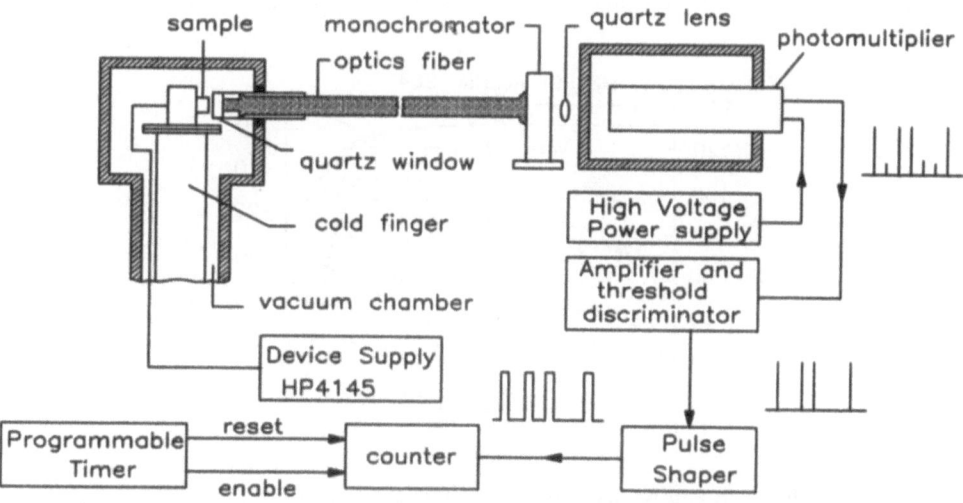

Fig. 2.Sketch of the experimental apparatus used for optical measurements, [4]

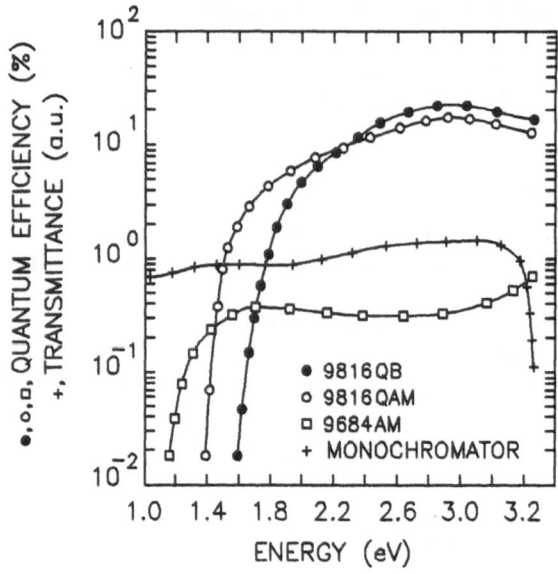

Fig. 3. Quantum efficiency of the photomultipliers adopted and transmittance band of monochromator, [4]

photomultiplier (1.7-6.2 eV energy band). Pulses deriving from multiple photon detection were rejected by a threshold discriminator (ORTEC 583). Shaped pulses were counted by a counter ORTEC 776. Experiments were performed in a dark ambient mostly at 300 K. The residual noise caused a constant count rate and was subtracted from the actual readings.

In Fig. 3 we report the quantum efficiency of the three photomultipliers adopted and the transmittance band of the monochromator. The quantum efficiency of the S20 photomultiplier was supplied by the manufacturer, while the S1 photomultiplier was characterized in our laboratory by using two different reference sources, a tungsten light and a black body [4].

Photon spectra in the 1.1-3.1 eV energy range have been obtained by means of the EMI 9816QAM and EMI 9684 photomultipliers. The two sets of data obtained in different energy ranges have been merged by normalizing the quantum efficiency of the two

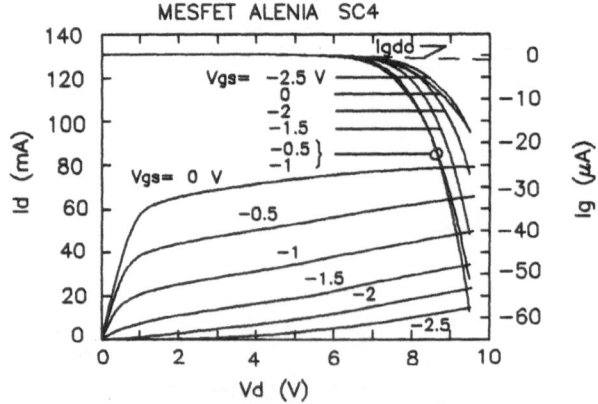

Fig. 4. Id and Ig characteristics vs Vds at fixed Vgs of a typical Alenia MESFET. Igdo is the gate-drain diode reverse current measured with the source kept floating, [18]

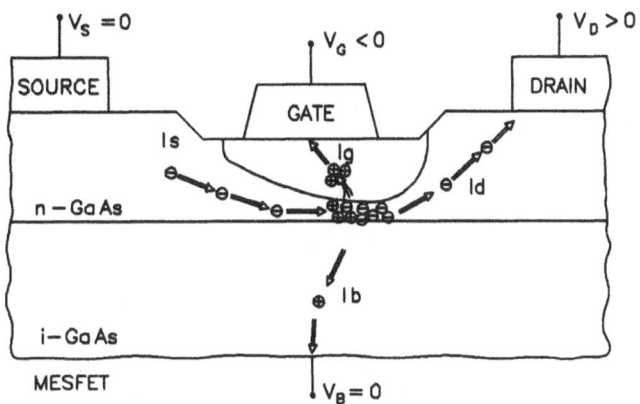

Fig. 5. Sketch of impact ionization phenomenon in a MESFET

phototubes in the energy range common to both phototubes (1.5-1.8 eV). Spectra have been corrected for the optical response of the components of the apparatus, for the transmittance of the monochromator and for the quantum efficiency of the phototubes.

The spectral sensitivity of the experimental apparatus was limited below 1.1 eV by the quantum efficiency of the 9684AM photomultiplier and above 3.1 eV by the monochromator transmittance.

The measurements of the integrated light intensity above 1.5 eV (or 1.7 eV) were performed by means of the 9816QAM (or 9816QB) photomultiplier, removing the monochromator filter and inserting or not low-pass or high-pass optical filters.

## III MESFETs

Since the first publication on GaAs MESFETs in 1966 by Mead [5] a good understanding of the principles of operation of these devices has been achieved. As the device dimensions become smaller, the increased electric field in the active region induces hot carrier effects such as breakdown phenomena and light emission. Breakdown is a significant limiting factor for the microwave power output of GaAs MESFETs and consequently has been extensively investigated. Solutions for improving the breakdown voltage based on recessed gate, highly doped $n^+$ drain contacts and lightly doped drain (LDD) structures, control of surface deep levels and use of buffer layers have been suggested and investigated [6-10]. Self-consistent two-dimensional device simulators, including the effect of tunneling and impact ionization and considering the influence of deep levels have been recently developed [6, 10-14]. Light emission of GaAs MESFETs close to breakdown conditions has been adopted to identify device weak areas and to study the effect of design variations on breakdown. Light emission has been observed in a wide energy range, both below and above the energy gap; nevertheless a close correlation of the light intensity and spectral distribution with impact ionization effects, device technology and their electrical characteristics has been only recently attempted [15-19]. A clear understanding of light emission mechanism has not yet achieved.

### III.a Samples

We have characterized several commercially available depletion-mode GaAs MESFETs (Mitsubishi, NEC, Toshiba) and devices prepared by different laboratories (Alcatel-Telettra, Alenia, ETH, Hewlett-Packard, Varian). For ease of description the

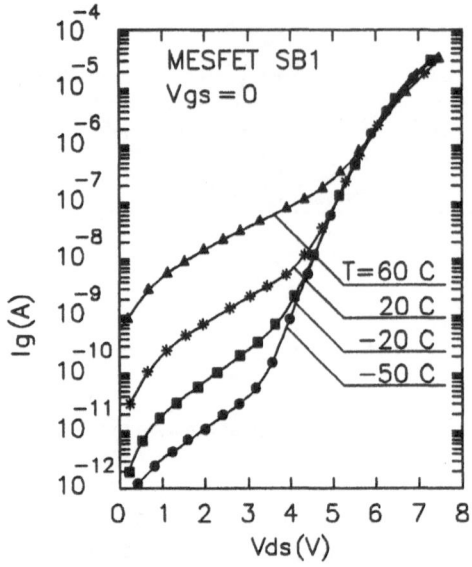

Fig. 6. Gate current |Ig| vs Vds at Vgs=0 V and at four
different temperatures for an Alenia MESFET, [22]

following discussion will refer mainly, as representative examples, to depletion mode GaAs MESFETs manufactured by Alenia and Mitsubishi (MGF1412).

Alenia MESFETs have been fabricated on semi-insulating <100> GaAs substrates, implanted with a $^{28}Si^+$ dose of $5x10^{12}$ cm$^{-2}$ at 100 keV (channel implant) and with $1x10^{13}$ cm$^{-2}$ at 40 keV (n+ shallow implant for the ohmic contact regions). Source and drain ohmic contacts were achieved by alloying a AuGeNi multilayer at 450 °C for 1 min., while the 0.5x300-μm$^2$ recessed gate electrode was based on thermally evaporated aluminium. Afterwards, a SiN passivating layer was deposited by plasma-enhanced chemical vapor deposition. The gate was aligned to the <110> direction, in which most MESFETs are fabricated. The channel implant under the gate has a peak concentration of $2x10^{12}$ cm$^{-3}$. The MESFET layout was defined with gate lenght Lg=0.5 μm, gate-source spacing Lgs=0.25 μm, and gate-drain spacing Lgd=0.25 μm.

### III.b Impact ionization effects

In this section we investigate the dependence of the gate current, Ig, on impact ionization phenomena. By means of experiments and numerical simulations we show that the main contribution to Ig comes from impact-ionization generated holes collected at the gate electrode. Ig is exponentially proportional to the negative inverse of the electric field, in agreement with impact ionization theory and Chynoweth's law [20, 21].

Drain, Id, and gate currents measured in a typical implanted Alenia MESFET as a function of drain, Vds, and gate, Vgs, biases are reported in Fig. 4. The reverse gate-drain current Igdo measured with source floating is reported as a dashed line. At high drain voltages (Vds>5V) a noticeable increase of gate current exiting from the device is clearly detectable. The fact that Igdo remains negligible in the whole range of Vds explored suggests that the increase in Ig observed when the device is operated as a transistor (source grounded) is due to the collection of holes generated by impact ionization and determined by multiplication of electrons flowing in the channel, as schematically shown in Fig. 5.

Carrier energies, however, are not so high to give rise to avalanche multiplication and the amount of generated hole-electron pairs is negligible compared to channel electrons; as a consequence, an increase of drain current is not detectable. The electrical bias condition in which substantial carrier multiplication occurs without avalanching is conventionally identified as "prebreakdown" conditions.

**Ig vs Vds at constant Vgs**

Figure 6 shows the gate current, Ig, of the Alenia MESFET #SB1 at Vgs=0V as a function of drain bias and at four different temperatures ranging from -50 °C to 60 °C, [22]. For drain voltages higher than a certain value, which increases with increasing temperature, Ig increases rapidly as a function of Vds. Such rapid increases of Ig are not present if the source is kept floating, and therefore cannot be due to the breakdown of the gate-drain junction. In agreement with Hui et al. [23], such sudden increase of Ig can be attributed to the onset of impact ionization. The sudden increase in Ig takes place when the contribution of generated holes dominates the reverse current of the gate diode. Figure 7 reports the dependence of Ig/Id versus 1/Vds at Vgs=0V for different temperatures. At high Vds, all data lie on a straight line, which extends over five orders of magnitude, confirming that at high Vds the ratio Ig/Id is proportional to an exponential of the negative inverse of Vds.

To obtain the electron ionization coefficient $\alpha_n$, the value of the electric field E is required. The simple model of Hui et al. [23], can be adopted thus assuming that

$$E\,max = (Vds-Vdsat)/Leff \tag{1}$$

where Vdsat is the saturation voltage, Emax is the maximum eletric field in the device, and Leff is the characteristic length of the high-field region. In the case of device #SB1 a value of Leff=0.35 µm was adopted and Vdsat was evaluated from device characteristics to be 0.5 V.

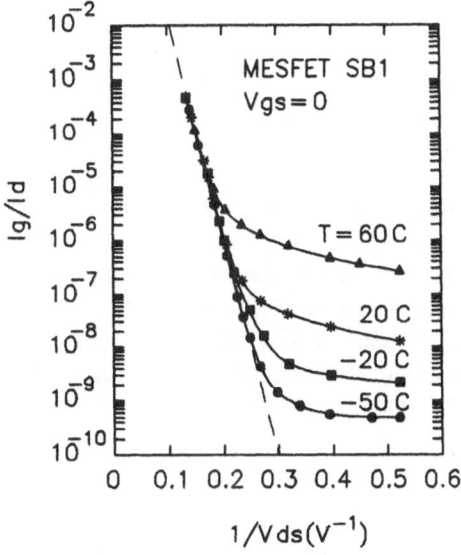

Fig. 7. Ig/Id vs 1/Vds at Vgs=0 V and at four different temperatures. Dashed line is a best fit of data obtained at higher Vds. Data refer to the same device of Fig. 6, [22]

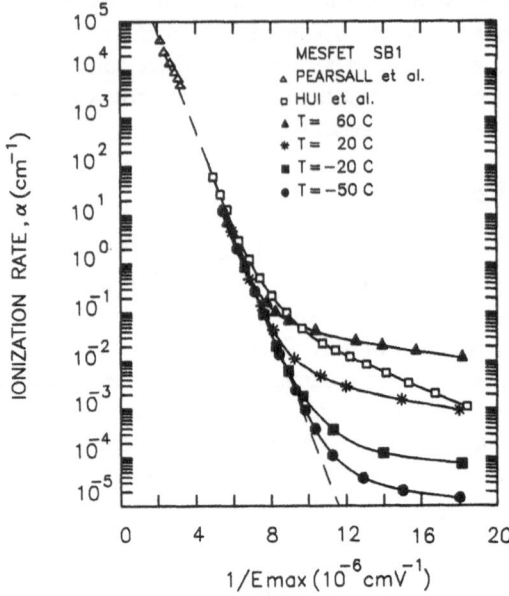

Fig. 8. Values of impact ionization coefficient in <110> GaAs obtained according to Eq. (2) from data of Fig. 7. The data from Pearsall et al [24] and from Hui et al. [23] have been included for comparison. The dashed line is a graphic plot of Eq. (3). Data refer to the same device of Fig. 6, [22]

In Fig. 8 the data reported in Fig. 7 was reorganized in order to obtain a plot of the ionization rate versus 1/Emax according to the equation

$$Ig/Id = \alpha_n(Emax) \cdot Leff \qquad (2)$$

The data on the impact coefficient in the <110> direction in GaAs measured by Pearsall et al. [24] at high electric fields and those obtained by Hui et al. [23] are also reported. The dashed line represents the equation

$$\alpha_n = 4.0 \times 10^6 \exp(-2.3 \times 10^6/E) \qquad (3)$$

Equation (3) is an excellent fit to the data obtained at high electric field, from $\alpha_n = 10^5$ down to $\alpha_n = 10^{-4}$ cm$^{-1}$, over nine orders of magnitude, [22].

At low electric fields, or better at low Vds, the experimental data departs from the exponential dependence given in (3) because, in this range, Ig is no more dominated by the collection of holes generated by impact ionization, but is mainly due to the reverse current of the gate diode, [22].

### Ig vs Vgs at constant Vds

When measured at constant and high Vds, in the impact ionization regime, Ig generally shows a non-monotonic behaviour, whose shape depends on device technology. Most commonly, we observe a "bell shaped" behaviour, as shown in Fig. 9, [18]. The position of the maximum usually varies from Vgs= -1 V to Vgs= +0.5 V, in particular

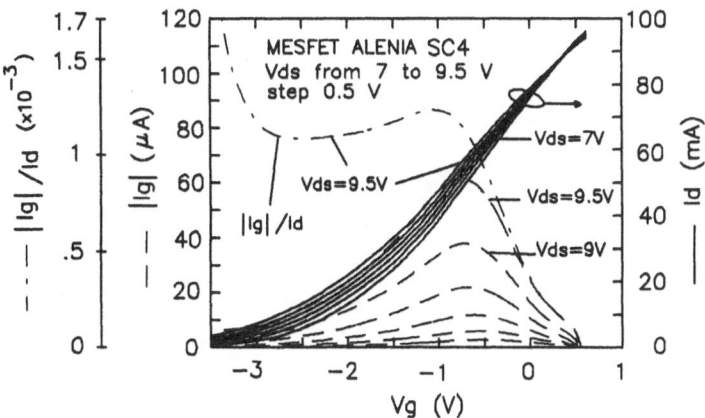

Fig. 9. Ig and Id as a function of Vgs at various Vds from 7 to 9.5 V, step 0.5 V measured in an Alenia MESFET. The ratio |Ig|/Id measured at Vds=9.5 V is also reported, [18]

depending on channel doping, recess profile and density of surface states. The Vgs value at which the maximum in Ig occurs can become so large that Ig is always increasing up to the bias point where the Schottky gate diode begins to be forward biased, as in the case of samples shown in Fig. 10, [16].

For the device showing a non-monotonic Ig behaviour, Fig. 9, the Ig/Id ratio is almost constant for -3 V<Vgs<-0.7 V and decreases for Vgs>-0.7 V; below -3V a sudden increase occurs due to the gate diode reverse current, which becomes dominant in these conditions. Drain current, Id, increases monotonically from Vgs= -0.3 V to Vgs= +0.5 V.

In Fig. 11 the ratio between the gate and the drain current, measured at two different temperatures, +20°C and -50°C, as a function of Vgs, for several values of Vds is reported [25]. It can be seen that the Ig/Id ratio increases in the whole range of Vgs and Vds explored as the temperature is lowered from +20 °C and -50 °C. This behavior confirms that Ig is caused by impact ionization; in fact an opposite behaviour would have been expected if the

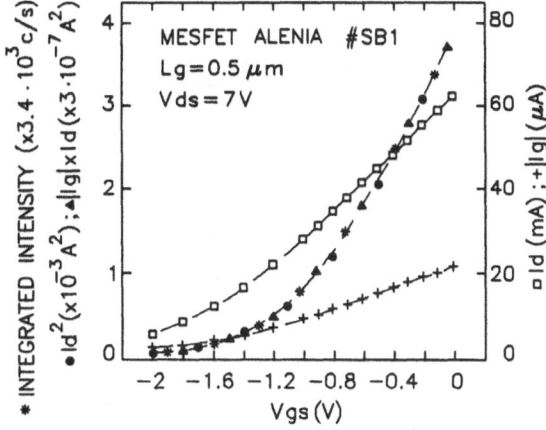

Fig. 10. Light intensity integrated over 1.5-3.1 eV energy range, Id, Ig Id$^2$ and Ig·Id vs Vgs at constant Vds=7 V for an Alenia MESFET, [16]

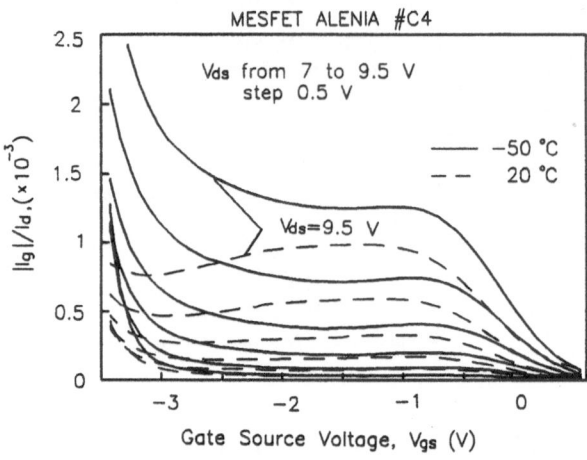

Fig. 11. The ratio |Ig|/Id versus Vgs at fixed Vds from 7 to 9.5 V, step 0.5 V, in an Alenia MESFET. Measurements obtained at T= -50 °C (continuous lines) and at T= 20 °C (dashed lines) are reported, [25]

contribution to Ig were related to tunneling through the gate Schottky barrier. A similar bell-shaped behavior was observed also for substrate current, [25].

The overall behaviour of Ig and Ib vs. Vgs is similar to that already reported for the substrate current Ib vs Vgs in Si MOSFETs, which also shows a "bell-shaped" curve as a function of Vgs [26].

The qualitative model proposed explains the bell-shaped behaviour of the genaration rate, as a superposition of two contrasting effects: as Vgs is moved from the pinch-off condition toward positive values, the number of electron drifting from source to drain increases, while their mean energy, due to the drop of the electric field in the channel, decreases. Initially the first mechanism dominates, so that the generation rate and Ig are roughly proportional to Id; eventually, since the ionization rate is exponentially dependent on the electric field, the generation rate and Ig reach a maximum and then drop.

This model is confirmed by numerical simulations performed by the two dimensional drift-diffusion HFIELDS simulator [11-14, 25]. In Fig. 12 the simulated Ig vs. Vgs curve, in good agreement with the measured one, is shown [25]. In the simulator thermal generation-

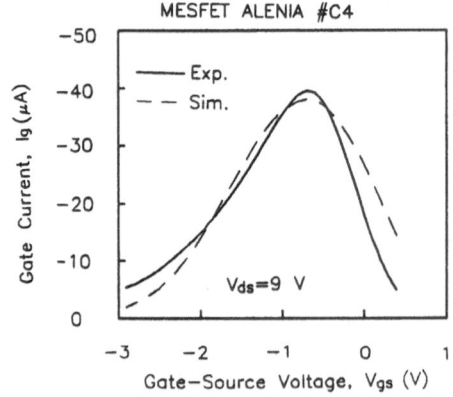

Fig. 12. Experimental (continuous line) and simulated (dashed line) Ig vs Vgs at Vds=9 V in an Alenia MESFET

recombination is accounted for through the standard Shockley-Read-Hall (SRH) model, [27], while generation by impact ionization is included in a self-consistent manner following the work of Laux et al., [11, 12]. Surface states are also taken into account through the introduction of a density $Nt=1.2x10^{12}$ cm$^{-2}$ of acceptor levels, modeled according the SRH theory, with energy 0.7 eV below the conduction band. The ionization rates for electron and holes are calculated according to the Chynoweth's law, [20]:

$$\alpha_{n(p)} = \alpha_{\infty} \exp\left[ -(B/F_{n(p)})^m \right] \qquad (4)$$

where the values of the parameters $\alpha_{\infty}=4x10^6$ cm$^{-1}$, B=2.3x10$^6$ V/cm, m=1 for electrons and $\alpha_{\infty}=3x10^5$ cm$^{-1}$, B=6x10$^6$ V/cm, m=3 for holes are taken from [23, 25].

The argument $F_{n(p)}$ is defined as the max(0, $E \cdot J_{n(p)}/|J_{n(p)}|$), where $E$ is the electric field and $J_{n(p)}$ is the electron (hole) current density.

### III.c  Light emission

Figures 13 and 14 show the photon energy distribution from 1.1 to 3.1 eV obtained in a typical Alenia MESFET and a Mitsubishi MGF1402 one respectively, taken at room temperature, at different Vds and at constant Vgs=0 V [16, 19]. Data are not corrected for self-absorption effects, which will be discussed in more detail in Section IV devoted to HEMTs.

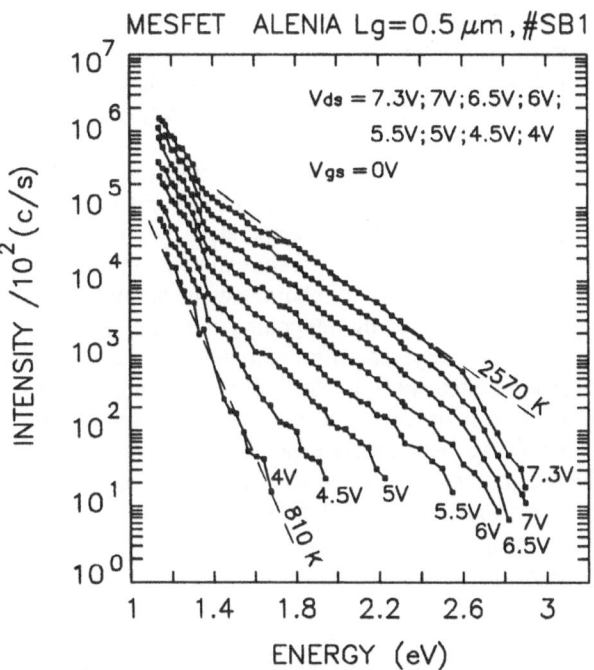

Fig. 13. Emitted light intensity as a function of energy at 300 K at various Vds and Vgs=0 V. Experimental data refer to Alenia MESFET. Electron temperatures extrapolated from the slope of the spectra (dashed lines) are shown. [16, 19]

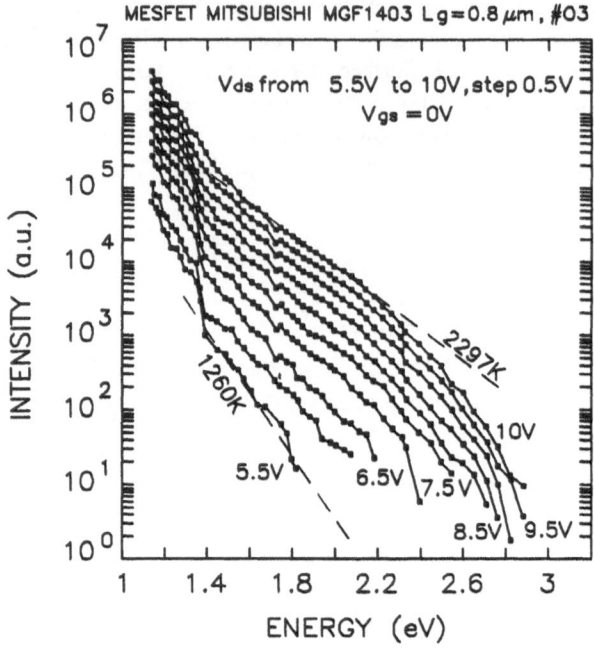

Fig. 14. Emitted light intensity as a function of energy at 300 K at various Vds and Vgs=0 V. Experimental data refer to Mitsubishi MESFET. Electron temperatures extrapolated from the slope of the spectra (dashed lines) are shown, [26, 19]

The spectra of both devices can be divided into three energy ranges. The first region spans approximately from 1.1 eV to 1.5 eV and is characterized by a kink at about 1.4 eV. In the second region, from 1.5 eV to 2.6 eV, the experimental data fit an exponential distribution (dashed lines). Finally the emitted light intensity steeply decreases at energies greater than 2.6 eV.

The central part of the spectrum is in good agreement with a maxwellian distribution. The equivalent temperatures of these distributions can be evaluated from the slope of the energy spectra and lie in the 800-2600 K temperature range. Both the intensity of the emitted light and the equivalent electron temperature increase at increasing drain voltages.

To study the emission mechanism of photons with energy higher than the energy gap, Eg, light intensity has been measured by integrating the light emitted over the 1.5-3.1 eV energy range and it has been compared with drain and gate currents. Figure 15 shows the behaviour of integrated light intensity (Iph), Ig, and the Ig·Id product as a function of Vds at constant Vgs=0 V. As it can be seen, when Ig is dominated by impact ionization, for Vds>5V, all curves exhibit the same nearly-exponential behaviour as a function of Vds, i.e. as a function of the longitudinal electric field. Since in the pre-breakdown regime Id is almost constant on increasing Vds, from Fig. 15 we can not discriminate whether light intensity is proportional to Ig or to the product Ig·Id. In Fig. 16 and in Fig. 10 we report the integrated light intensity, Ig, Id and the Ig·Id and Id$^2$ products as a function of Vgs at constant Vds, for two different MESFETs with different positions of the bell-shape maximum. Integrated light intensity does not appear to be simply proportional to Id or Ig. On the contrary a good correlation is generally found with both Id$^2$ or Ig·Id up to the maximum in Ig; for more positive Vgs values the correlation of Iph with Id$^2$ is lost, see Fig. 16, and integrated light intensity decreases proportionally to Ig·Id.

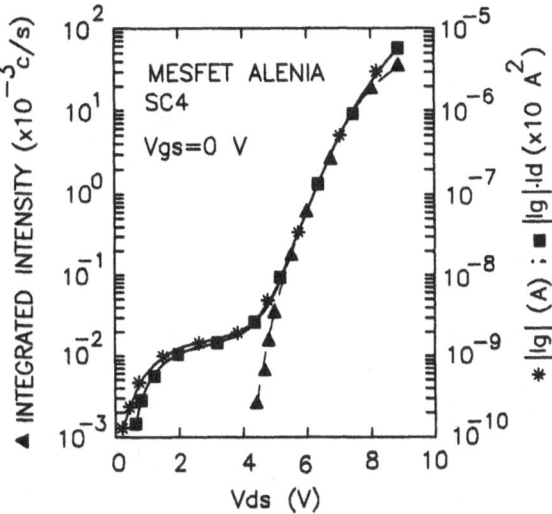

Fig. 15. Light intensity (integrated over 1.5-3.1 eV energy range), Ig and |Ig|·Id product as a function of Vds with Vgs=0 V in logarithmic scale. Ig and |Ig|·Id have been multiplied by constant coefficients. Normalizing constant have been used for graphical reasons

## III.d  Discussions

The correlation between integrated light intensity and $Id^2$ or the product Ig·Id can be better evaluated in Fig. 17 which shows the linear dependence of the integrated light intensity as a function of $Id^2$ and Ig·Id obtained varying Vgs from -2 V to 0 V at two fixed Vds values for the device shown in Fig. 10 [16]. The linear dependence of light intensity on $Id^2$ or Ig·Id has been verified for different Vds values, all in the pre-breakdown regime. This correlation can be explained by a simple model. According to Eq. (2) the current of holes generated by impact ionization can be approximated as Ihole = $\alpha_n$(Emax)·Leff·Id. If the emission of high energy photons is induced by the recombination of holes generated by

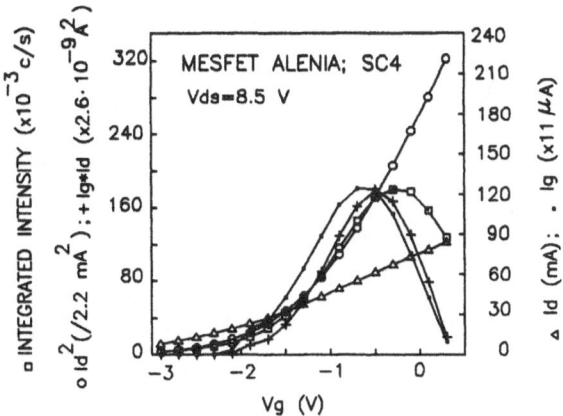

Fig. 16. Light intensity (integrated over 1.5-3.1 eV energy range), |Ig|·Id and $Id^2$ as a function of Vgs at constant Vds=8.5 V. Normalizing constants have been used for graphical reasons

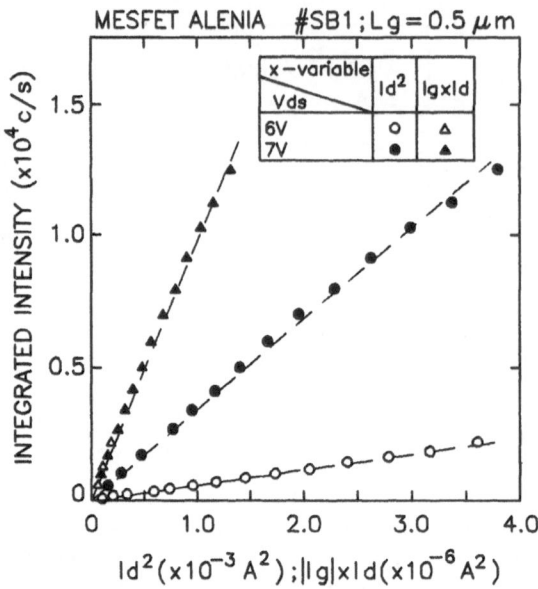

Fig. 17. Linear dependence of integrated light intensity on $Id^2$ and .Ig·Id obtained varying Vgs from -2 to 0 V at two fixed Vds values, [16]

impact ionization with channel electrons, the integrated light intensity should be proportional to the product Ihole·Id. When, at very high Vds and electric fields, the variation of $\alpha_n$(Emax)·Leff induced by changes in Vgs is negligible compared to the variation on Id, Iph∝Ihole·Id∝$Id^2$ as actually occurs for all Vgs values ranging from pinch-off to the maximum in gate current. For more positive Vgs values, the variation of $\alpha_n$(Emax)·Leff as a function of Vgs is no more negligible; as a consequence, Iph is only proportional to Ig·Id.

At increasing Vds, $\alpha_n$(Emax)·Leff increases thus increasing the slope of integrated light intensity against $Id^2$ curves as reported in Fig. 17. [16].

On the other hand, all holes generated by impact ionization are collected by the gate

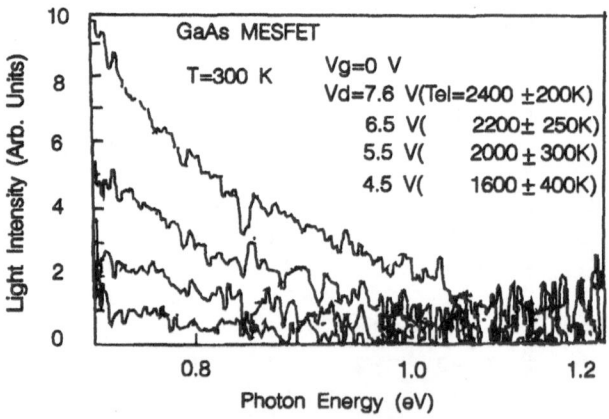

Fig. 18. Continuous background emission of a GaAs MESFET below the band-gap energy for several drain voltages. Tel gives the decay constant for a fit to an exponential decay, [31]

electrode, since the fraction of holes which recombines is negligible. Gate current is indeed of the order of $10^{-5}$ A which corresponds to $\approx 10^{14}$ holes collected per second, whereas the total number of photons detected per second is $< 10^{5}$. This suggests that the fraction of holes which recombine is negligible compared to Ig, even taking into account a very small radiative efficiency. If the gate current is dominated by the collection of holes and not by the gate diode reverse current, we expect therefore that the intensity of light would be also proportional to the product of Id and Ig as shown in Figs. 10, 16 and 17.

The correlation of light intensity with $Id^2$ and Ig·Id suggests that the dominant emission mechanism of visible light in GaAs MESFETs biased at high drain voltages in the prebreakdown region is the recombination of channel electrons with holes generated by impact ionization. On the contrary, emission of high energy photons due to bremsstrahlung would give rise to radiation emission proportional to the number of highly energetic electrons and therefore to Id or Ig separately.

For what concern the light emission with energy below the energy gap, few data are reported in literature [28-31]. The exponential energy dependence of the subgap spectrum, Fig. 18, has been explained by bremsstrahlung mechanism [31]. A strongly polarized component of electroluminescence, for $h\nu < Eg$, further substantiates the existance of radiative elastic collisions (bremsstrahlung) [29].

The measurement of gate current vs Vds and Vgs can be used to quantitatively evaluate impact ionization effects in GaAs MESFETs. Furthermore by simple analytical model [22, 23], electron impact ionization coefficients can be evaluated as a function of longitudinal electric field. However in order to achieve a deeper understanding of impact ionization, light emission and their correlation, a more detailed description of holes and electrons energy distribution function and of their distribution on different band minima is required, which can be obtained by means of suitable Monte Carlo simulations, [2, 3].

## IV. AlGaAs/GaAs HEMT

Very little data is available in the literature concerning impact ionization effects in AlGaAs/GaAs HEMTs. Experimental and theoretical studies have shown that tunneling is the dominant breakdown mechanism when the gate is made on a heavily doped AlGaAs layer [6]. In the case of a non-heavily doped AlGaAs layer, and in particular for delta-doped designs, breakdown is due to impact ionization. As reported in the following the question of determining whether impact ionization takes place in the channel or in the donor layer can be studied by measuring the bias dependence of the Ig/Id ratio. Spectral analysis of emitted light can provide further insights into the vertical (between the various device layers) distribution of electrons and holes when the transistor is operated at high biases.

### IV.a Samples

The devices studied in the present investigation were purchasable discrete depletion mode AlGaAs/GaAs microwave HEMTs, Toshiba S8901 and NEC NE20200. Figure 19 reports the 300 keV Transmission Electron Micrograph of the cross-section of the gate region of Toshiba HEMT, [4]. Toshiba devices were characterized by a recessed gate with gate length Lg= 0.3 μm and gate width W= 200 μm, and by gate-to-source and gate-to-drain contact spacing Lgs= 0.5 μm and Lgd= 1.5 μm, respectively. The micrograph shows the heterojunction formed by the AlGaAs layer, $\approx$ 30 nm thick, on the undoped GaAs layer, $\approx$ 350 nm thick. An AlGaAs/GaAs (approximately 30 nm thick) superlattice buffer is used for separating the active device from the semi-insulating substrate. Devices have recessed

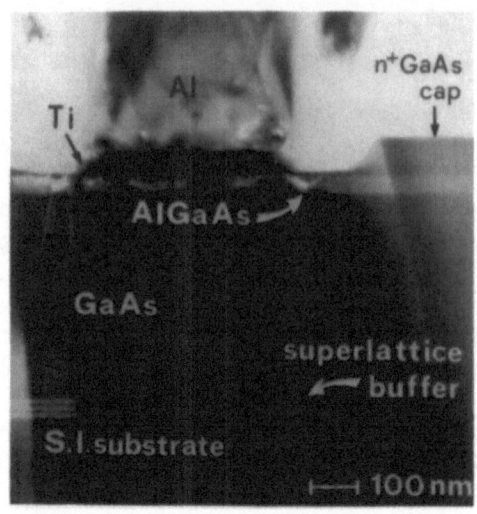

Fig. 19. 300-keV Transmission Electron Micrograph of the cross section of a
Toshiba S8901 AlGaAs/GaAs HEMT, [4]

Al/Ti gates and an $n^+$ GaAs cap layer, which is approximately 60 nm thick and extends from
the gate edge to the ohmic contact. The NEC devices show layout and technology similar to
the Toshiba ones, with Lg= 0.35 μm, Lgs= 0.6 μm and Lgd= 1.2 μm. All devices adopt
AuGeNi source/drain ohmic contacts, [4].

Typical drain and gate current characteristics of a Toshiba S8901 HEMT are shown
in Fig. 20. The reverse gate-drain current, Igdo, measured keeping the source floating, is

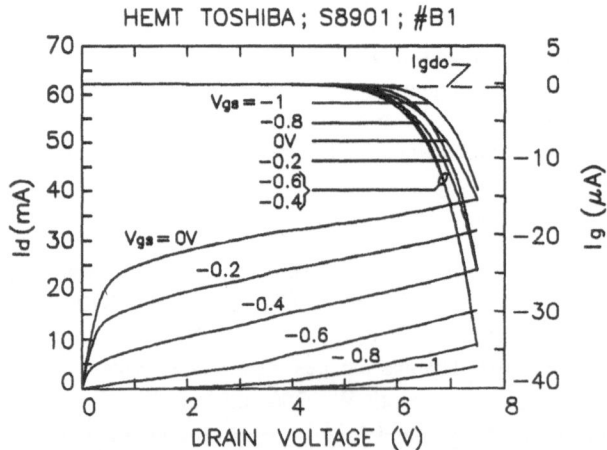

Fig. 20. Id and Ig characteristics vs Vds at fixed Vgs for a typical
Toshiba S8901 AlGaAs/GaAs HEMT. Igdo is the gate-drain diode
reverse current measured with the source kept floating, [4, 32]

reported as a dashed line in the same figures. When the devices are biased at high drain voltages, Vds>4 V, an increasing gate current (up to 40 μA for the Toshiba HEMT and 5 μA for the NEC one) exiting from the device is measured. As already observed in GaAs MESFET devices, Section III, this excess gate current is due to holes generated by impact ionization and collected at the gate electrode.

Despite the remarkable increase in Ig, however, avalanche does not take place up to Vds=7.5 V, and the amount of generated hole-electron pairs is negligible compared to the channel electron, see Fig. 20; consequently, the percentage increase of the drain current is very low, [4, 32].

### IV.b  Impact ionization

To investigate impact ionization phenomena the gate current has been studied as a function of Vds at constant Vgs and as a function of Vgs at constant Vds as for MESFETs, [Section III]. It has been shown that, as in GaAs MESFETs, also in AlGaAs/GaAs HEMT devices the gate current due to impact ionization can be approximated in the prebreakdown regions by Eq. (2), [32]. Accordingly, the dependence of $\alpha_n$ on the electric field can be evaluated by plotting Ig/(Id·Leff) vs 1/Emax=Leff/(Vds-Vdsat) in a semilogarithmic scale.

Figure 21 shows the plot of $\alpha_n$ versus 1/Emax obtained at four different temperatures between 60 °C and -50 °C. The Schottky gate reverse current dominates Ig at low drain voltages (i.e. at high 1/Emax values) leading to incorrect results. Reverse current can be reduced by lowering the temperature, so that the impact ionization coefficient can be correctly evaluated also at low electric fields, as shown in Fig. 21, [32].

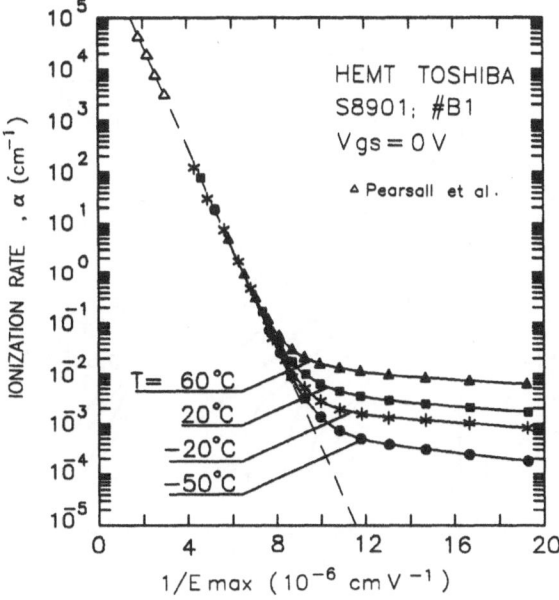

Fig. 21. Values of impact ionization rate obtained according to Eq. (2) for a AlGaAs/GaAs Toshiba HEMT at various temperatures. The data from Pearsall et al. [24] have been included for comparison. The dashed line is a graphic plot of Eq. (3), [32]

In order to evaluate $\alpha_n$ from the experimental data, we extracted Vdsat = 0.5 V from the device electrical characteristics, and used Leff as a fitting parameter in order to obtain a reasonable agreement with data reported in the literature either for GaAs [24, 33] or for AlGaAs [34, 35]. Figure 21 reports the data measured by Pearsall et al. [24] on <110> GaAs p-n junctions, and the fit of these data obtained by using the function $\alpha_n = a \cdot \exp(-b/E)$ with $a = 4 \times 10^6$ cm$^{-1}$ and $b = 2.3 \times 10^6$ V/cm (dashed line) [23, 39]. A good agreement between present data and those measured on <110> oriented GaAs [24, 33-35] can be obtained by using Leff = 0.25 $\mu$m, i.e. a value slightly lower than the nominal gate length Lg=0.3 $\mu$m, as shown in Fig. 21. On the contrary, since in the $0.2 < x < 0.3$ range the ionization rate of $Al_xGa_{1-x}As$ is one order of magnitude lower than that of GaAs [24, 33-35], unrealistic values of Leff would be required to fit the experimental data. This would suggest that impact ionization does not occur significantly in the AlGaAs layer, where the energy and drift velocity of electrons are smaller than in GaAs and the threshold energy for impact ionization is higher. Consequently, those electrons travelling in the AlGaAs layer as a result of parasitic MESFET effects [36] or real space transfer [37] do not contribute significantly to impact ionization, [32].

When measured at constant and high Vds in the impact ionization regime as a function of Vgs, Ig shows the same bell-shaped behaviour reported for MESFET in Fig. 9. The decrease of Ig when Vgs moves towards positive voltages can be ascribed to two different effects: (i) the decrease of the longitudinal electric field which takes place as Vgs is increased; (ii) the transfer of electrons to the AlGaAs, where they are not able to impact-ionize; this real-space transfer effect is enhanced at high Vgs, [37].

The presence of a non-monotonic behaviour of Ig also in conventional GaAs MESFETs without AlGaAs buffer or substrate layers [13, 25], suggests that (i) is the dominant effect both in MESFET and HEMT devices.

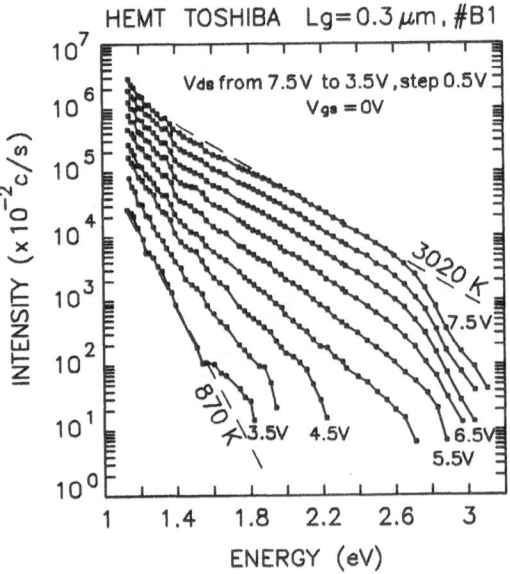

Fig. 22. Emitted light intensity as a function of energy at T=300 K at various Vds and Vgs=0 V. Experimental data refer to Toshiba HEMT. Electron temperatures extrapolated from the slope of the spectra (dashed lines) are shown, [4]

## IV.c Light emission

Figure 22 shows the photon energy distribution from 1.1 to 3.1 eV obtained in a typical Toshiba HEMT at room temperature by varying the drain voltage in the range 3.5 V - 7.5 V at Vgs=0 V [4, 38]. The energy spectra for a typical NEC device in the same bias conditions are reported in Fig. 23, [4, 38]. The optical transmittance of a GaAs layer is also reported at the top of Fig. 23 for two different layer thicknesses of 10 and 100 nm, respectively. The absorption coefficient data were taken from [39].

The spectra of both devices can be divided into three energy ranges. The first region spans approximately from 1.1 to 1.5 eV and is characterized by a kink or a peak at about 1.4 eV. The intensity of this peak increases with the drain-to-source voltage, and is much higher for the NEC device, Fig. 23, than for the Toshiba HEMT, shown in Fig. 22. In the second region from 1.5 to 2.6 eV, the experimental data fit an exponential distribution (dashed lines). Finally, the emitted light intensity steeply decreases at energies greater than 2.6 eV.

The peak in the first region is produced by the direct recombination of cold electrons, at the bottom of the conduction band, with cold holes, at the top of the valence band in GaAs (Eg=1.4 eV).

The central part of the spectra is in good agreement with a maxwellian distribution. Assuming the electron energy distribution to be reproduced by the photon energy distribution [40], the characteristics electron temperatures can be evaluated from the slopes of the energy spectra and lie in the 900-3100 K temperature range. Both the intensity of the emitted light and the equivalent electron temperatures increase at increasing drain voltages.

The shape of the spectra at high energy values (hv >2.6 eV) is distorted by light

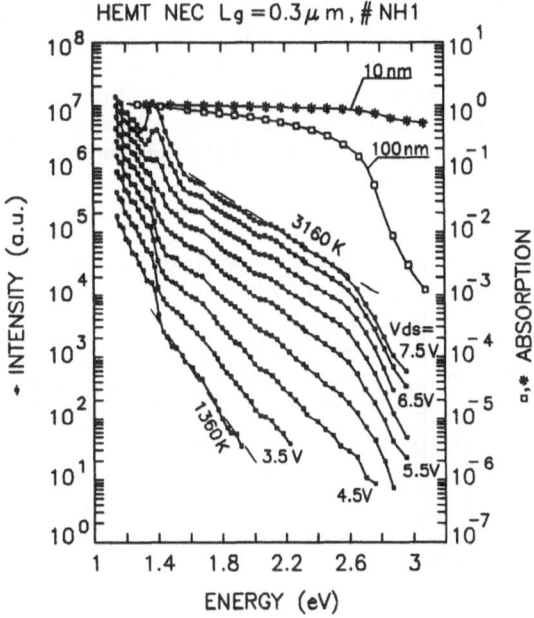

Fig. 23. Room temperature energy spectra of the emitted light at different Vds and Vgs=0 V fo a NEC HEMT. Optical transmittance of two GaAs layers, 10 and 100 nm thick, is shown at the top of the figure. Electron temperatures extrapolated from the slope of the spectra (dashed lines) are shown, [4]

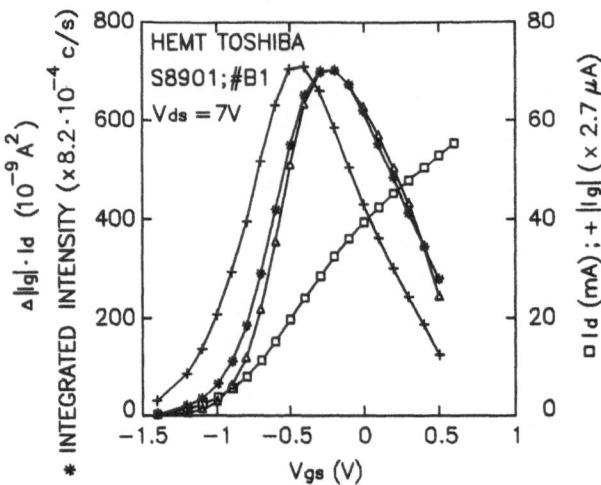

Fig. 24. Ig, Id, light intensity (integrated over the 1.7-3.1 eV
energy range) and |Ig|.Id as a function of Vgs with Vds=7
V measured in a Toshiba HEMT. Normalizing constants
have been used for graphical reasons, [4, 38]

absorption in the $n^+$GaAs cap layer. Direct observations in these devices indicate that light is
generated mainly in the gate-to-drain region, where the highest electric fields are present, as
reported also for MESFETs [7, 28] and MOSFETs devices [41]. In the measured HEMTs,
the emitted light passes through a $n^+$GaAs cap layer, about 100 nm thick, before being
detected, as shown in Fig. 19. For a 100 nm thick GaAs layer, which is comparable with the
thickness of the $n^+$GaAs cap layer in our samples, absorption can be considered negligible
for energies lower than 2.6 eV, while it suddenly increases at energies greater than 2.6 eV,
as reported in Fig. 23, producing the rapid fall of the light intensity detected in Figs. 22 and
23.

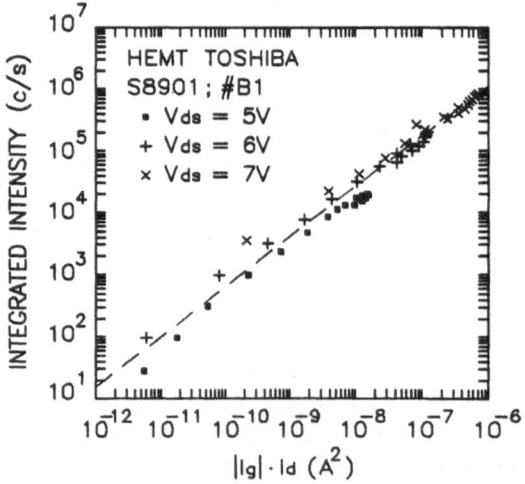

Fig. 25. Linear dependence of light intensity on
|Ig|.Id in a Toshiba HEMT. The device has been
biased as a function of Vgs for the three Vds
values reported in the figure, [4, 38]

The spectra do not show the presence of any peak in proximity of 1.7-1.8 eV, i.e. corresponding to the value of the energy gap of $Al_xGa_{1-x}As$ with x in the 0.2 - 0.3 range. The absence of this peak suggests that a detectable direct band-to-band recombination does not take place in the AlGaAs layer. Since a great amount of holes are generated by impact ionization and are collected at the gate electrode, thus crossing the AlGaAs layer, this would suggest that, at room temperature, AlGaAs may have a smaller radiative recombination probability than that of GaAs, or that very few electrons travel in the AlGaAs layer.

In order to discriminate the emission mechanism of photons with energy greater than Eg we integrated the light intensity, Iph, in the 1.7 - 6.2 eV energy range and analyzed it as a function of Vgs. Figure 24 shows |Ig|, Id, Ig·Id and Iph as a function of Vgs. Iph is not simply proportional to Ig while it seems proportional to Ig·Id. This correlation holds for various drain voltages and for more than five orders of magnitude, as Fig. 25 shows, [4, 38].

The behaviour of light intensity vs Vgs at different photon energies was also analyzed. Figure 26 shows |Ig|, |Ig|·Id and light intensity data as a function of Vgs for a Toshiba HEMT biased at Vds=7 V. Light intensity measurements were taken in a narrow band, about 0.01 eV, centered at 1.57 and 2.4 eV, using the monochromator as a filter, [4]. Maximum intensities were normalized to the same value by multiplying the values referring to the 2.4 eV curve by 18. As shown in Fig. 26, light intensity is never proportional to |Ig|, even taking into consideration photons emitted at high energy (2.4 eV). The light intensity vs Vgs curves measured at the two energies show only a slight difference at the most negative voltages, i.e. before the maximum. At increasing gate voltage towards positive values, the longitudinal electric field decreases [13, 25]; this decrease is reflected into a lower equivalent temperature of the photon spectra. As a consequence, the number of carriers having higher energy values is reduced and we observe a more steep decrease in light intensity as a function of Vgs in the curve taken at 2.4 eV with respect to the one taken at 1.57 eV.

## IV.d Discussions

In Si MOSFETs intensity of the radiation emitted in the hot-electron regime has been found to be proportional to the impact-ionization generated hole current, Isub, and light emission is usually attributed to the bremsstrahlung mechanism, [41-43].

On the contrary, in the HEMTs tested in the present work, light intensity has been found to be proportional to the product of the impact ionization generated current Ig with drain current Id, Fig. 25. Since the rate of recombination between electrons and holes would be also proportional to the product of the electron density (≈Id) with hole density (≈Ig), the

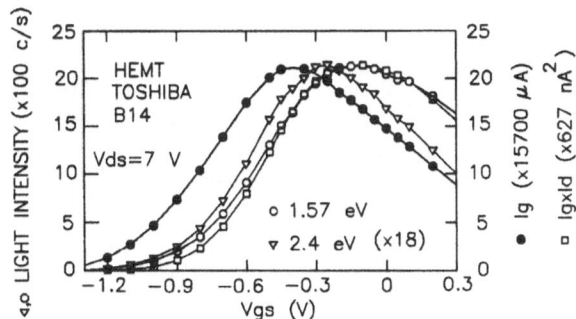

Fig. 26. Ig, |Ig|·Id product and the light intensity measured at 1.57 and 2.40 eV vs Vgs at Vds=7 V in a Toshiba HEMT. Normalizing constants have been used for graphical reasons, [4]

result shown in Fig. 25 suggests recombination as the dominant emission mechanism for high energy (>1.7 eV) photons in AlGaAs/GaAs HEMTs.

In HEMTs transistors, the experimental investigation of gate current and light emission can provide insights into vertical distribution of electrons and holes and of impact ionization events when devices are operated at high biases. In particular in uniformly doped AlGaAs/GaAs HEMTs, the data above reported suggest that impact ionization and radiative recombination occurs mainly in GaAs channel thus providing useful insights for the simulation of breakdown phenomena [44]. In δ-doped quantum well AlGaAs/GaAs HEMTs operated at 10 K and high Vds biases, the luminescence spectrum exibits a weak peak at 1.89 eV due to radiative recombination in $Al_{0.3}Ga_{0.7}As$; the intensity of this peak was used to evaluate the real space transfer effects vs Vds at low temperature in these devices [45, 46]. Furthermore the light emitted with energies <Eg was attributed to bremsstrahlung mechansim [45-47].


## V  Pseudomorphic AlGaAs/InGaAs HEMTs

The use of the AlGaAs/InGaAs heterojunction in PseudoMorphic High Electron Mobility Transistors (PM-HEMTs) provides several advantages over conventional AlGaAs/GaAs HEMTs because of the superior electronic transport properties of InGaAs and of the larger conduction band discontinuity. However power applications of conventional PM-HEMTs, having a constant and high doping level in the AlGaAs layer are limited by the occurrence of gate-drain breakdown due to tunneling effect at the gate contact [6]. The use of a δ-doped AlGaAs layer prevents this effect and also provides better transconductance and carrier confinement, higher channel electron density and drift velocity [6, 48]. The device current capability can be even further improved by inserting in the center of the InGaAs channel an additional planar doping layer [49].

Moreover the reduction of the InGaAs channel layer thickness to dimensions comparable to the electron wavelength increases the effective bandgap and therefore remarkably increases the breakdown voltage [50]. Recently it has been proposed to tailor the impact ionization coefficient in pseudomorphic GaAs-based and InP based structures by inducing a compressive strain in the channel layer [51].

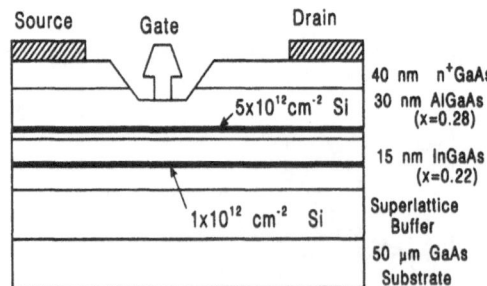

Fig. 27. Cross section of δ-doped and doped channel AlGaAs/InGaAs pseudomorphic HEMT [49]

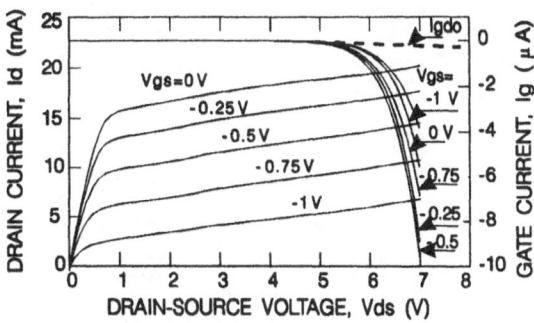

Fig. 28. Id and Ig current characteristics of a δ-doped and planar-doped channel AlGaAs/InGaAs PM-HEMT. Igdo is the gate-drain diode reverse current measured with the source electrode floating, [52]

## V.a Samples

The devices used in this work are high-power PM-HEMTs characterized by δ-doping in both donor and channel layer described in [49]. As shown in Fig. 27, they consist of an undoped superlattice buffer layer grown on top of semi-insulating GaAs substrate, followed by a 15 nm undoped $In_{0.22}Ga_{0.78}As$ channel with a planar doping of $10^{12}$ $cm^{-2}$ inserted in the center of channel, a 2 nm undoped $Al_{0.25}Ga_{0.75}As$ spacer layer, a silicon δ-doping of $5x10^{12}$ $cm^{-2}$, a 30 nm of undoped $Al_{0.25}Ga_{0.75}As$ and a 40 nm of $n^{+}GaAs$ cap doped at $6x10^{18}$ $cm^{-3}$.

## V.b Impact ionization

In the samples above described a remarkably high value of the gate-drain breakdown voltage, BVgdo=13 V at Igdo=6 µA (source floating), is measured, [52]. This value increases with increasing temperature, pointing out that the breakdown mechanism is avalanche multiplication [6]. Comparable values of BVgdo have been obtained by Bahl et al. in $InAlAs-n^{+}InGaAs$ HFETs [50] and were attributed to the enlargement of the effective energy gap caused by energy quantization in the InGaAs channel. However when the devices are biased in transistor configuration, impact ionization causes a noticeable increase of gate current already at Vds>5 V, limiting proper device operation. Typical drain and gate currents of PM-HEMT are shown in Fig. 28, [52]. At Vds>5 V, a remarkable increase of Ig is clearly detectable. As already observed in GaAs MESFETs [Section III] and AlGaAs/GaAs HEMTs [Section IV], this excess gate current is due to the collection of holes generated by impact ionization. Ig is linearly proportional to Id and exponentially proportional to the negative inverse of the electric field, in agreement with impact ionization theory [20, 21]. At Vds<7 V electric fields and carrier energies are not so high to give rise to avalanche multiplication and the amount of generated hole-electron pairs is negligible compared to the channel electrons; as a consequence, an increase of Id is not detectable. Only for Vds>7 V does breakdown occur, usually followed by device destruction, [52].

In impact ionization regime at high and constant Vds ranging from 5 to 7 V, Id, Ig and the ratio Ig/Id show a behaviour similar to the one reported for MESFET in Fig. 10, with the same dependence of Ig on electric field already discussed, [52].

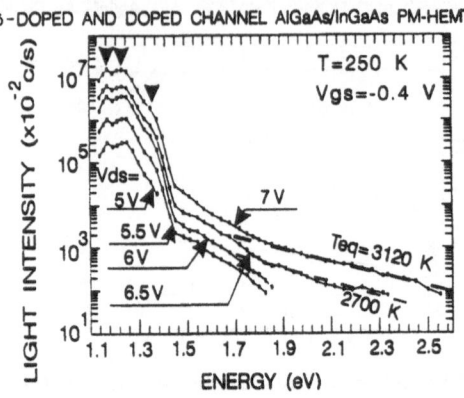

Fig. 29. 250 K emitted light intensity as a function of energy at different Vds values and at constant Vgs=-0.4 V in a AlGaAs/InGaAs PM-HEMT. Electron temperature extrapolated from the slopes of the spectra (dashed lines) are shown, [52]

## V.c Light emission

When biased in the impact ionization regime, devices emit light. Figure 29 shows the photon energy distribution in 1.1-2.6 eV range, measured at T=250 K for various Vds, from 5 to 7 V, and with Vgs= -0.4 V, [52]. Two peaks at approximately 1.18 and 1.24 eV can be observed in the spectra. Photoluminescence spectra at 10 K obtained on the same $In_{0.22}Ga_{0.78}As$ pseudomorphic structures [53], also show the presence of two similar peaks at 1.22 and 1.28 eV. The peak at higher energy has been associated with the quantum-well transition between the first allowable states in the conduction and valence bands, and the lower energy peak to a conduction band-carbon acceptor transition [53, 54]. The spectra in the 1.3-1.7 eV range show the presence of some shoulder, as the one at 1.38 eV (indicated by arrow), which could not be resolved, Fig. 29. They could possibly be attributed to recombination from higher levels in InGaAs quantum well, or to band-to-band recombination in the GaAs cap layer or in the superlattice buffer.

Emission at high energy values (greater than 1.7 eV) is due to hot-electrons and shows a nearly maxwellian distribution. The equivalent temperatures of the distribution increase on increasing Vds and are 3120 K for Vds=7 V and 2700 K for 6.5 V, respectively.

In order to discriminate the emission mechanism of high energy photons, light intensity integrated for hv>1.5 eV has been analyzed and compared with |Ig|, Id and the |Ig|.Id product in Fig. 30, [52]. On increasing Vgs from -0.5 V towards positive values, |Ig| and |Ig|·Id steeply decrease while light intensity shows only a slight decrease, with a minimum value which is approximately two thirds of light intensity maximum value. On the contrary, light intensity in MESFETs [Section III] and AlGaAs/GaAs HEMTs [Section IV] was found to be always proportional to the |Ig|·Id product, thus suggesting recombination of hot electrons with holes as the main emission mechanism. The emission of light when the concentration of holes is negligible (Ig≈0) suggests that in PM-HEMTs bremsstrahlung of electrons travelling in the channel results in a contribution to light emission larger than in MESFETs and conventional HEMTs, possibly due to the presence of the planar doping inside the channel which enhances the electron scattering with dopant impurities.

δ -DOPED AND DOPED CHANNEL AlGaAs/InGaAs PM-HEMT
Vds=6 V
T=300 K
• INTEGRATED INTENSITY (c/s)
60
40
20
0
-1.5
-1
-0.5
0
0.5
GATE-SOURCE VOLTAGE, Vgs (V)
•|Ig| (x60 μA); ◦Id (x3 mA)
◦|Ig|·Id (x5.5·10⁻⁹A²)

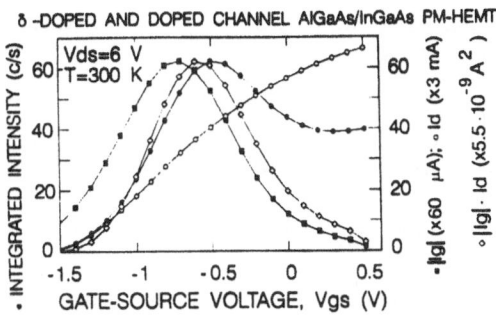

Fig. 30. |Ig|, Id and the |Ig|·Id product and the integrated (1.5- 6.2 eV) light intensity vs. Vgs at Vds=6 V measured in a δ-doped and planar-doped channel AlGaAs/InGaAs PM-HEMT. Normalizing constants have been used for graphical reasons, [52]

## VI AlGaAs/GaAs HBTs

There currently is great interest in III-V Heterojunction Bipolar Transistor (HBTs) for very high speed integrated circuits. Despite the fact that avalanche breakdown is a significant limiting factor of the microwave output power of AlGaAs/GaAs HBTs and markedly influences the design of collector composition and doping profile of these devices, only limited data on avalanche phenomena and impact ionization have been reported. An experimental and theoretical investigation of breakdown mechanism, microplasma, temperature dependance of avalanche phenomena is reported in [55, 56]. In particular the temperature dependence of breakdown occurring in common emitter configuration indicates that breakdown of AlGaAs/GaAs HBTs is primarily due to impact ionization occurring at the BC junction and not to tunneling effects, [56].

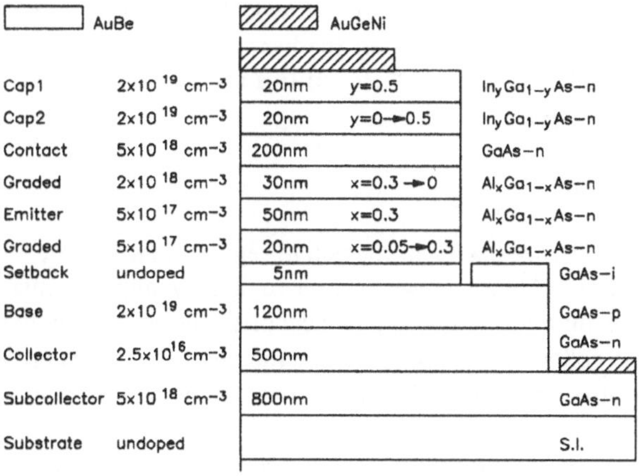

AuBe
AuGeNi
Cap1 2x10¹⁹ cm⁻³ 20nm y=0.5 InyGa1-yAs-n
Cap2 2x10¹⁹ cm⁻³ 20nm y=0→0.5 InyGa1-yAs-n
Contact 5x10¹⁸ cm⁻³ 200nm GaAs-n
Graded 2x10¹⁸ cm⁻³ 30nm x=0.3→0 AlxGa1-xAs-n
Emitter 5x10¹⁷ cm⁻³ 50nm x=0.3 AlxGa1-xAs-n
Graded 5x10¹⁷ cm⁻³ 20nm x=0.05→0.3 AlxGa1-xAs-n
Setback undoped 5nm GaAs-i
Base 2x10¹⁹ cm⁻³ 120nm GaAs-p
Collector 2.5x10¹⁶cm⁻³ 500nm GaAs-n
Subcollector 5x10¹⁸ cm⁻³ 800nm GaAs-n
Substrate undoped S.I.

Fig. 31. Schematic cross section of AlGaAs/GaAs npn HBT. An InGaAs layer was used on top of the emitter to improve ohmic contact characteristics. The emitter composition was graded towards the base to reduce conduction-band discontinuity at the junction, [60, 61]

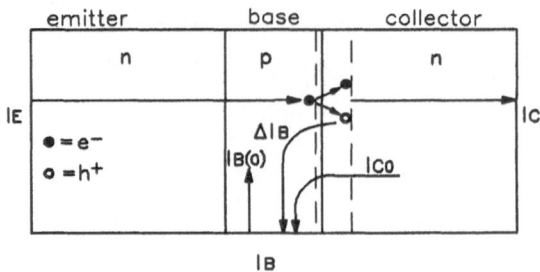

Fig. 32.Sketch of impact ionization phenomena in an HBT

A proper measurement and interpretation of light emitted from AlGaAs/GaAs HBTs may provide a deeper understanding of the physics of hot carriers in these devices. Unfortunately, very few data has been reported in the literature concerning electroluminescence in HBTs; moreover, this data mainly refer to breakdown of the base-collector, BC, junction [55, 57], or to recombination in the base-emitter, BE, junction at low electric fields [58, 59], rather than to the high electric field operation of the transistor biased in the active region.

In particular, the emission spectra of AlGaAs/GaAs HBTs measured with the BC junction in avalanche conditions have been shown to exhibit two peaks at 1.43 eV and 2.03 eV, whose intensity depends on the reverse current [57]. The high energy peak at 2.03 eV was attributed [57] to electrons excited by the high electric field to the upper conduction band, loosing their energy by impact ionizing electron-hole pairs, thus producing 2.03 eV photons, an energy value which corresponds to the threshold energy for impact ionization.

## VI.a Samples

Samples used in this work were npn single heterojunction AlGaAs/GaAs HBTs, grown by molecular beam epitaxy with the structure shown in Fig. 31 [60, 61]. The semiconductor layers consisted of a 800 nm GaAs-n layer ($5 \times 10^{18}$ cm$^{-3}$) which acts as the collector contact, followed by a 500 nm GaAs-n collector ($2.5 \times 10^{16}$ cm$^{-3}$). The GaAs-p base was 120 nm thick ($2 \times 10^{19}$ cm$^{-3}$) followed by an AlGaAs emitter ($5 \times 10^{17}$ cm$^{-3}$) compositionally graded immediately after the base to reduce conduction band discontinuity at the junction [60]. The epilayer was etched to form a two level mesa structure having an emitter diameter of 100 μm.

## VI.b Impact ionization

When npn HBTs are biased in the active region at high $V_{CB}$, electron-hole pairs are generated in the base collector space charge region by impact ionization. The generated electrons are swept into the collector, as schematically shown in Fig. 32, contributing a positive term to the collector current $I_C$, while the generated holes are injected into the base contributing a negative term to the base current, $I_B$, which increases on increasing $V_{CB}$ [61, 62].

Figure 33 shows $I_B$ measured as a function of $V_{CB}$ with the HBT biased in the common base configuration and driven at different emitter currents $I_E$. At $V_{CB}=0$ V impact ionization does not occur and $I_B$ is always positive. On increasing $V_{CB}$ the negative contribution to $I_B$, due to holes generated by impact ionization, becomes so high that $I_B$ decreases until it changes its sign and becomes negative in correspondence of the "dips" in

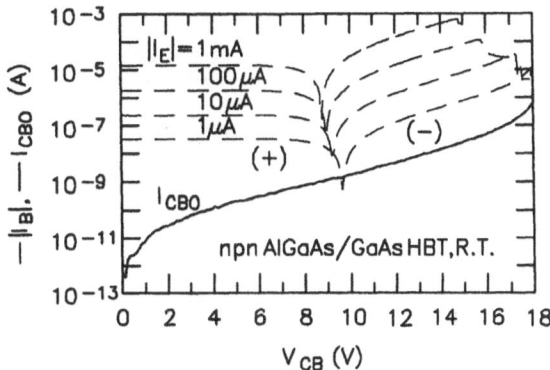

Fig. 33. Absolute value of the base current $I_B$ (dashed line) as a function of $V_{CB}$ with the device driven at different and constant $I_E$ in common base configuration. Data refer to AlGaAs/GaAs HBT. The dips are where $I_B$ changes its sign. The reverse current of the BC junction with open emitter, $I_{CBO}$, is also shown (continuous line), [63]

Fig. 33, [61, 63]. The quantity $\Delta I_B = I_B(V_{CB}=0) - I_B(V_{CB})$ is a measurement of the hole current due to impact ionization, neglecting the contribution of the base-collector junction reverse current $I_{CBO}$. $I_{CBO}$ remains negligible with respect to $I_B$ up to $V_{CB}$ values close to the breakdown voltage of the BC junction, $BV_{CBO}$, occurring abruptly at approximately 18 V [61].

Impact ionization effects can be quantitatively evaluated by measuring the M-1 coefficient, defined as the ratio of generated electron-hole pairs to the number of carriers injected in the collector:

$$M - 1 = \frac{\Delta I_B}{I_C - \Delta I_B} = \frac{I_{BO} - I_B(V_{CB})}{I_C(V_{CB}) - (I_{BO} - I_B(V_{CB}))} \tag{5}$$

where $I_{BO}$ is the base current without multiplication (i.e. $I_B$ at $V_{CB}$ = 0 V). This equation assumes that: i) the Early effect is negligible; ii) recombination current in the base does not change with $V_{CB}$; iii) holes generated by impact ionization would not end up in the emitter; iv) device self-heating is negligible [61, 62, 64]. The first two effects are negligible in our devices, while iii) and iv) are assumed to be small [61, 65]. M-1 can be written as [21, 62]:

$$M - 1 = \frac{1}{1 - \int\limits_{0}^{W} \alpha_n \cdot \exp\left[ -\int\limits_{0}^{x} (\alpha_n - \alpha_p)\, dx' \right] dx} - 1 \tag{6}$$

where W is the collector width and $\alpha_n$ and $\alpha_p$ are the electron and hole ionization coefficients respectively, which, according to Bulman et al. [66], follow the expression:

$$\alpha_n (E(x)) = 1.899 \times 10^5 \exp(-(5.750 \times 10^5 / E(x))1.82) \ (cm^{-1}). \tag{7}$$
$$\alpha_p (E(x)) = 2.215 \times 10^5 \exp(-(6.570 \times 10^5 / E(x))1.75) \ (cm^{-1}). \tag{8}$$

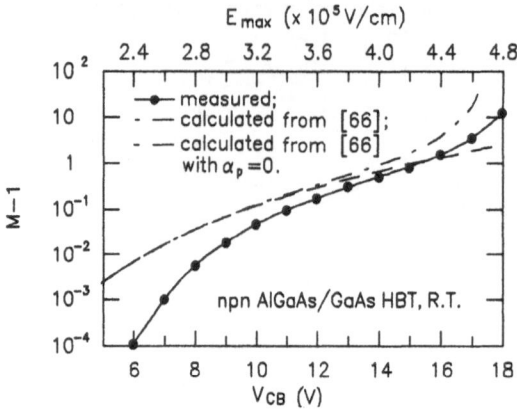

Fig. 34 Bias dependence of the multiplication factor in AlGaAs/GaAs HBT, [65]

To evaluate E(x), the Poisson's equation can be solved neglecting the free electron contribution in the space charge region, which is unimportant at the $I_E$ values considered. Fig. 34 shows the experimental values of M-1.

Experimental data were compared, Fig. 34, with the M-1 values calculated according to Eq. 6 excluding or not the contribution of the generated holes to impact ionization, i.e. assuming $\alpha_p$ (E (x)) = 0 (dashed line) or given by (8) respectively (dotted and dashed line), [61].

As it can be seen, both calculated curves largely overestimate the experimental values for $V_{CB}$<15 V. As demonstrated by Monte Carlo simulations of Si bipolar transistors [67], and confirmed for AlGaAs/GaAs devices [63, 65], electrons must travel a non-negligible distance within the collector before reaching the energy level sufficient to impact-ionize. In

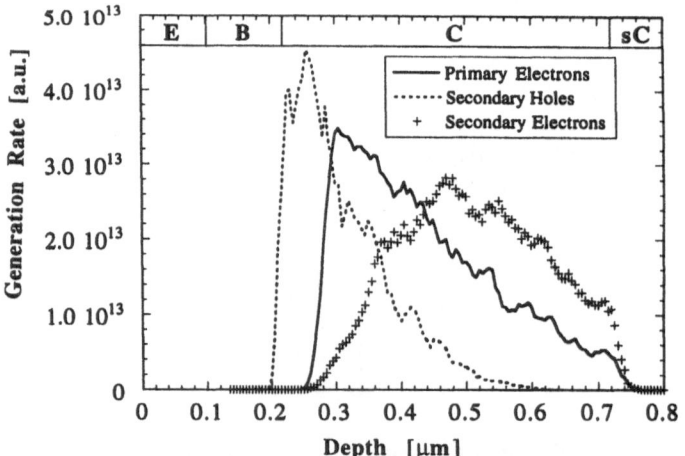

Fig. 35. Generation rate obtained by means of simulation in an AlGaAs/GaAs HBT at $V_{CB}$=18 V. The solid lines illustrates the contribution of primary electrons (those injected across the base from the E-B junction), while open circles and crosses refer to high-order contributions from holes and electrons repectively, [65]

fact this level is achieved deeper inside than where the maximum of the electric field occurs [63, 65, 67]. As a consequence, the M-1 values calculated as a function of the electric field overestimate the experimental data, since they assume a remarkable contribution to the integral in (2) from the region of maximum electric field.

In particular for the HBT structure reported in Fig. 31, Monte Carlo simulation shows that at $V_{CB}$=18 V electrons reach their maximum energy about 80 nm inside the collector [63, 65]. Most of them are found in the satellite L and X valleys, and move through the collector at saturated velocity. The ionization coefficient is further delayed due to finite time (and distance) required to reach the ionization threshold. As a consequence, ionization events do not occur in corrispondence to maximum electric field, nor to the maximum average energy, rather well into the collector. Such phenomenon, usually referred to as *Dead Space Effect*, is also evident in Fig. 35 where we plot the generation rate for both electrons and holes distinguishing between primary and secondary particles. At such high fields, the holes generated by primary ionization are able to ionize before reaching the base, thus creating electrons that will in turn ionize while drifting along the collector region. Such positive feedback marks, in fact, the onset of breakdown. The experimental breakdown is found to correspond to the intersection of generation rate of primary electrons with secondary carrier. The bias dependence of the generation rate is reflected in the behaviour of the multiplication factor M-1 illustrated in Fig. 34. The change in slope of the measured M-1 coefficient around 18 V can indeed attributed to the onset of higher-order ionization process. The Monte Carlo results agree well with the measured one.

## VI.c Light emission

For optical measurements a circular hole (30 μm diameter) was opened in the emitter metallization. The single photon counting system described in section II was adopted for measurements.

Figure 36 shows the spectrum emitted at room temperature by the base-collector junction forward biased at $I_C$ = -500 μA with open emitter (lowest curve), [63, 68]. A very intense peak at 1.44 eV and a broad and weak one at 2.09 eV can be observed. The same figure shows the spectra emitted by the HBT biased in the active region, in common base configuration at constant $I_E$ = -500 μA, for different $V_{CB}$.

At $V_{CB}$=0 V the emission spectrum is similar to that observed by forward biasing the BC junction, the only noticeable difference being the shift of the high intensity peak to a lower energy, i.e. 1.40 eV. We do not observe any change in the spectra on increasing $V_{CB}$ from 0 V up to 8 V. For $V_{CB}$ higher than 8 V, the intensity of the 1.40 eV peak remains practically unchanged, while the part of the spectra at hν>1.7 eV increases in intensity and changes its shape approximating a straight line, thus masking the presence of the peak at 2.09 eV. Moreover light intensity increases also for hν<1.3 eV.

In the whole energy range light intensity increases linearly with the current crossing the device.

To investigate the correlation between hot electrons and light emitted when the device is operating in the active region, we analyzed the dependence of the light intensity on $V_{CB}$ by integrating the emission over two different energy ranges; the first one from 1.38 eV to 1.55 eV (where the 1.40 eV recombination peak is dominant) and the second one from 1.77 eV to 2.06 eV (corresponding to the region of the spectrum mostly influenced by the increase in $V_{CB}$).

Figure 37 shows integrated light intensity in the two energy windows and $\Delta I_B$ as a function of $V_{CB}$. In the first energy window (upper curve) there is no relationship between

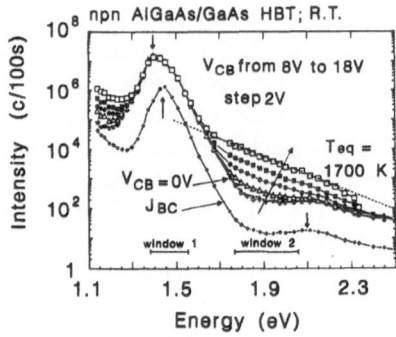

Fig. 36 Intensity of the light measured in an AlGaAs/GaAs HBT in the 1.1-2.5 eV energy range emitted at room temperature by: BC junction forward biased and emitter open ($I_C$=-500 mA, lowest curve, translated of a factor 1/100); HBT biased in the active region at different $V_{CB}$ and at constant $I_E$=-500 mA. Equivalent temperature $T_{eq}$=1700 K, extrapolated from the slope of the spectrum at $V_{CB}$=16 V is reported as a dashed line. Arrows indicate recombination peaks, [63, 68].

the emitted light intensity and $V_{CB}$ or $\Delta I_B$. In the second one the integrated light intensity follows closely the behaviour of the current generated by impact ionization, $\Delta I_B$, [63].

## VI.d  Discussions

When only low electric fields are present in the device, i.e. when the BC junction is forward biased with emitter floating, we observe a light emission distribution exibiting two peaks: a first one at 1.44 $e$V, very intense, and a second one at 2.09 $e$V, very broad and weak. The first one corresponds to direct recombination of cold electrons and holes occurring in the lightly doped collector region and corresponding to the energy gap of GaAs

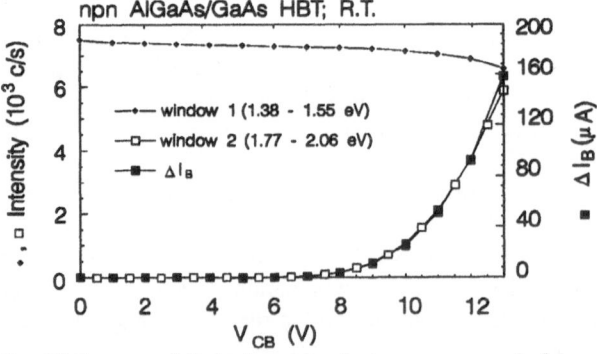

Fig. 37 Integrated light intensity in two energy windows (from 1.38 to 1.55 eV, and from 1.77 to 2.06 eV and $\Delta I_B$ as a function of $V_{CB}$ for an AlGaAs/GaAs HBT biased in the active region and driven at constant $I_E$= -500 mA, [63].

at room temperature (1.44 $eV$). In fact, owing to the high doping level of the base ($2x10^{19}$ $cm^{-3}$) with respect to the collector ($2.5x10^{16}$ $cm^{-3}$), the BC current is mainly due to holes diffusing from the base into the collector, where they recombine. This peak shifts to a lower energy (1.40 $eV$) when the HBT is operating in the active region; in this condition, in fact, the BC junction is reverse biased and electrons are injected from the emitter in the base, where part of them recombine. The shift in energy may be attributed to two effects: (i) band-gap narrowing due to the high p-type doping level in the base [69, 70], or (ii) recombination assisted by Be acceptor level. The absence of a luminescence peak at 1.7 $eV$ from the AlGaAs emitter confirms that the compositional grading at the BE junction preserves the holes confinement.

The broad peak at 2.09 $eV$ has an intensity which is about five orders of magnitude lower than that of the 1.44 $eV$ peak. A peak at 2.03 $eV$ was reported in [57] and attributed to recombination of hot electrons since its energy value corresponds to the threshold energy of electron-initiated impact ionization.

A peak at approximately 2.1 $eV$ was observed in photoluminescence spectra on heavily doped p-type GaAs [71, 72]. In particular in [72] it was shown that a peak at 2.1 $eV$ can be observed only in heavily doped p-type material independently of the nature of the dopant (Zn, Be, C) with an intensity which increases on increasing the doping level. This peak was attributed to a $L_{6c}$-$G_{7v}$ transition [72]. The presence of this peak in the spectrum of the forward biased BC junction where only very low electric fields are present, would exclude impact ionization or hot electron effects as emission mechanism, but suggests recombination between cold electrons and holes close to band minima as the most probable light emission mechanism, as reported in [72].

Finally, we observed that the energy value of the two peaks has a negative temperature coefficient as the energy gap.

When high electric fields are present in the device, i.e. when the device is operating in the forward active region at high $V_{CB}$, the following features are observed: (i) the intensity of the 1.40 $eV$ peak does not change significantly with $V_{CB}$, see Figs. 36 and 37. Only a slight (< 10%) decrease is observed, suggesting a lower radiative recombination in the base; (ii) for $V_{CB}$ < 8V we do not observe any change in the emitted spectra with respect to $V_{CB}$ = 0V; (iii) on increasing $V_{CB}$ above 8V a large increase of the emitted light is instead observed either for hν<1.3 $eV$ or hν>1.65 $eV$. In particular at high energies, hν > 1.7 $eV$, the light distribution becomes more and more maxwellian on increasing $V_{CB}$, giving an equivalent temperature Teq = 1700 K at $V_{CB}$ = 16V. Furthermore, the relative increase of light intensity integrated in the 1.77 $eV$ - 2.06 $eV$ energy range as a function of $V_{CB}$ is strongly correlated with the increase of the current $\Delta I_B$ generated by impact ionization, Fig. 37.

Despite the strong correlation observed between the light intensity and the impact ionization generated holes at $V_{CB}$ > 8V, we can not conclude that recombination between hot holes and hot electrons is the main emission mechanism. In fact, on increasing $V_{CB}$, we also observe an increase in the light intensity below 1.3 $eV$, which is difficult to explain with direct conduction to valence band recombination. Furthermore, it has been shown that, in order to explain the emission mechanism and light energy distribution in Si MOSFETs, other processes, such as direct and phonon assisted or impurity assisted conduction band to conduction band transition, may play a major role in the photon emission [1].

In conclusion, electroluminescence measurements on AlGaAs/GaAs HBTs in different bias conditions allowed us to observe and discriminate the light emitted owing to cold carrier recombination and to hot electron phenomena. Hot electron induced electroluminescence consists essentially in the emission of photons having a nearly exponential energy distribution, with an intensity strongly correlated with carrier generation due to impact ionization in the high field region at the BC junction.

## CONCLUSIONS

A proper measurement and interpretation of impact ionization phenomena and hot electron induced electroluminescence may provide a deeper understanding of the physics of hot carriers in submicron devices operated at high electric fields and suitable rules to design device structure in order to achieve better device performances.

The results presented above on GaAs based MESFETs, HEMTs, PM-HEMTs and HBTs can be summarized as follows.

The analysis of $I_g$ and $I_B$ current changes induced by impact ionization has demonstrated to be a powerful tool to investigate impact ionization in FETs ($I_g$) and bipolar ($I_B$) devices, respectively, and their dependence on device technology, presence of surface states, applied voltages, etc. Electron impact ionization coefficient vs electric field can be obtained by a simple model as reported for MESFETs and HEMTs. Results indicate that in AlGaAs/GaAs HEMTs impact ionization occurs in the GaAs channel and not in the AlGaAs/GaAs donor layer. To identify where impact ionization takes place in PM-HEMTs more detailed experimental data are required.

However, even if the reduction of InGaAs channel layer thickness to dimensions comparable to the electron wavelength and/or a strain induced in the channel layer have been demonstrated to increase gate-drain diode breakdown in PM-HEMTs, the limiting mechanism for high Vds applications when these devices are operated as transistor is the onset of impact ionization which induces a remarkable increase of the gate current. From $I_B$ changes, the multiplication factor M-1 can be evaluated in HBTs and Monte Carlo interpretation of experimental data and device simulation provide a detailed picture of carrier distribution function, of their location in different band minima and accurately predict the device breakdown demonstrating the fundamental role of secondary holes in determining the breakdown voltage.

When biased at high drain or collector voltages FETs and bipolar devices emit visible and infrared light. In general, the spectral distribution of the emitted radiation results from a superposition of: i) one or more peaks corresponding to conduction-band to valence-band recombination of cold electrons and holes in the various device layers, and ii) a nearly maxwellian tail due to hot carriers induced electroluminescence, whose intensity and equivalent temperature increase at increasing electric field. In FET devices and in particular in MESFETs and HEMTs the correlation of light intensity integrated above Eg with the product of electron and hole currents suggests that the main emission mechanism for visible light emission should be a two carriers radiative recombination (c-v transition) of channel electrons with holes generated by impact ionization, even though a contribution of ionized impurity-assisted (bremsstrahlung) or phonon assisted c-c or v-v radiation can not be ruled out. Moreover these mechanisms have been demonstrate to be the dominant scattering and radiative transitions in Si [1]. In particular the correlation between the integrated light intensity and $I_g \cdot I_d$ product indicates in channel doped PM-HEMTs a bremsstrahlung contribution larger than in MESFETs and AlGaAs/GaAs uniformly doped HEMTs possibly due to enhanced ionized impurity scattering in the doped channel. In HBT devices, the strong correlation observed at high electric fields between the light intensity and the impact ionization generated holes suggests a light emission mechanism mainly due to radiative transitions which involve only one type of carrier such as either conduction to conduction (c-c) or valence to valence (v-v) radiations even if we can not exclude that recombination between hot electrons and hot holes (c-v) can play a role in the high energy photon emission.

To achieve a deeper understanding of visible light emission in GaAs based advanced FET and bipolar devices, more sophisticated theoretical studies of all the important source of emission radiation must be considered and transport simulators including realistic band structures are required to obtain carrier distribution function and their location on the different bands [2,3, 73]. Unfortunately, such studies are today not available in literature.

## ACKNOWLEDGEMENTS

The authors would like to thank for many useful discussions and for sample provided: Federico Capasso and Roger Malik (AT&T Bell Labs.), Dwight C. Streit (TRW), Wallace T. Anderson (Naval Res. Lab.), Paolo Lugli (Univ. Roma), Antonio Cetronio (ALENIA); for assistance in optical measurements Stefano Bigliardi (Univ. Parma); for assistance in electrical measurements Paolo Telaroli, Paolo Pavan and Pietro Pisoni (Univ. Padova) and for Monte Carlo simulation Aldo di Carlo (W. Schottky Institute, Monaco).

## REFERENCES

[1] J. Bude, N. Samo, A. Yoshii, Hot carrier luminescence in Si, *Phys. Rev. B*, *45*, 5848 (1992).
[2] C. Jacoboni, P. Lugli, "The Monte Carlo method for semiconductor device simulation", Springer (1989).
[3] S. E. Laux, M.V. Fischetti, D.j. Franck, Monte Carlo analysis of semiconductor devices: the DAMOCLES program, *IBM J. Res. & Dev.*, *34*, 466 (1990).
[4] E. Zanoni, M. Manfredi, S. Bigliardi, A. Paccagnella, P. Pisoni, C. Tedesco, C. Canali, Impact ionization and light emission in AlGaAs/GaAs HEMTs, *IEEE Trans. Electron Devices*, *ED-39* (1992).
[5] C.A. Mead, Schottky barrier gate field effect transistor, *Proc. IEEE*, 307 (1966).
[6] Y. Crosnier, F. Temcamani, D. Lippens, G. Salmer, Avalanche and tunneling breakdown mechanisms in HEMT power structures, *J. de Physique, c4-563*, 49 (1988).
[7] J.M. Ashworth, H. Arnold, The gate bias dependency of breakdown location in GaAs Metal Semiconductor Field Effect transistors (MESFETs), *Japanese J. Appl. Phys.*, *30*, 3822 (1991).
[8] Y. Wada, M. Tomizawa, Drain avalanche breakdown in gallium arsenide MESFETs, *IEEE Trans. Electron Devices, ED-35*, 1765 (1988).
[9] H. Mizuta, K. Yamaguchi, S. Takahashi, Surface potential effects on gate-drain avalanche breakdown in GaAs MESFETs, *IEEE Trans. Electron Devices, ED-34*, 2027 (1987).
[10] J. Ashworth, P. Lindorfer, Analysis of the breakdown phenomena in GaAs MESFETs, *Int. Phys. Conf. Ser. No 112, Chapter 7, p. 395, 1990 and Proc. ESSDERC'90*, 241 (1990).
[11] S.E. Laux, R.J. Lomax, Numerical investigation on mesh size convergence rate of the finite element method in MESFET simulation, *Solid-State Electron*, *24*, 485 (1981).
12] S.E. Laux, B.M. Grossman, A general control volume formulation for modeling impact ionization in semiconductor transport, *IEEE Trans. Computer Aided Design, CAD-4*, 520 (1985).
[13] A. Neviani, C. Tedesco, E. Zanoni, C.U. Naldi, M. Pirola, Impact ionization phenomena in GaAs MESFETs: experimental results and simulation, *Int. Phys. Conf. Ser,. No. 120,: Chapter 5*, 267 (1992).
[14] G. Baccarani, R. Guerrieri, P. Ciampolini, M. Rudan, "HFIELDS: a highly flexible 2-D semiconductor device analysis program", Proc. 4th Int. Conf. on numerical Analysis of Semiconductor Devices and Integrated Circuits, J.J. Miller Ed., Boole Press (1985).
[15] E. Zanoni, S. Bigliardi, R. Cappelletti, F. Magistrali, M. Manfredi, A. Paccagnella, N. Testa, C. Canali, Light emission in AlGaAs/GaAs HEMTs and GaAs MESFETs induced by hot carriers, *IEEE Electron Device Lett., EDL-11*, 487 (1990).
[16] E. Zanoni, S. Bigliardi, M. Manfredi, A. Paccagnella, P. Pisoni, P. Telaroli, C. Tedesco, C. Canali, Correlation between impact ionization, recombination and visible light emission in GaAs MESFETs, *Electronics Lett.*, 29, 770 (1991).
[17] E. Zanoni, C. Tedesco, A. Paccagnella, C. Canali, S. Bigliardi, M. Manfredi, High energy photon emission in GaAs MESFETs and AlGaAs/GaAs HEMTs, *Microelectronics Engineering, 15*, 581 (1991).
[18] C. Tedesco, M. Manfredi, A. Paccagnella, E. Zanoni, C. Canali, Hot carrier induced photon emission in submicron GaAs devices, *IEDM Tech. Digest*, 437 ( 1991).
[19] E. Zanoni, C. Tedesco, M. Manfredi, M. Saraniti, P. Lugli, Hot carrier induced photon emission in submicron GaAs devices, *Semicond. Sci. Technol.*, 7, 1354 (1992).
[20] A.G. Chynoweth, Ionization rates for electrons and holes in silicon, *Phys. Rev.*, *109*, 1537 (1958).
[21] S. M. Sze, "Physics of semiconductor devices", 2nd Edition, J. Wiley & Sons (1981).
[22] C. Canali, A. Paccagnella, E. Zanoni, C. Lanzieri, A. Cetronio, Comments on Impact ionization in GaAs MESFETs, *IEEE Electron Device Lett., EDL-12*, 80 (1991).

[23] K. Hui, C. Hu, P. George, P.K. Ko, Impact ionization in GaAs MESFETs, *IEEE Electron Device Lett. EDL-11*, 113,(1990).

[24] T.P. Pearsall, F. Capasso, R. Mahory, M. Pollack, J. Cheliowsky, The band structure dependence of impact ionization by hot carriers in semiconductors:GaAs, *Solid-State Electron., 21*, 287 (1978).

[25] C. Canali, A. Neviani, C. Tedesco, E. Zanoni, A. Cetronio, C. Lanzieri, Dependence of ionization current on gate bias in GaAs MESFETs, to be published on IEEE Trans. Electron Devices.

[26] T.Y. Chan, P.K. Ko, C. Hu, A simple method to characterize substrate current in MOSFETs, *IEEE Electron Device Lett., EDL-5*, 505 (1980).

[27] W. Schottky,W.T. Read, Statistics of recombination of.holes and electrons, *Phys. Rev., 87*, 835 (1952).

[28] H. P. Zappe, C. Moglestue, Electroluminescence from Gunn domains in GaAs/AlGaAs heterostructure field-effect transistors, *J. Appl. Phys., 68*, 1501 (1990).

[29] H.P. Zappe, D.J. As, Mechanisms for the emission of visible light from GaAs field effect transistors, *Appl. Phys. Lett., 57*, 2919 (1990).

[30] H. P. Zappe, "Study of lower-dimensional transport by electroluminescence", Granular Nanoelectronics, D.K. Ferry Ed., Plenum Press, 507 (1991).

[31] M. Herzog, M. Schels, F. Koch, F. Moglestue, J. Rosenzweig, Electromagnetic radiation from hot carriers in FET devices, *Solid State Electronics, 32*, 1065 (1989).

[32] C. Canali, A. Paccagnella, P. Pisoni, C. Tedesco, P. Telaroli, E. Zanoni, Impact ionization phenomena in AlGaAs/GaAs HEMTs, *IEEE Trans. on Electron Devices, ED-38*, 2571 (1991).

[33] J.P.R. David, "Carrier ionization rates in GaAs", EMIS Datareview, RN=12287 (1984).

[34] V.M. Robbins, S.C. Smith, G.E. Stillman, Impact ionization in $Al_xGa_{1-x}As$ for x=0.1-0.4", *Appl. Phys. Lett., 52*, 296 (1988).

[35] J.P.R. David, J.S. Marsland, J.S. Roberts, The electron impact ionization rate and breakdown voltage in GaAs/AlGaAs MQW structures, *IEEE Electron Device Lett., EDL-10*, 294 (1989).

[36] P. Hendriks, E.A.E. Zwaal, J.G.A. Dubois, F.A.P. Blom, J.H. Wolter, Electric field induced parallel conduction in GaAs/AlGaAs heterostructures, *J. Appl, Phys., 69*, 302 (1990).

[37] T.S. Shawki, G. Salmer, O. El-Sayed, 2-D simulation of degenerate hot electron transport in MODFETs including DX center trapping, *IEEE Trans. Computer-Aided Des., CAD-11*, 1150 (1990).

[38] E. Zanoni, A. Paccagnella, P. Pisoni, P. Telaroli, C. Tedesco, C. Canali, N. Testa, M. Manfredi, Impact ionization, recombination and visible light emission in AlGaAs/GaAs High Electron Mobility Transistors, *J. Appl. Phys., 70*, 529 (1991).

[39] D.E. Aspnes, "Table of optical functions of intrinsic GaAs: refractive index and absorption coefficient vs energy (0-155 eV)", EMIS Datareview, RN=15437 (1985).

[40] M. Lanzoni, M. Manfredi, L. Selmi, E. Sangiorgi, R. Cappelletti, B. Ricco', Hot electron induced photon energies in n-channel MOSFETs operating at 77 and 300 K, *IEEE Electron Device Lett., EDL-10*, 173 (1989).

[41] S. Tam, C. Hu, Hot electron induced photon emission and photocarrier generation in silicon MOSFETs, *IEEE Trans. Electron Devices, ED-31*, 1264 (1984).

[42] A. Toriumi, Experimental study of hot carriers in small size Si-MOSFETs, *Solid State Electron, 32*, 1519 (1989).

[43] F. Ootsuka, The eveluation of the activation energy of interface state generation by hot electron injection, *IEEE Trans. Electron Devices, ED-38*, 1477 (1991).

[44] H.F. Chan, D. Pavlidis, K. Tomizawa, Theorical analysis of HEMT breakdown dependence on device design parameters, *IEEE Trans. Electron Devices, ED-38*, 213 (1991).

[45] H.P. Zappe, D.J. As, Spectrum of hot electron luminescence from high electron mobility transistors, *Appl. Phys Lett, 59*, 2257 (1991).

[46] H.P. Zappe, D.J. As, Carrier transport in HEMTs analyzed by high-field electroluminescence, *IEEE Electron Device Lett., EDL-12*, 590 (1991).

[47] R. Ostermeier, F. Koch, H. Brugger, P. Narozuy, H. Dambkes, Hot carrier light emission from GaAs HEMT devices, *Sem. Sci. Technol.,B5666*, 7 ( 1992).

[48] K.W. Kim. H.T. Tiam, M.A. Littlejohn, Analysis of delta doped and uniformly doped AlGaAs/GaAs HEMTs by ensemble MonteCarlo simulations, *IEEE Trans. Electron Devices, ED-38*, 737 (1991).

[49] K.L. Tam, D.C. Streit, R.M. Dia, S.K. Wang, A.C. Ham, P.D. Chow, T.R. Trinh, P.H. Liu, J.R. Velebir, H.C. Yen, High power V-band pseudomorphic InGaAs HEMT, *IEEE Electron Device Lett., EDL-12*, 213 (1991).

[50] S.R. Bahl. J.A. del Alamo. Breakdown voltage enhancement from channel quantization in InAlAs/n$^+$-InGaAs HFETs, *IEEE Electron Device Lett., EDL-13*, 123 (1992).

[51] J. Singh, The tailoring of impact ionization phenomena using pseudomorphic structures-appliations to InGaAlAs and GaAs and InP substrates, *Semicond. Sci. Technol., 7, B509* (1992).

[52] C. Tedesco, E. Zanoni, C. Canali, S. Bigliardi, M. Manfredi, D.C. Streit, W.T. Anderson, "Impact

ionization and light emission in high power pseudomorphic HEMTs", submitted to IEEE Trans. Electron Devices.

[53] L.P. Sadwick, D.C. Streit, W.L. Jones, C. W. Kim, R.J. Hwu, Device and material properties of pseudomorphic HEMT structures subjected to rapid thermal annealing, *IEEE Trans. Electron Devices, ED-39*, 50, (1992).

[54], Z. Xu, J. Xu, T.G. Anderson, Z. Chen, Photoluminescence of the residual shallow acceptor in $In_xGa_{1-x}As$ grown on GaAs (100) by molecular beam epitaxy, *Solid State Communications, 70*, 505, (1989).

[55] J.J. Chen, G.B. Gao, J.J. Chyi, H. Morkoc, Breakdown behaviour of GaAs/AlGaAs HBTs, *IEEE Trans,. Electron Devices, ED-36*, 2165 (1989).

[56] R.J. Malik, A. Feygenson, D. Ritter, R.A. Hamm, M.B. Panish, J. Nagle, K. Alewi, A.Y. Cho, Temperature dependence of collector breakdpwn voltage and output conductance in HBTs with AlGaAs, GaAs, InP and InGaAs collectors, *IEDM Tech. Digest*, 805 (1991).

[57] J. Chen, G.B. Gao, D. Huang, J.I. Chyi, M.S. Hunlu, H. Morkoc, Photon emisson from avalanche breakdown in the collector junction of AlGaAs/GaAs heterojunction bipolar transistors, *Appl. Phys. Lett., 55*, 374 (1989).

[58] J.R. Hayes, R.F. Leheny, H. Tenkin, A.C. Gossard, W. Wiegmann, Electroluminescence from a heterojunction bipolar transistor, *Appl. Phys. Lett., 45*, 537 (1984).

[59] A.F. Levi, J.R. Hayes, A.C. Grossard, J.H. English., Electroluminescence from the base of a GaAs/AlGaAs double heterojunction bipolar transistor, *Appl. Phys. Lett., 50*, 98 (1987).

[60] L.M. Lunardi, J.R. Malik, R.W. Ryan, P.R. Smith, S.C. Shunk, M.D. Fewer, T.R. Fullowan, "Characteristics of AlGaAs thin emitter heterojunctioon bipolar transistors", Proc. SPIE, Vol. 1288 (High Speed Electronics and Device Scaling), p. 44, (1991).

[61] E. Zanoni, R. Malik, P. Pavan, J. Nagle, A. Paccagnella, C. Canali, Negative base current and impact ionization phenomena in AlGaAs/GaAs HBT's, *IEEE Electron Device Lett., EDL-13*, 253 (1992).

[62] J.J. Liou, J.S. Yuan, An avalanche multiplication model for bipolar transistor, *Solid State Electronics, 33*, 35 (1990).

[63] P. Pavan, E. Zanoni, L. Vendrame, R.J. Malik, S. Bigliardi, M. Manfredi, A. Di Carlo, P. Lugli, C. Canali, Impact ionization and light emission phenomena in AlGaAs/GaAs HBTs, *Proc. GaAs and Related Compound Symp.* (1992).

[64] T.M. Liou, T.Y. Chiu, V.D. Archer, H.H. Kim, Characteristics of impact ionization current in the advanced self-aligned polysilicon emitter bipolar transistor, *IEEE Trans. Electron Devices, ED-38*, 1845, (1991).

[65] A Di Carlo, P. Lugli, P. Pavan, E. Zanoni, R. Malik, "Impact ionization phenomena in AlGaAs/GaAs HBTs", Proc. ESSDERC'92, 135 (1992).

[66] G.E. Bulman, V.M. Robbins, G. E. Stilman, The determination of impact ionization coefficients in (100) Gallium Arsenide by noise and photocurrent mutiplication measurements, *IEEE Trans. Electron Devices, ED-32*, 2454 (1985).

[67] E.F. Crabbe', J.M. Stork, G. Baccarani, M.V. Fischetti, S.E. Laux, The impact of non-equilibrium transport on breakdown and transit time in bipolar transistors, *IEDM Tech. Digest*, 463 (1990).

[68] P. Pavan, S. Bigliardi, R. Malik, M. Manfredi, E. Zanoni, "Hot electron induced electroluminescence and impact ionization in AlGaAs/GaAs HBTs", submitted to Appl. Phys. Lett.

[69] D. Olego, M. Cardona, Photoluminescence in heavily doped GaAs. I. Temperature and hole concentration dependence, *Phys. Rev. B, 22*, 886 (1980).

[70] M.C. Wu, Y.K. Su, K.Y. Cheug, C.Y. Chang, Electrical and optical properties of heavily doped Mg- and Te-GaAs grown by liquid phase epitaxy, *Solid State Electronics, 31*, 251 (1988).

[71] D. Olego, M. Cardona, Luminescence above the gap in heavily Zn-doped GaAs, *Solid State Communications, 32*, 1027 (1979).

[72] J. Sapriel, J. Chavignon, F. Alexandre, R. Azoulay, B. Sormage, K. Rao, M. Voos, Above bandgap luminescence of p-type GaAs epitaxial layers, *Solid State Communications, 79*, 534 (1991).

[73] I.C. Kizilyally, K. Hess: "Monte Carlo simulation of GaAs-AlxGa1-xAs Field Effect Transistors", Chapter III.2 in "Hot carriers in semiconductor nanostructures: physics and applications", J. Shah Ed. Academic Press Inc., 1992.

# NEGATIVE DIFFERENTIAL RESISTANCE, HIGH FIELD DOMAINS AND MICROWAVE EMISSION IN *GaAs* MULTI-QUANTUM WELLS

A. Straw, A. Da Cunha, N. Balkan and B.K. Ridley

University of Essex, Physics Department, Colchester,
United Kingdom

## Introduction

Experimental results presented in this work are concerned with high longitudinal electric field transport in n-type modulation doped *GaAs/AlGaAs* multiple quantum wells. Negative differential resistance with accompanying oscillations in the current have been observed at room temperature, and are attributed to real space transfer or $\Gamma$-L inter-valley transfer within the wells, the frequency of the oscillations is typically of the order of 1GHz. In samples, exhibiting negative differential resistance and current instabilities, high field domains have been observed, using field contrast techniques within a scanning electron microscope. The domains are shown to be either static or slow moving, propagating along the samples and collapsing before reaching the anode, in stark contrast to Gunn domains in bulk *GaAs*. These samples have exhibited microwave amplification in untuned circuits from 1.2GHz to 3.5GHz. Further experiments are required to determine the performance limits of these structures, but these results show the potential of such devices as high frequency components. In samples where breakdown occurs as a result of impact ionisation scanning electron microscope studies indicate the occurrence of current filamentation. The samples studied also emit light when biased. The threshold field required for light emission is somewhat smaller than that for the current instabilities. The spectral analysis of the electroluminescence indicates that it originates from the quantum wells. The relationship between the current instabilities and light emission is as yet not fully understood.

Previous work on *GaAs* 2d structures (Balkan and Ridley, 1988a; Balkan and Ridley, 1988b; Balkan et al, 1989a; and Balkan et al, 1989b) has shown current instabilities in low modulation doped samples, with frequencies as high as a few GHz, and current saturation in highly modulation doped samples at room temperature with the application of high fields. The latter phenomena has been explained in terms of the effect of non-drifting *hot phonons* on the momentum relaxation (Ridley, 1989; and Gupta and Ridley, 1989). This leads to the inhibition of negative differential resistance (n.d.r.) in heavily doped materials. In bulk material $\Gamma$-L inter-valley transfer occurs to give n-type n.d.r., and furthermore gives rise to current instabilities due to Gunn domains traversing the device. In *GaAs* quantum wells not only can $\Gamma$-L inter-valley transfer occur but also real space transfer

*Negative Differential Resistance and Instabilities in 2-D Semiconductors*
Edited by N. Balkan *et al.*, Plenum Press, New York, 1993

261

(Hess et al, 1979). Ridley (1988) has shown that real space transfer induced n.d.r should also give rise to high field domains and hence current instabilities. If there is no lateral dissipation of the domains than this process is very similar to the impurity-barrier process which produces slow domains in bulk *GaAs* as proposed by Ridley and Watkins (1961) and observed by Leach and Ridley (1978).

Current instabilities have been witnessed, not only at room temperature but also at lower temperatures, in a number of samples by us (Balkan and Ridley, 1988a; Balkan and Ridley, 1988b; Balkan et al, 1989a; Balkan et al, 1989b; Vickers et al, 1989) and others elsewhere (Keever et al, 1981; Sawaki et al, 1986; Xiao-Song et al, 1987; and Hendriks et al, 1991). The low temperature instabilities have been explained by recourse to a number of different mechanisms, not only inter-valley and real space transfer but also the acoustoelectric effect and carrier injection into the barriers and subsequent impact ionisation (Hendriks et al, 1991; and Schöll and Aoki, 1991). The common features of these instabilities are, firstly, that they give rise to continuous current oscillations with frequencies increasing with field and temperature up to 1GHz. Secondly, they can occur at low electric fields, a few hundred Vcm$^{-1}$, which are well below those needed for inter-valley and real space transfer. Thirdly, the oscillations are continuous: the shape and noise level changing arbitrarily from samples of the same material. Fourthly, the frequency of the oscillations increase with carrier density. It has been noted that the onset of current instabilities is accompanied by a pronounced non-uniformity of the field within the sample, suggesting the formation of static domains (Balkan et al, 1989a).

A possible explanation for this effect are the models proposed by Hendriks et al (1991) and Schöll and Aoki (1991). In the former model electrons, under the influence of a high electric field, tunnel into the barrier from the cathode. Likewise in the latter model electrons are injected into the barriers, but in this model the electrons proceed into the wells where under the influence of the applied electric field they become hot, ultimately some become hot enough to surmount the potential barriers presented by the the *GaAs/AlGaAs* interfaces, in other words real space transfer. Therefore in both models electrons are lost from the conducting process, leading to a lowering of the current drawn by the sample. Thus the charge neutrality is effected, ultimately leading to the re-injection of electrons into the wells, thus increasing the current and therefore completing the cycle of the current oscillation. Such a state of affairs leads to n.d.r. and to the possibility of static domains.

In the present work results of high electric field experiments on *GaAs/AlGaAs* multi-quantum wells (m.q.w.) are presented. N.d.r and current instabilities are observed, with associated amplification of microwaves. We also present results using the field contrast mode of a scanning electron microscope (s.e.m.) to image high field domains. Not only do these samples exhibit current instabilities but they also emit a strong electroluminescence signal from their surface. This phenomenon is very different to light emission which occurs in *GaAs/AlGaAs* m.e.s.f.e.t.s (Herzog et al, 1989; and Hans et al, 1990), where the fields needed to drive the light emission is in excess of $2.5 \times 10^4$Vcm$^{-1}$.


## Experimental

The structures studied were three differently doped ten period *GaAs/Al$_{0.3}$Ga$_{0.7}$As* m.q.w.s, two m.o.c.v.d. grown and the third m.b.e. grown. The structure schematics and summary of physical properties are shown in tables 1 and 2.

The samples were fabricated in the form of Hall bars and simple bars, ohmic contacts were formed by alloying the sample with *AuGe/Ni*. Six different sample lengths were fabricated 10, 100, 500, 1000, 1750 and 3500 μms. Current voltage characteristics were measured and negative differential resistance was observed at different voltages due to the different sample lengths. The fields were applied along the heterolayers, the highest of which were pulsed so as to minimise Joule heating, with a duty cycle less than 1%. To observe the high frequency current waveforms oscilloscopes with bandwidths of 1GHz were used. Some optical experiments were also performed on some of the samples, these were photoluminescence (p.l.) and electroluminescence (e.l.) measurements. The p.l. was excited with the 647nm line of a c.w. Krypton laser and recorded using standard lock-in techniques. Whilst the e.l. was measured with the sample in complete darkness and only during the duration of the field pulse.

**Table 1.** Sample parameters

| Sample | Growth | Doping (cm$^{-3}$) | Structure | $n_H$(300K) (cm$^{-2}$well$^{-1}$) | $\mu_H$(300K) (cm$^2$V$^{-1}$s$^{-1}$) |
|--------|--------|---------|-----------|-----------|-----------|
| CB520 | M.O.C.V.D. | n-4.5×10$^{17}$ | M.Q.W. (10×75Å) | 3.8×10$^{11}$ | 5.1×10$^3$ |
| CB535 | M.O.C.V.D. | n-1×10$^{17}$ | M.Q.W. (10×75Å) | 2.6×10$^{11}$ | 4.6×10$^3$ |
| A441 | M.B.E. | n-1×10$^{18}$ | M.Q.W. (10×100Å) | 1.1×10$^{12}$ | 7.0×10$^3$ |

**Table 2.** Schematic of samples

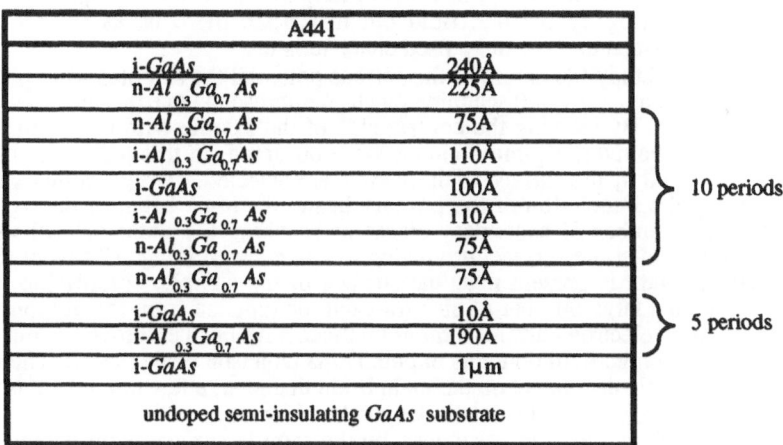

The electric fields required to cause n.d.r. in CB520, CB535 and A441 were, respectively, 3.5kVcm$^{-1}$, 3kVcm$^{-1}$ and 2kVcm$^{-1}$. These fields are typical for Γ-L inter-valley and real space transfer driven n.d.r.. However, the field required to achieve n.d.r in the 10μm samples was found to be much greater than for other sample lengths, for instance in CB520 the field is about 4kVcm$^{-1}$ and for CB535 about 5.2kVcm$^{-1}$. This is illustrated in

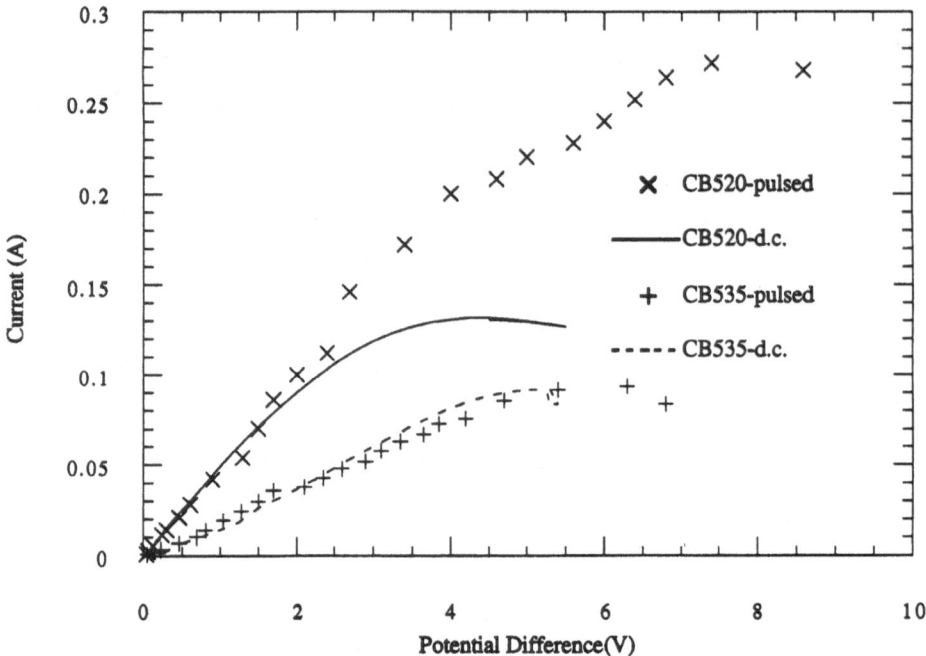

**Figure 1.** Current versus potential difference characteristics for the two 10μm samples, at room temperature.

figure 1 which shows the effects of Joule heating on these small samples. In CB535 there is no Joule heating by the d.c. field but in CB520 there is, as shown by the large discrepancy between the d.c. and pulsed characteristics.

In order to study the nature of the surface potential the samples were placed within a Cambridge Instruments S360 scanning electron microscope (s.e.m.) and a pulsed bias was applied to them. By viewing the contact side of the sample with the s.e.m. while it was biased within the n.d.r. régime domains were observed. In this technique the secondary electrons emitted by the surface of the sample are detected. These secondary electrons are produced by the interaction of the primary beam, the main beam of the s.e.m., with the surface. The number of secondary electrons reaching the detector will vary with the electrostatic field at the surface, producing a field contrast image (Straw et al, 1992). If a high field domain is present near the surface of the sample, comprising of a body of electrons, then a large number of electrons will be repelled from this region, producing a large secondary electron current. Similar methods have been employed to directly observe the dynamics of high field domains in bulk *GaAs* (Johnson et al, 1987a). The resolution of this method is only dependent on the main beam diameter, a few nm, which is many orders of magnitude smaller than any detail likely to be observed.

The secondary electrons produced by the sample surface were detected to give an image of the surface on a television monitor. Images could then be recorded permanently on 35mm film. Brighter regions indicating a higher secondary electron emission and therefore a more negative potential, and furthermore the position of any domains. So as not to damage the surface and affect the current oscillations (Johnson et al, 1987b) low beam currents, 130pA, and accelerating voltages, about 500V, were employed. With the experimental set up used only steady state measurements can be performed, that is only stationary and slow moving domains can be viewed. If fast domains are present in any sample then the high field would be observed over the whole sample.

To study the microwave properties of these structures, the samples were wire bonded onto a 50Ω microstrip waveguide that was fabricated on RT/duroid 6010.5 provided by Rogers Inc., the whole ensemble was then housed in an r.f. box. Two s.m.a. to microstrip transmission line transformers were used to connect the microstrip lines to the external co-axial cables. The microwave reflection coefficient was then measured at room temperature from the L band to the X band, 0.1GHz to 12.4GHz. If the reflection coefficient becomes greater than unity then amplification is occurring.

A Hewlett-Packard HP8410A network analyser was used which can measure quantities from 0.1GHz to 12.4GHz. Other components used were a microwave sweep oscillator, a reflection-transmission test set and a phase-magnitude display unit. The d.c. bias was introduced via a bias tee, the d.c. signal could then be superimposed upon the microwave signal from the r.f. sweeper, this combined signal was then applied to the sample test unit via s.m.a. co-axial cables. The bias tee also isolated the reflection-transmission test set from the d.c. voltage. Figure 2 shows the microwave test circuit. The reflected microwave signal was then down converted by a Hewlett-Packard HP8411A harmonic frequency converter to 278kHz, this signal then formed the input to the network analyser, which could then be displayed on the phase-magnitude display unit. From the phase-magnitude display unit the reflection coefficient was then plotted versus frequency on a x-y recorder for further analysis.

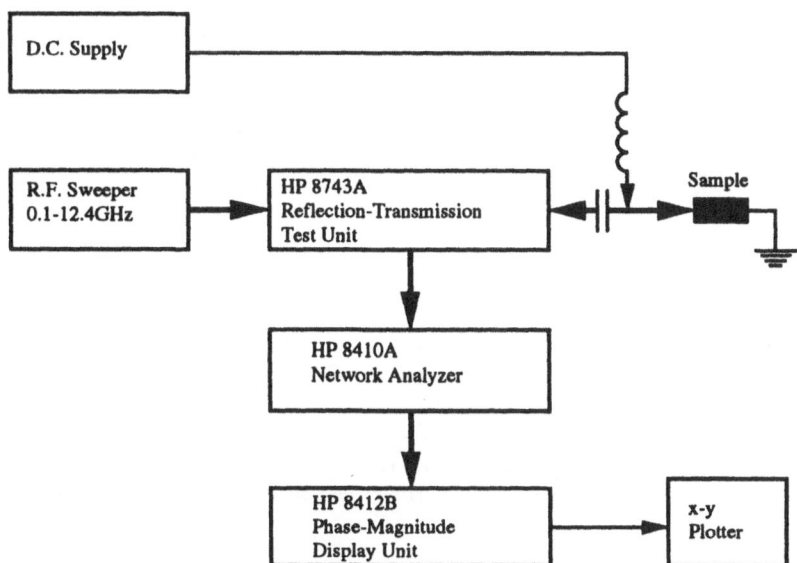

**Figure 2.** A schematic diagram of the microwave reflection coefficient measurement set up.

To calibrate the system fixed impedances were used from Omni-Spectra, a precision short circuit and a 50Ω termination. Calibration data was then obtained over the whole frequency range of interest. During this procedure and the later experiments the input microwave power was set at 5mW. After the initial calibration the experimental set up was checked again and the error was found to be within 1%. The phase difference between the short circuit and the 50Ω termination was established as $180°$, within $1°$.

## Results

High field drift velocities for the three samples measured at room temperature are shown in figure 3. The drift velocities for CB520, CB535 and A441 saturate at values of $1.02\times10^7\,\mathrm{cms}^{-1}$, $1.08\times10^7\,\mathrm{cms}^{-1}$ and $6.41\times10^6\,\mathrm{cms}^{-1}$ respectively. N.d.r. has been observed in all three samples with associated current instabilities. The current instabilities take the form of sinusoidal oscillations, typically with frequencies of the order of 1GHz, which is the limit of our experimental set up, as shown in figure 4. The oscillations occur up to the highest electric fields applied, typically $4\mathrm{kVcm}^{-1}$. However, the oscillation frequency doesn't depend on the electric field, in contrast to results obtained at lower temperatures (Balkan et al, 1989a). There is an indication of a reduction in the saturation drift velocity with increasing carrier concentration, in accord with Balkan et al (1989b).

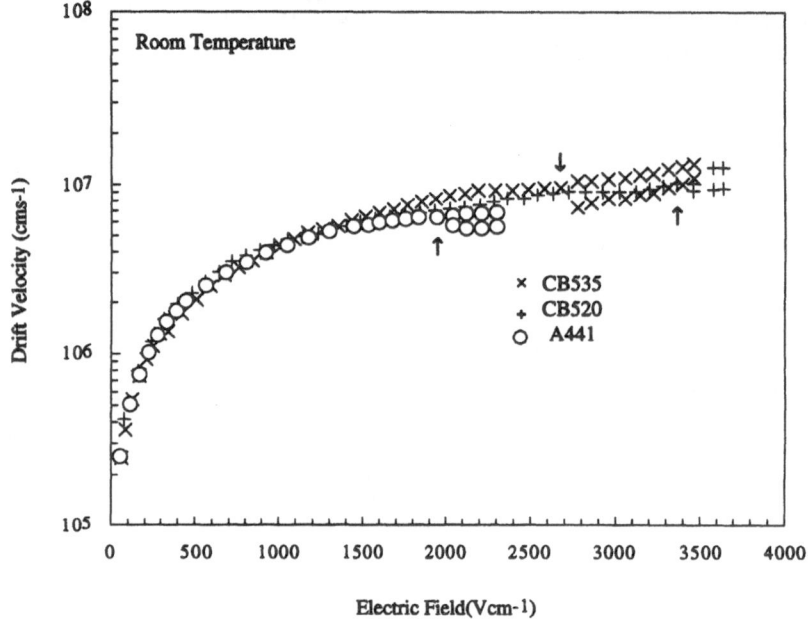

**Figure 3.** The measured electron drift velocity versus electric field measured at room temperature, the arrows indicate the points at which current instabilities occur.

Each sample when viewed with the s.e.m. exhibited unique characteristics. With sample CB520 a series of bright fringes were observed as shown in figure 5(a), we believe that this is due to the presence of a mobile domain. When time averaged the brightness is smeared out over the surface, as expected if a moving domain is present. Assuming that each fringe marks the position of the same domain as it progresses through the sample, then it is moving with a velocity of approximately $8\times10^{-4}\,\mathrm{cms}^{-1}$, which is very slow compared to the velocity of Gunn domains. However, when CB535 and A441 were viewed, stationary domains were observed in the middle of the sample in the case of CB535 and near to the cathode in the case of A441, as shown in figures 5(b) and 5(c). With further increases in the bias applied to A441 current filamentation occurred as shown in figure 5(d). These features persisted even after the field was removed, eventually disappearing after a few minutes. When this occurred the sample was permanently damaged.

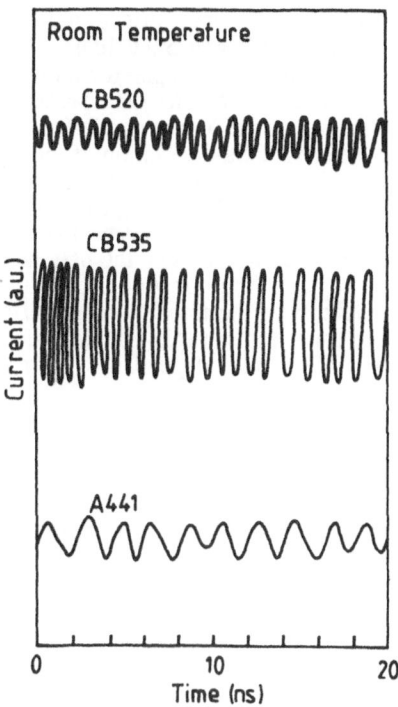

**Figure 4.** Typical current oscillations at threshold for the three samples at room temperature, in all cases the sample length was 100μm.

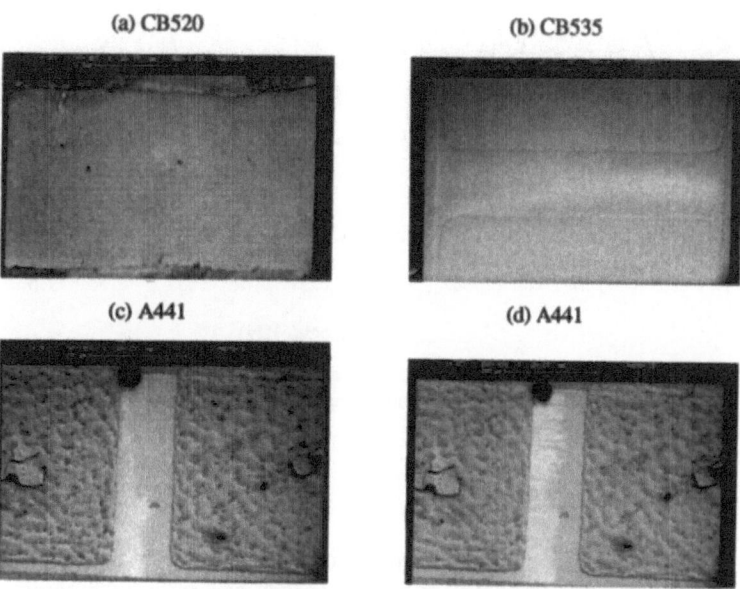

**Figure 5.** Field contrast images of the three samples, with each sample biased above oscillation threshold. (a) shows CB520, which exhibits a series of bright fringes. (b) shows CB535, which has a static domain in the middle of the sample. While (c) and (d) both show A441, in (c) a static domain has occurred near the cathode and in (d) current filamentation has occurred.

Figure 6 shows the results of the microwave reflection coefficient measurements for the 10μm sample of CB535 with 2V, 4V and 5.2V bias. When there is no bias the reflection coefficient varied from 0.95 to 0.80 compared with the short circuit termination, this represents the return loss of the device. The data presented in figure 6 has been corrected for this return loss. When a bias of 2V was applied to the sample, the sample still in the ohmic régime, there was no or very little variation of the reflection coefficient over the whole frequency range from the zero bias situation. However, when the bias was increased to 4V, where current saturation is beginning to occur, some amplification was observed, where the reflection coefficient becomes greater than unity, as shown in figure 6. The amount of amplification increased further when the bias was increased to 5.2V, the sample now well within the n.d.r régime. Amplification occurs in two clear bands, 1.2GHz to 2.3GHz and 2.7GHz to 3.6GHz, with the maximum amplification at a frequency of 1.6GHz. The voltage required to cause n.d.r in this sample was 5.2V, but even a voltage of 4V causes some amplification with a similar dependence on frequency.

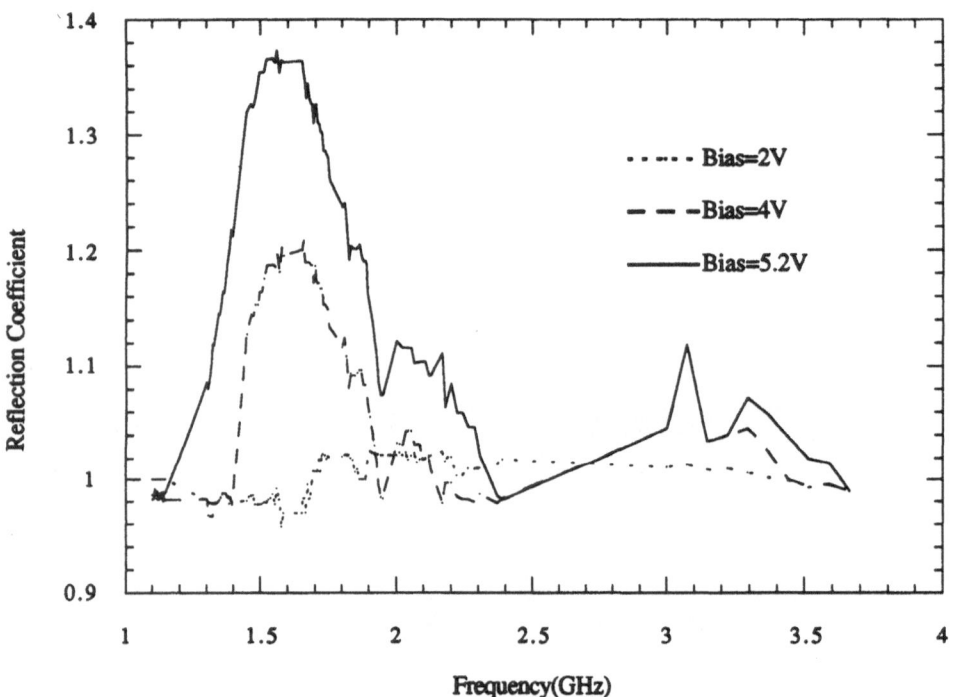

**Figure 6.** The measured reflection coefficient magnitude versus frequency of CB535 at room temperature, sample length was 10μm.

Figure 7 shows the p.l. and e.l. spectra for CB520 and CB535 at an applied field of 2.2kVcm⁻¹ and 2.0kVcm⁻¹ respectively, both recorded at a lattice temperature of 60K. These fields are just below the oscillation threshold field. The peak intensity of the two spectra, in both cases, are identical, within experimental error, indicating that they originate within the wells. The occurrence of e.l. indicates the presence of holes within the structures, which are predominantly n-type as shown by Hall measurements.

## Conclusions

The results presented in this work were obtained from modulation doped n-type *GaAs* multi-quantum wells. At room temperature high frequency current instabilities have been observed with an applied electric field of between 2.0kVcm$^{-1}$ and 3.5kVcm$^{-1}$, with frequencies of the order of 1GHz. The field contrast mode of a s.e.m. has successfully been used directly, and simply, to observe the resulting high field domain dynamics. These domains are found to be static, or slow moving, in contrast to the more familiar Gunn domains in bulk *GaAs*. Possible explanations of these results can be found, we believe, in the models proposed by Hendriks et al (1991) and Schöll and Aoki (1991). These novel mechanisms for n.d.r can both explain the observed instabilities, their frequencies and the stationary nature of the high field domains. These structures have also exhibited an ability

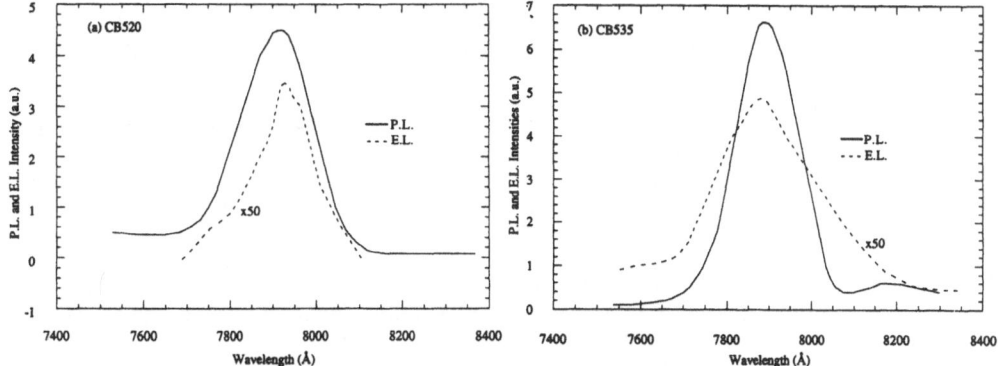

**Figure 7.** Photoluminescence and electroluminescence spectra for (a) CB520 at a field of 2.2kVcm$^{-1}$ and (b) CB535 at a field of 2.0kVcm$^{-1,}$ both at a lattice temperature of 60K. The e.l. peak occurs at a slightly higher energy than the p.l. peak in both cases.

to amplify microwave signals. An amplification of 1.4 has been achieved at a frequency of 1.6GHz. This result is meagre when compared with state of the art microwave devices, but they do compare favourably with other m.q.w. devices. E.l. has also been observed in these samples at low temperatures, which persist up to room temperature. The peak intensity of the e.l. is found to occur at a similar wavelength as the p.l. peak, we therefore conclude that they both originate in the wells.

## Acknowledgements

The authors would like to thank the help and expertise provided by Doctor K. Moulding for the s.e.m. studies; Doctor G. Hill (University of Sheffield) for producing the 10μm samples used in this work; and Doctors C. Button (University of Sheffield) and D. Ritchie (University of Cambridge) for growing the material.We would also like to thank the

S.E.R.C. for financial support during this work, and the *Junta Nacional de Investigação Científica e Tecnologica* for providing a scholarship for A. Da Cunha.

## References

Balkan, N., and Ridley, B.K., 1988a, Hot electron capture in *GaAs* m.q.w.s: n.d.r. and photo-emission, *in*: "Properties of Impurity States in Superlattice Semiconductors", Plenum Press, New York.

Balkan, N., and Ridley, B.K., 1988b, Electrical instabilities in *GaAs/AlGaAs* quantum wells: acoustoelectric effect, *Semicond. Sci. Technol.* 3:507-510.

Balkan, N., Ridley, B.K., and Roberts, J.S., 1989a, Current instabilities in *GaAs/AlGaAs* single and multi-quantum wells, *Superlatt. and Microst.* 5:4, 539-544.

Balkan, N., Gupta, R., Ridley, B.K., Emeny, M., Roberts, J.S., and Goodridge, I., 1989b, Hot phonons and instabilities in *GaAs/AlGaAs* structures, *Solid-State Elect.* 32:1641-1646.

Gupta, R., Ridley, B.K., 1989, High field transport with hot phonons in degenerate semiconductors, *Solid-State Elect.* 32:1241-1246.

Hans, P.Z., and Moglestue, C., 1990, *J. Appl. Phys.*, 68:2501-2503.

Hendriks, P., Zwaal, E.A.E., Dubois, J.G.A., Blom, F.A.P., and Wolter, J.H., 1991, Electric field induced parallel conduction in *GaAs/AlGaAs* heterostructures, *J. Appl. Phys.*, 69:1, 302-306.

Herzog, M., Schels, M., and Koch, F., 1989, 32:1065-1069.

Hess, K., Morkoc, H., Shichijo, H., and Streetman, B.G., 1979, Negative differential resistance through real space electron transfer, *Appl. Phys. Lett.*, 35:6, 469-471.

Johnson, D.A., Maracas, G.N., Myhajlenko, S., Goronkin, H., Edwards, J.L., and Roedal, R.J., 1987a, Direct observation of long range potentials in semi-insulating *GaAs* and *GaAs:In*, *in*: *Inst. Phys. Conf. Ser.*, 91:161-164, Adam Hilger, Bristol.

Johnson, D.A., Myhajlenko, S., Edwards, J.L., Maracas, G.N., Roedal, R.J., and Goronkin, H., 1987b, Imaging of deep level domains in semi-insulating *GaAs* by voltage contrast, *Appl. Phys. Lett.*, 51:15, 1152-1154.

Keever, M., Shichijo, H., Hess, K., Banerjee, L., Withowski, L., Morkoc, H., and Streetman, B.G., 1981, Measurements of hot-electron conduction and real-space transfer in *GaAs/AlGaAs* heterojunction layers, *Appl. Phys. Lett.*, 38:1, 36-38.

Leach, M.F. and Ridley, B.K., 1978, On the origin of slow domains in *GaAs:O*, *J. Phys. C: Solid State Phys.*, 11:2265-2280.

Ridley, B.K., and Watkins, T.B., 1961, The dependence of capture rates on electric field and the possibility of negative resistance in semiconductors, *Proc. Phys. Soc.*, 85:710-715.

Ridley, B.K., 1988, Space charge waves and the piezo-electric interaction in 2d semiconductor structures, *Semicond. Sci. Technol.*, 3:542-545.

Ridley, B.K., 1989, Hot phonons in high-field transport, *Semicond. Sci. Technol.*, 4:1142-1150.

Ridley, B.K., 1990, Hot-electron percolation, *Solid-State Elect.*, 33:859-861.

Sawaki, N., Suziki, M., Takagaki, Y., Goto, H., Akasaki, I., Kano, H., Tanoka, Y., and Hashimoto, M., 1986, Photo-luminescence studies of hot electrons and real space transfer effect in a double quantum well superlattice, *Superlatt. Microst.*, 2:281-285.

Schöll, E., and Aoki, K., 1991, Novel mechanism of a real-space transfer oscillator, *Appl. Phys. Lett.*, 56:12, 1277-1279.

Straw, A., Özturk, E., Guerreiro, P.T., and N. Balkan, 1992, to be published.

Vickers, A.J., Straw, A., and Roberts, J.S., 1989, *Semicond. Sci. Technol.*, 4:743-746.

Xiao-Song, J., Li-Sheng, Y., Shu-Min, W., Hong-Xun, L., and Bei, Z., 1987, Observation of real space transfer in *GaAs/AlGaAs* heterostructures grown by l.p.e., *Solid State Commun.*, 62:9, 597-598.

# ON NEGATIVE DIFFERENTIAL RESISTANCE AND SPONTANEOUS DISSIPATIVE STRUCTURE FORMATION IN THE ELECTRIC BREAK-DOWN OF p-Ge AT LOW TEMPERATURES

Joachim Peinke,[1] Wilfried Clauss,[2] Achim Kittel,[3] Jürgen Parisi,[4] Uwe Rau,[3] and Reinhard Richter[3]

[1]CRTBT, C.N.R.S., BP 166, 38042 Grenoble-Cédex 9, France
[2]Institut für Angewandte Physik, Universität Tübingen, D-7400 Tübingen, Germany
[3]Physikalisches Institut, Universität Tübingen, D-7400 Tübingen, Germany
[4]Physik Institut, Universität Zürich, CH-8001 Zürich, Switzerland

## ABSTRACT

Experimental evidence on self-generated electrical structures in space and time is given for the electric breakdown of p-germanium at low temperatures. The existence of NDR in the breakdown region is shown. By means of scanning electron microscopy current filaments are visualized. Spontaneous temporal oscillations are discussed also on the background of non-linear dynamics and chaos.

## INTRODUCTION

The electronic charge transport in solids can be described by the equation

$$j(E) = \sigma(E) \, E,$$

where j denotes the current density, $\sigma$ the electric conductivity, and E the electric field. Non-linearities arise, if the conductivity changes with the electric field. Usually, the conductivity changes within a finite transition region from one constant value to another one. Such a transition belongs to the class of non-equilibrium phase transition, because the electric charge transport in solids is a dissipative process (exemption the superconductivity).

For an increasing electric field, there are two possibilities, either the conductivity increases or it decreases. For each case the phase transition may give rise to negative differential j-E characteristics. A system driven into the negative differential part of the

*Negative Differential Resistance and Instabilities in 2-D Semiconductors*
Edited by N. Balkan *et al.*, Plenum Press, New York, 1993

261

characteristic is instable against perturbations, like a ball sitting on the top of a hill. This instability leads to the spontaneous formation of dissipative structures. For extended systems whose sizes are larger than characteristic lengths, like the mean free path length or the diffusion length of mobile charge carriers, one may observe in space a spontaneous phase separation of the system into different regions with different conductivity values. In time, on the other hand, the phase (i.e. the conductivity) may start oscillating. Such an oscillation can be denoted as a temporal structure. These spontaneous structure formations cause some practical problems if one wants to verify the existence of a negative differential j-E characteristic.[1] In a current-voltage characteristic, to some extent one always measures some spatially and temporally averaged values of the current density j and the electric field E. Only if j and E are homogeneous in space and time, one can measure the j-E characteristic by a current-voltage characteristic. The presence of spatial or temporal structures over which one has averaged will often display positive differential current-voltage characteristics, although the j-E characteristic is negative differential. If a negative differential part in the current-voltage characteristic is found, it is evident that the j-E characteristic is also negative differential.

For the case of a transition from a low conducting state into a high conducting one, the negative differential characteristic has an S-shape. Here, current filaments are expected as a typical dissipative structure is space.[2] Current filaments are non uniform current distributions perpendicular to the externally applied field. Either an alternating switching on and off of a filament or a temporal change of the form of a filament can lead to temporal oscillations.

For the case of a transition to a low conducting state, N-shaped negative differential characteristics may appear. Here, electric field domain formation in space is typical.[2] The motion of these field domains from one contact to the other gives rise to temporal oscillations.


## ELECTRIC AVALANCHE BREAKDOWN

In the following, we focus on breakdown effects leading to S-shaped characteristics. More precisely, the electric avalanche breakdown of homogeneously doped semiconductors with shallow impurities at low temperatures is regarded. As an exemplary material, p-germanium with an acceptor concentration of about $10^{14}$ cm$^{-3}$ and with a compensation ratio smaller than $10^{-2}$ was used. For further details of this material see Ref. 3. Comparable results were obtained also for doping concentrations of p-Ge between $10^{13}$ cm$^{-3}$ and $10^{15}$ cm$^{-3}$ and for epitaxically grown GaAs and InP samples.[4]

On homogeneously doped samples, ohmic contacts were fabricated by ion implantation or by allowing technique. The contact distances were usually in the mm range. The arrangement of the contacts, the typical sample geometry, and the experimental set up are shown in Fig. 1.

In such a material at low temperatures (T < 10 K) the charge carriers freeze out at the impurities. An ionization breakdown takes place, if an electric field of some V/cm is applied (4.5 V/cm for p-Ge with $10^{14}$ cm$^{-3}$ acceptor concentration). The avalanche breakdown yields to a nonlinear transport characteristic, see Fig. 2. At a certain threshold value of the electric field, mobile charge carriers bound to impurity atoms are activated. This happens by the auto-catalytic process of impact ionization.[5] As a consequence, the electric conductivity of the semiconductor raises by orders of magnitude. Moreover, in a certain interval above the critical electric field, both the low and the high conducting state are stable. The coexistence of two possible values of the current density j directly leads to spatial pattern formation inside the sample, characterized by high conducting filaments, and to spontaneous current oscillations.

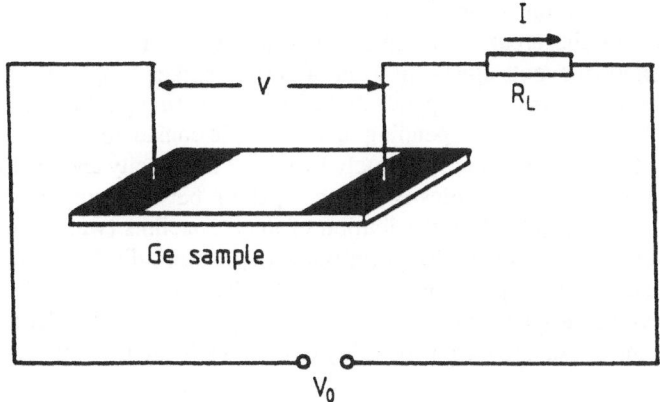

**Figure 1**. Schematic illustration of the experimental set-up.

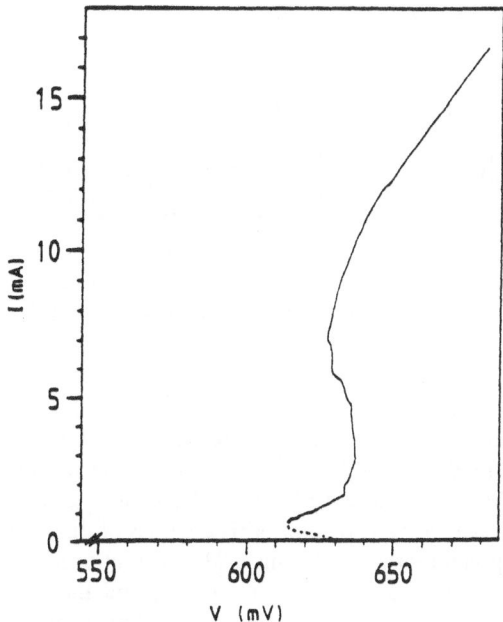

**Figure 2.** Current-voltage characteristic of a p-Ge sample at the temperature of 4.2 K (contact distance 1.5 mm).

## SPATIO-TEMPORAL STRUCTURES

Next, we summarize the scenario of different spatio-temporal structure formations which were detected in the breakdown region by increasing the current value. For a theoretical treatment of these phenomena we refer to Schöll.[6]

Approaching the breakdown region by going up the I-V characteristic, first some random of breakdown events appear in form of current spikes (Fig. 3a). The current spikes start from a current level that corresponds to the low conducting phase. The amplitudes range from some µA to some 100 µA depending on the sample considered. The form of the spikes is influenced by the external circuitry, namely by the cable capacity and the load resistor. One remarkable feature of this dynamics is that the spacing between single breakdown events shows criticality. The probability distribution of these spacings (waiting times) displays a power law behavior. The Fourier transform (power spectrum) displays a $1/f^\alpha$ noise regime. In contrast to usual critical phenomena, this $1/f^\alpha$ noise seems to be self-organized, i.e. it could be observed in a whole range of the characteristic (30 mV). Further details and a discussion in terms of self-organized criticality of a sand pile are given in Ref. 7.

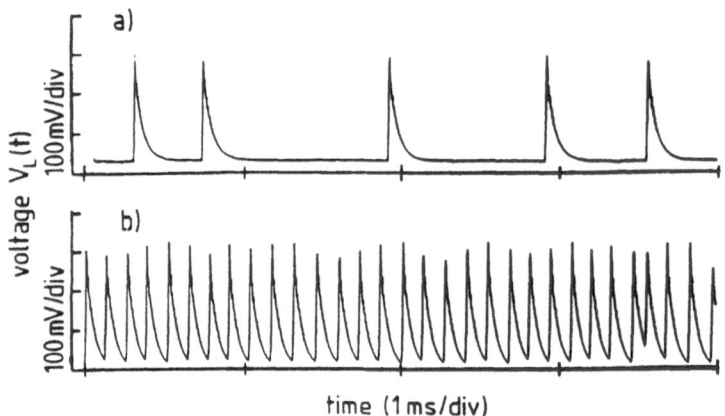

**Figure 3.** Spontaneous current oscillations observed at different current values corresponding to first negative differential part of the characteristic marked by a broken line, compare with Fig. 2 (temperature 4.2 K, load resistor 100 kΩ).

Going up the characteristic, these random breakdown spikes become more frequent. They start to overlap and finally a periodic switching oscillation is found (see Fig. 3b). This periodic oscillation can be explained as follows[8] : The system is driven into the negative differential part of the I-V characteristic by appropriate values of the load resistor and the bias voltage. A capacitance is given by the used cables. The current oscillation is now due to the charging process of the capacity over the load resistor, and the successive discharging process of the capacity over the breakdown of the sample. Due to the fact that the load line intersects the I-V characterisitc in the negative differential part, no temporally stable situation can be reached. The oscillation is thus the periodic charging and discharging process.

At the end of the first negative differential part of the characteristic (marked by a broken line) the switching oscillations vanish and a transition to new dissipative structures takes place. On the basis of a power balance argument, it can be shown that at this point of the characteristic a current filament can be stabilized for the first time.[9] The power loss over the surface becomes here smaller than the power dissipation inside the current filament caused by the resistive current transport. In Fig. 4 two examples of current filaments are shown. By means of low temperature scanning electron microscopy, it is possible to detect the boundaries of a current filament as regions which are sensitive to the electron beam.[10] Increasing the current, first one filament can be detected (like Fig.4a), then more and more filaments appear. Filaments may branch or split up into two. For higher current values, complex patterns are found like shown in Fig. 4b. Finally, the semiconductor becomes homogeneously high conducting.

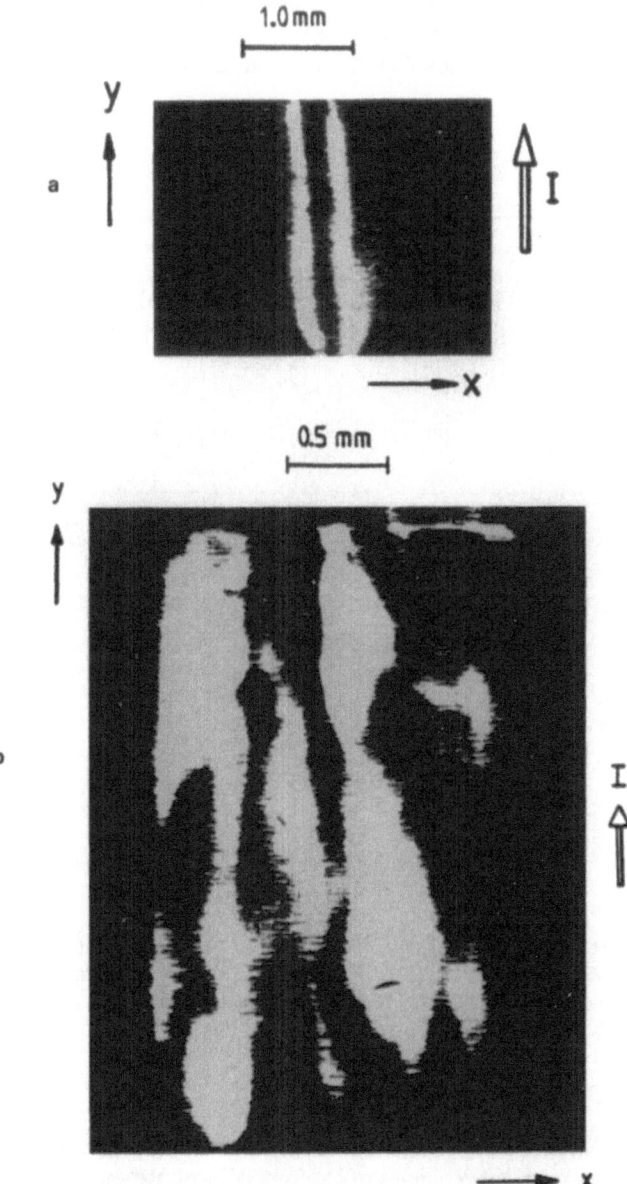

**Figure 4.** Brightness-modulated images of current filaments in p-Ge samples. a) The dark area between the two bright regions corresponds to one filament (parameter : I = 50 μA, T = 4.2 K. b) Complex filamentary structure for the parameters : load resistor $R_L = 1 \, \Omega$, $V_0 = 2.0$ V, T = 4.2 K.

For these filaments we found two further temporal instabilities. For small current values, corresponding to one filamentary state, an external magnetic field perpendicular to the electric field causes a transverse motion.[11] This leads to an oscillation of periodically switching on and off of a filament. First, a filament is created. Due to the Lorentz force, it moves to one side of the sample and extinguishes. At the other side of the sample, a new filament is created, and so on.

In contrast to these current oscillations discussed above, which have amplitudes corresponding to single breakdown effects, a different kind of oscillation is found at higher current values (see Fig. 5a). These oscillations are now superposed on a permanently flowing current of some mA. The typical amplitude of this oscillation mode is in the percent range of the flowing current. This indicates that this oscillation is due to relatively small structural

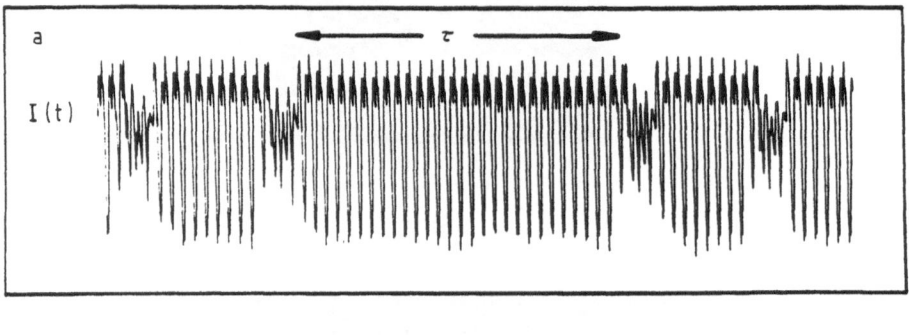

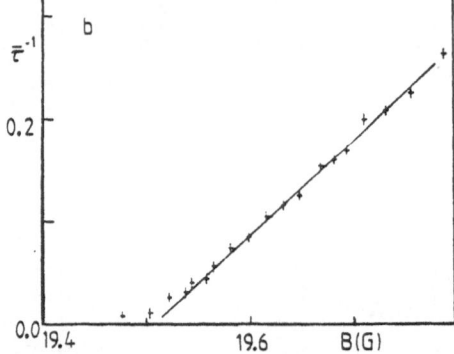

**Figure 5.** a) Time trace of the current signal for B = 19.581 G. The amplitude is in the μA range ; the total time span is 15 ms. With τ one laminar length is marked. b) Inverse of the mean laminar length $\bar{\tau}$ as a function of the magnetic field. I = 3.45 mA and T = 1.88 K were kept constant.

changes of the current filament configuration. An other remarkable feature of this oscillation mode is its sensitivity to minute changes of external control parameters, like the bias voltage, bath temperature, and an external magnetic field. The changes of the form of the oscillation under the variation of control parameters can be best explained in terms of scenarios known from nonlinear dynamics. Thus, universalities of different bifurcations have been found in the present system.[12]

The oscillation displayed in Fig. 5a has a remarkable structure. There are parts of nearly periodic oscillations and parts of irregular oscillations. These different temporal phases change randomly. Changing the control parameter (bias voltage for example), one finds changes of the lengths of these phases. In Fig. 5b the mean value of the laminar (periodic) phase length is shown as a function of the magnetic field. The chosen presentation (inverse of the laminar length versus magnetic field) shows a clear scaling behavior. This scaling is typical for type III intermittency for a transition into chaos. A more detailed discussion of these semiconductor instabilities, especially on the background of nonlinear dynamics, is given in Ref. 13.

## CONCLUSION AND ACKNOWLEDGEMENTS

The electric breakdown regime of p-Ge at low temperatures displays a variety of different instabilities which are linked to different spontaneous structure formations in space and time. Going through the breakdown regime with increasing power dissipation, we found the following sequence. First, a stochastic firing mode of breakdown events sets in, which continuously transforms in a periodic switching mode. At a well defined threshold value stable filaments become possible. A transverse magnetic field may destabilize these filaments. For filamentary current flow, oscillations displaying nonlinear dynamical phenomena are found. Finally, for higher dissipation values the sample becomes homogeneously high conducting.

For helpful discussion we thank A. Aoki, R.P. Huebener, O.E. Rössler and R. Stoop. J. Peinke wants to thank the Deutsche Forschungsgemeinschaft for financial support.

## REFERENCES

1. J. Peinke, D.B. Schmid, B. Röhricht, J. Parisi, Z.Phys. B 66:65 (1987).
2. B.K. Ridley, Proc. Phys. Soc. 82:954 (1963).
3. J. Peinke, J. Parisi, B. Röhricht, K.M. Mayer, U. Rau, W. Clauss, R.P. Huebener, G. Jungwirt, W. Prettl, Appl. Phys. A 48:155 (1989).
4. K.M. Mayer, J. Parisi, R.P. Huebener, Z. Phys. B 71:171 (1988) ; K. Aoki, U. Rau, J. Peinke, J. Parisi, R.P. Huebener, J. Phys. Soc. Jpn 59:420 (1990) ; K.D. Morhard, R.P. Huebener, W. Clauss, J. Peinke, J. Appl. Phys. 71: 3336 (1992).
5. K. Seeger, Semiconductor Physics (Springer, Berlin 1989) ; J. Parisi, U. Rau, J. Peinke, K.M. Mayer, Z. Phys. B 72:225 (1988).
6. E. Schöll, D. Drasdo, Z. Phys. B 81:183(1990).
7. W. Clauss, A. Kittel, U. Rau, J. Parisi, J. Peinke, R.P. Huebener, Europhys. Lett. 12:423 (1990) ; U. Rau, K.D. Morhard, J. Peinke, W. Clauss, A. Kittel, J. Parisi, Phys. Lett. A 153:385 (1991).
8. A. Kittel, U. Rau, J. Peinke, W. Clauss, M. Bayerbach, J. Parisi, Phys. Lett. A 147:229 (1990) ; U. Rau, W. Clauss, A. Kittel, M. Lehr, M. Bayerbach, J. Parisi, J. Peinke, R.P. Huebener, Phys. Rev. B 43:2255 (1991).
9. K.M. Mayer, R.P. Huebener, U. Rau, J. Appl. Phys. 67:1412 (1990).
10. J. Peinke, W. Clauss, R.P. Huebener, A. Kittel, J. Parisi, U. Rau, R. Richter, in Spontaneous Formation of Space-Time Structures and Criticality, ed. T. Riste and D. Sherrington (Kluwer Academic Publishers, Dordrecht 1991) p. 145.
11. W. Clauss, U. Rau, J. Peinke, J. Parisi, A. Kittel, M. Bayerbach, R.P. Huebener, J. Appl. Phys. 70:232 (1991).

12. J. Peinke, U. Rau, W. Clauss, R. Richter, J. Parisi, *Europhys. Lett.* 9:743 (1989) ; J. Peinke, J. Parisi, R.P. Huebener, M. Duong-van, P. Keller, *Europhys. Lett.* 12:13 (1990) ; R. Richter, J. Peinke, W. Clauss, U. Rau, J. Parisi, *Europhys. Lett.* 14:1 (1991) ; R. Richter, U. Rau, A. Kittel, G. Heinz, J. Peinke, J. Parisi, R.P. Huebener, *Z. Naturforsch.* 46a:1012 (1991).

13. J. Peinke, in *Non-Linear Dynamics in Solids*, ed. H. Thomas (Springer, Berlin 1992) p. 52 ; J. Peinke, J. Parisi, O.E. Rössler, R. Stoop, *Encounter with Chaos - Self-Organized Hierarchical Complexity in Semiconductor Experiments* (Springer, Berlin 1992).

# DISSIPATIVE STRUCTURES IN BISTABLE ELECTRONIC AND OPTOELECTRONIC SEMICONDUCTOR DEVICES

Ralf Symanczyk, Steffen Knigge, and Dieter Jäger

FG Optoelektronik
Universität - GH - Duisburg
D-4100 Duisburg, Germany

## INTRODUCTION

There is a growing interest world-wide in the physical mechanisms and the potential applications of dissipative structures in semiconductor materials and devices, for recent reviews see Refs. 1,2,3,4. Phenomenas such as self-generated oscillations, hysteresis and bi- and multistability have been observed experimentally. Additionally, pronounced switching properties and different chaotic scenarios have been demonstrated by using various semiconductor components, where electronic elements are studied as well as optical and optoelectronic devices.

From the technical point of view, the commercially available Gunn-diode makes already good use of the propagation of field domains to generate microwave or millimeter wave electrical power. In contrast, the switching behaviour of power thyristors can be influenced by the formation of current density filaments which leads to an undesired time delay and the possibility of an enhanced degradation and irreversible breakdown[5]. As a further example, the study of the nonlinear dynamics of laser diodes should include the distributed nature of the waveguiding structure, i.e. wave propagation effects, in order to obtain a quantitative description of the device especially for operation in the GHz- or ps-range[6]. It should finally be noted, that optical or optoelectronic bistability are regarded to play an important role on future systems of information technology such as optical signal processors.

Physically, the occurence of dissipative structures in semiconductors is traced back to a negative differential conductivity (NDC) of the material, see also Refs. 2,3,7, where different mechanisms may be responsible. One commonly can distinguish between two types of the current density ($j$) versus electric field ($E$) characteristic[7,8]: The N-shaped characteristic is connected with the formation of electrical field domains and the S-shaped behaviour is expected to lead to the generation of current density filaments. Up to now, the physics of current density filaments is not well understood and practically no approach has been published to combine N- and S-shaped properties. Moreover, there

*Negative Differential Resistance and Instabilities in 2-D Semiconductors*
Edited by N. Balkan *et al.*, Plenum Press, New York, 1993

269

is a need to transfer the results of electronic devices to the corresponding behaviour of optical or optoelectronic bistable elements.

The aims of this paper are twofold. Firstly, the present status of research work in the field of current density filaments is summarized with special emphasis on the results of recent experimental investigations. Secondly, selected examples of bistable optoelectronic elements are discussed with respect to their characteristics arising from an optically induced S-NDC or N-NDC behaviour. In both cases, the physical background is described as well as the areas of potential applications.

## CURRENT DENSITY FILAMENTS

The experimental observation of current density filaments started in the fifties[9]. In the following years several semiconductor materials and devices were studied using different methods[8,10,11]. Recently, the spatio-temporal pattern formation in bulk Ge and thin n-GaAs layers at low temperatures were investigated, where the S-NDC is due to impact ionisation of shallow impurities[11,12]. Two models have been developed to explain the filamentation process. The first one is based on nonlinear generation - recombination kinetics of electrons[2,13] and has been applied to oscillating filaments[13,14]. In the second one, the device is devided into two layers with different electrical properties and an activator - inhibitor system is used to describe the electrical potential and current distribution at the interface of the layers[15]. In spite of these investigations a complete and comprehensive understanding of the phenomena in connection with S-shaped NDC is lacking until now. There exists several open questions concerning for example the physical mechanism and the microscopic picture leading to this kind of pattern formation. The influence of the boundaries and contacts as well as the stability and instability of the structures have still to be analysed.

We have investigated in detail the spatio-temporal pattern formation due to current filaments in pin diodes at room temperature. In these devices the double injection of electrons and holes is responsible for a S-type NDC and several experimental techniques have been employed to observe the current distribution between the contacts[16-18]. In the following, we propose theoretical models which can describe both the $j - E$ characteristics of the material and the current inhomogeneities in the diodes. Furtheron, experimental results are presented showing the properties of stable and unstable filaments in Au doped Si and Cr doped GaAs diodes. A comparison is finally carried out between the measured results and the predictions of the models revealing a good agreement.

### Theoretical Description

The theoretical treatment is twofold: (i) numerical calculations on the basis of the Shockley-Read-Hall (SRH) model yield the $j - E$ characteristic in forward biased pin-diodes with homogeneous current density distribution[8] and (ii) a distributed two-dimensional equivalent circuit can explain the formation of filaments in large area diodes.

Starting with the Poisson equation, the continuity equation and the SRH-model, describing the nonlinear generation and recombination processes of charge carries via deep impurity centers, one can calculate the voltage drop $V$ and the electric field as a function of the current density in homogeneous pin diodes[19,20]. In the case of Au doped Si diodes there is an excellent agreement between experimental $j - V$ characteristics and calculations[20]. We determined numerically the $j - E$ behaviour in the material at different contact distances using the parameters of our devices. The results have shown that Si:Au pin diodes simultaneously exhibit two regions with different electrical properties, a bistable, S-shaped behaviour near the p-contact and in the center region and a monostable characteristic without NDC close to the n-contact[21].

In the following this model is applied to Cr doped GaAs by using capture cross sections of the Cr accepter level as mentioned in the literature[22]. Fig. 1(a) shows the calculated j-E characteristics at the contacts and in the center of a GaAs:Cr pin diode with an effective donor concentration $N_D = 1 \cdot 10^{15}\,cm^{-3}$ and a contact distance $d = 73\,\mu m$. The concentration of the deep Cr level is used as a fit parameter to achieve the same low field resistivity as measured in our diodes.

To take into account this nonuniformity of the electrical properties along the direction of current flow we have developed a two-dimensional equivalent circuit for a thin pin-diode with current inhomogeneities perpendicular to the current flow[21] as represented in Fig. 1(b). The diode is assumed to be sufficiently thin and therefore

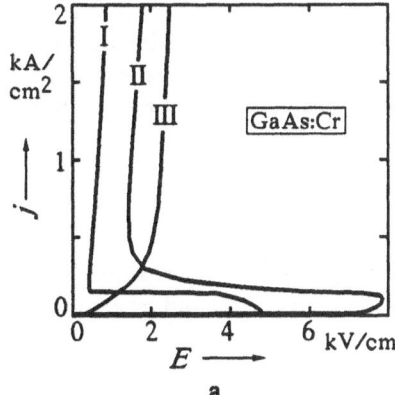
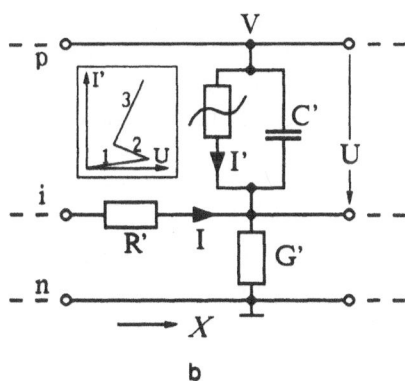

a                                                     b

**Figure 1.** (a) Calculated $j - E$ characteristic for different contact spacings of a homogeneous GaAs:Cr pin diode (I = p-contact, II = between the contacts, III = n-contact).
(b) Equivalent circuit for a two-dimensional pin-diode representing the electrical properties.

homogeneous in the direction perpendicular to the paper-plane. The elements are distributed per unit length in $x$-direction and transversal, i.e. horizontal coupling is achieved by the resistance per unit length $R'$. The monotonic region close to the n-contact is approximated by the conductance per unit length $G'$, and the capacitance per unit length $C'$ allows for dynamical properties. The region with S-shaped $j - E$ characteric is represented by a nonlinear resistance per unit length and the relationship between the current per unit length $I'$ and the voltage drop $U$ may be simplified in form of three piecewise linear sections $i = 1, 2, 3$ as indicated in the inset of Fig. 1(b):

$$U = a_i I' + b_i \quad ; \quad b_1 = b_3 = 0, b_2 > 0\,,\ a_1 > a_2 > 0, a_2 < 0 \qquad (1)$$

Using Kirchhoff's laws one easily derives the following equations

$$\frac{\partial U}{\partial x} = I R' \quad , \quad \frac{\partial I}{\partial x} = I' - (V - U)G' + C'\frac{\partial U}{\partial t} \quad .$$

Eliminating the current $I$ we obtain a differential equation for the voltage drop $U$:

$$\partial^2 U/\partial x^2 = R'I' - (V - U)G'R' + C'R'\partial U/\partial t \quad . \qquad (2)$$

Because $I'(U)$ is not a mathematical function this diffusion equation has no unique solution. As a key point of the following discussion, however, we write $U$ as a function of the current per unit length through the nonlinear resistor $U = f(I')$, which is a unique function. This leads to a differential equation for $I'$ and an analytical solution for this equation can be found if the S-shaped nonlinearity is approximated by piecewise linear relations. Accordingly, combining eqs. (1) and (2) yields:

$$a_i \partial^2 I' / \partial x^2 = R' I' - (V - f(I')) G' R' + C' R' a_i \partial I' / \partial t \quad . \tag{3}$$

Generally, this equation describes the spatial and time-dependent distributions of the current density in a thin pin diode. For constant $a_i$ eq. (3) has the form of the FitzHugh-Nagumo (FHN) equation which gives well known stationary solutions in form of solitary or kink structures[23].

For a further analytical study the phase portrait of the stationary dynamic system described by eq. (3) has been investigated[21]. Current filaments correspond to trajectories on a homoclinic orbit in the $(I', \partial I'/\partial \xi)$-plane with $\xi = \sqrt{G'R'}x$. The condition for the existence of filamentary solutions can be stated as an equal areas rule[2] which determines simultaneously the maximum current in the center of the filament. It turns out, that for our model filaments exist in a fixed range of the parameters and the profile $I'(x)$ can be solved explicitely. This yields the width of the filament $\Delta x_F$ as a function of the slopes $a_i$ and the resistance elements in the equivalent circuit:

$$\Delta x_F = \frac{2}{\sqrt{G'R'}} \frac{1}{\alpha_2} \mathrm{artanh} \left( \frac{|a_2| \alpha_2}{a_1 \alpha_1} \right) \tag{4}$$

and $\alpha_i^2 = 1 + (G'a_i)^{-1}$. For our pin diodes the following relations hold, which has been confirmed by numerical calculations on the basis of the SRH-model and by experimental study of the $I - V$ characteristics:

$$G' \gg \frac{1}{a_2} \gg \frac{1}{a_1} \quad .$$

Therefore, the argument of the artanh-function in eq. (4) can be opproximated by $|a_2|/a_1 \ll 1$ and $\mathrm{artanh}(...) \approx |a_2|/a_1$ is valid. The equation for the width simplifies now to the expression

$$\Delta x_F \approx \frac{2}{\sqrt{G'R'}} \cdot \frac{|a_2|}{a_1} \quad .$$

It should be noted at this point, that $\Delta x_F$ is independent of $V$ and therefore likewise from the total current through the device. For a comparison with experimental data it is profitable to know the dependence of $\Delta x_F$ on the contact distance $d$. If the doping level is not changed in the case of Si:Au the calculations show that the fraction $|a_2|/a_1$ is independent on the contact distance. The elements $G'$ and $R'$ can roughly be estimated by $G' \approx 1/(\rho_1' df)$ and $R' \approx \rho_2'/d$ where $\rho_1', \rho_2'$ are effective surface resistivities in Ohm and $f$ is the fraction of the extension of the monotonic region as compared to the contact distance. Thus we get

$$\Delta x_F \propto d\sqrt{f} \tag{5}$$

for the width of current filaments in our model.

### Experimental Results and Comparison

In our experiments we used Au doped Si and Cr doped GaAs pin diodes at room temperature. The samples are made from commercial Si wafers with resistivities

$\rho = 10...100 \ \Omega cm$ and GaAs:Cr wafers with $\rho = 10^7...10^9 \ \Omega cm$. An Au diffusion process is carried out in the case of Si in order to get compensated high-resistivity material. Highly doped $n^+$- and $p^+$-layers are fabricated on the surfaces of the wafers, respectively. Finally, samples are cut from the wafer with the following typical dimensions (see inset of Fig. 2): contact distance $d \approx 80 \ \mu m$(GaAs), 100...500 $\mu m$(Si) and length$\gg d \approx$width. For details of the preparation see Ref. 16. Due to these relations the samples can be regarded as long, thin layered diodes as used in our two-dimensional model.

The device is forward biased by applying a dc voltage via a load resistor. A typical current $(I)$ - voltage $(V)$ characteristic of a GaAs:Cr pin diode is shown in Fig. 2 and the Si:Au pin diodes exhibit the same features. As a key result a pronounced

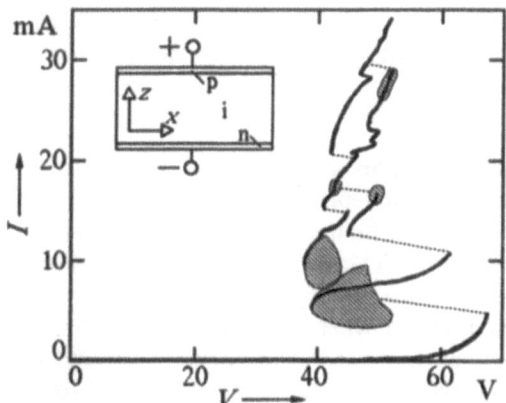

**Figure 2.** Measured $I - V$ characteristicristic of a GaAs:Cr pin diode. The hatched areas marke selfgenerated oscillations. The inset shows a sketch of the device.

multistability is observed which is due to the S-shaped NDC of the material and the jumps are associated with the formation of current filaments[8,16,17]. In the vicinity of the jumps we often observe selfgenerated oscillations of the current and the voltage[8,18]. To obtain detailed information about the properties of the filaments we have carried out several spatially and temporally resolved measurements between the contacts. First we will describe stationary patterns. In Fig. 3(a) the intensity of the emitted recombination radiation and in Fig. 3(b) the result of a potential probe measurement on the surface of the same GaAs:Cr diode are shown. In both cases a working point beyond the first jump in the $I - V$ characteristic was choosen and the signals at both contacts are compared. The recombination radiation as a measure for the product of the density of electrons and holes reveals an inhomogeneity with a pulse-shaped distribution, whereas the surface potential clearly depends on the $z$-coordinate: At the n-contact the hump is single-peaked and near the p-contact one can observe a W-shaped characteristic. This behaviour reproduces the different electrical properties of the material. The monotonic relation between $j$ and $E$ close to the n-contact and the S-type NDC at the p-contact lead to the observed potential in case of a pulse-shaped current density distribution. This is in good agreement with our numerical calculations and it should be pointed out, that the same results can be obtained with the Si pin diodes[21].

For a further comparison between the experimental observations and theoretical predictions we have determined quantitatively the extension of the filaments in $x$-direction. The FWHM-value of the inhomogeneities show the following characteristics:

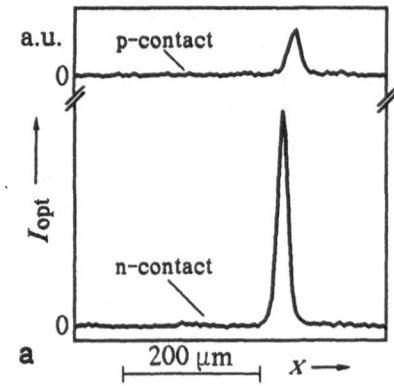 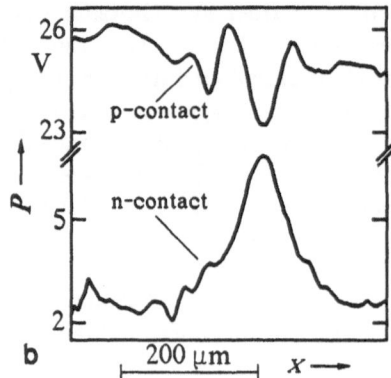

**Figure 3.** Spatially resolved measurements showing one current filament in a GaAs:Cr pin diode. Emitted recombination radiation $I_{opt}$ (a) and surface potential $P$ (b) perpendicular to the current flow.

Different filaments in the same sample have nearly the same widths thereby slightly increasing from p- to n-contact. The independence on the working point is demonstrated in Fig. 4(a). The FWHM-value of a filament in a Si pin diode is plotted as a function of the current through the device. The results from the potential probe signals were taken at a $z$-coordinate, where the W-shaped characteristic passes over to a single-peaked hump. The IR-signal on the other hand yields the width of the same filament close to the n-contact due to an increased absorption caused by free charge carriers[17]. The $z$-coordinate at the transition from the monotonic region to the S-shaped NDC

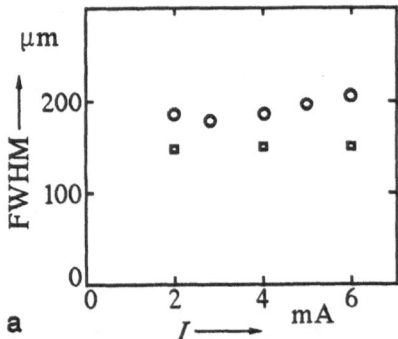 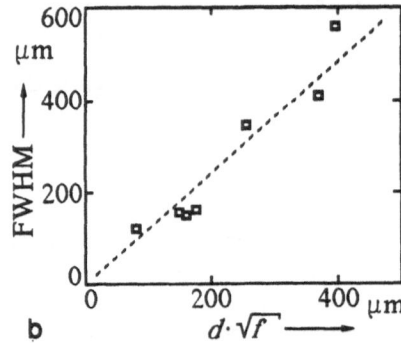

**Figure 4.** Width of a filament perpendicular to the current flow in Si pin diodes for different currents (a) and varying contact distances (b). □ = determined from potential probe measurements, ○ = determined from IR-absorption signals. In (a) one sample is investigated, whereas in (b) different samples are compared.

material permits to calculate the factor $f$ provided that the contact distance is known. So, we can compare likewise the widths of filaments in different samples. In Fig. 4(b) the FWHM of potential inhomogeneities in several Si samples is plotted versus the value $d\sqrt{f}$. The dashed line indicates the proportionality as specified by eq. (5). In summary, the experimental observations reveal the same properties of the filaments concerning their widths as predicted by our two-dimensional model.

Let us now look at temporally unstable patterns. As mentioned above, we often observe selfgenerated oscillations of the current and the voltage. The shape of the signals versus time can vary with the external dc voltage and the frequencies range from 10 kHz to 1 MHz. There occur simple periodic oscillations as well as signals with odd or even numbers of maxima during one period and even nonperiodic oscillations[8]. The current may show pulse-like or more sinusoidal oscillations. These instabilities are attributed to the dynamical behaviour or the filaments. To proof this assumption we have carried out temporally resolved potential measurements using an active probe[18]. During simple periodic oscillations one can observe spatially limited regions on the surface of the sample with an oscillating potential in synchronism with the external current. The extension in the plane of the contacts and the position of this region is comparable to those of a stable filament. Consequently, this result can be interpreted as a switching current density. In Fig. 5(a) a Si pin diode is operated in such a way that the current shows two maxima during one period. One can see two spatially separated regions on the sample with simple periodic behaviour but their frequencies differ by a factor of two. Other measurements at operating points with several maxima during one period show a phase shift of simple periodic oscillations in different regions[18]. These results demonstrate a mechanism for the formation of complex integral current and voltage signals: Period multiplication is caused by a correlated interaction of two or more filaments, each acting in a simple periodic switching manner.

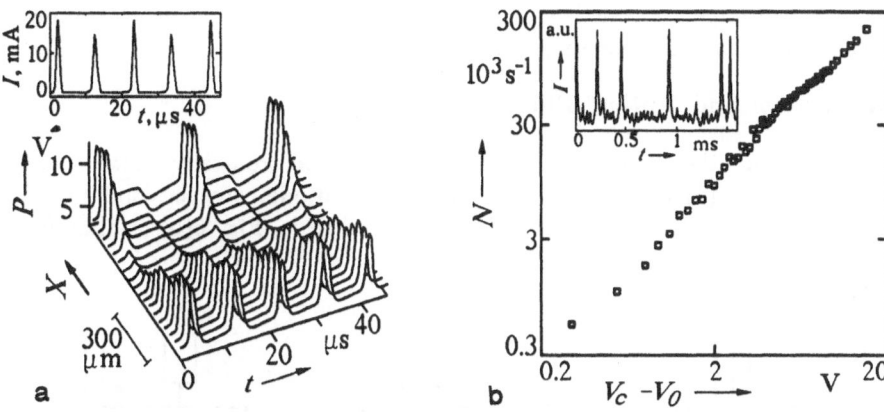

**Figure 5**. Behaviour of a Si pin diode during selfgenerated oscillations. (a) Surface potential at different $x$-positions while the external current shows periodic oscillations with two maxima. (b) Scaling of the number of pulses $N$ per time as a function of the external voltage during aperiodic spiking of the current.

The nonperiodic oscillations often indicate a deterministic behaviour and obey universal scaling rules. Because of the complexity of the signals and the variety of methods this cannot be treated here in detail. To conclude, however, we present one example to demonstrate the scaling behaviour. In the inset of Fig. 5(b) a current versus time signal is shown, where the pulses have nearly the same amplitude but the distances between successive spikes differs in a chaotic manner. The number of spikes per time, $N$, varies with the external dc voltage $V_0$ as plotted in Fig. 5(b) ($V_c$ is the value of $V_0$ at the onset of oscillations). This scaling behaviour seems to be an universal property of special nonlinear systems and can be observed for example in microwave resonators[24].

In the following we discuss special bistable optoelectronic devices which exhibit an optically induced S-NDC or N-NDC. It is shown that a close similarity of these devices to the above pin diodes exists experimentally as well as theoretically. As a result, several features of the already observed and described dissipative structures can directly be transfered to the optoelectronic devices. Additionaly, some potential applications are foreseen.

## OPTICAL AND OPTOELECTRONIC BISTABILITY

Since the discovery of optical bistability by Szöke et al.[25], a huge amount of work has been dedicated to the development and optimization of suitable devices with technically interesting properties. Preliminary work on the occurence of dissipative structures has also been described[26]. Recently, optoelectronic elements have become of major interest because of their inherent large optical nonlinearities and the flexibility of technological realization[27,28]. The basic properties of such hybrid devices are due to an appropriate combination of optoelectronic photodetector and electrooptic modulator characteristics where the latter can preferably be realized by a laser diode or LED function[27,29]. Special socalled SEEDs (self electro-optic effect devices) have been developed where the same device operates simultaneously as a photodetector and a modulator[30,31]. In the following, special hybrid switching devices and SEEDs are discussed where transversal or longitudinal dissipative structures are predicted. Particularly, optical filaments are expected in large area SEEDs or pnpn four-layer diodes and optical solitons are foreseen to occur in nonlinear Bragg reflectors, respectively.

### SEED Devices and Optical Filaments

Figure 6 shows the sketches of three different optoelectronically bistable switching devices. In Fig. 6(a) an upside down Schottky photodetector is combined with a thermooptical Fabry-Perot-modulator. The structure in Fig. 6(b) is that of a pin

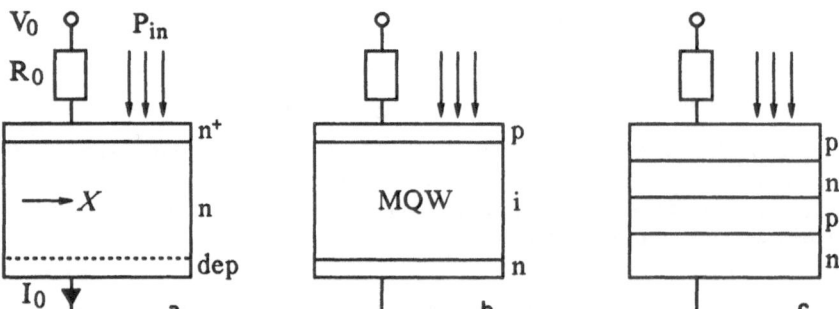

**Figure 6.** Hybrid optical switching devices. (a) Thermooptical SEED[29]; (b) electrooptical MQW-SEED[30]; (c) active pnpn photothyristor[32]. The depletion layer is marked by 'dep'; MQW denotes the multiple quantum well material.

photodiode and an absorptive modulator based on the quantum confined Stark effect. The four-layer diode of Fig. 6(c) consists of a pnp phototransistor vertically integrated on top of a pn LED where only one p-type layer is used for both functions. In case of distributed diodes, the devices and the optical beams are supposed to be extended in x-direction (length $L$). From experimental observations and theoretical studies the

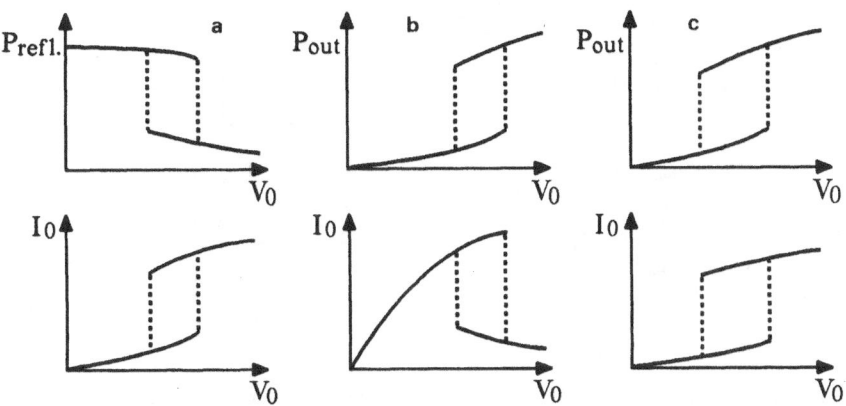

**Figure 7.** Schematic electrooptical and electrical characteristics of the corresponding switching devices (a), (b), (c) in Fig. 6. $P_{refl}$ and $P_{out}$ are the reflected and the optical output power, respectively.

switching behaviour of the devices in Fig. 6 are known. Fig. 7 summarizes the results in a schematic way useful for the model to be developed.

For the following analysis, the electrical behaviour of the devices in Fig. 6 are important. For that purpose, in a first step homogeneous current flow with respect to the x-direction is assumed. Fig. 8(a) shows the corresponding equivalent circuit when dynamical effects are neglected. As can be seen, the circuit consists of a series connection of the external resistance $R_0$, an internal resistance $1/G$ of the doped layers and the nonlinear part from $NL$. Note that $NL$ can be represented by an S-shaped or N-shaped current-voltage characteristic in cases of Figs. 6(a) and (c) or Fig. 6(b), respectively. This situation is sketched in Figs. 8(b) and (c) where the load lines are also included in a schematic way. From Figs. 8(b) and (c) the external $I_0$ vs. $V_0$ characteristic for a given load can be derived, cf. Fig. 7.

Similar to the procedure discussed in the first chapter, the distributed, i.e. travelling-wave devices of Fig. 6 can electrically be respresented by the equivalent circuit of Fig. 9. Here $R'_0 = R_0 \cdot L$, $G' = G/L$, $C' = C/L$ and $I' = I_0/L$ are elements per unit length

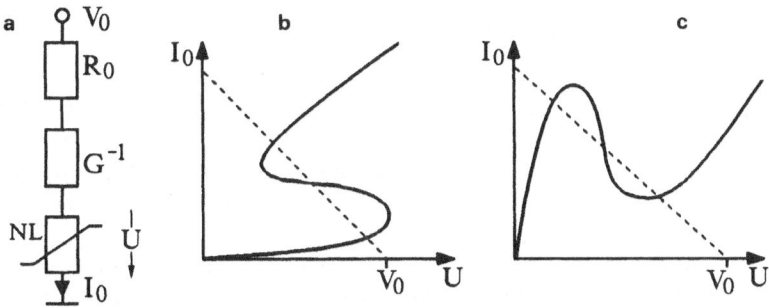

**Figure 8.** (a) Static equivalent circuit for the electrical properties of the devices in Fig. 6. (b) S-shaped and (c) N-shaped characteristics of the nonlinear resistance (NL) of (a). The slopes of the load lines are $(R_0 + 1/G)^{-1}$.

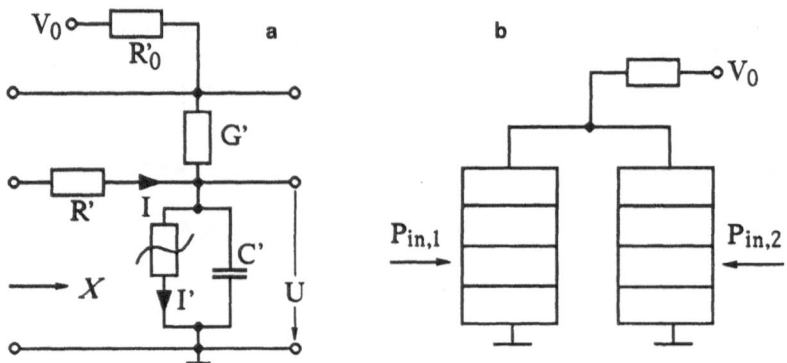

**Figure 9.** (a) Equivalent circuit of a distributed SEED or pnpn device of Fig. 6 (For $I' = I_0/L$ see Figs. 8(b) and (c)) and (b) schematic configuration of an optical comparator.

$dx$ where $L$ is the length of the device. Note the capacitance $C$ has been introduced to represent relaxation phenomena (cf. Fig. 1). The coupling between the elements is taken into account by $R'$ (see above).

For the equivalent circuit of Fig. 9(a) the equations for wave propagation, $U(x,t)$ or $I(x,t)$, can be drived. In case of the S-shaped current-voltage characteristic, $R_0 = 0$ can be used to derive basic properties. As a result, the circuit of Fig. 9(a) is that of the pin diode of the first chapter and eqs. (2) and (3) hold also in case of the Schottky SEED or pnpn device. Consequently, current density filaments $I'(x,t)$ are expected with properties similar to the above discussion. From Figs. 8 and 9 one can further recognize that $I'$ is connected with the voltage drop $U$ which determines the local optical power density $P'_{refl}$ or $P'_{out}$. In particular, the current filament in the Schottky SEED is connected with a dark region in the reflected power and a bright region in the transmitted power. These optical filaments can propagate along the device, depending on the operating point. Besides solitary solutions, kink structures are also imaginable leading to switching waves converting the reflective state into a transmissive state and vice versa. In the distributed pnpn device the current filament determines the optical filament of $P'_{out}$. In this case, a propagating current filament corresponds to an optical beam scanner.

The case of the MQW-SEED (Fig. 7(b)) must be treated separately because of the N-shaped characteristic (Fig. 8(c)). Consequently, the wave equation can easily be derived for the parameter $U$ which gives:

$$\frac{1}{R'C'}\frac{\partial^2 U}{\partial x^2} = \frac{\partial U}{\partial t} + \frac{1}{C'}I'(U) \tag{6}$$

Note that $1/G = 0$ has been used to obtain basic results and that eq. (6) is again the above FHN equation. Now filaments or kink structures in the potential distribution $U(x,t)$ are expected under stationary conditions. However, the filaments are now combined with W-shaped current density distributions and therefore W-shaped optical output power density, $P'_{out}(x,t)$.

Because the optical input power determines the electrical operating point of the device, i.e. the current - voltage characteristic, interesting new phenomena are expected in large area devices with spatially dependent optical input power. In the following, some examples are briefly discussed where the fundamentals of dissipative structures are applied to signal processing methods.

## Dissipative Structures and Logic Functions

When switching devices such as the pnpn photothyristor of Fig. 6(c) are used with inhomogeneous optical input different logical functions can be implemented. The basic concept is that of connecting two parallel pnpn diodes with two optical inputs in series with the load[32], see Fig. 9(b). The circuit provides the properties of a differential optical comparator, i.e. only the switch with the higher input power is turned on and emits light. When choosing a proper load resistance, NOR or NAND gates can be realized or an inverter when only one optical input is used. It should be mentioned that similar results can be obtained in distributed structures (see Fig. 6) where signal processing in adjacent regions is accomplished and the information is simultaneously propagated along the structure. A further very interesting image processing function - pattern edge extraction - has been realized in a chain of parallel - connected pnpn switches[33].

As can be seen, the use of inhomogeneous optical input power, e.g. from an image, can lead to interesting signal processing properties of distributed optoelectronic switching devices. However, this field of research work is still at the very beginning.

## Bragg Reflectors and Optical Solitons

Another type of dissipative structures is predicted to occur in a Bragg reflector consisting of a chain of bilayers of a linear and a nonlinear film[34]. Again a hybrid struc-

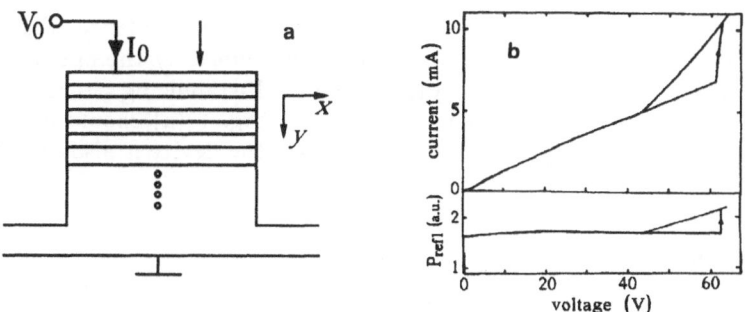

**Figure 10.** (a) Bragg reflector consisting of bilayers of GaAs and AlGaAs[35,36]; (b) measured electrical and optoelectronic bistability[37].

ture has been studied experimentally and optical bistability has been found recently[37] near the Bragg frequency, see Fig. 10. Model experiments in the microwave range have pointed out that the bistability is connected with the occurence of static kink structures of the excitation density and solitons of the wave envelope in $y$-direction[38]. Additionally, switching is traced back to discontinuous jumps of the structures along the optical beam direction.

The general analysis in Ref. 38 has shown that the excitation density $n = U$ can be described by eq. (6) where $I'(n)$ is N-shaped and depends on the optical power through $I'(n) \sim P(y)$. Furtheron, $P(y)$ is determined by a nonlinear Schrödinger equation (NLS) for the complex optical field amplitude $E$ according to

$$i\left(\frac{\partial}{\partial t} + u_g\frac{\partial}{\partial y}\right)E + \frac{1}{2}u_g'\frac{\partial^2 E}{\partial y^2} - \kappa|E|^2 E = 0 \qquad (7)$$

Here $u_g$ denotes the group velocity and $u_g'$ the group velocity dispersion. $\kappa$ is a parameter of nonlinearity. Most important for the present analysis is that $u_g$ can become zero near the cut-off frequency and eq. (7) yields static solitons which separates the device into two domains, an optically transparent and an optically reflecting region, see Fig. 11. The transition region is determined by the kink structure of the excitation density $n$, which can be the temperature in a photothermal device or the photocarrier

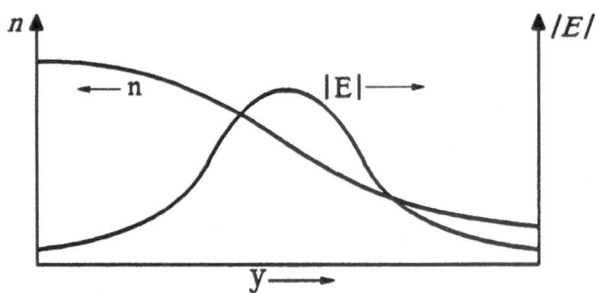

**Figure 11.** Gap soliton and excitation density in a nonlinear Bragg reflector (schematic).

density when the optical nonlinearity arises from charge carriers. Note that in this case the hybrid device consists of two electrical field domains similar to what happens in the Gunn diode. Since the operating point is determined by the optical power, it varies with the depth according to $P(y)$ leading to a self-stabilizing structure. Finally, it should be pointed out that the reflector shown in Fig. 10(a) should exhibit additional dissipative structures in form of filaments when the device is also distributed in $x$-direction.

## CONCLUSION

The formation of dissipative structures in bistable electronic and optoelectronic semiconductor devices is studied. First a GaAs:Cr pin diode is investigated where different experimental techniques are utilized to detect the properties of current density filaments in a quantitative way. A comparison with theoretical results based upon a suitable equivalent circuit is carried out. Most interesting are the results on spatially resolved oscillations. In a second chapter it is shown that current filaments can also occur in large area optoelectronic switching devices such as SEED elements and pnpn four-layer diodes. Theoretically, pattern formation is described by the FitzHugh-Nagumo (FHN) equation as in the case of the double injection pin diode. Besides these transversal inhomogeneities longitudinal domains are found in a nonlinear Bragg reflector similar to the Gunn effect. Again the formation of domains is given by the FHN equation. The corresponding optical field amplitude is determined by the nonlinear Schrödinger equation (NLS). It is foreseen that further phenomena will most probably be discovered leading to a variety of potential technical applications in image processing.

As a final result, it is found that static patterns in NDC-materials can be described by only one basic nonlinear active diffusion equation, the FHN equation. And the innovative concept is that this holds for the generation of domains as well as filaments. Additionally, the NLS equation is discovered to control the behaviour of the optical input power as a second important physical parameter. Note that both, the FHN- and the NLS-equation, provide stationary solutions which are independent of time. Further work on the physics of such phenomena including experimental investigations are in progress.

## ACKNOWLEDGEMENT

This research work was financially supported by the Volkswagen-Stiftung within the two joint programs "Synergetik" and "Photonik".

## REFERENCES

1. H. Thomas, editor. "Nonlinear Dynamics in Solids", Springer, Heidelberg (1992).
2. E. Schöll. "Nonequilibrium Phase Transitions in Semiconductors", Springer-Verlag, Heidelberg (1987).
3. V.L. Bonch-Bruevich, I.P. Zvyagin, and A.G. Mironov. "Domain Electrical Instabilities in Semiconductors", Studies in Soviet Science. Consultants Bureau, New York (1975).
4. J. Pozhela. "Plasma and Current Instabilities in Semiconductors", Pergamon Press, Oxford (1981).
5. M. Stoisiek and R. Sittig, Power thyristors using ic-technology, in: "Festkörperprobleme, Advances in Solid State Physics", vol. 26, P. Grosse, ed., Vieweg, Braunschweig (1986).
6. X. Pan, H. Olesen, B. Tromberg, and H.E. Lassen, Analytic description of the standing wave effect in DFB lasers, IEE Proc.-J 139:189 (1992).
7. F. Stöckmann, Elektrische Instabilitäten in Halb- und Photoleitern, in: "Festkörperprobleme, Advances in Solid State Physics", vol. 9, O. Madelung, ed., Vieweg, Braunschweig (1969).
8. D. Jäger and R. Symanczyk, Current density filaments in semiconductor devices, in: "Nonlinear Dynamics in Solids", H. Thomas, ed., Springer, Heidelberg (1992).
9. R. Newman, Visible light from a silicon p-n junction, Phys. Rev. 100:700 (1955).
10. A.M. Barnett, Current filament formation, in: "Injection Phenomena", vol. 6 of "Semiconductors and Semimetals", R.K. Willardson and R.C. Beer, eds., Academic Press, New York (1970).
11. J. Peinke, Current instabilities in the interplay between chaos and semiconductor physics, in: "Nonlinear Dynamics in Solids", H. Thomas, ed., Springer, Heidelberg (1992).
12. A. Brandl and W. Prettl, Current filaments and nonlinear oscillations in n-gaas, in: "Festkörperprobleme, Advances in Solid State Physics", vol. 30, U. Rössler, ed., Vieweg, Braunschweig (1990).
13. E. Schöll, Current instabilities in semiconductors: Mechanisms and self-organized structures, in: "Nonlinear Dynamics in Solids", H. Thomas, ed., Springer, Heidelberg (1992).
14. E. Schöll and D. Drasdo, Nonlinear dynamics of breathing current filaments in n-GaAs and p-Ge, Z. Phys. B - Condensed Matter 81:183 (1990).
15. C. Radehaus, R. Dohmen, H. Willebrand, and F.-J. Niedernostheide, Model for current patterns in physical systems with two charge carriers, Phys. Rev. A 42:7426 (1990).
16. D. Jäger, H. Baumann, and R. Symanczyk, Experimental observation of spatial structures due to current filament formation in silicon pin diodes, Phys. Lett. A 117:141 (1986).

17. R. Symanczyk, E. Pieper, and D. Jäger, Optical study of current filaments in pin diodes, *Phys. Lett. A* 143:337 (1990).

18. R. Symanczyk, S. Gaelings, and D. Jäger, Observation of spatio-temporal structures due to current filaments in Si pin diodes, *Phys. Lett. A* 160:397 (1991).

19. W.H. Weber and G.W. Ford, Double injection in semiconductors heavily doped with deep two-level traps, *Solid-St. Electron.* 13:1333 (1970).

20. I. Dudeck and R. Kassing, The influence of recombination center data on the I-V characteristics of silicon p-i-n diodes, *J. Appl. Phys.* 48:4786 (1977).

21. R. Symanczyk, D. Jäger, and E. Schöll, Equivalent circuit model for current filamentation in p-i-n diodes, *Phys. Lett. A,* 59:105 (1991).

22. G.M. Martin, A. Mitonneau, D. Pons, A. Mircea, and D.W. Woodard, Detailed electrical characterisation of deep Cr acceptor in GaAs, *J. Phys. C* 13:3855 (1980).

23. A.C. Scott, The electrophysics of a nerve fiber, *Rev. of Modern Phys.* 47:487 (1975).

24. D. Jäger, A. Gasch, and H.G. Schuster, Emergence of aperiodic spikes and a boundary crisis in nonlinear microwave resonators, *Phys. Rev. A* 33:1451 (1986).

25. A. Szöke, V. Danen, J. Goldhar, and N.A. Kurnit, Bistable optical element and its applications, *Appl. Phys. Lett.* 15:376 (1969).

26. C. Klingshirn, J. Grohs, and M. Wegener, Optical instabilities in passive semiconductors, *in:* "Nonlinear Dynamics in Solids", H. Thomas, ed., Springer, Heidelberg (1992).

27. D. Jäger, Large optical nonlinearities in hybrid semiconductor devices, *J. Opt. Soc. Am. B* 6:588 (1989).

28. D. Jäger, Optical bistability: Semiconductor devices for digital optical signal processing, *in:* "Nonlinear Optical Material", *SPIE* 1127:2 (1989).

29. D. Jäger, Hybrid semiconductor devices for optical and optoelectronic signal processing, *SPIE ICOESE '90* 1230:707 (1990).

30. D.A.B. Miller, D.S. Chemla, T.C. Damen, A.C. Gossard, W. Wiegemann, T.H. Wood, and C.A. Burrus, Novel hybrid optically bistable switch: The quantum well self-electro-optic effect device, *Appl. Phys. Lett.* 45:13 (1984).

31. D. Jäger, F. Forsmann, and B. Wedding, Low power optical bistability and multistability in a self-electro-optic silicon interferometer, *IEEE J. Quant. Electron.* 21:1453 (1985).

32. K. Hara, K. Kojima, K. Mitsunaga, and K. Kyuma, Differential optical comparator using parallel connected AlGaAs pnpn optical switches, *Electron. Lett.* 25:43 (1989).

33. K. Hara, K. Kojima, K. Mitsunage, and K. Kyuma, Pattern edge extraction using parallel connected AlGaAs - GaAs pnpn optical switches, *IEEE Photon. Techn. Lett.* 3:62 (1991).

34. W. Chen and D.L. Mills, Gap solitons and the nonlinear optical response of superlattices, *Phys. Rev. Lett.* 58:160 (1987).

35. S. Zumkley, G. Wingen, G. Borghs, F. Scheffer, W. Prost, and D. Jäger, Electrooptical modulation in AlGaAs/GaAs distributed feed-back structures, *in:* "High Speed Phenomena in Photonic Materials and Optical Bistability", *SPIE* 1280:202 (1990).

36. S. Zumkley, G. Wingen, F. Scheffer, W. Prost, and D. Jäger, Photonic switching and SEED effect in AlGaAs/GaAs DFB structures grown by MOVPE, *in:* "Photonic Switching II", K. Tada and H.S. Hinton, eds., Springer, Heidelberg (1990).

37. S. Knigge, S. Zumkley, G. Wingen, O. Humbach, C. Chaix, and D. Jäger, Experiments on optoelectronic bistability in distributed AlAs/GaAs-Bragg reflectors, *in:* "Proceedings of the 22nd ESSDERC '92", Elsevier Science Publisher, Amsterdam (1992).

38. D. Jäger, A. Gasch, and K. Moser, Intrinsic optical bistability and collective nonlinear phenomena in periodic coupled microstructures: model experiments, *in:* "Optical Schwitching in Low-Dimensional Systems", H. Haug and L. Banyai, eds., Plenum Publishing Corp. (1989).

# NDR, HOT ELECTRON INSTABILITIES AND
# LIGHT EMISSION IN LDS

A. Da Cunha, A. Straw and N. Balkan

University of Essex, Physics Department, Colchester,
United Kingdom

## 1. INTRODUCTION

### 1.1 Instabilities

It is well known that negative differential resistance, NDR, in bulk material can occur as a result of mixed scattering[1,2], non-parabolicity[2,3], intervalley transfer[4,5], impurity barrier capture[6,7] and other esoteric mechanisms[8,9]. All these mechanisms give rise to voltage controlled NDR or, in reference to the shape of the current-voltage characteristic, n-type NDR. A second type is current controlled NDR, s-type, which is associated with impact ionisation in special conditions of one form or another[10]. Both types give rise to electrical instabilities, the former generating propagating high field domains, the latter, high current filaments[11]. The formation of travelling domains in bulk $GaAs$ is well known, namely the Gunn effect[12], and equally well known are the many applications of the effect in microwave electronics. In contradistinction, the formation of filaments in practical devices is usually something to be avoided, heralding, as they do, irreversible destruction of the device through excessive Joule heating.

There is no shortage of proposals for novel voltage controlled NDR in the field of LDS. Indeed, the research into superlattices was initially motivated by the prediction of NDR associated with vertical transport, that is transport perpendicular to the wells and the barriers forming the superlattice[13]. Novel mechanisms for n-type NDR associated with longitudinal transport, in other words transport parallel to the layers, include real space transfer[14], electron-electron scattering in the presence of an abrupt optical-phonon emission threshold[15], and inter-well tunnelling[16] as well as the conventional bulk mechanisms. Although the observation of NDR in, particularly, $GaAs/Ga_{1-x}Al_xAs$ quantum well structures has been widely reported[17-20], very little work on the accompanying electrical instabilities has appeared in the literature as yet. Experimental work on the topic has been reported by us[21-27] and by a few other research groups[28-33]. Our work at Essex on high field longitudinal transport in $GaAs$ quantum well structures has shown that there exists a number of electrical instabilities associated with hot electron transport. These can be categorized as follows.

1. Acoustoelectric like instabilities. These occur typically at and above 77K in long specimens with electron densities less than $2\times10^{11}cm^{-2}well^{-1}$. The instabilities are in the form of damped, or continuous, oscillations at a frequency corresponding to the transit time associated with transversely polarised acoustic (TA) modes[21]. The electron drift velocities at the thresholf field for the instabilities are found to be slightly higher than the TA

*Negative Differential Resistance and Instabilities in 2-D Semiconductors*
Edited by N. Balkan *et al.*, Plenum Press, New York, 1993

283

2. Carrier depletion. In nominally undoped material at lattice temperatures, $T_l$, less than 10K current collapse and carrier depletion occurs at electric fields, F, of about 600Vcm$^{-1}$, followed by current oscillations of frequency, f, of the order of 3MHz[22,23].

3. High frequency low temperature oscillations. These occur in samples with 'intrinsic' *GaAlAs* buffers and electron densities, $n_{2d}$, between $1 \times 10^{12}$cm$^{-2}$ and $2 \times 10^{11}$cm$^{-2}$, below 200K and at electric fields of a few hundred Vcm$^{-1}$. The threshold field, however, does vary by as much as factor of 2 in different samples made from the same wafer, indicating an association with the uniformity of the samples. The instabilities prevail in the form of continuous current oscillations with frequencies increasing with temperature and electric field up into the GHz régime. The frequency of the oscillations is independent of the length of the specimen, but increases with electron densities[24,25]. In samples with *GaAs* buffer layers, however, similar instabilities occur at higher threshold fields, typically about 2kVcm$^{-1}$ at $T_l$=60K, which generally increases with increasing lattice temperature. Typically $n_{3d} \le 4 \times 10^{17}$cm$^{-3}$ in the samples exhibiting instabilities.

4. High field oscillations. These occur typically at room temperature in samples containing 3d electron densities, $n_{3d} \le 4 \times 10^{17}$cm$^{-3}$. The threshold field for the oscillations depends on the carrier concentration and the structural parameters such as the well width and barrier height, but typically the field is between 2.2kVcm$^{-1}$ and 4kVcm$^{-1}$. The frequency of the oscillations is independent of the field and the length of the samples[26,27].

5. Destructive instabilities. In samples containing large electron densities, $n_{2d} > 10^{12}$cm$^{-2}$, current filamentation occurs as observed by scanning electron microscopy, accompanied by instabilities in the form of current oscillations, ultimately leading to the irreversible destruction of the sample[28].

Table 1 gives a summary of all the different types of instability observed at Essex. Categories 1,2 and 3 are discussed in detail at this workshop by our colleagues R. Gupta and A.J. Vickers, as well as by J.H. Wolter who has reported similar observations to type 2 and 3 instabilities[32-34]. It is however worth noting that types 1,2 and 3 instabilities occur at low lattice temperatures and at electric fields well below those required for the conventional NDR mechanisms in LDS. Type 4 is probably caused by either of the conventional electron transfer mechanisms, namely real space transfer or intervalley transfer. What makes this type of instability markedly different from those in bulk material is the absence of propagating high field domains. Also the inhibition of NDR in samples where $n_{3d} >> 4 \times 10^{17}$cm$^{-3}$. The inhibition of NDR has been attributed to non-drifting hot phonon effects[27,28]. Type 5 is likely to be associated with impact ionisation as it is usually followed by the appearance of a static high field domain near the cathode, where the local fields can exceed $10^4$Vcm$^{-1}$.

## 1.2 Light Emission

Since the inception of LDS, much attention has been directed towards the application of size quantisation in tailoring the wavelength of light produced by semiconductor lasers for use in optical communications and optical information read out systems. The relevant devices, which attracted most attention, are quantum well heterostructure lasers, whose operation is based on vertical transport and the capture of charge carriers within the quantum well[35]. Electroluminescent devices based on hot carrier transport parallel to the layers however, have not received much attention until recently. Earlier observations of tunable wavelength light emission were reported in periodically doped superlattices (nipi structures)[36]. When an electric field is applied along the superlattice layers, spatially separated electrons and holes are heated, these carriers can then surmount the respective space charge potential barriers via thermionic emission to ultimately recombine radiatively across the *indirect gap in real space*. So far problems associated with the growth of high quality nipi structures have hindered progress in this area. More recent light emitters utilizing parallel transport are based upon real space transfer. One of these involves the injection of hot minority electrons across a semiconductor heterojunction into a p-type collecting layer[37]. The device, in principle, is a three terminal NDR field effect transistor (more commonly referred to by the acronym NERFET) where the source-drain field heats up the channel electrons transferring them, in real space, into a biased complementary p-type collector. NDR in the source-drain circuit is then accompanied by radiative recombination from the collector. Another electroluminescent device, based upon NERFET operation, involves the injection, in real space, of majority electrons into a biased n-type collector, causing the creation of holes by impact ionization[38]. High energy photon emission ($h\nu > E_g$) has also been observed in *GaAs* HEMT structures under heavy bias, this has been attributed to the recombination of the channel electrons with holes generated by impact ionization in the Gunn domains, where local electric fields in excess of $10^4 \text{Vcm}^{-1}$ exist[39]. Detailed discussions of electroluminescent devices based on hot electron longitudinal transport are given by S. Luryi, T.G. Higman and C. Canali in the relevant chapters of this book. We shall therefore concentrate on a new type of tunable electroluminescence that has been observed at Essex in n-type modulation doped *GaAs/Ga$_{1-x}$Al$_x$As* multiple quantum well structures[22-25]. To our knowledge no other observation of light emission similar to ours has been reported in the literature. This is not surprising however, as one does not usually expect to see, and moreover to look for, electroluminescence in modulation doped n-type 2d material.

The aim of the work presented here is to give a brief summary of our observations of type 3 and 4 instabilities together with hot electron electroluminescence studies. Possible mechanisms to describe the observed phenomena will also be discussed and compared with the experimental results.

**Table 1.** Summary of the electrical instabilities in 2d *GaAs* (longitudinal transport) observed at Essex

| Type | Acousto-electric Like | Carrier Depletion | High Frequency Low Temperature (a)*GaAlAs* buffer | (b)*GaAs* buffer | High Field High Temperature (a)*GaAlAs* buffer | (b)*GaAs* buffer | Current Filamentation |
|---|---|---|---|---|---|---|---|
| Temp. | ~77K | <10K | <220K | <300K | >280K | <300K | |
| Carrier Density | $n_{2d}$~2×10$^{11}$ cm$^{-2}$well$^{-1}$ | Undoped | 2×10$^{11}$<$n_{2d}$<1×10$^{12}$cm$^{-2}$well$^{-1}$ | $n_{3d}$<4×10$^{17}$ cm$^{-3}$ | $n_{3d}{}^*$≤1×10$^{17}$ cm$^{-3}$ | $n_{3d}{}^+$<4×10$^{17}$ cm$^{-3}$ | $n_{2d}$>10$^{12}$cm$^{-2}$ |
| Oscillation | Damped | Current Collapse | Continuous | | Continuous | | Irreversible destruction of sample |
| Frequency | f=1/Δ | 3MHz | f=f(T$_l$,n$_{2d}$,F) f$_{max}$=1GHz | f=f(T$_l$) | f≠f(T$_l$,F,l) f=1GHz | | |
| Threshold Field | →$v_d$>$v_s$ | ~600 Vcm$^{-1}$ | ~200Vcm$^{-1}$ | ~1.5kVcm$^{-1}$ | ≥4kVcm$^{-1}$ | F$_{thres}$=f(T$_l$) ~2.3kVcm$^{-1}$ at 300K | |

* When $n_{3d}$>1×10$^{17}$cm$^{-3}$ then there are no instabilities.

+ When $n_{3d}$>4×10$^{17}$cm$^{-3}$ then there are no instabilities.

where Δ is the TA transition time, $v_d$ is the drift velocity, $v_s$ is the sound velocity, F$_{thres}$ is the threshold field, f is the frequency, T$_l$ is the lattice temperature, F is the electric field and l is the sample length.

## 2. EXPERIMENTAL RESULTS AND DISCUSSION

The parameters and the structural diagrams of the samples are shown in table 2. The samples were grown by both MOCVD and MBE techniques and fabricated into simple bar or Hall bar shaped specimens. Ohmic contacts were formed by alloying *AuGe,In* or *NiAuGeCu*. All high field measurements were carried out in pulsed field conditions where electric field pulses of a few microseconds in duration, with a duty

**Table 2.** Sample parameters and structural diagrams.

| Sample | Growth | %Al | $L_Z$ /Å | $L_B$ /Å | $n_H$(4.2K) /cm$^{-2}$ | $\mu_H$(4.2K) /cm$^2$V$^{-1}$s$^{-1}$ | $n_{3d}$ /cm$^{-3}$ | Instabilities | Light Emission |
|---|---|---|---|---|---|---|---|---|---|
| MV280* | MOCVD | 45 | 2×40 | 100 | $4.00\times10^{11}$ | $3.8\times10^4$ | $1.00\times10^{18}$ | Types 1&4 | Yes |
| MV281 | MOCVD | 45 | 40 | - | $9.60\times10^{11}$ | $4.3\times10^4$ | $2.40\times10^{18}$ | Type 5 | Yes |
| OC82 | MOCVD | 30 | 100×75 | 100 | $1.55\times10^{12}$ | $5.0\times10^3$ | $2.07\times10^{16}$ | Types 1&4 | Yes |
| CB520 | MOCVD | 30 | 10×75 | 185 | $3.06\times10^{12}$ | $1.0\times10^4$ | $4.08\times10^{17}$ | Types 3&4 | Yes |
| CB535 | MOCVD | 30 | 10×75 | 185 | $1.66\times10^{12}$ | $9.0\times10^3$ | $2.77\times10^{17}$ | Types 3&4 | Yes |
| C568 | MBE | 30 | 10×100 | 185 | $1.18\times10^{13}$ | $3.5\times10^4$ | $1.18\times10^{18}$ | Type 5 | Yes |

* Most carriers are in one well.

**MV280**

| i-GaAs | 90Å | |
|---|---|---|
| n-$Al_{0.45}Ga_{0.65}As$ | 580Å | |
| i-$Al_{0.45}Ga_{0.65}As$ | 120Å | |
| i-$Al_{0.45}Ga_{0.65}As$ | 100Å | x2 |
| i-GaAs | 40Å | |
| i-$Al_{0.45}Ga_{0.65}As$ | 0.5μm | |
| i-GaAs | 0.5μm | |
| undoped semi-insulating GaAs substrate | | |

**MV281**

| i-GaAs | 90Å |
|---|---|
| n-$Al_{0.45}Ga_{0.65}As$ | 580Å |
| i-$Al_{0.45}Ga_{0.65}As$ | 120Å |
| i-GaAs | 40Å |
| i-$Al_{0.45}Ga_{0.65}As$ | 0.5μm |
| i-GaAs | 0.5μm |
| undoped semi-insulating GaAs substrate | |

**OC82**

| i-GaAs | 100Å | |
|---|---|---|
| i-$Al_{0.45}Ga_{0.65}As$ | 580Å | |
| i-GaAs | 75Å | x100 |
| i-$Al_{0.45}Ga_{0.65}As$ | 100Å | |
| i-$Al_{0.4}Ga_{0.6}As$ | 0.5μm | |
| i-GaAs | 100Å | |
| undoped semi-insulating GaAs substrate | | |

**CB520,CB535**

| i-GaAs | 240Å | |
|---|---|---|
| n-$Al_{0.3}Ga_{0.7}As$ | 225Å | |
| n-$Al_{0.3}Ga_{0.7}As$ | 75Å | |
| i-$Al_{0.3}Ga_{0.7}As$ | 110Å | |
| i-GaAs | 75Å | x10 |
| i-$Al_{0.3}Ga_{0.7}As$ | 110Å | |
| n-$Al_{0.3}Ga_{0.7}As$ | 75Å | |
| i-GaAs | 0.2μm | |
| undoped semi-insulating GaAs substrate | | |

**C568**

| i-GaAs | 240Å | |
|---|---|---|
| n-$Al_{0.3}Ga_{0.7}As$ | 225Å | |
| n-$Al_{0.3}Ga_{0.7}As$ | 75Å | |
| i-$Al_{0.3}Ga_{0.7}As$ | 110Å | |
| i-GaAs | 100Å | x10 |
| i-$Al_{0.3}Ga_{0.7}As$ | 110Å | |
| n-$Al_{0.3}Ga_{0.7}As$ | 75Å | |
| n-$Al_{0.3}Ga_{0.7}As$ | 75Å | |
| i-GaAs | 10Å | x5 |
| i-$Al_{0.3}Ga_{0.7}As$ | 190Å | |
| i-GaAs | 0.5μm | |
| undoped semi-insulating GaAs substrate | | |

cycle less than 1%, were applied along the 2d layers. Current waveforms were observed on a 1GHz oscilloscope and plotted by using a box car x-t plotter assembly. Photoluminescence (PL) was excited with the 647nm line of a CW Krypton laser and recorded using standard lock-in techniques. The electroluminescence (EL) was measured when the sample was in complete darkness. The EL signal was detected during the longitudinal field pulse duration, it was then dispersed and plotted by using simple gating techniques as described elsewhere[40]. The spatial distribution of the EL was observed and recorded in the steady state. A TV frequency infrared camera, fitted with zoom lenses and a video printer assembly was employed in these measurements.

## 2.1 Samples with $Ga_{1-x}Al_xAs$ Buffer Layers

Figure 1 shows the I-V characteristics of the double quantum well sample MV280 at a lattice temperature, $T_l$, of 4.2K. Instabilities in the form of noisy oscillations in the current pulse occurred with a bias of 3.2V applied, corresponding to an electric field, F, of 320Vcm⁻¹ and an electron temperature, $T_e$, (as calculated from mobility-lattice temperature and mobility-field mapping[29]) of 100±5K. The drift velocity, $v_d$ at the instability threshold was $9.8 \times 10^6$cms⁻¹.

Increasing the applied field above 320Vcm⁻¹ caused a slight increase in the frequency of the oscillations and a decrease in the Ohmic conductivity. Subsequent heating of the sample up to room temperature however, resulted in the recovery of the Ohmic conductivity. Noisy oscillations occured at lattice temperatures below 220K. However, the threshold field for the instabilities increased and the amplitude of the oscillations decreased with temperature. At $T_l$=77K the threshold field and the drift velocities were found to be 1.38kVcm⁻¹ and $1.1 \times 10^7$cms⁻¹ respectively. At temperatures above 220K current saturation occured at electric fields of about 3kVcm⁻¹. Figure 2 shows the I-V characteristics of the sample at $T_l$=300K. Here the current saturates at F=3.5kVcm⁻¹, corresponding to a saturation drift velocity of about $9 \times 10^6$cms⁻¹. Increasing the field further, up to F=8kVcm⁻¹, did not alter the shape of the I-V characteristic: no NDR was observed and there was no instabilities in the current pulse.

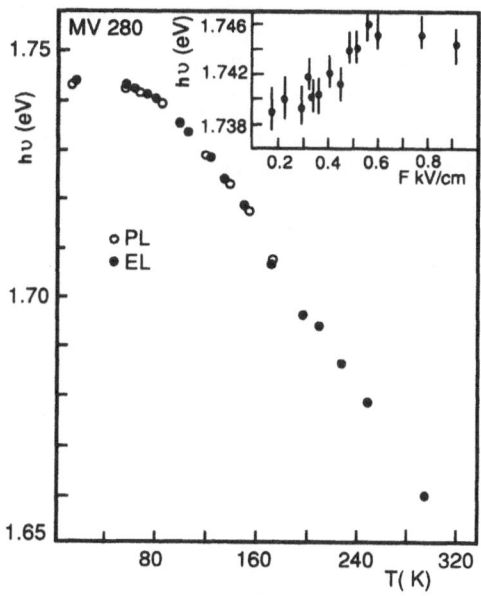

**Figure 1.** Current-voltage characteristics of the double quantum well sample, MV280, at 4.2K. Instabilities in the form of current oscillations occur at an applied electric field of 320Vcm⁻¹.

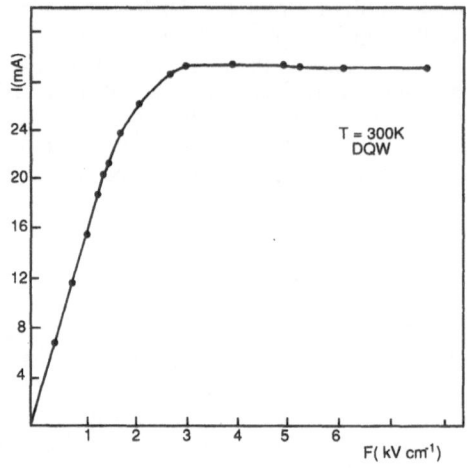

**Figure 2.** Current-voltage characteristics of the double quantum well sample at $T_l=300K$. No NDR and no instabilities are observed up to an applied field of $8kVcm^{-1}$.

A strong light emission, hereafter referred to as Electroluminescence (EL), was observed from the surface of the sample over the whole measured temperature range, when electric fields in excess of a few hundred $Vcm^{-1}$ were applied along the layers. The rise time of the EL (the incubation time) was much slower than the risetime of the field pulse and was found to be a decreasing function of the applied electric field. We did not measure this dependence in MV280, but in previous studies on other samples indicate an inverse exponential dependence. The decay time of the EL pulse, on the other hand, was comparable with that of the field pulse and it was independent of the applied field[24]. The EL had a similar spectral dependence to that of the quantum well PL. The peak energy of the EL spectrum increased slightly with increasing electric field, whilst the peak energy of the PL increased with increases in the excitation intensity, but they both saturated at the same energy. Figure 3 shows the EL and the PL spectra of MV280 and figure 4 shows the temperature dependence of the peak energies. It is evident from figure 3 and 4 that both the EL and the PL are of the same origin, that is quantum well luminescence. The EL persisted up to room temperature while the instabilities disappeared at $T_l=220K$, suggesting that they were of different sources. It should be noted, however, that the reduction in the Ohmic conductivity upon the application of electric fields in excess of the instability threshold were also accompanied by a reduction in the intensity of the (zero field) PL.

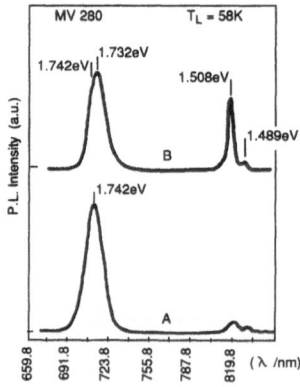

**Figure 3.** PL(B) and EL(A) spectra of sample MV280 at a lattice temperature of 58K

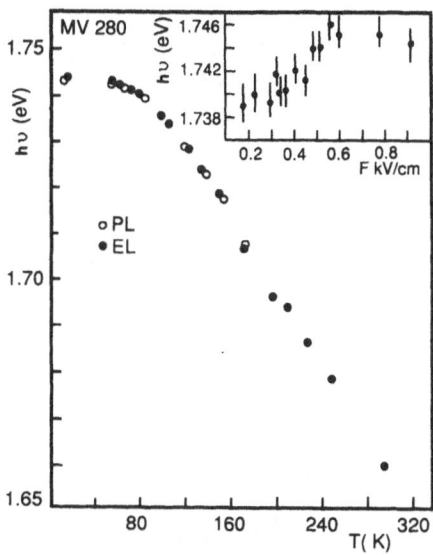

**Figure 4.** The lattice temperature dependence of the peak energy of the PL spectra (open circles) and the EL spectra (closed circles). The inset shows the shift of the EL peak with applied field.

The EL intensity was found to be a rapidly increasing function of the applied electric field as depicted in figure 5. In the figure the EL intensity is plotted against the applied field on a logarithmic scale at $T_l=4.2K$, 56K and 294K. At lower ranges of the electric field the EL intensity, I, has a power law dependence on F of the form $I \alpha F^s$. Where the index, s, is 6.3, 5.8 and 4.7 at $T_l=4.2K$, 56K and 294K respectively. The EL intensity tends to saturate at high electric fields. However, it is difficult to attach any significance to the field dependence. This is because the EL is collected from the whole surface of the sample. Our investigations of the spatial distribution of the EL signal on the sample surface indicate that at low electric fields only some microscopic regions in the sample emit light strongly. As the electric field is increased, however, many more active spots or channels appear, as well as an increase in the brightness of the light. Thus the field dependence not only represents the intensity but also the widening of the active area. We therefore investigated how the overall light output changed with the average energy (or the temperature) of the electrons.

The high energy tail of the EL spectra exhibited a Maxwellian distribution of carriers, characterized by an electron temperature, $T_e$, greater than $T_l$[25]. We could therefore extract $T_e$ from the applied electric field, and therefore investigate the EL intensity as a function of $T_e$. In these measurements the sample was kept immersed in liquid *He*, and hence the lattice temperature was constant, $T_l=4.2K$. In figure 6 the EL intensity is plotted against the inverse electron temperature, as obtained from the high energy tail of the EL spectra[40].

It is evident from the figure that at $T_e>105K$ the EL intensity is thermally activated with an activation energy, $\Delta E=97meV$. However, at $T_e<105K$ there is a deviation from the thermally activated behaviour. The reason for this will be discussed below. However, the figure shows clearly that the observed EL is directly associated with hot electrons.

In a double quantum well (DQW) structure it is difficult to predict if the EL is due to either of the layers or both. We therefore investigated a single quantum well (SQW) sample with an identical well width and with very similar growth parameters to the DQW structure, namely MV281. We also studied a multiple quantum well (MQW) structure to make sure that the observed phenomena was not an artefact of some unknown growth fault in the single or DQW structures.

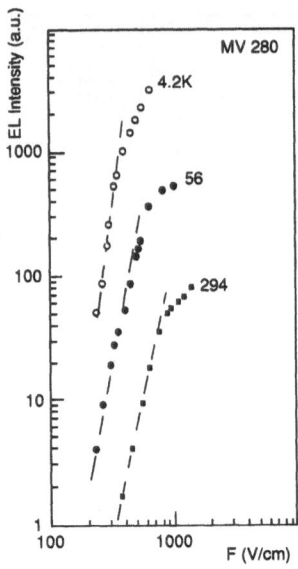

**Figure 5.** Electric field dependence of the EL intensity at the indicated lattice temperature.

Instabilities and light emission, similar to that of the DQW sample, was observed from both the SQW and MQW samples. The EL spectra for these samples are plotted together with the PL spectra in figure 7, which also indicates that both the EL and the PL are of the same origin. The peak energy of the EL spectra also had an identical lattice temperature dependence to that of the PL spectra in these samples. Instabilities in the SQW sample persisted up to $T_l=250K$. When $T_l>250K$ the current saturated at electric fields of about $3kVcm^{-1}$ and there were no instabilities or NDR up to the highest fields applied, $8kVcm^{-1}$. However, in the MQW samples NDR did occur, at $F=4.5kVcm^{-1}$ followed by high frequency oscillations in the current pulse $(f\sim1GHz)$[40]. SEM studies of the high field domains in this sample indicate the formation of a static (hanging) domain in the middle of the sample, perpendicular to the path of the electrons. The current instabilities *could not therefore be associated* with the propagation of high field domains.

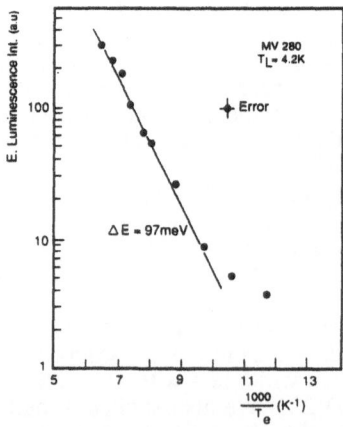

**Figure 6.** EL intensity versus inverse electron temperature for the DQW sample at Tl=4.2K. The solid line over the experimental points at high electron temperatures gives an activation energy of 97meV.

290

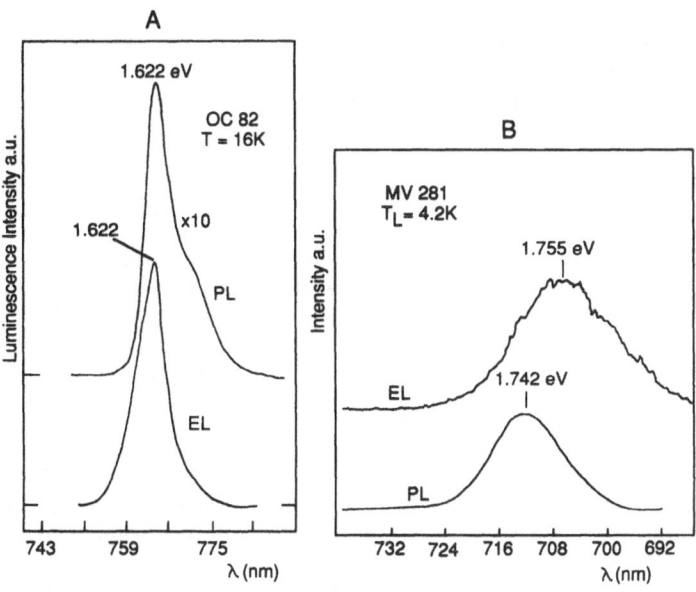

**Figure 7.** The PL and EL of the MQW (A) and the SQW (B) samples.

One possible reason for the observed light emission might be the recombination of quantum well electrons with holes generated by impact ionisation. However, the threshold fields, as observed in our samples, are far too low for impact ionisation in the wells to occur. Another explanation for the observed EL is the possible existence of one or more p-type channels. Indeed, in order for the EL intensity to be persistent (the intensity does not change during the duration of a long train of field pulses) there must be a continuous supply of minority carriers. The Hall measurements yield an n-type conduction as shown in table 2. This suggests that the conduction along the layers occurs via a parallel combination of hole and electron channels with electrons dominating the conduction, at least at low fields, whilst the supply of holes is maintained along the continuous conduction paths for the holes. The most obvious candidate for the presence of the p-type channels is the carbon impurities in the undoped *GaAlAs* material. The samples mentioned above have either an undoped or nominally n-type doped thick *GaAlAs* cladding layer at the bottom of the mesa. It is well established that such material contains a large density of non-uniformly distributed carbon impurities. The local density of carbon in microscopic clusters can be as high as $10^{17} cm^{-3}$[22,41]. It is then possible that the donation of holes to the lowest lying quantum well from such p-type regions along the bottom interface of the layers can give closely spaced pockets or channels of large potential fluctuations along the layers in the neighbourhood of the bottom interface. Further potential fluctuations associated with interface roughness, well width fluctuations, alloy clustering, non-uniform distribution of the n-type doping in the top barriers could also be present.

The electrons primarily in the n-type material would then occupy and conduct along percolation channels in real space. The hole conduction between pockets of holes can be achieved directly or via hopping[22]. A possible explanation for the observed instabilities at low temperatures and at low fields may therefore be found in hot electron percolation as predicted by Ridley[42]. According to this model electrons proceeding along a percolation path can gain energy from the electric field, become hot and can then surmount the potential barriers to be captured at localised states. This novel NDR would therefore give rise to the observed instabilities at low temperatures. At high lattice temperatures, however, the regeneration rate of the captured electrons might become comparable with the capture rate and hence the instabilities should be expected to diminish as observed. The transfer of hot electrons over the barriers into the spatially separated hole channels, on the other hand, would give rise to the observed EL.

In view of the considerations above, the model can be applied to our structures as shown in figure 8. Here a *GaAlAs* buffer layer can either be uniformily p-type doped (figure 8a) or alternatively it could contain p-type clusters randomly distributed in a nominally n-type material (figure 8b). In the former case the structure of the MQW sample consists of modulation doped wells in an n-type *GaAlAs* host material and depleted wells in the vicinity of the p-type *GaAlAs* buffer layer. Thus *GaAlAs* forms a p-n junction in the growth direction where some quantum wells are placed in the depletion region. In very extreme cases, the MQW structure itself can incorporate a p-n junction. When the electric field is applied parallel to the layers hot electrons and holes proceeding along their respective channels diffuse into the depletion region where they recombine radiatively in the quantum wells contained in this region. At low electric fields the greatest contribution to the recombining electrons is from free electrons within the barrier regions of the modulation doped wells. As the field is increased injection of electrons into the n-type barriers and holes into the p-type channel increases as expected if a mechanism as described by Hendriks et al[33] is operating. Also at high fields the carriers may become energetic enough to impact ionize their parent donors (and acceptors). Thus, an avalanche multiplication in the number of available carriers for recombination occurs. Therefore a very strong dependence of the emitted light intensity with electric field is expected as observed in our samples. Furthermore, electrical instabilities may also occur as a result of this mechanism[33]. At high $T_l$, when the majority of donors are ionized, the increase in the number of electrons available for recombination occurs as a result of real space transfer of the quantum well electrons at sufficiently high fields.

Since the quantum wells are placed on a potential ramp in the depletion region, emitted light intensity can be expected to exhibit a thermally activated behaviour with electron temperature. However, when the carriers become hot the potential profile at the junction is expected to change, this change involves some complicated hot carrier dynamics. Therefore it is difficult to attach a significant meaning to the thermal activation energy, but to a first approximation it should represent the energy difference between the n-type *GaAlAs* conduction band and the energy of the ramp at the highest lying quantum well in the depletion region. This is because the electron temperature at a given field would be much higher than the hole temperature due to the disparity in their effective masses, thus at high fields most recombination should occur in the quantum wells in the depletion region at the top of the ramp.

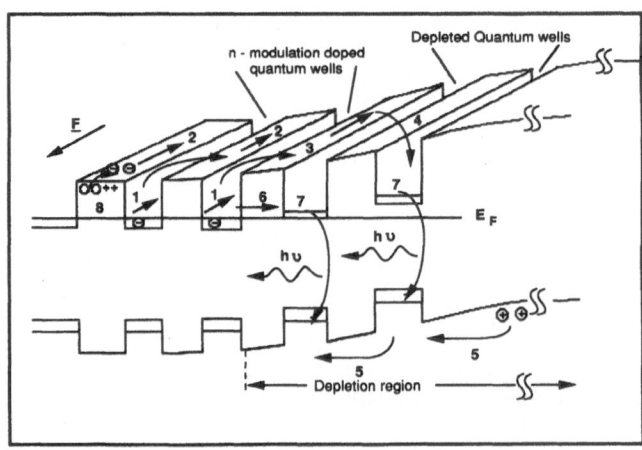

**Figure 8(a).** Schematic representation of the MQW structure in the vicinity of the *GaAlAs* n-p junction. The arrows indicate the possible hot carrier processes when the electric field is applied along the layers. 1) Drift of quantum well electrons. 2) Drift of barrier electrons. 3-4) Real space transfer, diffusion and capture of hot electrons by the depleted quantum wells. 5) Diffusion and capture of holes. 6) Tunnelling of hot quantum well electrons through the barriers into the depleted wells. 7) Radiative recombination. 8) Impact ionization of shallow donors in the barriers.

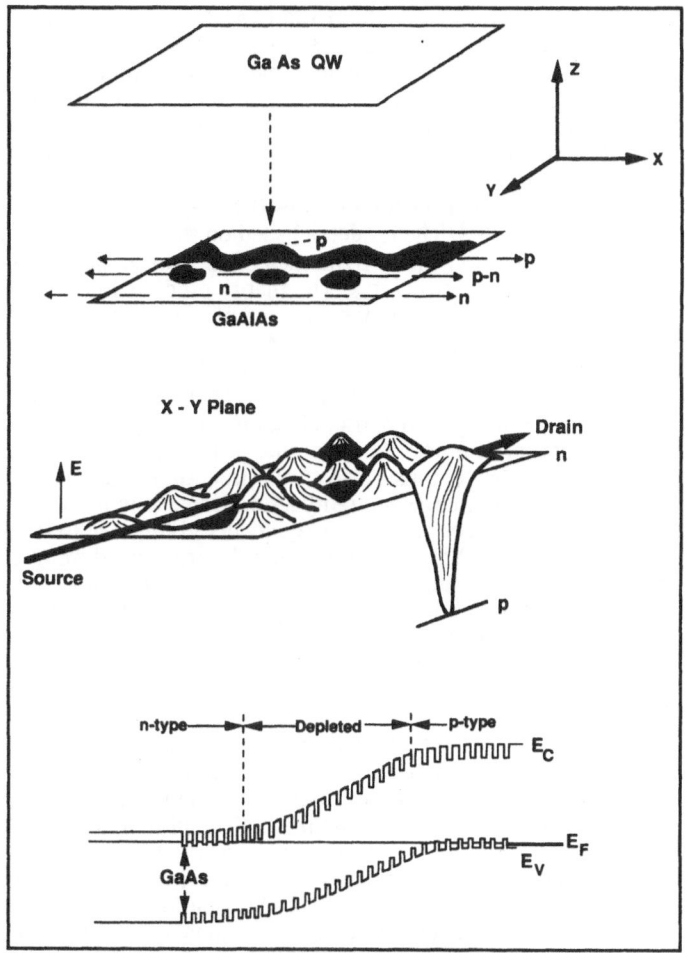

**Figure 8(b)** The schematic diagram of the process which gives rise to p and n-type and depleted quantum wells in the xy plane.

Although this mechanism can explain the light emission in a MQW structure (and probably in a DQW) it fails to explain why there is light emission in a SQW. The alternative model, as illustrated in figure 8b, however, suggests that light emission can occur in a SQW sample. If the *GaAlAs* buffer layer has p-type and n-type regions, as explained above, there will be randomly distributed p-n junctions in the plane of the buffer layer. The quantum well layer grown on the buffer layer in real space may, therefore, contain electrons in the vicinity of the n-type regions of the buffer as well as holes in the vicinity of the p-type regions in the adjacent buffer layer. In order to simplify the picture we can assume a single channel of p-type buffer continuous along the sample between the contacts, whilst the rest of the buffer layer is n-type. We can then represent the conduction channels along the quantum well layers as if they were made up of multiples of identical quantum wells, some n-type, some depleted and others p-type, as illustrated in figure 8(b). Obviously the real situation is far more complicated due to the randomness of the distribution of the p-type and n-type regions of the buffer layer. With such a random distribution of potential fluctuations, the existence of n-type and p-type channels as well as the depleted channels can indeed give rise to both NDR and also light emission as predicted by the hot electron percolation model. In view of this model we can explain the thermally activated behaviour of the EL intensity with $T_e$ as well as why the light emitting *'active'* spots and channels in the

293

sample get wider with increasing applied field. At low $T_e$ the deviation from the thermally activated behaviour merely represents a random distribution of activation energies as would be expected from the random nature of the processes involved in the model. Increasing the field and hence the electron temperature results in the activation of more p-type channels which are separated from the electron channels by higher barriers. The limiting height would be that represented by the single thermal activation at the higher electron temperatures as observed.

Another possible explanation for the observed (type 3) instabilities has been proposed by Hendriks et al[31,33]. The model is based on the theoretical work by Schöll and Aoki[43] concerning the non-linear dynamics of the space charge in the *GaAlAs* layers of the *GaAs/GaAlAs* 2d structures. The theory predicts intrinsic self generated oscillations. The mechanism involves the transfer of electrons (either by real space transfer or by contact injection) into the *GaAlAs* layers and the delayed dielectric relaxation of the interface space charge potential. According to the proposed mechanism involving the contact injection of electrons into the *GaAlAs* layers; at low temperatures, with increasing electric field, injected barrier electrons may become energetic enough to impact ionize their parent donors. Thus the charge neutrality at the *GaAs/GaAlAs* interface, particularly in the regions close to the injecting source contact, changes. Thus, in order to preserve the charge neutrality a time dependent relaxation of the space charge potential and the 2d electron gas density occurs giving rise to instabilities. The factor which determines the threshold field should vary with doping concentration (sample dependent) and with temperature. At high enough temperatures when the majority of donors are ionized, however, one may expect the instabilities to diminish as observed. Such a model, however, fails to explain light emission, unless either of the models for light emission, as outlined above, is also operative.

As for high field instabilities which occur at high temperatures in the lowly doped sample OC82 (and CB520 and CB535 as presented below), both the threshold fields and electron temperatures are high enough to cause real space transfer or intervalley transfer NDR, as reported by us earlier[40], and in this book by Straw et al[27]. The oscillation frequency of 1GHz, the limit of our experimental set up, is independent of the sample length. Considering the fact that the high field domain in OC82 is a static one we believe that the oscillations in the current pulse, unlike in bulk material, is not the result of the transit of propagating Gunn like domains but, rather the time dependent dielectric relaxation of the space charge, at the quantum well-barrier interface, associated with the lateral dissipation of the QW domains. Electron temperatures in the high field domain can be high enough to completely deplete the local space charge via real space transfer. Thus an oscillatory instability occurs as predicted by Schöll and Aoki[43]. The reason for the inhibition of NDR and current instabilities in samples with high 3d electron concentration is, on the other hand, believed to be associated with the non-drifting hot phonon effect[44]. Most theoretical treatments of the non-equilibrium (hot-phonon) effects consider a forward displaced distribution of the non-equilibrium phonons in momentum space which arises from the drift of hot electrons[25,26]. Consequently phonon reabsorption reduces the energy relaxation rates as generally accepted, however, it has only a small effect on the momentum relaxation rate, and therefore on the electron drift velocity at high fields. We have, however, shown both experimentally and theoretically that the momentum relaxation rates of hot electrons in *GaAs/GaAlAs* quantum wells increase with the decreasing energy relaxation rates and hence with the increasing 3d electron density[25]. In figure 9 the experimentally obtained high field saturation drift velocities are shown versus the 3d carrier concentration. The reduction in the saturation drift velocity with increasing carrier concentration is evident from the figure. This is believed to be due to the increasing hot phonon effects at large electron densities. The reduction in the drift velocity arises as a result of a randomization of the hot phonon distribution via the elastic scattering of phonons interacting with, for example, interface imperfections and well width fluctuations, and thus as a result of the enhanced momentum relaxation rate at high fields. We have observed high field NDR and instabilities in samples with 3d electron densities of $n_{3d} \leq 4 \times 10^{17} \mathrm{cm}^{-3}$, where hot phonon effects are negligible. This is shown in figure 10 where the drift velocity is plotted as a function the applied field at $T_l = 300K$ for the samples studied in the current work and previous work[23,26]. Samples with 3d electron concentrations greater than this do not show any prominent NDR or instabilities.

Since the observed low temperature, and low field, instabilities and light emission are shown to be primarily associated with the existence of nominally undoped or very lightly n-type doped *GaAlAs* buffer layers in our samples we investigated another set of samples with intrinsic *GaAs* buffer layers to see if the phenomena existed in this type of structure as well.

## 2.2 Samples with *GaAs* Buffer Layers

In order to prevent spurious effects associated with the unintentional non-uniform p-type doping of the *GaAlAs* layers we investigated samples with nominally intrinsic (slightly p-type) *GaAs* buffer layers. Figure 11 and 12 show the I-V characteristics of two samples, CB520 and CB535, at a lattice temperature of 60K. The channel length of the samples were 1mm and 1.75mm for CB535 and CB520 respectively. The threshold

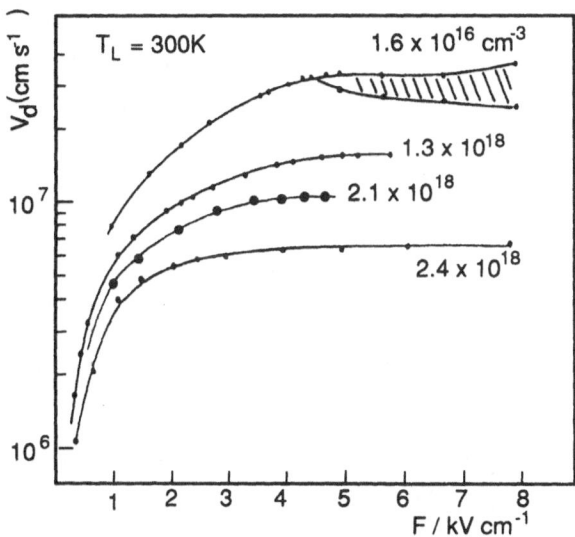

**Figure 9.** Drift velocity versus applied electric field in different QW samples with varying 3d carrier concentrations, $T_l$=300K.

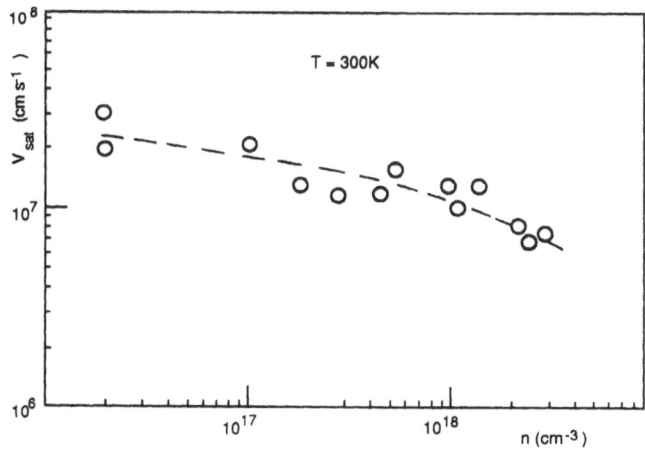

**Figure 10.** Saturation drift velocities measured at room temperature versus the 3d electron density in a number of samples investigated at Essex over a period of 5 years.

field for the instabilities at $T_l=60K$ is therefore identical for both samples and is $2.1\pm0.2kVcm^{-1}$. Corresponding drift velocities are in the range of about $1\times10^7cms^{-1}$. Instabilities in these samples are in the form of continuous current oscillations with a constant frequency of the order of 1GHz, which is the upper limit of our measurement set-up. The frequency of the oscillations, however, did not depend on the lattice temperature, applied field or the length of the samples within our measurement range, indicating that the oscillations were not associated with the propagation of charge domains along the channels.

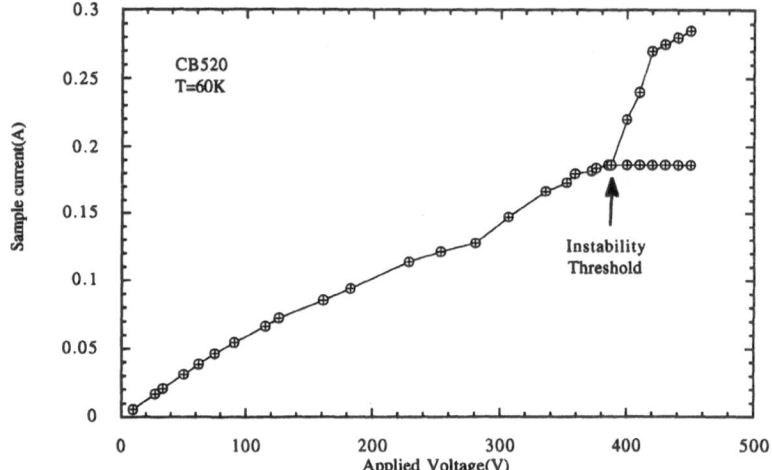

**Figure 11.** Current-voltage characteristic of CB520 measured at $T_l=60K$.

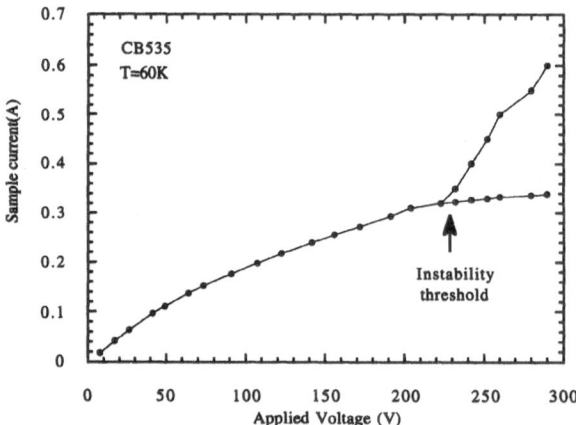

**Figure 12.** Current-voltage characteristic of CB535 measured at $T_l=60K$.

Unlike the samples with *GaAlAs* buffer layers, the threshold field for the instabilities did not show any appreciable variation in different samples made from the same wafer. This indicates that the instabilities could not be associated with non-uniformities as described above. Further evidence for the disparity in the underlying physical mechanisms, in the two sets of samples studied, comes from the actual magnitude of the threshold field. In samples with *GaAs* buffer layers the threshold field is much larger than the corresponding values in samples containing *GaAlAs* buffer layers. In figure 13 the threshold field for the instabilities is plotted as a function of lattice temperature, for both the samples with *GaAs* buffers. The data is rather scattered but the general trend is very clear. At lattice temperatures below 120K the threshold field decreases, whilst above 120K it increases with increasing lattice temperature. It should be noted, however, that in a mechanism which involves thermionic emission of hot carriers over potential barriers, as is the case in conventional NDR mechanisms, the important factor which determines the threshold for the instabilities is not the applied field itself but the average energy input per electron per unit time, and hence the electron temperatures at the threshold fields. In figure 13 the power input per electron, p, from the external field is plotted as a function of the lattice temperature as calculated from:

$$p = e\mu F_{th}^2 \qquad\qquad (1),$$

where $\mu$ is the electron mobility at the threshold field, $F_{th}$. It is evident from this figure that the threshold power input per electron decreases monotonically with increasing lattice temperature.

At the lattice temperatures and electric fields of interest we can assume that the dominant energy relaxation is via the emission and absorption of longitudinal optical phonons. It is therefore relatively straight forward to obtain the electron temperatures at the instability threshold via[21]:

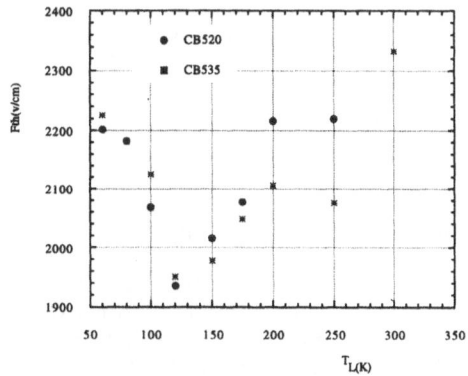

**Figure 13.** The threshold electric field for the instabilities as a function of lattice temperature, for the two samples shown in figures 11 and 12.

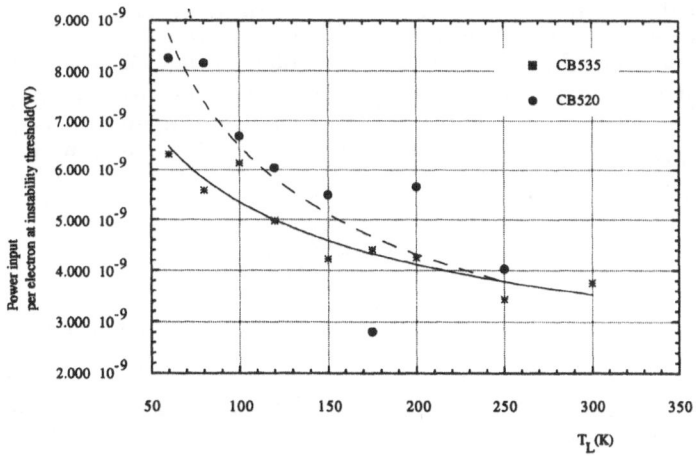

**Figure 14.** The power input per carrier at the instability threshold versus lattice temperature.

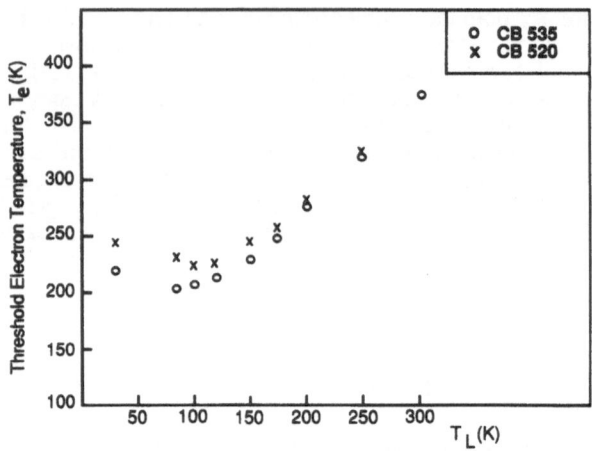

**Figure 15.** Electron temperature at the instability threshold obtained from the power input curves in figure 14, by using the energy balance equation (2), versus lattice temperature.

$$p = e\mu F_{th}^2 = \frac{\hbar\omega}{\tau}\left[\exp\frac{-\hbar\omega}{kT_e} - \exp\frac{-\hbar\omega}{kT_l}\right] \tag{2},$$

where $\hbar\omega$, $\tau$, $T_e$ and $T_l$ are the LO phonon energy (36meV), the electron-LO phonon scattering time (130fs), the electron and lattice temperatures respectively.

Electron temperatures as obtained from the solid lines in figure 14, using equation 2, are plotted as a function of lattice temperature in figure 15 for both samples studied. The threshold electron temperature for the onset of instabilities is constant within about 10% at lattice temperatures below 100K. At higher lattice temperatures the electron temperature increases rapidly, suggesting a larger population of carriers are needed to be excited to higher energies to trigger off the instabilities. The behaviour can be explained by invoking a mechanism for the instabilities similar to that proposed by Schöll and Aoki[43]. According to this mechanism self generated current oscillations occur as a result of the transfer of 2d electrons into the *GaAlAs* barriers via the thermionic emission followed by the delayed dielectric response of the interface space charge as depicted in figure 16. At low lattice temperatures the free electron density in the *GaAlAs* barrier, $n_2(T_l)$, will be altered by the real space transfer of carriers from the quantum well, because of the finite carrier temperature, $T_e \gg T_l$. To a first approximation, at low lattice temperatures when $n_2(T_l)$ is much smaller than the electron density within the quantum well, we can say that the minimum electron temperature required - corresponding to a minimum hot electron population with energies greater than $\Delta E_C - E_F$ (where $\Delta E_C$ is the conduction band discontinuity and $E_F$ is the Fermi level) - such that the minimum number of carriers transferred into the barriers appreciably alters the space charge potential, is given by the thermionic emission equation:

$$\Delta n_{min} = A\exp\frac{-(\Delta E_C - E_F)}{kT_{e_{min}}} \tag{3},$$

where A and $T_{e_{min}}$ are the appropriate Richardson constant and the minimum electron temperature to satisfy the conditions as described above. At higher lattice temperatures, however, when the free carrier density in the barrier is no longer negligible we should take account of the ratio of the number of transferred carriers to those within the barrier. When this ratio exceeds a critical value the space charge potential will be disturbed (in fact it will be diminished) enough as to cause an increase in the backward thermionic

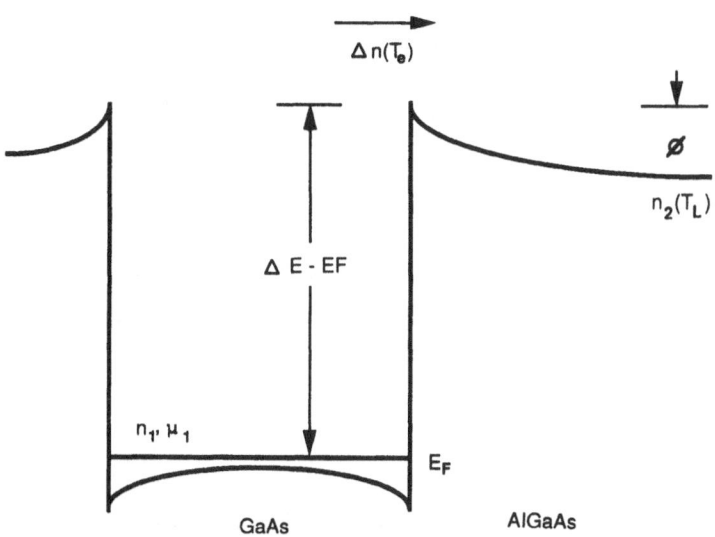

**Figure 16.** Schematic representation of the energy band diagram of a *GaAs/GaAlAs* heterostructure. The space charge potential barrier depends on $\Delta n(T_e)$ and $n_2(T_l)$ as described in the text.

emission, which in turn triggers off the instabilities. We can express the critical ratio for the instability as:

$$C_{th} = \frac{\Delta n(T_e)}{n_2(T_l)} \tag{4}.$$

The onset of the instabilities is therefore determined by both the temperature of the 2d electron gas and by the lattice temperature, via the number of thermally excited free carriers in the barriers. If we consider the lattice temperature dependence of the sheet carrier concentration, $n_s$, as shown in figure 17, we see that the sheet carrier concentration is constant and represents the quantum well electrons below about 80K, for both samples. At higher lattice temperatures the Hall measurements indicate an increase in the measured sheet carrier concentration which is primarily due to the parallel conduction in the *GaAlAs* layers where an increase in the free carrier concentration in the *GaAlAs* barriers occurs, with increasing lattice temperature as a result of the thermal excitation of shallow donors. An appreciable change in $n_s$ occurs at lattice temperatures above 100K. The denominator in equation 4, $n_2(T_l)$, increases rapidly at these temperatures. In order, therefore, to achieve the critical ratio, $C_{th}$, more carriers have to be transferred from the quantum wells into the barriers, requiring higher electron temperatures for the onset of instabilities. The behaviour as observed in figure 15 can therefore be understood in terms of this simple model. At lattice temperatures less than 100K, when the free carrier density in the barriers is negligible compared to those in the quantum wells, a minimum electron temperature, $T_e \sim 225K$ is sufficient to transfer enough quantum well carriers to disturb the space charge potential to start the instabilities. At high lattice temperatures, however, equation 4 will be satisfied. The electron temperature required to trigger off the instabilities are therefore expected to increase with increasing $n_2(T_l)$. A detailed analysis of this argument would require the numerical solution of the time dependent Shrödinger and Poisson equations to give direct knowledge of the free carriers in the barriers and the wells. However, the simple analysis given here suggests that the instabilities observed in our samples may be associated with the same mechanism as that proposed by Schöll and Aoki[43]. Further evidence for this comes from the good agreement between the predicted and the observed threshold fields and the drift velocities presented above.

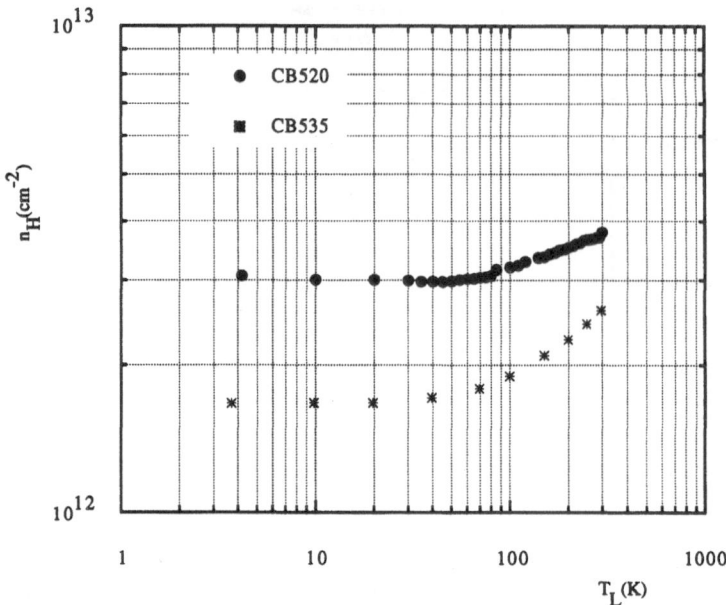

**Figure 17.** Hall electron density versus lattice temperature in the samples studied.

When electric fields are applied along the layers of the quantum wells, light emission is also observed in the samples with *GaAs* buffers. The spectral dependence of the EL and the temperature dependence of the peak emission wavelength were similar to those of the PL spectra as shown for sample CB535 in figure 18. The EL peak was, however, much broader than the PL peak at the same lattice temperature. It should also be noted that the EL signal could be detected with equal efficiency from all the facets of the cleaved sample. Unlike the samples with *GaAlAs* buffer layers the light emission in these samples have the following additional features.

1. The light emission began at electric fields of the order of 1kVcm$^{-1}$ at all temperatures.

2. The field dependence of the emitted light intensity had a sub-linear dependence at electric fields below the threshold field for instabilities and a super-linear dependence above threshold, of the form $I_{EL} \alpha F^s$ where the index s, at $T_l$=60K, was greater than 6, as shown in figure 19.

3. At a fixed wavelength the emitted light pulse has a much longer incubation time than the pulse widths of the applied electric field pulses, which were usually a few μs. The light emission persisted for a number of μs after the electric field pulse returned to zero. The radiative lifetime of the carriers were therefore much longer than those in the samples with *GaAlAs* buffers.

4. In the super-linear field dependence régime the high energy tail of the EL spectra indicated that the electron temperatures were high enough to cause a substantial proportion of the quantum well electrons to be transferred thermionically over the ~200meV barriers. In the sub-linear field dependence régime, however, the electron temperatures were not high enough to transfer a significant number of carriers over the potential barriers into the *GaAlAs* layers.

Measurements on light emission in these samples are at present at an early stage. However, coupled with our observations of a carbon peak in the *GaAs* PL spectra and the long incubation and radiative life times of the EL suggests, strongly, that the observed light emission occurs as a result of the recombination (indirect in real space) in the p-type *GaAs* layers as depicted in figure 20.

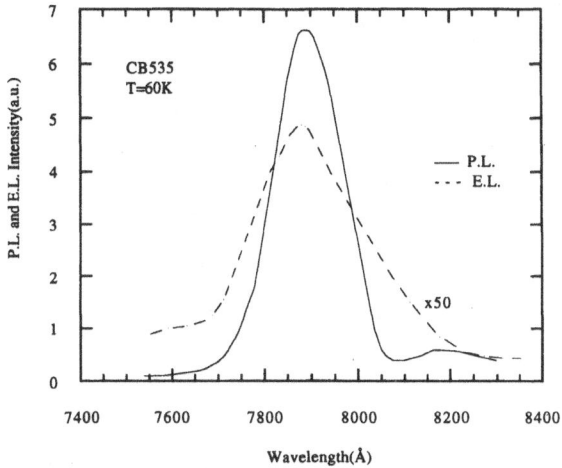

**Figure 18.** EL and PL spectra of sample CB535 at $T_l$=60K.

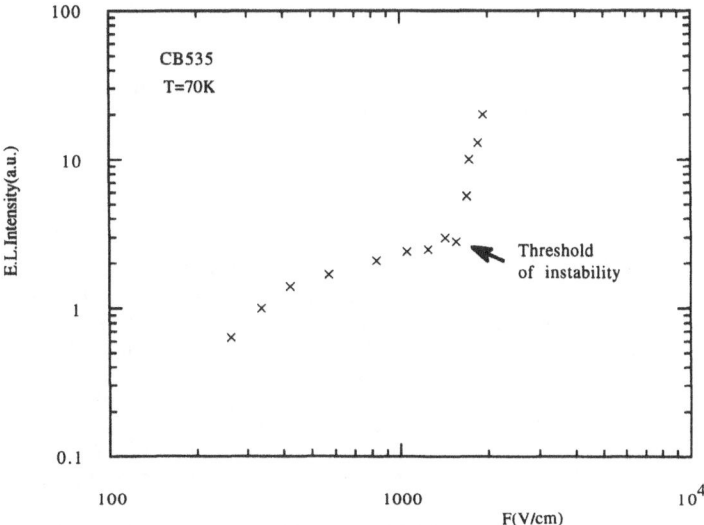

**Figure 19.** EL intensity versus applied electric field at $T_l$=70K.

The structure shown in this figure is fundamentally a p(*GaAs* buffer)-n(*GaAlAs* barrier) junction with quantum wells situated in a *GaAlAs* host material. The transfer of electrons from the *GaAlAs* form an inversion layer in the *GaAs* buffer layer. When an electric field is applied along the layers electrons in the quantum wells, in the vicinity of the junction, can gain sufficient energy to tunnel through the thin *GaAlAs* layers into the inversion layer. This transfer of negative charge will induce positive holes to drift towards the junction to recombine radiatively with the electrons. As the field is increased the electron temperature becomes high enough to enhance the transferred electron population via thermionic emission over the potential barriers, thus the emission intensity increases super-linearly with the applied field. It is interesting to note that the super-linear dependence of the emitted light intensity on the applied field begins at the threshold field for the instabilities, and hence, at high enough electron temperatures for thermionic emission to be the dominant carrier transfer mechanism.

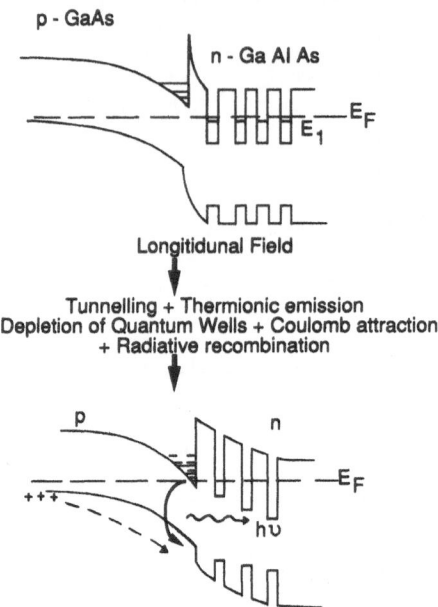

**Figure 20.** Schematic representation of the processes involved at the n-*GaAlAs* and p-*GaAs* interface for light emission to occur.

## 2.3 Summary and Conclusions

A number of instabilities associated with high field transport parallel to the layers in *GaAs/GaAlAs* single, double and multiple quantum well structures are reported. A classification of the instabilities is made according to the underlying phenomena. It is shown that in quantum wells with poor quality *GaAlAs* buffer layers low temperature and low threshold electric field instabilities may occur as a result of potential fluctuations associated with non-uniform p-type doping in the buffer layer. In special circumstances these potential fluctuations may be high enough to give rise to p and n-type quantum well conduction channels in the growth plane. Light emission, associated with the hot electron-hole recombination can therefore be observed in such structures at high fields.

In order to eliminate the spurious effects associated with potential fluctuations we carried out the experiments at room temperature. The results of our high field drift velocity measurements strongly suggest that in samples where the 3d carrier density is greater than $10^{17}$cm$^{-3}$ then non-drifting hot phonon effects inhibit NDR, via either real space transfer or intervalley transfer. In samples where the 3d carrier concentration is less than $10^{17}$cm$^{-3}$, however, NDR and instabilities in the form of current oscillations do occur. The frequency of the oscillations is in the GHz régime and independent of sample length. This observation coupled with the SEM study of high field domains - which indicate either static or slow moving domains - suggest that the instabilities are not due to propagating Gunn domains.

Samples with nominally intrinsic (slightly p-type) *GaAs* buffer layers were also studied. These samples present a more coherent picture of the instabilities. Threshold electric fields and drift velocities as well as the lattice temperature dependence of the threshold electron temperature indicate that a mechanism based on the hot electron real space transfer, and the consequent time dependent interface space charge relaxation, a mechanism proposed by Schöll and Aoki[43] could explain the observed instabilities. It is also shown that at the interface between the n-type *GaAlAs* and p-type *GaAs* buffer layers an inversion layer is likely to occur which, in turn, can give rise to light emission at high longitudinally applied electric fields. This happens as a result of indirect recombination, in real space, of electrons transferred into the inversion layer with holes attracted to the interface by the Coulomb interaction.

## ACKNOWLEDGEMENTS

We would like to thank the financial support provided by the SERC and the *Junta Nacional de Investigação Científica e Technologica* (Portugal), who provide a scholarship for A. Da Cunha.

## REFERENCES

1. C. Hilsum and J. Welborn, *J. Phys. Soc. Jpn* 21:532 (1966).
2. J.J. Harris and B.K. Ridley, *J. Phys. Chem. Solids* 34:197 (1973).
3. G. Persky and D. Bartelink, *IBM J. Res. Dev.* 13:607 (1969).
4. B.K. Ridley and T.B. Watkins, *Proc. Phys. Soc.* 78:293 (1961).
5. C. Hilsum, *Proc. IRE* 50:185 (1962).
6. B.K. Ridley and T.B. Watkins, *Proc. Phys. Soc.* 78:710 (1961).
7. B.K. Ridley and R.G. Pratt, *Phys. Lett.* 4:300 (1968).
8. J. Pozhela. "Plasma and Current Instabilities in Semiconductors", Pergamon Press, Cambridge (1981).
9. M. Kikuchi and Y. Abe, *Phys. Soc. Jpn* 17:1268 (1962).
10. J.J. O'Dwyer, *J. Appl. Phys.* 40:3887 (1969)
11. B.K. Ridley, *Proc. Phys. Soc.* 82:952 (1963).
12. J.B. Gunn, *Solid State Commun.* 1:88 (1963).
13. L. Esaki and R. Tsu, *IBM J. Res. Dev.* 14:61 (1970).
14. K. Hess, *J. Physique Coll.* 42:C7, 3.
15. B.K. Ridley, *J. Phys. C: Solid State Phys.* 15:5899 (1982).
16. S.W. Kirchoefer, R. Magno and J. Comes, *Appl. Phys. Lett.* 44:334 (1983).
17. P.D. Coleman, J. Freeman, H. Morkoc, K. Hess, B.G. Streetman and M. Keever, *Appl. Phys. Lett.* 40(6):493 (1982).
18. K. Tsubaki, A. Livingston, M. Kawashima, H. Okamoto and K. Kumabe, *Solid State Commun.* 46(7):517 (1983).
19. A. Kastalksy and S. Luryi, *IEEE Elect. Dev. Lett.* 4:334 (1983).
20. S. Luryi and M.R. Pinto, *Semicond. Sci. Technol.* 7:B520 (1992).
21. N. Balkan and B.K. Ridley, *Semicond. Sci. Technol.* 3:507 (1988).
22. N. Balkan and B.K. Ridley, "Properties of Impurity States in Superlattice Semiconductors", C.Y. Fang, I. Batra and S.C. Ciraci (eds.), NATO ASI Series B183:229 (1988).
23. N. Balkan, B.K. Ridley and J.S. Roberts, *Solid State Elect.* 31:799 (1988).
24. N. Balkan and B.K. Ridley, *SPIE*, Aachen, 1361:927 (1990).
25. N. Balkan, B.K. Ridley and J.S. Roberts, *Superlatt. Microst.* 5:539 (1989).
26. R. Gupta, N. Balkan, B.K. Ridley and M. Emeny, *SPIE*, Aachen, 1362:798 (1990).
27. R. Gupta, N. Balkan and B.K. Ridley, *Semicond. Sci. Technol.* 7:274 (1992).
28. A. Straw, A. Da Cunha, N. Balkan and B.K. Ridley. In this book.
29. N. Sawaki, M. Suziki, Y. Tagayaki, H. Goto, I. Akasaki, H. Kano, Y. Tanoka and M. Hashimoto, *Superlatt. Microst.* 2:281 (1986).
30. J. Xiao-Song, Y. Lisheng, W. Shu-Min, L. Honz-Xun and Z. Bei, *Solid State Commun.* 62:597 (1987).
31. A.J. Vickers, A. Straw and J.S. Roberts, *Semicond. Sci. Technol.* 4:743 (1989).
32. P. Hendriks, F.J.M. Schitxeler, J.E.M. Haverkort, J.H. Wolter K. de Kort and W.G. Wiegmann, *Appl. Phys. Lett.* 45:1763 (1989).
33. M. Keever, H. Shichijo, K. Hess, S. Banerjee, L. Witkowski, H. Morkoc and B.G. Streetman, *Appl. Phys. Lett.* 38:36 (1981).
34. P. Hendriks, E.A.E. Zwaal, J.E.M. Haverkort and J.H. Wolter, *SPIE*, Aachen, 1362:217 (1990).
35. P. Blood, E.D. Fletcher and K. Woodbridge, *Appl. Phys. Lett.* 47:193 (1985).
36. K. Köhler, G.H. Döhler and K. Ploog, *Superlatt. Microst.* 2:279 (1992).
37. S. Luryi, *Appl. Phys. Lett.* 58:1727 (1991).
38. T.K. Higman, M.S. Hagadon and J. Chen, *Appl. Phys. Lett.* 60:1342 (1992).

39. E. Zanoni, S. Bigliardi, R. Capaletti, P.Lugli, F. Magistrati, M. Manfredi, A. Pacagnallo, N. Testa and C. Canali *IEEE Elect. Dev. Lett.* 11:487 (1990).
40. N. Balkan, R. Gupta, B.K. Ridley, M.Emeny, J.S. Roberts and I. Goodridge,*Solid State Elect.* 32:1641 (1989).
41. N. Balkan, B.K. Ridley, J. Frost, D.A. Andrews, I. Goodridge and J.S. Roberts, *Superlatt. Microst.* 2:357 (1986).
42. B.K. Ridley, *Solid State Elect.* 33:859 (1990).
43. E. Schöll and K. Aoki, *Appl. Phys. Lett.* 58:1277 (1991).
44. R. Gupta, N. Balkan and B.K. Ridley, *Phys. Rev.* 46:7745 (1992).

# LIGHT EMISSION AND DOMAIN FORMATION IN REAL-SPACE TRANSFER DEVICES

T.K. Higman, Jihong Chen, M.S. Hagedorn, and R.T. Fayfield

Department of Electrical Engineering

University of Minnesota

200 Union St. S.E.

Minneapolis, Minnesota 55455

## INTRODUCTION

Light emission from real-space transfer devices can be brought about by several mechanisms. For device application the most promising of these mechanisms is the injection of minority electrons across a semiconductor heterojunction into a p-type collecting layer. So far, problems associated with the growth of p-type collecting layers have hindered efforts in this area, resulting in inefficient light output. Our group has reported observing light emission caused by the real-space transfer of majority electrons into an n-type collecting layer. We propose a mechanism to account for this light which consists of hole creation by impact ionization of real-space transferred electrons. Microscopic observation of the spatial distribution of the light along the length and width of the channel (variations have been observed in both directions), along with its spectral distribution, gives insight into the mechanisms involved in high field domain formation in the channel of real-space transfer FETs and the contribution of high field domains to negative differential resistance. New device structures which incorporate a p-type collecting layer as the top layer of the device in order to decrease the problems associated with p-type doping in real-space transfer devices will also be discussed.

Recent results are also presented on devices incorporating thin AlAs barriers which show that extremely large source-collector currents can be initiated by real-space transfer. In these devices, real-space transfer current causes an apparent shift of the electric field distribution resulting in high fields in the vicinity of the source. This results in the large source-collector current which is independent of drain bias. This switching is similar to the switching mechanism in the heterostructure hot electron diode, which has been discussed previously by the authors.

*Negative Differential Resistance and Instabilities in 2-D Semiconductors*
Edited by N. Balkan *et al.*, Plenum Press, New York, 1993

# LIGHT EMISSION

## Light Emission From Majority Carrier Real-Space Transfer Devices

In this section, we report on the observation of light emission from majority carrier real-space transfer devices of the negative resistance field effect transistor (NERFET) type. The devices were grown by molecular beam epitaxy (MBE). The growth was performed with on-axis (100) orientation GaAs:Si substrate having a net donor concentration of $1 \times 10^{18}$ cm$^{-3}$. The entire device structure was grown at 600°C and consists of a 5000Å GaAs:Si buffer layer ($1 \times 10^{18}$ cm$^{-3}$), a 1000Å GaAs:Si spacer layer ($1 \times 10^{16}$ cm$^{-3}$), and a 2000Å undoped $Al_{0.5}Ga_{0.5}As$ barrier layer followed by    a 2000Å undoped GaAs channel layer. To improve the quality of the top GaAs/AlGaAs interface where the conduction channel is located, $(Al_{0.5}Ga_{0.5}As)_{10}/(GaAs)_1$ short-period superlattices with growth interruptions were used. Using this method, extremely high mobility has been observed in inverted selectively doped GaAs/AlGaAs heterojunctions.[1] 500µm wide devices were then defined using standard photolithographic techniques.

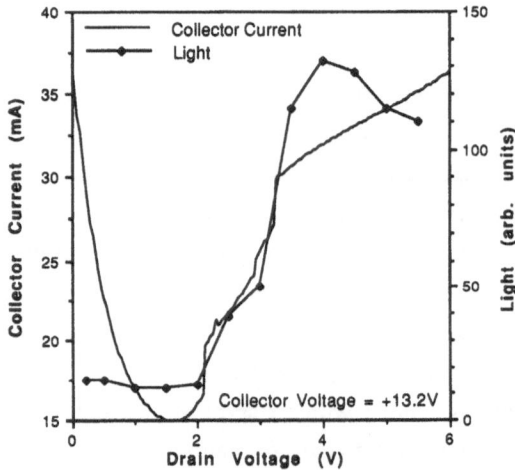

Fig. 1 - Collector current and light output vs. drain voltage at a substrate temperature of 150 K. The light present at low drain bias is due to source and drain leakage. The decreasing light output at higher bias is believed to be the result of heating effects at higher drain voltages.

While these devices are structurally similar to earlier AlGaAs NERFETs,[2] they differ in one important respect; there is a 1000 Å lightly doped GaAs layer located directly beneath the barrier.[2] We will refer to this layer as the spacer layer. The additional potential drop across this spacer layer increases the required substrate voltage for RST. This extra potential has the net effect of raising the substrate threshold voltage for drain circuit NDR to approximately 12V from the typically observed 8V in a device without a spacer layer. This can be seen in the I-V results of Fig. 1. This potential drop also provides additional energy to the real-space transferred (RST) electrons which subsequently impact ionize at an enhanced rate due to the barrier[4] in the spacer layer. The subsequent recombination of holes created by this process produces a measurable amount of light. The light we observed was not emission from the mesa sidewall but rather light emitted from the top surface of the wafer in the narrow stripe defined by the channel. This light was observed at room temperature, but low temperature operation greatly increases its intensity. At 150 K, the light coming from the source-drain gap was easily visible with the unaided eye. Fig. 1 provides a trace of light output and collector current vs. drain voltage. By comparing the light and collector current traces of Fig. 1, one can see that except for a small amount of light at low drain bias attributed to leakage current, the light output intensity is proportional to collector current. It is important to note that the leakage current creates a measurable amount of light by the same mechanism as outlined above. While easily visible from the side of the device, this light is effectively blocked from our detector (located above the device) by the metallic source and drain contacts. Therefore the detected light is predominantly that which is emitted from the source-drain gap. Note that this light output saturates when the RST process begins to saturate at approximately 3.5V on the drain. This indicates that the light is due to RST current rather than the competing effect of impact ionization in the channel. The coincidence of a sudden increase in substrate current and a sudden decrease in drain current at $V_D = 2V$ indicates that a high field domain is being created in the channel, which can be a source of impact ionization, however, other evidence suggests that the dominant source of the light is the RST mechanism outlined above. If channel impact ionization dominated, the light would continue to increase as drain bias (and thereby the spatial extent of the high field domain) increased. In addition to the correlation between substrate current and light emission, ensemble Monte-Carlo simulations of NERFETs by Kizilyalli and Hess[5] have shown that for AlGaAs barriers the number of electrons which gain enough energy to impact ionize while still in the channel region is negligible.

## Light Emission from 'Inverted' Real-Space Transfer Devices with p-type Collectors

One of the more interesting and potentially useful features of the NERFET is the fact that two terminals (the source and drain) are used to inject channel carriers into a collector region (often the substrate) over a wide range of collector bias. While present devices have been primarily limited to majority electron injection into n-type collectors, with limited examples of hole injection,[6] there also exists the possibility of minority carrier injection. Since the RST of electrons is more efficient than that of holes, we can only practically consider the case of minority electrons injected into a p-type collector. Several structural variations of a light emitter based on this concept have been previously discussed by Luryi.[7]

One of the problems with the minority carrier injection described in the above paragraph is that the most commonly fabricated structure of the NERFET is the collector down structure which uses the positive collector potential to create an accumulation layer of two-dimensional channel electrons.[8] Since the collector is then only separated from the source and drain contacts by a thin (typically 2000Å) barrier, leakage current is a problem. While low leakage devices have been built in the GaAs/AlGaAs,[8] and InP/InGaAs/InAlAs[3] systems, problems have been encountered when the collector is p-type and the doping is switched to beryllium.[9] The barrier, which must withstand electric fields between source and collector on the order of 350,000 V/cm without leaking suffers from the beryllium diffusing into it. This greatly increases the tunnelling sites in the barrier. P-type collectors also create another problem. While the leakage current in the n-type collector device consists solely of electrons tunnelling from source (and drain, depending on drain bias) to the collector, in the p-type collector device, an additional source of leakage current is holes tunnelling from collector to source and drain. This undesired effect has inhibited the performance of both GaAs/AlGaAs devices (attempted by the author and unpublished) and InP/InGaAs/InAlAs devices.[9]

One possible solution to barrier degradation by beryllium diffusion is to limit the diffusion during growth by inverting the structure, in other words going to a 'collector up' structure. If the principles of operation of the 'standard' NERFET (collector down) are to be simply extended to a collector up structure a problem immediately becomes apparent. Since the collector down structure uses the collector potential to accumulate a two-dimensional electron gas in the channel, a collector up structure that works in the same manner would require some sort of self-aligned structure in order precisely align the collector all the way to the source and drain contacts. Any gap between source or drain and collector would cause a gap in the induced channel. A pictorial conception of such a device is given by Luryi.[7] An alternative is to build the collector up device depletion-mode with a doped or modulation doped channel. Majority carrier devices with lightly doped n-type collectors and n-doped pseudomorphic InGaAs channels built in this manner have demonstrated current gain to 60 GHz when operated in charge injection transistor (CHINT) mode.[10]

We have constructed a collector-up device similar to the one detailed above but incorporating a p-type collector. This device is schematically detailed in Fig. 2. For this structure we have used the pseudomorphic InGaAs/AlGaAs/GaAs material system. It was grown by molecular-beam epitaxy (MBE) at 590°C on a semi-insulating GaAs substrate. The sequence of epitaxially grown layers was as follows: first, a 1500Å-thick graded AlGaAs layer was grown and the mole fraction of Al was varied from 0 to 0.4 as the growth proceeded. Next a 100Å-thick $In_{0.25}Ga_{0.75}As$ channel layer doped with Si ($2*10^{18}$ cm$^{-3}$) was grown, followed by 2000Å-thick undoped $Al_{0.15}Ga_{0.85}As$ barrier layer, and finally 2000Å-thick GaAs collector layer doped with Be ($2*10^{18}$ cm$^{-3}$). Conventional processing techniques were used to form the device structure. Mesa etching was carefully done to ensure good step coverage of the collector metal. Then 2500Å SiO$_2$ layer was deposited to prevent electrical shorts. An active device area was obtained by opening a window in the SiO$_2$. Source and drain contacts were Au/Ge/Ni based, alloyed at 420°C for 20 sec, which diffuse deep enough to make connection to the channel. The collector contact was a Cr/Au Schottky barrier which was defined as a stripe across the 300μm width of the channel. The separation between source and drain was 6μm. Finally, in order to reduce leakage, the areas between drain and collector (and source and collector) were etched to remove a portion of the p-type material (see Fig. 2) using the Cr/Au collector as a self aligned etch mask .

The electrical characteristics of the p-type collector up device are shown in Fig. 3. Note that the while negative differential resistance is present in the drain circuit, there is very little corresponding collector (RST) current. Increasing the collector bias only increased the

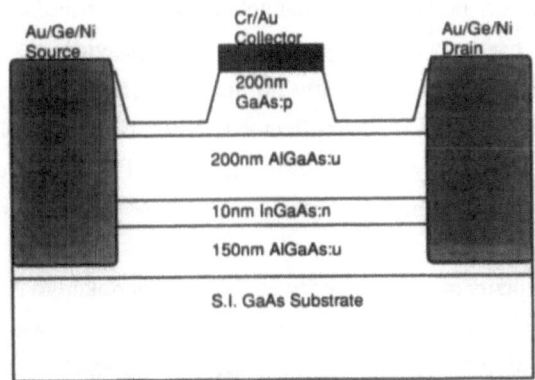

Fig. 2 - Schematic illustration of collector up NERFET incorporating a p-type collector.

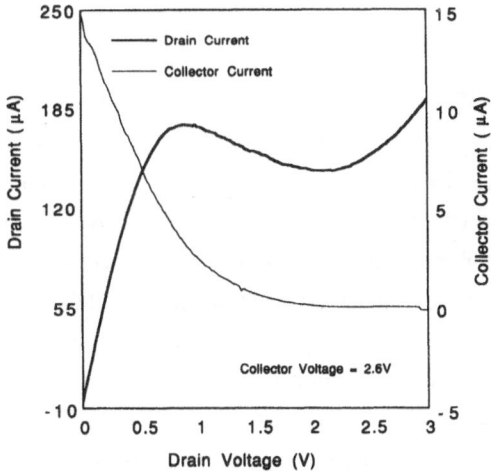

Fig. 3 - Current-voltage trace of the drain and collector currents in the p-type collector up NERFET structure.

amount of collector leakage without increasing the RST current. It is not known at this time whether the collector leakage is dominated by electron or hole tunnelling. Increasing the drain voltage above the range shown in Fig. 3 when the collector was increased to approximately 5 volts did give a slight increase in RST current, unfortunately this is also accompanied by evidence of substantial impact ionization in the channel (light emission near the drain contact along with a substantial increase in drain current) so the higher voltage effects are suspect at best.

The lower voltage results contrast with the results of Hueschen et al.[10] whose n-type collector up devices showed large collector injection with very little negative differential resistance in the drain circuit (the opposite of the present situation). Since the operating voltage range of our devices is similar to that of Hueschen et al.,[10] and since the structures are so similar, we suspect that the origin of the inhibited RST current is the additional barrier caused by the p-type doping in the collector. If this is the case, then the negative differential resistance in the drain may be due to hot channel electrons thermionically emitting into the low mobility AlGaAs barrier where they cool down and provide less channel conduction. These heated channel electrons apparently do not have enough energy to overcome the additional pn junction potential of the collector and contribute to collector current. This proposed mechanism is actually very close to the original RST concept of Hess et al.,[11] which has previously been observed only as a very weak effect[12] in modulation-doped superlattices.

Current experiments in our laboratory are being undertaken to reduce the barrier hole leakage by increasing the valence band offset. It is hoped that this will allow us to increase collector bias without increasing leakage and thereby increase RST current.


## DOMAIN FORMATION IN AlAs BARRIER DEVICES

Most GaAs/AlGaAs NERFETs fabricated to date have incorporated AlGaAs with Al mole fractions around 35% to 45%. This provides a $\Gamma$-point barrier to the GaAs channel between 0.27 and 0.35 eV. Barriers in this range are low enough to allow significant RST transfer of hot channel electrons but high enough to retard tunnelling leakage currents (although the crystal growth of high quality insulating AlGaAs barriers can be quite difficult). The authors have not been able to discover any reports of NERFET-type structures containing AlGaAs barriers of higher Al mole fraction than 50%.

Several interesting transport properties of hot electrons in NERFET structures can be exploited if the barrier Al mole fraction is increased above 45%. If this mole fraction is increased the (X -point) bandgap of the AlGaAs becomes indirect and stays more or less constant at around 2 eV (with a conduction band offset to GaAs of approximately .35eV, the $\Gamma$-point bandgap continues to increase to nearly 3 eV for AlAs and the $\Gamma$-point conduction band offset is 1 eV. In this section we report results on such a device. The crystal was grown by MBE at 590°C on a (100) oriented GaAs:Si substrate having a net donor concentration of $1*10^{18}$ cm-3. The device consists of a 2000Å GaAs:Si buffer layer ($2*10^{18}$ cm-3), a 500Å GaAs:Si spacer layer ($5*10^{16}$ cm-3), a 2000Å undoped $Al_{0.25}Ga_{0.75}As$ barrier layer, a 200Å GaAs undoped layer and 150Å AlAs/GaAs/AlAs double barrier, with a 1500Å undoped GaAs channel layer on the top. Standard photolithgraphic techniques were used. Mesa etching provided isolation and an $SiO_2$ layer was deposited. Shallow ohmic contacts were formed using alloyed Ge/Au/Ag. The separation between source and drain varied from 1.5µm to 3µm. This device is ilustrated in Fig. 4.

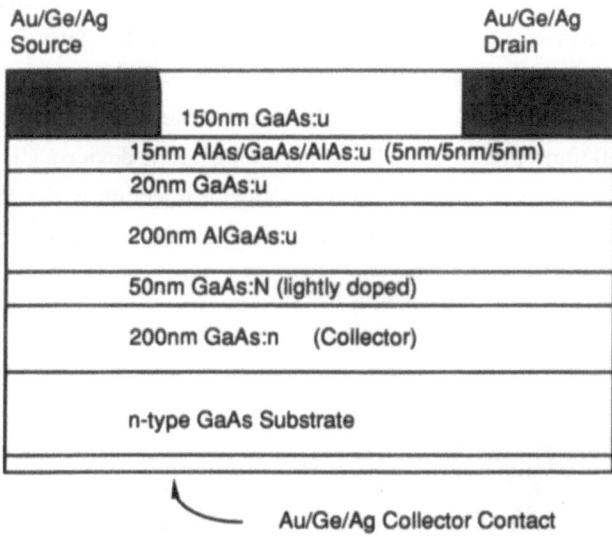

Fig. 4 - Schematic illustration of AlAs barrier NERFET.

Given the above, if the barrier in a NERFET is made of AlAs, the lattice temperature two dimensional electron gas in the channel will be confined by a large, 1 eV, Γ-point barrier. When the two-dimensiional electron gas is heated by a source-drain field the electrons have been shown to become three dimensional,[13] and a large percentage of them scatter into the indirect valleys.[5] These heated electrons then see virtually no barrier, since the X and L points in AlAs are energetically very close to the X and L points in GaAs. In this manner the bandstructure of the GaAs/AlAs heterojunction can be exploited to give extremely low source and drain leakage while still allowing RST to occur (albeit by the indirect valleys, which may be a slower process). This mechanism was demonstrated to enhance negative differential resistance in a similar device, the heterostructure hot electron diode (HHED).[14,15] Rather than make an entire barrier out of AlAs, only the top layers are alternating 50 Å layers of AlAs/GaAs. This alternating structure was designed in order to facilitate interfacial smoothing by utilizing pauses during GaAs growth,[1] not specifically to create confined subbands. It is felt by the authors that the high electric fields present in the device negate the effect of transport via the subbands confined by the AlAs layers. (When this device is operated at very low collector voltage, with correspondingly low collector currents, there is evidence of thermally assisted subband tunnelling,[16] but the main thrust here is the operation at a much larger collector voltage).

The current-voltage trace of this device is shown in Fig. 5. Note that currents are very small until the drain voltage is approximately 10 volts, at which point the drain and collector currents dramatically increase. The drain current present at drain voltages less than 10 volts is considerably lower than what is expected for a collector bias of 7.4 volts. This is probably due to unintentional p-type doping in the AlGaAs portion of the barrier, which causes a delay in the formation of a two dimensional electron gas until the collector voltage has depleted the p-type AlGaAs barrier. In this device an offset of 7 volts before channel formation corresponds to barrier doping of approximately $N_A = 1x10^{17}$. Note that the shape of the collector current vs. voltage trace follows the shape of the drain current trace but its level is approximately 10 times that of the drain current, making the device a CHINT with a gain of 10. We suggest that the mechanism for this collector injection is similar to the switching mechanism in the HHED. In the following paragraph we will give a brief review of the switching mechanism of the HHED.

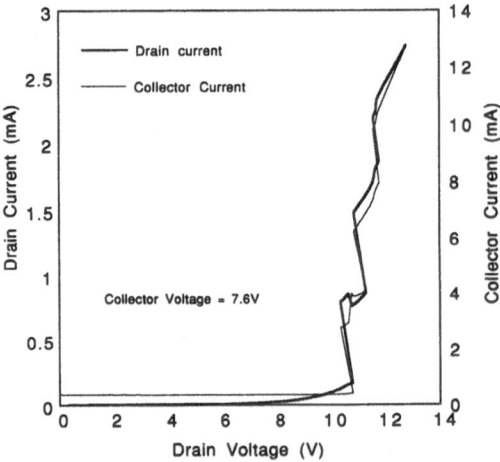

Fig. 5 - Room temperature Current-voltage trace of AlAs barrier NERFET.

The HHED (heterostructure hot electron diode) is a two terminal device exhibiting s-shaped negative differential resistance in its current-voltage trace. In this device two possible conduction regimes exist. At low bias voltage, the electric field in the device is dropped mainly across a wide bandgap heterostructure barrier layer, henceforth the barrier, and current is limited to tunnelling through the barrier (see Fig.6 (a)). At higher bias, electrons drifting towards the tunnelling barrier gain energy in the lightly doped narrow bandgap drift region, henceforth the drift region, which is adjacent to the barrier (Figs. 1(b) and 1(c)). Eventually these electrons are heated to the extent that thermionic emission becomes the dominant conduction mechanism(Fig. 1(d)). Since the thermionic emission mode requires less electric field in the barrier for a large current density compared to the tunnelling mode, the transition between the two modes has been shown to result in a region of s-shaped negative differential resistance.[14,15]

Given the above description, it can now be stated that the channel region of the NERFET, which lies adjacent to the barrier, gives a structure with physical similarities to the

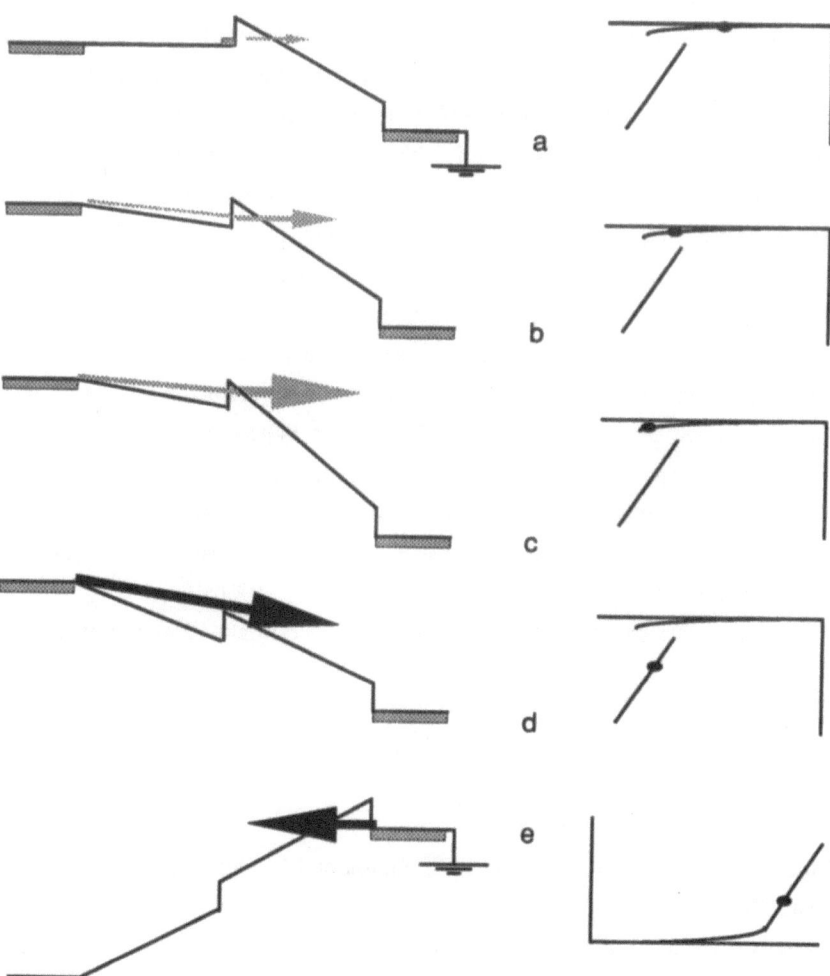

Fig. 6 - Conduction band diagram with arrows indicating electron flow and the corresponding current-voltage characteristic for an HHED under applied bias showing (a) initial low tunnelling current in negative bias, (b) and (c) increasing negative bias with increasing electric field in the drift region and increased tunnelling current, (d) the "on" state showing predominantly thermionic emission current, and (e) positive, nonswitching bias.

HHED. The fundamental difference between the two being the orientation of the electric field which heats the carriers. In the NERFET the heating field is from source to drain, parallel to the barrier interface, and in the HHED the heating field is perpendicular to the interface. In the AlAs barrier NERFET devices, the source-collector or drain-collector diodes, when contacted independently as two terminal devices, exhibit s-shaped negative differential resistance, acting as an HHED. One posible explanation for this is fringing electric fields in the channel region near the contacts causing electron heating and subsequent thermionic emission over the barrier and into the collector. This effect in the NERFET source-collector circuit is not observed in AlGaAs barrier devices, only AlAs barrier devices (this does not preclude the possibility of other high Al content barrier devices exhibiting s-shaped negative differential resistance, but to our knowledge no one has tested such a device).

The effect we have observed is also similar to the s-shaped negative differential resistance observed in four-terminal NERFET structures by Kastalsky et al.[17] This structure consists of a modulation doped NERFET with a Schottky gate placed between source and drain. When the device is operated with a large positive bias on the gate, avalanche currents from the reverse biased Schottky gate initiate an HHED type instability near the source, which causes large collector currents and s-shaped negative differential resistance in the source-collector circuit. Since avalanche is present and the resulting holes decrease the field across the barrier, the voltage peak to valley ratio of the s-shaped region is as high as 10, as opposed to 1.2 in the two terminal HHED. We feel that the inclusion of AlAs barriers, with their unique transport properties, in our device allows a source-drain electric field to initiate the avalanche, triggering the source instability, in a similar fashion to the four terminal device of Kastalsky et al.[17] (it is also interesting to note that the NERFETs in Ref. 17 contain barriers of $Al_{0.5}Ga_{0.5}As$, pointing to the possibility that a heterostructure barrier with a high $\Gamma$-point conduction band offset and a low X-point offset are critical).

This work was supported by the National Science Foundation, Grant no. NSF/ECS-9110221.

## REFERENCES

1) H. Shtrikman, M. Heiblum, K. Seo, D. E. Galbi, and L. Osterling, *J. Vac. Sci. Technol.* B6:670(1988).

2) A. Kastalsky, S. Luryi, A. C. Gossard, and R. Hendel, *IEEE Electron Device Lett.* EDL-5:57(1984).

3) Piotr M. Mensz, Paul A. Garbinski, Alfred Y. Cho, Deborah L. Sivco, and Serge Luryi, *Appl. Phys. Lett.* 57:2558(1990).

4) K. Brennan, T. Wang, and K. Hess, *IEEE Electron Device Lett.* EDL-6:199(1985).

5) I. C. Kizilyalli and K. Hess, *J. Appl. Phys.* 65:2005(1989).

6) Piotr M. Mensz, Serge Luryi, John C. Bean, and Conor J. Buescher, *Appl. Phys. Lett.* 56:2663(1990).

7) S. Luryi, *Appl. Phys. Lett.* 58:1727(1991).

8) A. Kastalsky, R. Bhat, W. K. Chan, and M. Koza, *Solid-State Electronics* 29:1073(1986).

9) M. Mastrapasqua, F. Capasso, S. Luryi, A.L. Hutchinson, D.L. Sivco, and A. Y. Cho, *Appl. Phys. Lett.* 60:2415(1992).

10) M.R. Hueschen, N. Moll, and Alice Fischer-Colbrie, *Appl. Phys. Lett.* 57:386(1990).

11) K. Hess, H. Morkoc, H. Shichijo, B.G. Streetman, *Appl. Phys. Lett.* 35:469(1979).

12) M. Keever, K. Hess, and M. Ludovise, *IEEE Electron Device Lett.* EDL-3:297(1982).

13) T.K. Higman, S.J. Manion, I.C. Kisilyalli, M.A. Emanuel, K. Hess, and J.J. Coleman, *Phys. Rev. B* 36:9381(1987).

14) T.K. Higman, L.M. Miller, M.E. Favaro, M.A. Emanuel, K. Hess, and J.J. Coleman, *Appl. Phys. Lett.* 53:1623(1988).

15) D. Arnold, K. Hess, T. Higman, J.J. Coleman, G.J. Iafrate, *J. Appl. Phys* 66:1423(1989).

16) Submitted to *Superlattices and Microstructures*.

17) A. Kastalsky, M. Milshtein, L.G. Shantharama, J. Harbison, and L. Florez, *Appl. Phys. Lett.* 54:2452(1989).

# HOT ELECTRONS IN δ-DOPED GaAs

Marion Asche

Paul-Drude-Institut für Festkörperelektronik
O - 1086 Berlin, Hausvogteiplatz 5 -7

## INTRODUCTION

Advanced epitaxial growth techniques allow to fabricate δ-doped multilayers of high quality in GaAs. Whereas the dopants are distributed arbitrarily in the layers the distance between the δ-doping planes can be chosen intentionally and a partially ordered arrangement of impurities is obtained in this way. The aim of the present report is to review hot electron transport in such systems.

When the doping is strong and the distance between the layers greater than several times the effective Bohr radius the donors create a V-shaped potential confining the electrons to the neighbourhood of these δ-layers consequently, supposed they are not energetically excited. On the other hand if the average distance between the dopants equals the distance between the layers (aspect ratio 1) the multilayers resemble homogeneously doped material.

When the aspect ratio is greater 1 some similarities to the transport properties in square potential wells of heterostructures are expected. However, due to the lower barrier height in δ-doped GaAs in comparison to $GaAs/Al_xGa_{1-x}As$ with x about 0.3 for instance the real space transfer is enhanced for a given excitation, whereas the upper levels already allow a certain carrier motion perpendicular to the well due to the larger spatial extensions of the states. Consequently, a change from transport in two-dimensionally confined to extended states is expected at comparatively low fields in contrast to rectangular quantum wells. A further important difference is given by the fact that the transfer into the interlayer space is connected with a remarkable increase of the mobility due to the low residual doping instead of a decrease on account of higher effective masses for transfer into AlGaAs.

Furthermore the electron-phonon interaction appears to be describable by three dimensional GaAs phonons for the δ-doped systems instead of the mixed scattering on bulk, confined and interface modes in heterostructures. Besides, it is assumed that the momentum conservation implies less restrictions to phonon emission perpendicular to the quantum wells, probably with exception of the scattering in the lowest level, for which the form factor plays a more decisive role.

*Negative Differential Resistance and Instabilities in 2-D Semiconductors*
Edited by N. Balkan *et al.*, Plenum Press, New York, 1993

When the donor concentration is low and the electron wave functions do still not overlap within the δ-layers properties are expected similar to those of isolated impurities in bulk material. However, the partial ordering of the dopants in planes leads to potential contours which evoke preferential carrier movement along the δ-layers. We therefore also expect differences to homogeneously doped samples, as long as the aspect ratio is high due to the influence of partial ordering of the impurities on mobility and carrier heating. Consequently, impact ionisation and recombination should quantitatively change the low temperature breakdown in comparison to homogeneously doped GaAs with regard to the static characteristics as well as to oscillations, filamentation and transition to chaos.

Experimental results are reviewed for low lattice temperatures, at which, however, the carrier motion is not determined either by variable range hopping on the "isolator side" of doping concentration or by manyparticle quantum interference effects on the "metallic side".

## FIELD INDUCED REAL SPACE TRANSFER

In this section high doping concentrations in distinctly separated δ-layers are regarded leading to the formation of carrier states extended in two dimensions (2D electrons) with potential wells parallel to the doping planes. While the electrons in the energetically lower subbands of such stacks of layers (resembling a superlattice) are confined to the single wells the strong overlap of the wave functions belonging to the highest subbands manifests itself as a broad miniband with a high electron density between the layers, leading to a quasi-free mobility perpendicular to the doping planes, consequently (quasi-3D behaviour of electrons).

At low lattice temperatures there is only a small fraction of carriers in the uppermost subband and their contribution to transport properties may be negligible. By energetic excitation like increasing lattice temperature, carrier heating by electric fields or light absorption the 2D electrons of the lower subbands are transferred to the uppermost subbands as well as the states above the barriers (3D electrons).

Such a real space transfer from the wells into the interlayer regions by electric fields applied along the layers was investigated in[1] for GaAs with a Si-concentration in the δ-layers corresponding to the dilute metallic limit and an unintentionally p-type doped background. The parameters are reported in table 1.

The current voltage characteristics were measured immersing the Hall bar samples in liquid Helium, and the differential conductivity was obtained numerically. The results are shown in Fig. 1.

Regarding the zero field limit it can be seen that in the multilayers with aspect ratio 1 lowering the dopant concentration (sample B to D) leads to smaller conductivity values since the influence of carrier density dominates the increase of mobility due to reduced impurity scattering as expected for homogeneously doped material. However, the conductivities of

Table 1. Sample parameters.

| | Sample | | | |
|---|---|---|---|---|
| | A | B | C | D |
| number of $\delta$-layers | 8 | 10 | 10 | 12 |
| sheet doping concentration $[10^{11} cm^{-2}]$ | 5 | 3.4 | 1.7 | 1.35 |
| layer spacing $d$ $[nm]$ | 100 | 17.1 | 24.3 | 27.1 |
| equivalent volume doping $[10^{16} cm^{-3}]$ | 5 | 20 | 7 | 5 |
| average distance of dopants in a sheet $\langle a \rangle$ $[nm]$ | 14.1 | 17.1 | 24.3 | 27.1 |
| aspect ratio $r = \langle a \rangle / d$ | 0.14 | 1 | 1 | 1 |

samples A and B exhibit nearly the same value in spite of higher volume doping in B in comparison to the equivalent volume doping density in sample A. It was assumed that this effect can be explained by a strongly enhanced mobility due to partial ordering of the dopants in $\delta$-multilayers.

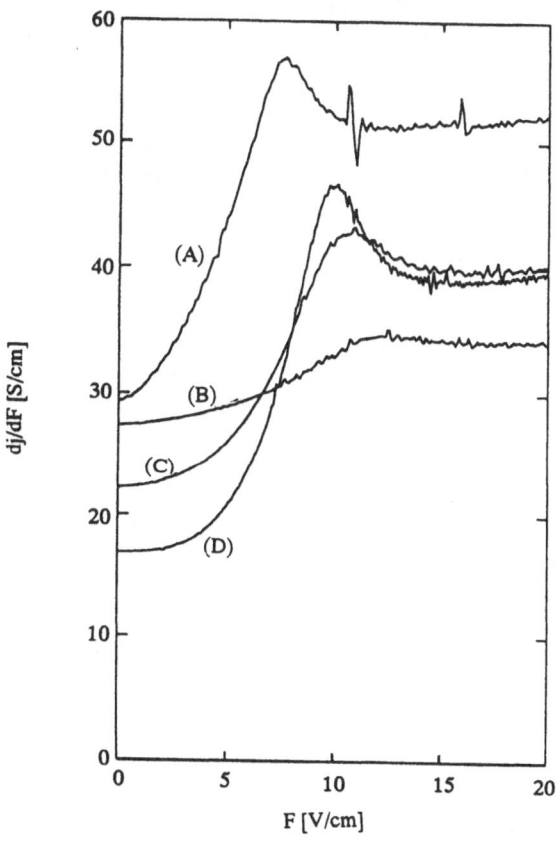

**Figure 1.** Differential conductivities at 4.2 K of the samples of Table 1 versus the electric field strength.

In the weak field region up to 2 V/cm (region I) the differential conductivity for sample A increases, whereas the curves of the other samples with aspect ratio 1 remain constant. With further rising fields (region II) both samples with the same weak effective volume doping (A and D) behave similarly, their differential conductivities show a significant increase up to 8 and 10 V/cm, respectively, while at higher doping concentration the increase is weaker pronounced, because carrier heating becomes less effective with increasing scattering at ionised impurities. This fact also explains the shift of the maximum of differential conductivity towards higher fields. The subsequent drop of differential conductivity with rising fields observed for samples A, D, and C is assumed to be due to the growing interaction with phonons at heating electric fields as well known from hot electrons in bulk semiconductors, whereas in the heavily doped sample B impurity scattering dominates. A specific feature of the investigated samples is the existence of an extended field region III, in which the differential conductivity as well as the magnetoresistivity and the Hall effect remain almost constant (see[1] for details).

As this happens in the δ-multilayers with clearly separated wells (A) as well as in sample D with aspect ratio 1 this behaviour seems to be characteristic for the 3D electrons.

Since the ratio $n^{(3D)}/n^{(2D)}$ increases with growing lattice temperature, this influence can be used to pronounce more details of the described situation. The zero field conductivity (Fig. 2) clearly reflects that the percentage of quasi-3D and 3D carriers with higher mobility than those values in the lower subbands of the wells grows with lattice temperature as mentioned in[2]. Furthermore the maximum value of the differential

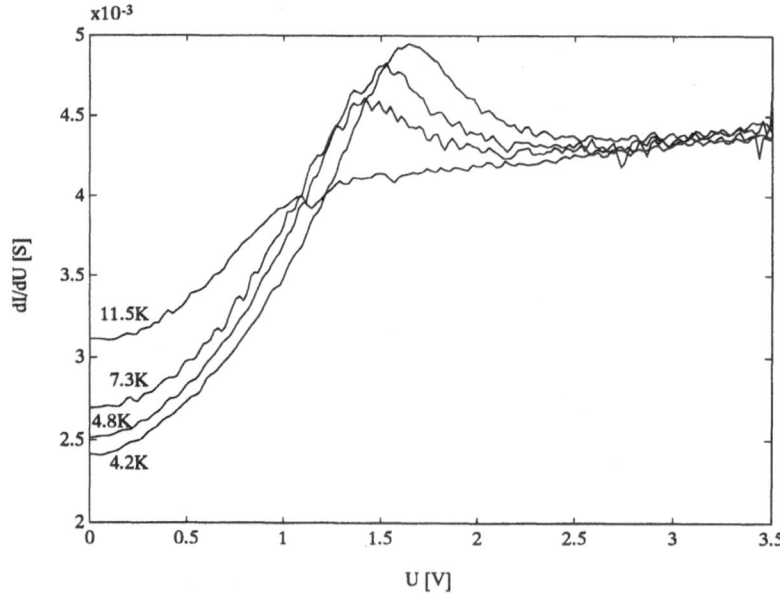

**Figure 2.** Temperature dependence of differential conductivity for sample A as a function of the electric field strength.

conductivity decreases with growing lattice temperature and shifts towards smaller fields, mirroring a smoother increase of $n^{(3D)}/n^{(2D)}$ with applied fields, because the thermal excitation of the electrons becomes more and more important and the interaction with phonons rises in comparison to scattering on ionised impurities. At high fields (region III), of course, the average carrier energy is enhanced up to a value, for which a remarkable fraction of electrons is able to emit optical phonons.

The experimental results are acompanied by numerical calculations supporting the interpretation[1].

The theoretical description approximates the real potential and the charge distribution of sample A by a rigid superlattice of wells with one quantised electron level 18 meV below and extended states above the barriers.

Furthermore the electron-electron collisions within both subsystems (2D and 3D electrons, respectively) are assumed to play the main role in comparison with all the other scattering mechanisms, therely allowing to describe the carrier distributions by shifted Fermi functions with their own temperatures and the drift velocities obtained by the balance method, consequently, after transforming the coordinates into the centre of mass and relative coordinates[3]. This proposal includes that the electron electron collisions between both subsystems are regarded as being weak. A further simplification is introduced by a common Fermi energy, which strictly is only valid for a single electron temperature. However, it permits to replace the dynamical par-

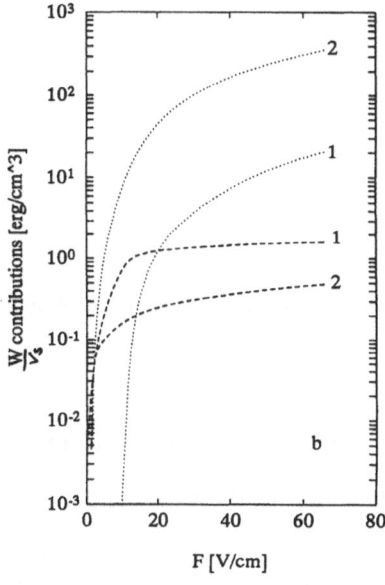

**Figure 3.** Energy losses w normalised to the sound velocity v as functions of the electric field strength for interaction with acoustic (- - - dashed lines) and polar optic phonons (... dotted lines) at 4.2 K. Curves 1 and 2 for 2 D and 3D electrons, respectively.

ticle balance between the subsystems by the conservation of the particles, and may be justified for a low transfer rate into 3D states in order to describe the experimentally observed features qualitatively.

The various contributions to the friction forces and energy dissipation corresponding to the interaction processes are calculated as functions of carrier heating. As an example Fig. 3 demonstrates the contributions to energy loss (normalised to the sound velocity) as functions of the electric field. The following features should be reminded: the dissipation of the energy per carrier to acoustic phonons (including piezoelectric and deformation potential interaction) is an order of magnitude lower for the 3D electrons than for 2D electrons because of the stronger heating in the interlayer space and higher emission of optical phonons, consequently.

Fig. 4 describes the strongly pronounced temperature rise at comparatively weak fields and the subsequent low increase for higher fields, when the emission of optical phonons becomes efficient.

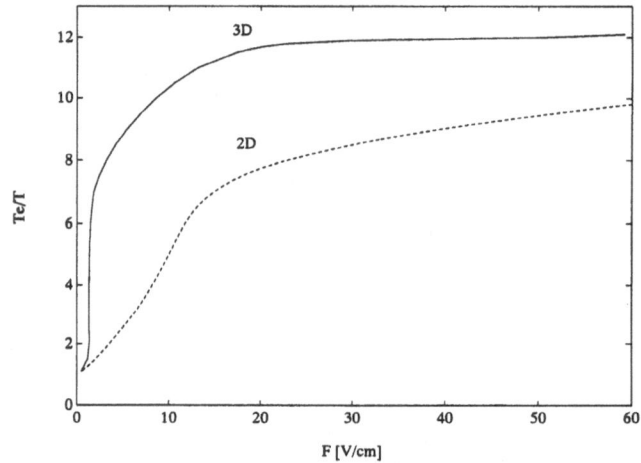

**Figure 4.** Electron temperatures normalised to 4.2 K versus electric field strength.

The differences of temperatures with respect to 3D and 2D carriers will be still more flagrant if a lower residual doping between the layers is assumed (the data of Fig. 4 are obtained for $5 * 10^{11}$ cm$^{-2}$ in the layers and $10^{15}$ cm$^{-3}$ in the interspace).
The real space transfer due to carrier heating is shown in Fig. 5. It leads to a remarkable increase of the 3D carrier concentration in the region of the steep rise of the electron temperature (region I and II in Fig. 1) and remains almost constant at those fields which lead to the slow increase of electron temperature (according to region III).

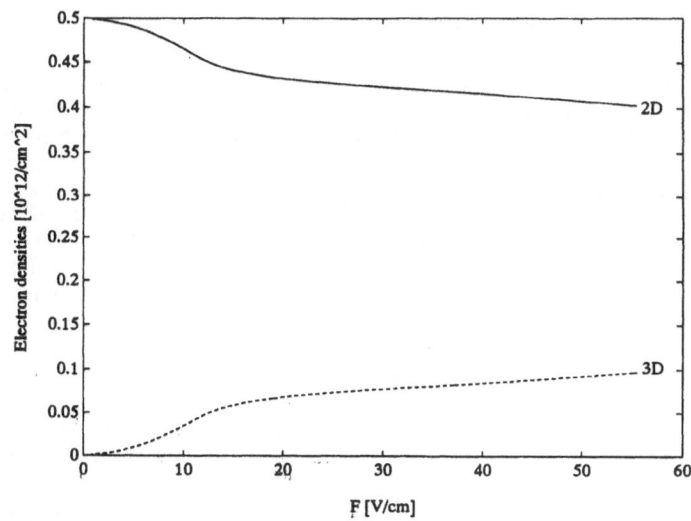

**Figure 5.** Electron densities as functions of electric field strength for $n_s$ = 5 * $10^{11}$ cm$^{-2}$, d = 100 nm, a = 10 nm, $N_J$ = $10^{15}$ cm$^{-3}$ and $E_c$ = 18 meV from ground level to barrier height.

The current contribution of the hot 3D electrons due to their high mobility dominates the total current density by more than one order of magnitude at about 10 V/cm for the parameters chosen in the simulation. The resulting differential conductivity (Fig. 6) qualitatively agress with the experimental results and mirrors the specific features of transport for sample A with its partially ordered distribution of dopants.

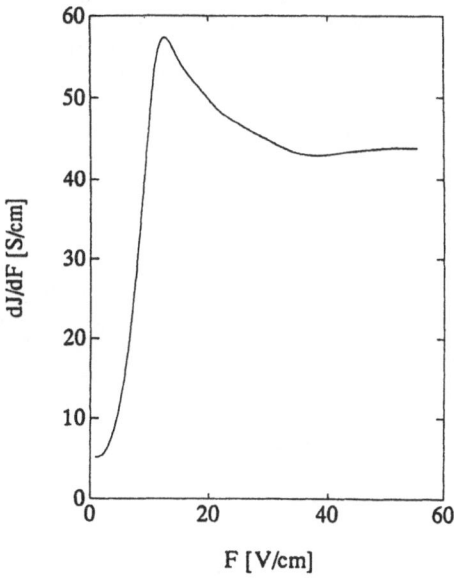

**Figure 6.** Differential conductivity dJ/dV versus applied voltage for the parameters of Fig. 5

## PHONON FLUX FROM HOT CARRIERS

As already described the energy gain by the electric field is balanced by the loss due to phonon emission. Therefore, phonon spectra originated by hot electrons reflect the kinetics of carrier heating. In time of flight measurements phonons below some $10^{11}$ cps propagate through the semiconductor without scattering by isotopes. As known ballistically moving phonons become self-focused in crystallographic directions of high symmetry. The flux arriving on a surface of such an orientation therefore consists of low frequency acoustic phonons directly emitted by electrons as well as phonons created within a time interval of 100 ns by repeated decay and conversion processes starting from optical phonons emitted by high energy carriers (e.g.[4,5]).

Investigations were performed in a broad range of electric fields using an epitaxial GaAs structure grown on a thick slide of GaAs with (100) orientation[6]. The MBE structure consists of 1 μm buffer layer, 2 δ-layers doped with 1.35 and 1.2 * $10^{12}$ cm$^{-2}$ Si atoms, respectively, in a distance of 100 nm, a 150 nm undoped layer and a 150 nm layer doped with 1 * $10^{17}$ cm$^{-3}$ Si on top. Au:Ge current contacts in a distance of 2 to 3 mm were prepared at 420 C in a $N_2/H_2$ atmosphere.

An Indium bolometer was evaporated on the back side with a linear sensitivity for the whole range of signals.

The sample was immersed in superfluid He for the measurements (2 K). The voltage was applied in the form of square pulses of 11, 30, and 65 ns duration, respectively. The current was measured as the voltage drop on a 50 Ω load resistor connected right next to the sample in the dewar. The current values for fixed fields did not differ remarkably with respect to the pulse duration.

The conductivity calculated from the current voltage characteristics is demonstrated in Fig. 7. It exhibits two specific features: a kink at 10 V/cm and a drop above 1 kV/cm. The sudden increase of conductivity at 10 V/cm can be assumed to be caused by carrier heating in the wells and consequent real space transfer into the extended states with their significantly enhanced

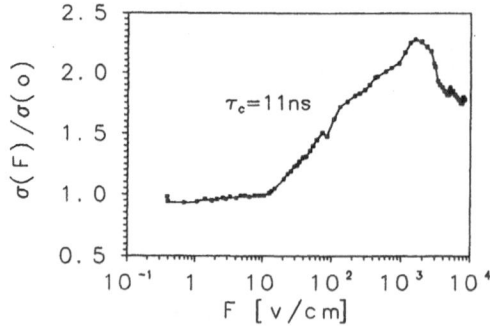

**Figure 7.** Normalised conductivity versus electric field strength at 2 K (pulse duration 11 ns).

mobility due to a low impurity density similar to the results in multi δ-layers reported in the previous section. Though the data differ they are in accordance having in mind the higher doping densities in the present structure leading to a shift of carrier heating to higher field strength. A mere influence of carrier heating on the mobilities of each single level should not exhibit such a kink, but lead to a smoother rise.

The drop of the conductivity above 1 kV/cm resembles the Gunn effect in bulk material (whereas in HEMTS the electron transfer from the $\Gamma$ to the L band edge in GaAs is mixed with real space transfer into the AlGaAs barriers), although a negative differential conductivity[7] is not obtained, but only a severe drop to one third between 1.1 and 3 kV/cm for the sample considered. This can be explained by the parallel contributions of several subensembles of carriers, when the fraction of electrons remaining in the confined states of the wells and those being perhaps present near to the top layer are not heated strongly enough for a $\Gamma$-L-transition at these fields, thus shunting the effect of decreasing conductivity of that fraction of electrons from the layers which move quasifree in the extended states outside the wells.

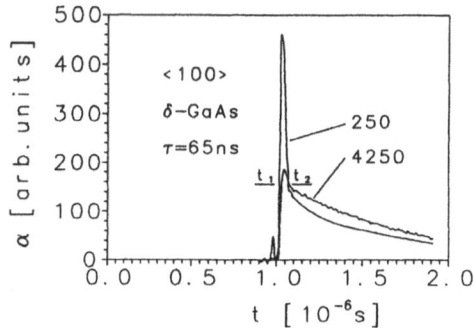

**Figure 8.** Bolometer signal normalised to input power as a function of the time of flight at 2 K.

Concerning the time of flight spectra a steep front at $t_1 = 1$ μs after the voltage pulse is observed according to the arrival of TA phonons, while LA phonons are not focussed along the <100> direction (Fig. 8). The narrow peak between $t_1$ and $t_2$ is explained by the acoustic phonons directly emitted by electrons. Such peaks were not observed in GaAs/AlGaAs[8]. Of course, it already overlaps with the signal of the broad band of "later arriving" phonons connected with the decay of the optical phonons emitted by hot electrons. The curves shown in Fig. 8 present the phonon fluxes normalised to the power dissipated in the sample. It is to be seen that for higher heating fields the fraction of phonons in the narrow peak is diminished in correspondence with an increase of phonons arriving in the broad time interval as expected due to an enhanced optical phonon emission with rising carrier energy. Because of the low sensitivity of the bolometer at weak phonon fluxes (irrespective of an accumulation of 1000 pulses) the amplification of the short pulse signals was somewhat increased in comparison to the measurements with 65 ns pulse duration. Of course, for all measurements the linear dependence on the electric field was verified.

Fig. (9) demonstrates the time integrated phonon fluxes per input power $\alpha(F)$ as functions of the electric field with pulse duration as a parameter for the narrow peak (lower curves) and the broad time interval of arriving phonons (upper curves). Evidently, two regions have to be considered: fields up to 1 kV/cm and those above this value. Regarding the moderate fields the decrease of the value for the narrow phonon peak could be associated with a redistribution of the dissipated energy between acoustic and optic phonon emission as functions of carrier heating (see remark in context with Fig. 8). However, the growing fraction of energy dissipation with respect to optical phonons normalised to the input power should be reflected by

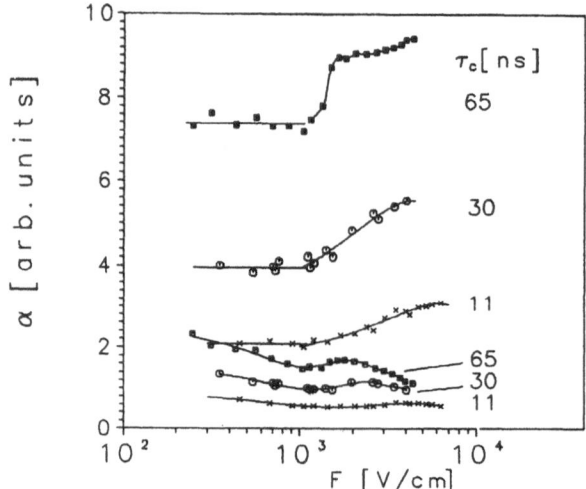

**Figure 9.** Bolometer signals normalised to input power as functions of electric field for the narrow peak (lower curves) and the signal in the broad time of flight interval (upper curves), respectively.

an increase of their decay products presented in the upper curves then. Regarding the 11 and 30 ns measurements such conclusions can not be drawn within the limitation given by the scattering of experimental values. (The horizontal lines are drawn for the eye only, they do not present a numerical fit to the data). However, the results of the measurements with 65 ns pulses beyond any doubt do not exhibit an increase of the normalised signal in the broad time interval in correspondence to the decrease of the normalised narrow peak. Yet this unexpected behaviour can be explained by a diminishing number of electrons confined to the wells by growing carrier heating, remembering that emission of acoustic phonons by 2D electrons is higher than by 3D electrons for a fixed field strength (Fig. 3).

The significant increase of the normalised phonon fluxes at fields above 1 kV/cm is clearly demonstrated in all curves, i.e. a new channel of phonon creation becomes active. This situation has to be regarded in connection with the conductivity drop at about the same field strengths. The proposed electron transfer from the $\Gamma$ to the L band edge is phonon assisted and explains the increase, because the probability of this process is greater than for carrier interaction with polar optical phonons due to the deformation potential scattering type. One has to remember that the electrons in the L valleys have approximately lattice temperature in contrast to the elevated temperatures of hot electrons in the $\Gamma$ minimum.

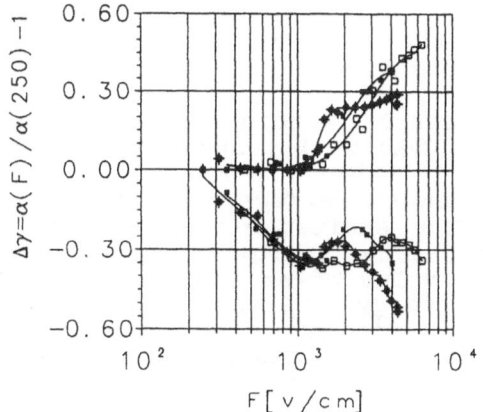

**Figure 10.** Changes of the bolometer signal normalised to input power as functions of electric field (see figure 9) for pulse durations of 11 ns, 30 ns, 65 ns.

When the $\alpha(F)$ curves are normalised to their values at 250 V/cm $\gamma(F)$ curves are obtained for both types of fluxes present. The curves are unified in the moderate field region, but exhibit differences for the high field region (Fig. 10) with respect to the duration of applied voltage pulses. Regarding the 11 ns data the lower branch presents a saturation region between 1 and 3 kV/cm (in contrast to the drop at lower fields) and a rise for further increasing fields, while the upper branch increases already at 2 kV/cm.

Introducing the power into the sample by longer pulses is clearly reflected by the phonon spectra in contrast to the current voltage characteristics. This circumstance can be described by inelastic phonon-phonon interaction when the nonequilibrium phonon densities reach a critical value. For more detailed information the reader is referred to Fig. 7 in[6].

## IMPURITY INDUCED BREAKDOWN

As mentioned in the introduction for small impurity concentration at low lattice temperatures $\delta$-doped systems will behave similar to homogeneously doped semiconductors, i.e. most of the electrons are bound to the donors. As well known the application of an electric field leads to a drop of the recombination and an increase of impact excitation as well as ionisation, resulting in breakdown at a critical electric field strength. The probabilities of these processes depend on the density of populated and free states envolved in this generation-recombination reaction kinetics, the cross sections for the regarded interactions and the energy distributions of the conduction electrons. While only the densities of donors and acceptors determine the numbers of neutral and ionised centres, the local arrangement of the impurities, too, influences the mobility and energetic distribution of the conduction electrons. Therefore the parameters of the generation-recombination reactions are expected to show specific features for partially ordered structures besides the general characteristics of breakdown.

In[10] multi $\delta$-layers were investigated and compared with homogeneously doped GaAs with respect to donors and acceptors. The data are given in table II. The distance of the donors is evidently larger than the effective Bohr radius, whereas the wave functions of the excited states already overlap. Due to potential fluctuations the energy levels of the donors are distributed around the value of an isolated centre, but it can be assumed[11] that this energy interval is still small at the chosen concentration of $7 * 10^{10}$ cm$^{-2}$ Si Atoms and the "bands" of the ground and the first excited states are well separated from the conduction band for the structures A to C.

**Table 2.** Sample parameters.

| sample | donors | | acceptors | |
|--------|--------|-------------|-----------|-------------|
| | density | arrangement | density | arrangement |
| A | $7 \cdot 10^{10} cm^{-2}$ | $\delta$, 100nm | $2 \cdot 10^{10} cm^{-2}$ | $\delta$, 100nm |
| B | $7 \cdot 10^{10} cm^{-2}$ | $\delta$, 100nm | $2 \cdot 10^{15} cm^{-3}$ | homogeneous |
| C | $7 \cdot 10^{15} cm^{-3}$ | homogeneous | $2 \cdot 10^{15} cm^{-3}$ | homogeneous |
| D | $1 \cdot 10^{11} cm^{-2}$ | $\delta$, 100nm | $3 \cdot 10^{10} cm^{-2}$ | $\delta$, 100nm |

The static current voltage characteristics were measured in a four point arrangement paying attention to eliminate effects of the side probes of the Hall bars, which else can reduce the effective sample length due to their finite size if not carefully mesa etched.

The current voltage characteristics in all three arrangements of dopants A to C exhibit a S-shape (Fig. 11): a low conductivity branch (I), negative differential conductivity (ndc) for the current above the critical field for breakdown, followed by a subsequent steep increase over an order of magnitude (II) and a region with weaker increase of current (III). The breakdown field in samples with donors arranged in nominally monoatomic planes (type A,B) is lower than in a homogeneously doped sample (type C) with the same effective donor concentration and compensation ratio, whereas the arrangement of acceptors seems to have a minor influence. The branch with ndc, which is seen in these current controlled measurements, is shorter in δ-layered than in homogeneously doped samples.

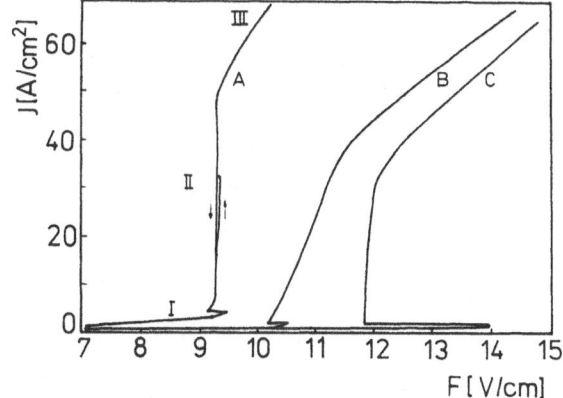

**Figure 11.** Current densities as functions of field strength at 4.2 K for the samples A, B, C of Table 2.

With concern to generation-recombination reaction kinetics in the two-donor level model, paying the main attention to the terms proportional to $n^2$ in determining ndc[12,13] no direct relation to the local arrangement of the donors is given, but only to their global density. In a less simplified treatment of the probalities of the generation and recombination processes taking an influence of impurity density on carrier heating via scattering and screening into account[14,15] a nonlinear dependence on doping density is obtained for other terms, too. In a detailed numerical treatment the partial ordering of the dopants in comparison to a random distribution should be of importance with respect to breakdown and ndc.

In contrast to the structures A to C a sample with a higher impurity concentration but the same compensation ratio (type D) only exhibits a steep current increase at the breakdown field instead of a branch with ndc. This is specific for generation-recombination reaction kinetics, in which the terms being directly nonlinear with respect to concentration play no decisive role. Such a case was implemented for instance in GaAs with about $8 * 10^{15}$ cm$^{-3}$ donors investigated in[16], where the impact ionisation of the first donor level was excluded from the balance on account of merging with the conduction band states.

With respect to the characteristics with ndc all the samples of types A to C show circuit limited oscillations in this region when the measurements are current controlled with a load resistance above a few kΩ.

These oscillations set in at a certain critical point of the branch with ndc. They are reproducible for each sample, but the current region (step height in the branch with ndc) differs from sample to sample. Such circuit limited oscillations had been observed in homogeneously doped semiconductors before(e.g. for thick GaAs layers in[17]).

However, more detailed investigations[18] demonstrate that the observed form of the ndc branch significantly depends on the measuring circuit. With the load resistor right next to the sample, reducing the parasitic capacitance to a minimum (about 1pF) the shape of the J-V-characteristics is depicted by curve II in Fig. 12. It exhibits two jumps between stable sections. As to be seen at the critical breakdown field the slope changes abruptly its sign. The following negative differential resistivity is smaller than the value of 1 MΩ of the load resistor and

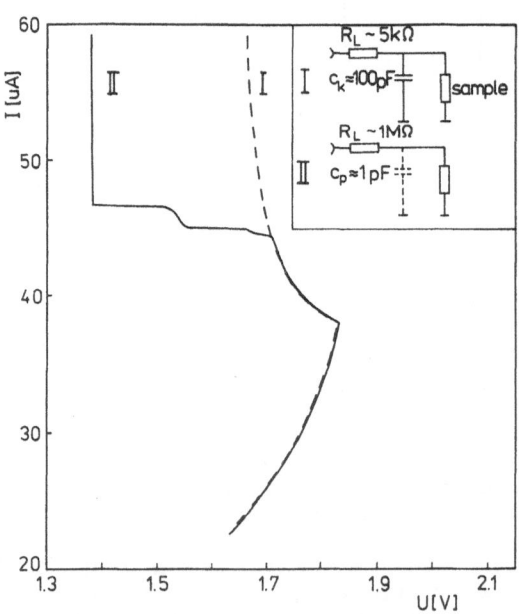

**Figure 12.** Current voltage characteristics (region II) for different load resistors

becomes continously reduced with increasing current. At a certain point the I-V-characteristics jumps to the next stable branch with negative differential resistance. The point to which it is switched is determined by the load resistor. Whether the transition corresponds to a negative differential resistivity, which is higher than 5 MΩ, or even to a positive value could not be decided experimentally. A similar jump is repeated at somewhat higher currents leading to the branch of stable steep increase, yet at a lower current value than in the case of a high parasitic capacitance, which can not be so quickly discharged. Furthermore it should be mentioned that now the current voltage characteristics shows no hysteresis and no oscillations, in the regarded region.

Furthermore hysteresis loops are observed in the region of steep current increase, they are reproducible with a given sample, too. In Fig. 11 at significantly higher currents than for the circuit limited oscillations such a loop is exhibited. In the characteristics of this sample only a single loop appears in contrast to the staircase observed in[19] because of the narrow sample width of d = 20-100 μm for the investigations reported in[10] compared to 1 mm for the GaAs results in[17].
According to the conventional explanation these hysteresis effects are due to current filamentation and the number of loops is given by the minimum and maximum filament diameters then. Besides in a δ-layered sample of about 2.6 mm width for instance, consequently more regions of hysteresis were observed, too.

The presence of magnetic fields does not only shift the static current voltage characteristics as expected[9,10], but induces now regular oscillations, not becoming realised in the absence of B[20]. This situation is similar to homogeneously doped GaAs layers[21] in contrast to p-Ge, in which they always can be detected. In dependence on magnetic and electric fields the temporal structures change, covering many types: intermittence type I, frustrated chaos, locking modes of as well stable character as with chaotic bursts[20].

In order to obtain more insight into the dynamics in[10] the voltage was applied by square pulses with 5ns rise time, while the resulting current was measured as a voltage drop on a 50 Ω resistor connected right next to the sample. The results for sample A are shown in Fig. 13. In the pre-breakdown region the current exactly follows the voltage pulse without any delay. When the pulse height corresponds to the static breakdown field at 3.22 V it takes 7 μs until the expected current increase turns up. This means that such a long time is necessary until avalanche sets in. With an increase of 10 mV this time is shortened to about 1 μs and by such a further voltage increase to 0.5 μs already, demonstrating the complex processes of carrier heating and its influence on recombination, impact excitation and ionisation.

Already at 3.23 V a second enhanced increase of the current can be noticed at about 10 us. This effect is accelerated with increasing voltage up to 3.3 V, too. This feature seems to be connected with the current filamentation (presented on the static characteristics by a loop in Fig. 11), i.e. with respect to dynamic measurements this formation of an inhomogeneous carrier distribution in the sample is supposed to take place even if the load line intersects the static current voltage characteristics only in the beginning of the post-breakdown region (III).

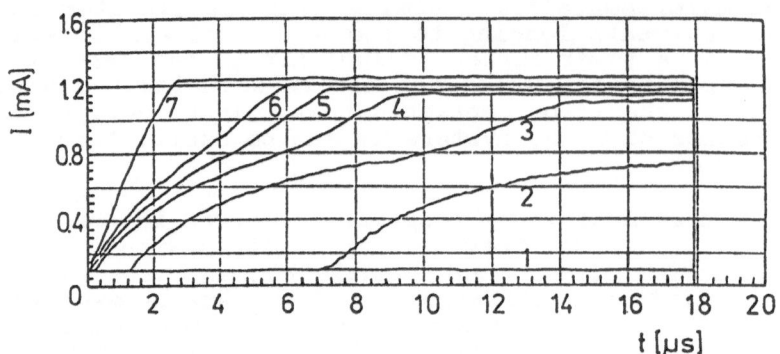

**Figure 13.** Current in sample A as a function of time for applied voltages of 3.20, 3.22, 3.23, 3.25, 3.29, 3.30, and 3.31 V for 1-7, respectively.


On the other side about 3.45 V a quick current rise on the sub μs time scale sets in, proceeding again faster with voltage increase. It is assumed that now the avalanche leads to the end state on the static characteristics without filamentation as a preliminar state.

Such dynamic investigations can be complemented by measurements of the relaxation processes by double pulse techniques as pointed out in[9]. A full understanding, however, is not possible without theoretical calculations.


# REFERENCES

1. H. Kostial, Th. Ihn, P. Kleinert, R. Hey, M. Asche, and F. Koch
   Phys. Rev. B, to be published
2. Xiao-mei Feng, J.J. Mares, M.E. Raikh, Fr. Koch, D. Grützmacher, and A. Kohl,
   Surface Science 263, 147 (1992)
3. X.L. Lei and C.S. Ting, Phys. Rev. B 32, 1112 (1985)
4. B.A. Danilchenko, V.N. Poroshin, M.J. Slutskii, and M. Asche
   phys. stat. sol (b) 136, 63 (1986)
5. P. Hawker, A.J. Kent, M. Hennini, and O.G. Hughes
   Sol. State El. 32, 1755 (1989)
6. B.A. Danilchenko, S. Roshko, M. Asche, R. Hey, M. Höricke, and H. Kostial
   Semicond. Science Technol. to be published
7. Y. Balynas, A. Krotkus, T. Lideikis, A. Stalnionis, and G. Treideris
   El. Letters 27, 2 (1991)

8. J.K. Wigmore, M. Erol, M. Sakraoni-Takar, C.D.W. Wilkinson, J.H. Davies, and C. Stanley
   Semicond. Science Technol. $\underline{6}$, 837 (1991)
9. H. Kostial, Th. Ihn, M. Asche, R. Hey, K. Ploog, and F. Koch
   Proc. SSDM92, Tsukuba, 1992, p. 287
10. H. Kostial, Th. Ihn, M. Asche, R. Hey, K. Ploog, and F. Koch
    Jap. Journ. Appl. Phys., to be published
11. Qui-yi Ye, B. J. Shklovskii, A. Zrenner, F. Koch, and K. Ploog
    Phys. Rev. B $\underline{41}$, 8477 (1990)
12. A. A. Kastalskii,
    phys. stat. sol. $\underline{15}$, 599 (1973)
13. E. Schöll
    Zeitschrift für Physik B, Condensed Matter $\underline{46}$, 23 (1982)
    Nonequilibrium Phase Transitions in Semiconducters, Springer Verlag (1987)
14. J. Yamashita
    Journ. Phys. Soc. Jap. $\underline{16}$, 720 (1961)
15. T. Kurosawa
    Journ. Phys. Soc. Jap. $\underline{20}$, 1405 (1965)
16. R. S. Crandall
    Phys. Rev. B1, 730 (1970)
    Phys. Chem. Sol. $\underline{31}$, 2069 (1970)
17. U. Rau, K. M. Mayer, J. Parisi, J. Peinke, W. Clauss, and R. P. Huebener
    Sol. State El. $\underline{32}$, 1365 (1989)
18. K. M. Mayer, R. P. Huebner, and U. Rau,
    Journ. Appl. Phys. $\underline{67}$, 1412 (1990)
19. H. Kostial
    private communication
20. W. Prettl
    private communication
21. J. Spangler, A. Brandl, and W. Prettl
    Appl. Phys. A $\underline{48}$, 143 (1989)
22. A. Brandl and W. Prettl
    PRL $\underline{66}$, 3044 (1991)
    Festkörperprobleme, Advances in Solid State Physics $\underline{30}$, (1990) p. 371

# OPTICAL PHONON MODES IN SEMICONDUCTOR QUANTUM WELLS AND SUPERLATTICES

M.P. Chamberlain

Department of Physics
University of Essex
Colchester CO4 3SQ
United Kingdom

## INTRODUCTION

There has been considerable interest in the properties of optical phonons in low dimensional heterostructures over a number of years since it has been possible to grow alternative layers of two semiconductors using techniques such as molecular beam epitaxy. The vibrational modes of GaAs/AlAs superlattices were first investigated by Merz et al. (1977) using Raman and infrared spectroscopy. They found modes at acoustic frequencies which were attributed to the folded dispersion relation of the acoustic modes. The effects of the superlattice on optical modes were first clearly seen and described in terms of zone folding by Jusserand et al. (1983, 1984, 1985), Colvard et al. (1985a,b) and Sood et al. (1985). For backscattering in the growth direction from superlattices with small layer thicknesses longitudinal optic (LO) modes are observed that are shifted down in frequency from the bulk. This shift in frequency of the modes is evidence of confinement to one of the two constituent layers in the superlattice. Quantum wells and superlattices also exhibit interface phonon-polariton vibrations which extend into both the layers with exponentially decreasing amplitudes. Sood et al. (1985) observed a broad band of interface modes in a backscattering geometry from the (001) surface when no in-plane phonon momentum should be generated. This was attributed to a breakdown of momentum conservation under resonant conditions. The first direct evidence of interface phonons was seen by Nakayama et al. (1988, 1990) in a GaAs/AlAs heterostructure using a quasi-backscattering configuration. A range of in-plane interface phonon wavevectors were generated by varying the incident angle with respect to the normal to the sample surface. Fuchs et al. (1991) have seen interface modes with a well-defined in-plane wavevector, induced by an 80Å period grating on the surface of the GaAs/AlAs superlattice. A detailed investigation of interface modes in GaAs/AlAs superlattices was performed by Huber et al (1991), Heβmer et al. (1992) using micro Raman scattering where the spot size of the incident laser beam is small enough to be able to use scattering geometries where all the momentum transfer is in the in-plane direction.

*Negative Differential Resistance and Instabilities in 2-D Semiconductors,*
Edited by N. Balkan *et al.*, Plenum Press, New York, 1993

335

In parallel with the experimental work there has been considerable theoretical investigation into the properties of phonons in heterostructures. The lattice vibrations can be described either by a microscopic model, where the atomic arrangement and the forces between the atoms near an interface are treated explicitly, or via a continuum model where the interface is a boundary dividing two dissimilar continua. The microscopic approach is computationally demanding and difficult to incorporate into theories of electron-phonon interactions or Raman scattering from phonons in heterostructures. Continuum models can treat all but very small layer width heterostructures, say ($\leq 20\text{Å}$) and are readily used to calculate electron-phonon interactions and Raman scattering intensities.

A review of the early work on superlattices was undertaken by Klein (1986). Reviews of phonons in superlattices have also been undertaken by Menéndez (1989) Jusserand and Cardona (1989) and Cardona (1990). In this review we shall give an outline of the dielectric continuum and hydrodynamic models. In particular we shall emphasise the shortcomings of these models and the need to consider coupling of confined and interface modes. We also review the hybrid continuum mode theories which agree very well with microscopic models. We compare the results of the different models in calculating electron-phonon transition rates. In addition Raman scattering data will be compared with the phonon model dispersion relations.

## DIELECTRIC CONTINUUM MODEL

Until recently there were two proposed continuum models for optical phonons in semiconductor heterostructures, the dielectric continuum model and the hydrodynamic model. In the dielectric continuum model (or sometimes called slab model) (Fuchs and Kliewer 1965, Huang and Zhu 1988) the Born and Huang equations for the ionic displacement field along with Maxwells equations are used to determine the longitudinal optic modes. The relative ionic displacement **u** is defined as the displacement of the positive ions relative to the negative ions. The dependence of the mode frequency on wavevector (dispersion) is neglected in this model and the displacement is coupled to the electric field **E** and the polarisation field **P** by:-

$$\ddot{\mathbf{u}} = -\omega_T^2 \mathbf{u} + [(\varepsilon_s - \varepsilon_\infty)/4\pi]^{1/2} \omega_T \mathbf{E} \qquad (1)$$

$$\mathbf{P} = [(\varepsilon_s - \varepsilon_\infty)/4\pi]^{1/2} \omega_T \mathbf{u} + [(\varepsilon_\infty - 1)/4\pi]\mathbf{E} \qquad (2)$$

where $\omega_T$ is the limiting bulk transverse optical (TO) phonon frequency and $\varepsilon_s$ and $\varepsilon_\infty$ are the static and high frequency dielectric constants. It is also required to use Maxwell's equations:-

$$\nabla.\mathbf{D} = 0 \qquad (3)$$

$$\nabla \times \mathbf{H} = \frac{1}{C}\dot{\mathbf{D}} \qquad (4)$$

$$\nabla.\mathbf{H} = 0 \qquad (5)$$

$$\nabla \times \mathbf{E} = \frac{1}{C}\dot{\mathbf{H}} \qquad (6)$$

The dielectric continuum model considers the non-retarded limit (i.e. the speed of light $c \rightarrow \infty$) and the right hand sides of equations (4) and (6) become equal to zero. The electric field **E** associated with the optical phonons is found to be proportional to the ionic displacement **u**

$$E = \left[ 4\pi/(\varepsilon_s - \varepsilon_\infty)\omega_T^2 \right]^{1/2} \left( \omega_T^2 - \omega^2 \right) u \qquad (7)$$

For LO modes the electric field can be written in terms of a scalar potential $\Phi$ such that $E = -\nabla\Phi$. It is this scalar potential which interacts with the electrons via the Fröhlich coupling when considering electron-phonon interactions.

Equation (3) can be rewritten in terms of $\Phi$ as

$$\varepsilon_i(\omega)\nabla^2\Phi = 0 \qquad (8)$$

where $\varepsilon_i(\omega)$ is the frequency dependent dielectric constant in each material. The material is considered to be isotropic and the direction perpendicular to the interfaces (growth direction) is taken to be the z-axis. Let us consider the direction of propagation of the phonons to be along the x-axis. The dielectric continuum model employs electrostatic boundary conditions such that

$E_x$ is continuous $\qquad\qquad (9)$

and

$\varepsilon E_z$ is continuous $\qquad\qquad (10)$

at the interfaces. There are two solutions of equation (8); the solution that $\varepsilon_1(\omega) = 0$ in the well gives rise to confined modes. The form of the dielectric function is

$$\varepsilon_i = \frac{\varepsilon_{\infty i}(\omega_{Li}^2 - \omega^2)}{\omega_{Ti}^2 - \omega^2} \qquad (11)$$

For $\varepsilon_1 = 0$ in the well implies $\omega = \omega_{L1}$ so in the barriers at this frequency $\varepsilon_2(\omega_{L1}) \neq 0$. In order to satisfy (8) in the barriers $\nabla^2\phi = 0$. Since $\varepsilon_1 = 0$ in the well $D = 0$ in the well on both sides of the barrier, and the continuity of the perpendicular component of the electric displacement $D_z$ requires that in the barrier layers $E_z$ must vanish on both boundaries. Thus a potential satisfying $\nabla^2\phi = 0$ must vanish in the barriers. Due to the continuity of the tangential component of the electric field at an interface, $E_x = 0$ in the well which leads to solutions of the scalar potential in the well with nodes at the interfaces of the form

$$\phi = \begin{cases} e^{iq_x x}\cos(q_z z) & q_z = \dfrac{n\pi}{d} \quad n = 1,3,5,\ldots \\[2ex] e^{iq_x x}\sin(q_z z) & q_z = \dfrac{n\pi}{d} \quad n = 2,4,6,\ldots \end{cases} \qquad (12)$$

where d is the thickness of the confining layer with interfaces at $z = -d/2$ and $z = d/2$.

The second solution of (8) that $\nabla^2\phi = 0$ in both the well and barriers leads to the Fuchs and Kliewer (1965) interface modes. For a single quantum well in the non-retarded limit the interface phonon dispersion relation is

$$\frac{\varepsilon_2}{\varepsilon_1} = \begin{cases} -\coth(q_x d/2) \\[1ex] -\tanh(q_x d/2) \end{cases} \qquad (13)$$

The scalar potential $\phi$ and the z-component of the ionic displacement for the confined and interface modes of the slab model is shown in Figure 1.

# THE HYDRODYNAMIC MODEL

The hydrodynamic (or guided mode) model is derived from Maxwell's equations (3-6) and a modified Born and Huang model that includes dispersion in the equation of motion (1) Babiker (1986), Ridley (1989). The equation of motion is

$$\ddot{\mathbf{u}} = -\omega_T^2 \mathbf{u} + [(\varepsilon_s-\varepsilon_\infty)/4\pi]^{1/2}\omega_T\mathbf{E} - (\beta_L^2-\beta_T^2)\nabla(\nabla.\mathbf{u}) - \beta_T^2\nabla^2\mathbf{u} \qquad (14)$$

For LO modes $\nabla \times \mathbf{u} = 0$ so (14) can be rewritten as

$$\ddot{\mathbf{u}} = -\omega_T^2 \mathbf{u} + [(\varepsilon_s-\varepsilon_\infty)/4\pi]^{1/2}\omega_T\mathbf{E} - \beta_L^2\nabla^2\mathbf{u} \qquad (15)$$

where $\beta_L$ is the parameter describing dispersion and is taken to be the velocity of the LA modes in the (001) direction for each material.

Since the hydrodynamic model includes dispersion equation (8) can be satisfied for $\varepsilon_i(\omega) = 0$ in both materials. Hydrodynamic boundary conditions are applied at the interfaces which have been shown to conserve the energy flux vector across an interface (Chamberlain 1992) when the LO and TO modes are considered to be decoupled. The boundary conditions applied are the continuity of the pressure field

$$\beta_L^2\nabla.\mathbf{u} \qquad \text{continuous} \qquad (16)$$

and the continuity of the normal component of the velocity

$$\frac{\partial u_z}{\partial t} \qquad (17)$$

Applying these boundary conditions gives rise to confined modes with opposite symmetry to those of the dielectric continuum model

$$\Phi = \begin{cases} e^{iq_xx}\cos(q_zz) & q_z = \dfrac{n\pi}{d} & n = 2,4,6,\ldots \\[2mm] e^{iq_xx}\sin(q_zz) & q_z = \dfrac{n\pi}{d} & n = 1,3,5,\ldots \end{cases} \qquad (18)$$

These are the modes obtained for material systems where the zone centre LO frequencies of each material are quite different as is the case for GaAs/AlAs. For the GaAs/Ga$_{1-x}$Al$_x$As material system Ga$_{1-x}$Al$_x$As has two LO vibrations one close to GaAs LO (GaAs-like) and the other close to AlAs LO (AlAs-like). For $x = 0.3$ the difference between the GaAs and GaAs-like frequencies is only 15 cm$^{-1}$ and application of hydrodynamic boundary conditions predicts modes that are not completely confined to the GaAs layer. Modes in this frequency range have scalar potentials of the form (18) in the well and decaying exponentials in the barriers penetrating up to 30% of d into the barriers. These modes are called guided modes.

For GaAs/Ga$_{1-x}$Al$_x$As the hydrodynamic model also predicts purely mechanical interface modes which satisfy the dispersion relation

$$\frac{(\omega_2^2-\omega^2)\rho_2q_1}{(\omega_1^2-\omega^2)\rho_1q_2} = \begin{cases} -\coth(q_1L/2) \\ -\tanh(q_1L/2) \end{cases} \qquad (19)$$

$\rho_i$ is the mass density of each material and $q_1(q_2)$ are the z components of the phonon wavevector in the well (barrier).

These mechanical interface modes have a very shallow dispersion relation and lie within 3 cm$^{-1}$ of the bulk GaAs LO phonon frequency. This would make them virtually impossible to distinguish from the bulk GaAs phonons of the substrate in Raman scattering experiments.

The hydrodynamic model also allows solutions of equation (8) where $\nabla^2\Phi=0$ in both materials giving rise to the Fuchs and Kliewer interface modes. The scalar potential $\Phi$ and the ionic displacement $u_z$ for the confined modes, Fuchs and Kliewer interface modes and mechanical interface modes are shown in Figure 1. In this model the fact that $\varepsilon \neq 0$ and $\nabla.\mathbf{u} = 0$ is taken to imply that these interface modes are transverse. They then satisfy the electrostatic boundary conditions and exhibit the same dispersion as the slab model interface

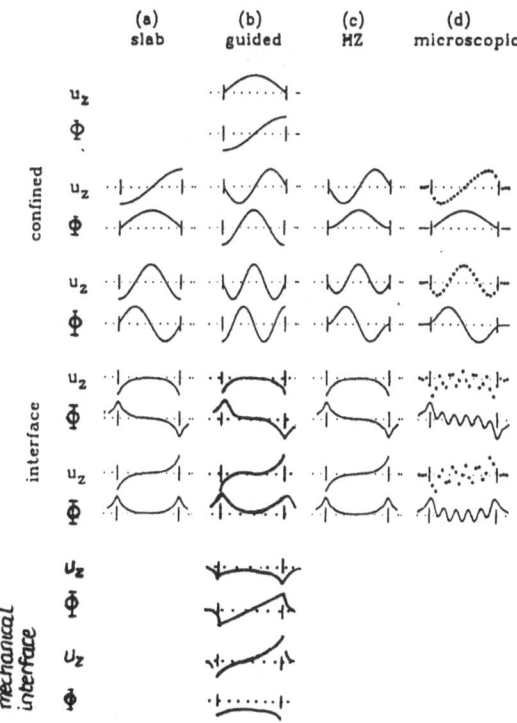

Figure 1. $u_z$ and $\Phi$ for GaAs-like optical modes in a GaAs quantum well. Microscopic model from Rücker et al (1992).

modes in equation (13). The hydrodynamic model emphasises the transverse nature of these modes and the fact that they should interact with electrons via an $\mathbf{A.p}$ interaction, where $\mathbf{A}$ is the interface mode vector potential and $\mathbf{p}$ is the momentum of the electron. In the slab model the interface modes interact with the electrons via an $e\Phi$ coupling. A calculation of the electron interaction with TO modes in bulk GaAs using an $\mathbf{A.p}$ coupling was found to be small (Chamberlain and Babiker 1989) and it was thought that the electrostatic interface modes would contribute little to electron-phonon interactions. However, calculation of electron -interface phonon interactions in a GaAs/AlAs quantum well using the $\mathbf{A.p}$ minimal coupling were found to be very significant (Al-Dossary et al 1992) and similar to those obtained using an $e\Phi$ coupling (Rudin and Reinecke 1990,1991).

# SHORTCOMINGS OF THE DIELECTRIC CONTINUUM AND HYDRODYNAMIC MODELS

One of the major problems of the dielectric continuum model is that it predicts modes with finite displacements at the interfaces which is in contradiction with the results of microscopic models (Rücker et al 1992, Chang et al 1991) and Raman Scattering experiments (Cardona 1990, Samson et al 1992). With this in mind Huang and Zhu (1988) compared the dielectric continuum model with a microscopic model and were able to modify the phonon mode potentials $\Phi$ such that both $\Phi$ and $u_z$ vanished at the boundaries. The Huang and Zhu potentials are

$$\Phi = \cos(n\pi z/d) - (-1)^{n/2} \qquad n = 2, 4, \dots \qquad (20)$$

$$\Phi = \sin(\mu_n n\pi z/d) + c_n z/d \qquad \mu_3 = 2.86, \mu_5 = 4.92, \dots \qquad (21)$$

where $c_n$ are constants. These modified potentials were obtained from the microscopic model in the limit of zero dispersion. In Figure 1 the microscopic displacements are taken from Rucker et al (1992) for $q_x = 0.15$ Å$^{-1}$ who found that at these larger wavevectors the dielectric continuum model agrees more closely with the microscopic model.

The hydrodynamic model includes dispersion and predicts zero displacements at the interfaces but it disagrees with microscopic models in that it predicts discontinuities in the scalar potentials at the interfaces. Since the scalar potential is non-zero at the interface the hydrodynamic guided modes do not agree with the microscopic models when considering electron-phonon interactions.

Enderlein (1991) has developed a macroscopic theory of optical modes which leads to the Fuchs and Kliewer interface modes and two sets of confined mode solutions equivalent to the hydrodynamic and slab model confined modes. He choses the slab model confined modes to be the correct ones since they are orthonormal to the interface modes.

A common shortcoming of all these continuum models is that they predict confined modes and interface modes that are uncoupled. It was first pointed out by the microscopic model of Huang and Zhu (1988) that the confined and interface modes of the same symmetry intermix when $q_x$ is finite. The intermixing is particularly strong for antisymmetric modes where for increasing $q_x$ the first confined mode $\omega_{LO1}$ mixes strongly with the interface mode and at large $q_x$ becomes a mode vibrating with the $\omega_{LO3}$ frequency. Neither the hydrodynamic nor the slab model can take account of this intermixing and more sophisticated continuum models have been required. The modified Huang and Zhu model described in (20) and (21) agrees with micrcoscopic models for small $q_x$ but cannot take into account the variaton in the amount of confined and interface mode mixing as $q_x$ increases.

Recently Trallero-Giner et al (1992) have given a detailed description of how continuum models of optical phonons can be derived which result in intermixing of confined and interface modes. They consider exactly the same equations describing the optical phonons as the hydrodynamic model but stress that the displacement consists of a longitudinal and a transverse field.

$$\mathbf{u} = \mathbf{u}_L + \mathbf{u}_T \qquad (22)$$

where $\nabla \times \mathbf{u}_L = 0$ and $\nabla.\mathbf{u}_T = 0$. In the hydrodynamic model $\mathbf{u}_L$ and $\mathbf{u}_T$ are always decoupled but Trallero-Giner et al (1992) point out that $\mathbf{u}_L$ is in general affected by the coupling to $\mathbf{u}_T$. They include the coupling of mechanical and electrical fields explicitly in the electrostatic field equation by noting that the source of $\Phi$ is the polarisation charge and writing (3) as

$$\nabla^2 \Phi = 4\pi \nabla.\mathbf{P} \qquad (23)$$

This leads to two simultaneous differential equations coupling $\mathbf{u}$ and $\Phi$ which require $\mathbf{u}$ and $\Phi$ are continuous at the interfaces. They also show that the other required boundary conditions are continuity of derivatives of the stress tensor, equivalent to the continuity of hydrostatic pressure, and continuity of the electric displacement field. They have also generalised their formalism to include anisotropy where three parameters are required to describe dispersion in each principal direction, their values obtained by fitting of experimental bulk dispersion curves.


## HYBRID OPTICAL PHONON MODELS

A number of models have been proposed recently which couple together confined and interface modes. Zianni et al (1992) have combined LO and interface modes including the effects of dispersion. The scalar potential is taken to be of the form

$$\Phi = Ae^{iq_L z} + Be^{-iq_L z} + Ce^{q_x z} + De^{-q_x z} \tag{24}$$

in the GaAs wells and decaying exponentials in tha AlAs barriers. They apply the electrostatic boundary conditions that $\Phi$ and $D_z$ are continuous at the interfaces along with the boundary condition that $u_z$ is also continuous . The dispersion relation for the odd and even modes are:

$$n\tan(nd)\alpha = \frac{q_x \tanh(q_x d)(\varepsilon_2 a_1 - \varepsilon_1 a_2)}{\varepsilon_1 + \varepsilon_2 \tanh(q_x d)} \tag{25}$$

$$n\cot(nd)\alpha = \frac{q_x \coth(q_x d)(\varepsilon_2 a_1 - \varepsilon_1 a_2)}{\varepsilon_1 + \varepsilon_2 \coth(q_x d)} \tag{26}$$

where

$$a_j = \frac{(\varepsilon_{sj} - \varepsilon_{\infty j})^{1/2} \omega_{Tj}}{4\pi(\omega_{Tj}^2 - \omega^2)} \tag{27}$$

and

$$\alpha = \frac{(\varepsilon_{s2} - \varepsilon_{\infty 2})^{1/2} \omega_{T2}}{4\pi(\omega_{T2}^2 - \omega^2 - b^2 q^2)} \tag{28}$$

Figure 2 illustrates the dispersion relation for a 50Å GaAs quantum well. It can be seen particularly for the odd modes that there is anticrossing between the interface and confined mode constituents. This is exactly the behaviour that was first pointed out by Huang and Zhu and has also been seen in the microscopic calculations of Chang et al (1991) and Molinari et al (1992). Constantinou et al (1992) have applied a similar model to GaAs/AlAs superlattices and the results are discussed in these proceedings. The above hybrid phonon models are restricted to the optical modes whose frequencies lie close to the bulk LO frequency. It is seen from the microscopic model of Molinari et al (1990) that the symmetric interface mode intermixes strongly with the lowest order confined TO modes.

Ridley (1992) has proposed a triple hybrid model which combines LO, TO and interface phonon polaritons. The displacement of the phonons is taken as

$$\mathbf{u} = \mathbf{u}_L + \mathbf{u}_T + \mathbf{u}_I \tag{29}$$

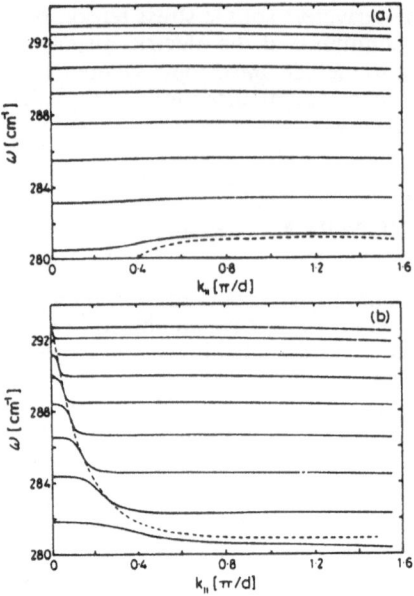

Figure 2. The dispersion curves (full-time) for (a) even modes (b) odd modes from Zianni et al (1992).

in the well and zero in the barriers as is appropriate for GaAs/AlAs structures. The LO and interface modes have associated electric fields which are non-zero in the barriers. The boundary conditions applied are $u_x$ and $u_z$ continuous at the interfaces along with the continuity of $E_x$ and $\varepsilon E_z$. Ridley (1992) has obtained results for a GaAs/AlAs quantum well which are similar to those of Zianni et al (1992) and to the dispersion shown in Figure 2. Chamberlain and Ridley (1992) have applied this model to GaAs/AlAs superlattices. The dispersion relation obtained for GaAs width a and AlAs, width b is given by

$$\tan(q_L a/2) = \frac{q_L r s d}{q_x(\alpha+\beta)}$$

$$\left[ -\left(1 + \frac{q_x^2}{4q_L^2 r^2 s^2 d^2}\ (\beta^2 - \alpha^2 + \gamma^2)\right) \pm \left\{\left[1 + \frac{q_x^2}{4q_L^2 r^2 s^2 d^2}\ (\beta^2 - \alpha^2 - \gamma^2)\right]^2 - \frac{q_x^2}{q_L^2 r^2 s^2 d^2}\ (\beta^2 - \alpha^2)\right\}^{1/2} \right] \quad (30)$$

where  $\gamma^2 = r^2 \sinh^2(q_x a)\ \sin^2(q_z(a+b))$

$\alpha = \sinh q_x b\ \cosh q_x a + r\ \cosh q_x b\ \sinh q_x a$

$\beta = \sinh q_x b + r \sinh q_x a\ \cos(q_z(a+b))$

$d = \cos(q_z(a+b)) - z\ \sinh q_x a\ \sinh q_x b - \cosh q_x a\ \cosh q_x b$

$$Z = \frac{1+r^2}{2r}\ , \qquad r = \frac{\varepsilon_\infty}{\varepsilon_1}\left(\frac{\omega^2 - \omega_{LO}^2}{\omega^2 - \omega_{TO}^2}\right) \qquad s = \frac{\omega^2 - \omega_{TO}^2}{\omega_{LO}^2 - \omega_{TO}^2} \qquad (31)$$

Note that the dispersion relation depends on d which when set equal to the zero is the interface phonon-polariton dispersion relation. The dispersion relation for an a = 56Å, b = 56Å GaAs/AlAs superlattice with $q_z(a+b) = \pi/4$ is shown in Figure 3. Again it exhibits the property of mixing confined and interface modes. Figure 4 shows the dispersion relation

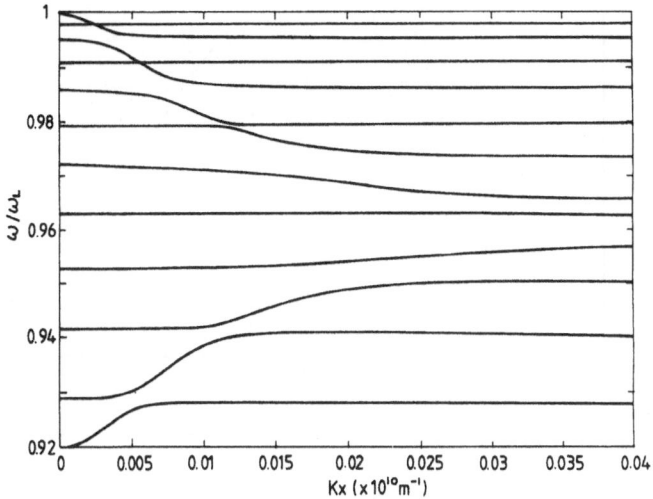

Figure 3. The dispersion relation for a = 56Å, b = 56Å superlattice with $k_z(a+b) = \pi/4$.

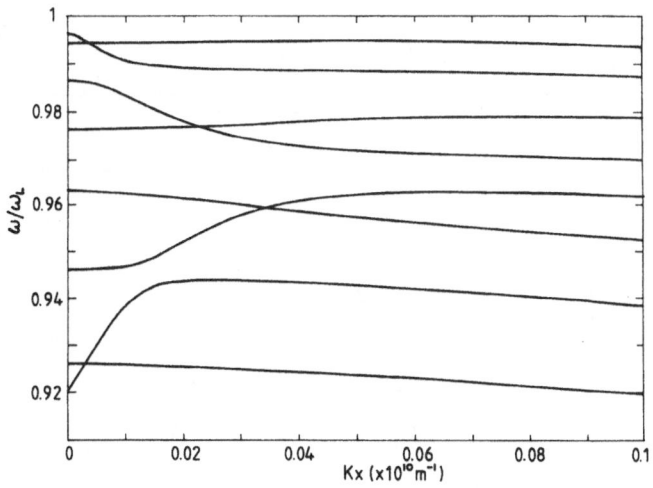

Figure 4. The dispersion relation for a = 50Å, b = 1000Å superlattice with $k_z = 0$.

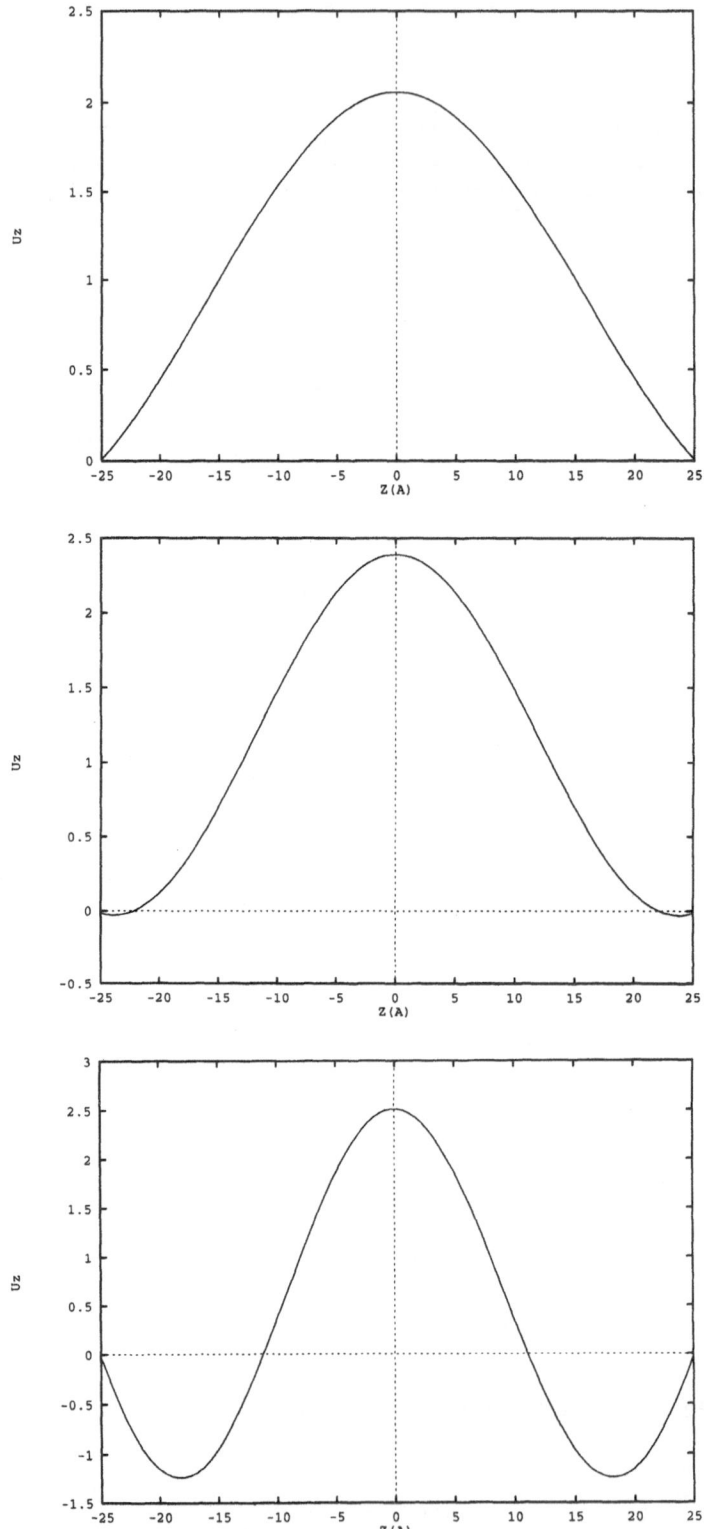

Figure 5. The z-component of the atomic displacement for a = 50Å, b = 1000Å superlattice with (a) $k_x = 1 \times 10^4\,m^{-1}$, (b) $k_x$ $5 \times 10^7\,m^{-1}$, (c) $k_x$ $5 \times 10^8\,m^{-1}$.

for an a = 50Å, b = 1000Å superlattice with $q_z = 0$. The first confined LO1 mode mixes strongly with the odd interface mode and at large $q_x$ becomes an LO3 mode. This is illustrated in Figure 5 where $u_z$ is plotted as a function of z for the highest frequency mode at three different values of $q_x$. The nature of the mode is seen to change from LO1 at $q_x = 0$ to LO3 at $q_x = 5 \times 10^8 m^{-1}$. To simplify the algebra the limit that $k_T$ is large was taken which is true for frequencies close to $\omega_L$. Further work will consider how the dispersion relation is modified around $\omega_T$ where this limit is inappropriate.

Gerecke and Bechstedt (1991,1992) have developed a continuum model of optical phonons using differential equations for the envelopes of the atomic displacement derived from a rigid ion model. The envelopes are seen to automatically satisfy the electrostatic boundary conditions and the additional mechanical boundary conditions are extracted from the underlying microscopic equations of motion. Their results are equivalent to the above hybrid models. Nash (1991) has also developed a hybrid model of optical phonons but this is as yet unpublished.

## ELECTRON-PHONON INTERACTIONS

A number of authors have calculated electron-phonon interactions in quantum well structures comparing different phonon models. Rudin and Reinecke (1990,1991) calculated electron-phonon interactions in an infinitely deep GaAs/GaAlAs quantum well for the dielectric continuum model, hydrodynamic model, and the Huang and Zhu model. They obtained good agreement between the dielectric continuum model and the Huang and Zhu model. Weber et al (1991) undertook a similar calculation for a finite potential GaAs/GaAlAs quantum well and also obtained good agreement between the same two models. Neither of these calculations included the contribution from the mechanical interface modes of the hydrodynamic model which give large contributions to the intrasubband transitions for GaAs/GaAlAs quantum wells (Chamberlain et al 1991) but do not exist for the GaAs/AlAs system.

Haupt and Wendler (1991) have calculated the electron-phonon interaction using a modified dielectric continuum model where the LO phonon fields are required to be orthogonal . They point out that the phonon modes of the Huang and Zhu (1988) model are not orthogonal and thus give incorrect results for the scattering rates. Haupt and Wendler found that when the results of their modified model are compared to the standard dielectric continuum model there is a strong redistribution of scattering channels from low-index LO phonons to higher index phonons but the sum of all contributions to the scattering rate is almost equal. Tsuchiya and Ando (1992) have calculated the electron-phonon interaction in GaAs/AlAs superlattices using an envelope function approximation which reproduces the modes for long wavelength optical phonons calculated in a microscopic model. It is found that the dielectric comtinuum model is quite accurate in superlattices with wide layers but fails to describe individual modes for narrow layers. They also point out that although the contributions of each mode are quite different the total scattering rate is the same.

Molinari et al (1992) and Rücker et al (1992) have compared the electron-phonon interaction in a GaAs/AlAs quantum well for the three continuum models with a microscopic model (Figure 6). They find that the best agreement with the microscopic model comes from the dielectric continuum model and that there is little agreement with the hydrodynamic model. They also find that the contribution of the AlAs interface modes to the scattering rate is very important.

Ridley (1992) has compared the electron transition rates in an infinite quantum well of the hybrid phonon model with those of the dielectric continuum, hydrodynamic and Huang and Zhu models. In his hybrid model the interface modes are taken to be the retarded limit of the interface phonon-polaritons. As such the hybrid mode interaction Hamiltonian is

$$H = -e\Phi + \frac{eA.p}{m^*} \tag{32}$$

The electron hybridon interaction is part Fröhlich-like and part electromagnetic. The scalar potential $\Phi$ comes from the LO modes and the vector potential A from the interface phonon-polaritons. Figure 7 shows a comparison of intrasubband transition rates with varying well

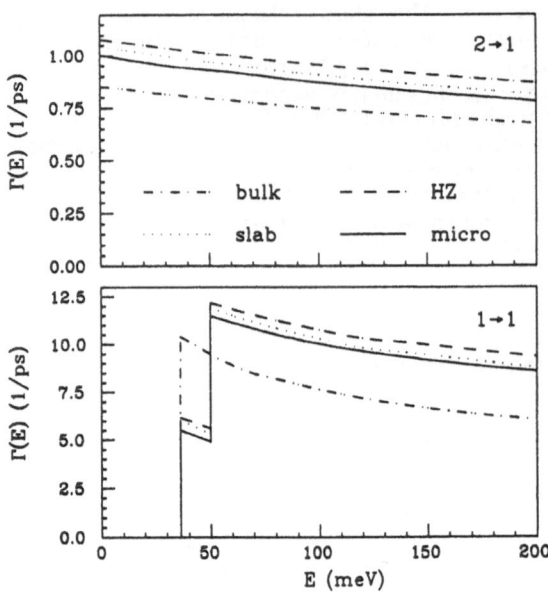

**Figure 6.** Comparison of total scattering rates (GaAs - and AlAs-like modes) for microscopic and macroscopic models. From Rücker et al (1992).

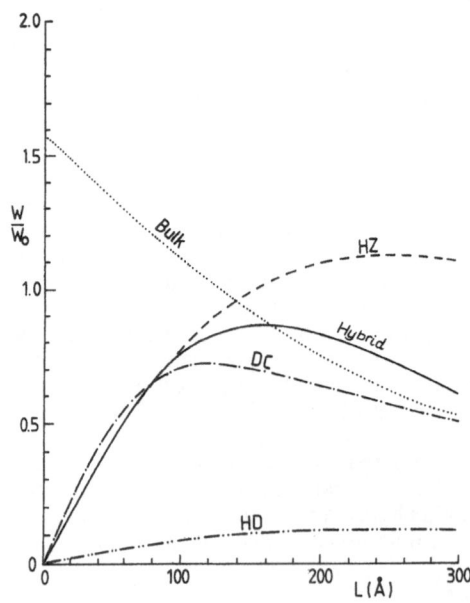

**Figure 7.** Comparison of intrasubband scattering rates of GaAs-like modes with the hybrid model. From Ridley (1992).

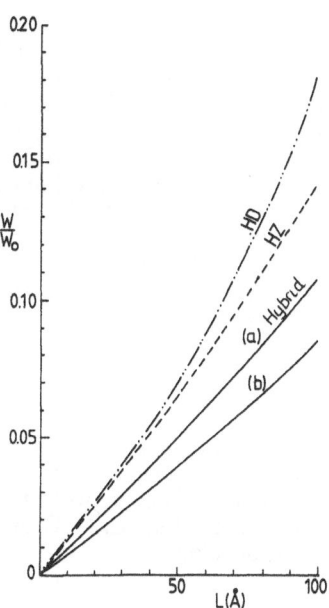

**Figure 8.** Comparison of intersubband scattering rates of GaAs-like modes with the hybrid model. From Ridley (1992).

width for the different phonon models, considering the initial electron to have just enough energy to emit an optical phonon. In Figure 8 the different intersubband transition rates as a function of well width are shown for an initial electron state at the bottom of the first excited subband. For well widths up to 80Å the transition rates for the hybrid model lie just below those of the Huang and Zhu model. The results of the electron-hybrid phonon interaction would agree well with those of the microscopic model of Molinari et al if the AlAs interface modes are also considered.

Time resolved Raman scattering has been used to probe intra- and intersubband electron energy relaxation via optical phonon emission in quantum well structures (Tatham and Ryan 1982). However the resolution of the laser pulse is 1 ps and these experiments are only able to place an upper limit on the relaxation rates. Grahn et al (1990) have used sequential resonant tunnelling of electrons to acheive occupation of higher subbands in GaAs/AlAs superlattices. Photoluminescence is used to determine the relative occupation of the second subband at resonance. Together with the electrically measured transport time between adjacent wells they have determined the $2 \rightarrow 1$ intersubband relaxation time for a = 123Å, b = 21Å and a = 118Å, b = 16Å superlattices to be between 0.57 and 0.65 ps. This agrees well with the intersubband transition rates obtained with the hybrid model of Ridley.

## CONCLUSIONS

An overview of the continuum models describing optical phonons in GaAs/AlAs quantum wells and superlattices has been presented. It was seen that neither the hydrodynamic nor the dielectric continuum models give a good description of the phonon modes. Although the dielectric continuum model gives similar scalar potentials to microscopic models for wide layers and large in-plane wavevectors, it fails to describe individual modes for narrow layers. Both of these models consider the confined and interface modes to be uncoupled but it has been shown by microscopic models that there is strong mixing of these modes of the same symmetry.

More sophisticated continuum models have been suggested which couple together LO and interface modes and in some cases TO modes as well. These hybrid models give similar results to microscopic models for dispersion and individual mode displacements, The values obtained from the hybrid phonon models for the strength of the electric phonon interaction also agree with the limited number of experimentally determined transition rates.

A few minor points still remain to be resolved. Recent micro-Raman experiments (Huber et al 1991, Heβmer et al 1992) have observed interface modes for a number of diferent in-plane wavevectors. The observed modes can be explained solely by the interface phonon dispersion relation and no intermixing with confined modes is needed. The experiments were undertaken for relatively large layer widths where confined modes are not normally observed in Raman scattering. For large period superlattices the large number of modes are not so easily resolved and their effect on the interface modes is not so pronounced. However, micro-Raman results on ultra-thin layer superlattices have observed the decrease in the $LO_1$ frequency as $k_x$ is increased, which is predicted theoretically (Scamacio et al 1992)

All continuum models ignore the fact that the GaAs/AlAs superlatices are anisotropic. Akera and Ando (1989) have pointed out that mechanical boundary conditions on the ionic displacement do not reflect the real situation at the interfaces , though they are a reasonable approximation provided the differences between force constants and mass ratios is small. Continuum hybrid phonon models for the GaAs/GaAlAs system, where the barrier has two bulk LO modes have yet to be developed.

## Acknowledgements

The author would like to thank B.K.Ridley, N.C.Constantinou, M.Babiker and O.Al-Dossary for a number of useful discussions. He is also grateful to the Science and Engineering Research Council for the provision of a Research Fellowship.

## References

Akera H and Ando T (1989) Phys. Rev. B **40** 2914
Al-Dossary 0,Babiker M and Constantinou N C (1992) Semicond. Sci. Tech. **7** B91
Babiker M (1986) J.Phys. C: Solid State **19** 683
Cardona M (1990) Superlattices and Microstructures **7** 183
Chamberlain M P (1992) Semicond. Sci. Tech. **7** 478
Chamberlain M P and Babiker M (1989) J. Phys: Condes. Matter **1** 1181
Chamberlain M P, Babiker M and Ridley B K (1991) Superlatt. Microstruct. **9** 227
Chamberlain M P and Ridley B K (1992) Proceedings of NATO ARW on Phonons in Nanostructures, 15-18 Sept. St-Feliu de Guixols, Spain
Chang Y-C, Ren S-F and Chu (1991) Superlatt. Microstruct. **9** 383
Colvard C, Gant T A, Klein M V, Merlin R, Fischer R, Morkoc H and Gossard A C (1985a) Phys. Rev. B **31** 2080
Colvard C, Fischer R, Gant T A, Klein M V, Merlin R, Morkoc H and Gossard A C Morkoc H and Gossard A C (1985b) Superlatt. Microstruct **1** 81
Constantinou N C and B K Ridley (1992) Proceedings of NATO ARW on NDR and Instabilities in 2-D Semiconductors
Enderlein R (1991) Phys Rev B **43** 14513
Fuchs R and Kliewer K L (1965) Phys Rev **140A** 2076
Fuchs H D, Mowbray D J, Cardona M, Chalmers S A and Gossard A C (1991) Solid State Commun. **79** 223
Gerecke H and Bechstedt F (1991) Phys Rev B **43** 7053
Gerecke H and Bechstedt F (1992) Semicond. Sci,. Technol. **7** B80
Grahn H T, Schneider H, Ruhle W W, von Klitzing K and Ploog K (1990) Phys. Rev. Lett. **64** 2462
Haupt R and Wendler L (1991) Phys. Rev. B. **44** 1850

Heβmer R, Huber A, Egeler T, Haines M, Tränkle G, Weimann G and Abstreiter G (1992) a Phys. Rev. B **46** 4071
Huang K and Zhu B (1988) Phys. Rev. B **38** 13377, 2183

Huber A, Egeler T,Ettmüller W,Rothfritz H, Tränkle G and Abstreiter G (1991) Superlatt. Microstruct. **9** 309

Jusserand B and Cardona M (1989) in Light Scattering in Solids V (Springer, Heidelberg) p203

Jusserand B, Paquet D and Regreny A (1984) Phys. Rev. B **30** 6245

Jusserand B, Paquet D and Regreny A (1985) Superlatt. Microstruct. **1** 61

Jusserand B, Paquet D and Regreny A  and Kervarec J (1983) Solid State Commun. **48** 499

Klein M V (1986) Journal of Q.E. QE22 **9** 1760

Menendez J (1989) J. Luminescence **44** 285

Merz J L, Barker A S and Gossard A C (1977) Appl. Phys. Lett. **31** 117

Molinari E, Baroni S, Giannozzi P and de Gironcoli (1990) in Proceedings of the 20th International Conference on the Physics of Semiconductors, edited by J.D. Joannopulos and E. Anastasakis (World Scientific, Singapore)

Molinari E, Bungaro C, Gulia M, Lugli P and Rucker H (1992) Semicond. Sci. Tech. **7** B67

Nakayama M, Ishida M and Sano N (1988) Phys. Rev. B **38** 6348

Nakayama M, Ishida M and Sano N (1990) Surf. Sci. **288** 131

Nash K (1991) Condensed Matter and Materials Physics Conference, Dec. 1991, Birmingham, U.K.

Ridley B K (1989) Phys. Rev. B **39** 5282

Ridley B K (1992) Proc. SPIE  Compound Semiconductor Physics and devices, Somerset, New Jersey **1675** 492 and submitted to Phys. Rev. B

Rucker H, Molinari E and Lugli P (1992) Phys. Rev. B **45** 6747

Rudin S and Reinecke T L (1990) Phys. Rev. B  41 77143; (1991) Phys. Rev. B **43**  9298

Samson B, Dumelow T, Hamilton A A, Parker T J, Smith S R P, Tilley D R, Foxon C T, Hilton D, Moore K J (1992) Phys. Rev. B **46** 2375

Scamarcio G, Haines M, Absteiter G, Molinari E, Baroni S and Ploog K (1992) submitted to Phys. Rev. B.

Sood A K, Menendez J, Cardona M and Ploog K (1985) Phys. Rev. Lett. **54** 2111

Tatham M C and Ryan J F (1992) Semicond. Sci. Technol. **7** B 102

Trallero-Giner C, Garcia-Moliner F, Velasco V R and Cardona M (1992) Phys. Rev. B. **45** 1194

Tsuchiya T and Ando T (1992) Semicond. Sci. Tech. **7** B73

Weber G, de Paula A M and Ryan J F (1991) Semicond. Sci. Tech. **6** 397

Zianni X, Butchet P N and Dharssi I (1992) J. Phys: Condens. Matter **4** L77

# PLASMONS ON LATERALLY DRIFTING 2DEGs

H P Hughes, R E Tyson, L C Ó Súilleabháin, and R J Stuart

University of Cambridge
Cavendish Laboratory
Madingley Road
Cambridge
CB3 0HE, U.K.

## INTRODUCTION

The plasmon modes of a two dimensional electron gas (2DEG) formed at a semiconductor heterojunction or in a quantum well have received much attention over the past few years. The dispersion relation for 2D plasmon modes in the absence of lateral drift has been derived by Stern[1], neglecting retardation:

$$\omega_0{}^2 = \frac{N_s e^2}{2m^* \varepsilon_0 \bar{\varepsilon}} k_x \qquad \text{Eqn (1)}$$

Here $\omega_0$ is the plasmon frequency, $k_x$ the plasmon wavevector in the plane of the 2DEG, $N_s$ the areal density of charge carriers each with charge $e$ and effective mass $m^*$, and $\bar{\varepsilon}$ is an effective dielectric constant which includes the dielectric screening effects of the layers of media around the 2DEG. Mode frequencies are typically in the range 20~70 cm$^{-1}$, so relevant optical techniques are far infra-red (FIR) and Raman spectroscopies. The properties of the drifting electron gas are also of interest, partly because of the possibility, in principle, of plasmon instabilities at high drift velocities; there have been several theoretical studies of the drifting gas and its plasmon modes[2,3], but experimental studies[4] of such systems are uncommon. In the simplest description, a Doppler-type shift in the plasmon frequency is expected as a result of the lateral drift of the carriers, and this should be proportional to the drift velocity $v_d$ and the wavevector $k$ of the plasmon mode itself:

$$\omega_v{}^2 = (\omega_0 + k \cdot v_d)^2 \qquad \text{Eqn (2)}$$

If $k$ and $v_d$ are both along the $x$-axis, this becomes, for the modes travelling up- and downstream :

$$\omega^{\pm}(k_x) = \omega_0(k_x) \pm k_x v_d \quad \text{or} \quad \omega^+(k_x) - \omega^-(k_x) = 2k_x v_d \qquad \text{Eqn (3)}$$

*Negative Differential Resistance and Instabilities in 2-D Semiconductors*
Edited by N. Balkan *et al.*, Plenum Press, New York, 1993

The plasmon wavelength is always longer than that of freely propagating electromagnetic radiation of the same frequency when retardation is taken into account, so direct radiative coupling of photons to plasmon modes of 2DEG's cannot occur since both energy and momentum cannot be simultaneously conserved. For Raman scattering, the photon is not absorbed, and momentum and energy can be simultaneously conserved:

$$\omega_0(k_x) = \omega_L - \omega_R \qquad k_x = k_L \sin\theta_L + k_R \sin\theta_R \approx k_L(\sin\theta_L + \sin\theta_R) \qquad \text{Eqn (4)}$$

where $\omega_0(k_x)$ is the plasmon frequency for wavevector $k_x$, and $\omega_L$, $\omega_R$ and $k_L$, $k_R$ are the frequencies and wavenumbers of the incident laser and Raman scattered photons, making angles $\theta_L$ and $\theta_R$ to the sample normal respectively in the plane of incidence; by changing the scattering geometry, the dispersion of the modes for lateral momentum transfers up to $\sim 2k_L$ can be mapped out[5].

Direct optical coupling can occur, however, by using a coupling structure, typically an overlaid metal grating close to the 2DEG, to match the wavelength and frequency of the plasmon to that of the photon, as first used to couple FIR to 2D plasmons in Si MOS systems[6,7]. A grating of period $d$ generates spatial harmonics with wavevectors $k_x = nG$, where $G = 2\pi/d$, and these couple directly to the plasmon modes; equivalently, in $k$-space, the grating periodicity sets up a reciprocal lattice and the dispersion curve is reduced to the first Brillouin zone, introducing modes with zero reduced wavevectors which couple directly to the radiation (Figure 1).

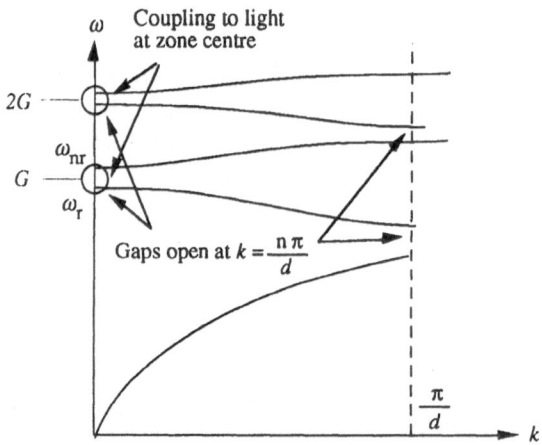

**Figure 1.** Schematic of the 2D plasmon dispersion relation in the reduced zone scheme.

FIR studies thus give information at these discrete wavevectors. Band gaps occur where branches of the dispersion relation would otherwise cross, arising from the different screening by the grating of the two plasmon modes at these points; the two modes are standing waves with oscillating charge density profiles which are either antisymmetric (and hence radiative) or symmetric (and hence non-radiative) about the centre of a grating finger[8]. For the first order plasmon (n = 1), the radiative mode ($\omega_r$) is screened more effectively by the grating than is the non-radiative mode ($\omega_{nr}$), and therefore has the lower frequency. As we shall see, the periodic screening effect of the grating also means that the plasmon frequency shift is no longer linear with drift velocity. The harmonics are evanescent modes and decay away from the grating more rapidly with increasing n; coupling to the n = 4 wavevector plasmon has been reported[9], but here we consider only the first order plasmon.

Grating-enabled FIR emission has been observed[10] and was attributed to radiative decay of plasmons excited by the applied current; the mechanism by which the plasmons were generated was shown to be impurity scattering since increased emission was observed in lower mobility systems. However, if sufficiently high drift velocities can be obtained, other mechanisms for plasmon growth have been calculated in high mobility systems, involving streaming plasma instabilities, leading to plasmon growth and increased radiative emission[11]. Studies of the shift of the plasmon resonance with applied current is an important preliminary in understanding the possible development of plasma instabilities.

## EXPERIMENTAL AND COMPUTATIONAL STUDIES

Here we report Raman and FIR measurements of plasmon shifts for 2DEGs resulting from lateral drift. What is ideally desired is a classically cold carrier velocity distribution which is uniformly shifted in velocity, but this is experimentally unrealisable in the solid state. The samples investigated here were fabricated from MBE-grown, modulation-doped, GaAs/AlGaAs heterostructure wafers; very high mobilities are desirable to minimise heating as the 2DEG is drifted. To obtain uniform current flow when laterally biased, electrically isolated mesas were defined by wet etching, and Au/Ge/Ni ohmic contacts were evaporated and sintered on each end of the mesas to contact the 2DEG; ohmic contacts along the side of the mesa were also included to allow the voltage to be probed when passing drift current through the main contacts. The FIR results are interpreted with the aid of a computational model for the electromagnetic response of a multilayer system including a metallic lamellar grating. Different samples were used for each experimental technique, and the results will be described separately.

### Raman Spectroscopy

The sample used for Raman studies includes a p-type $\delta$-doped layer below the heterojunction; the wavefunctions of the virtual photoholes produced during the Raman process are localised near these acceptors, enhancing the Raman cross section. The optical system consists of a Kr ion laser pumping a dye laser operating at ~1.597eV; the horizontal incident laser beam is focused on the sample in a back-scattering geometry ($\theta_L \approx \theta_R = \theta$) with the scattered light collected by a lens and passed to a Dilor XY triple grating spectrometer with parallel detection electronics and an overall resolution of 2cm$^{-1}$. The sample was mounted in a liquid helium cooled cold finger cryostat and maintained at a constant temperature of 8K. The momentum transfer ($k_x = 2k_L \sin\theta$) can be varied continuously, either parallel or anti-parallel to the plasma drift direction. $N_s$ was measured (under LED illumination) using Shubnikov-de Haas techniques to be $2.47\times10^{11}$cm$^{-2}$, and was also determined, using Eqn (1) and Raman measurements of the zero drift plasmon frequencies for various $k_x$, to be $3.17\times10^{11}$cm$^{-2}$; the latter situation most closely parallels the experimental situation for the drift measurements, and this value for $N_s$ was adopted in subsequent analysis. The lateral drift velocity under particular bias conditions was obtained from the measured drift current and $N_s$, assumed to be unchanged under drift conditions. The electron mobility for the Raman sample was $0.3\times10^6$cm$^2$V$^{-1}$s$^{-1}$.

Figure 2 shows a selection of (Stokes shift) Raman spectra for several drift velocities for one value of $k_x$; there is an upward and a downward shift of the plasmon energy, depending on whether $k_x$ is parallel or anti-parallel to $v_d$. Figure 3 summarises data for three values of $k_x$, and for $k_x = 1.32\times10^5$cm$^{-1}$ with the lateral momentum transfer perpendicular to the drift velocity; the data were obtained under similar illumination conditions in order to minimise any shifts arising from possible light-induced changes in $N_s$.

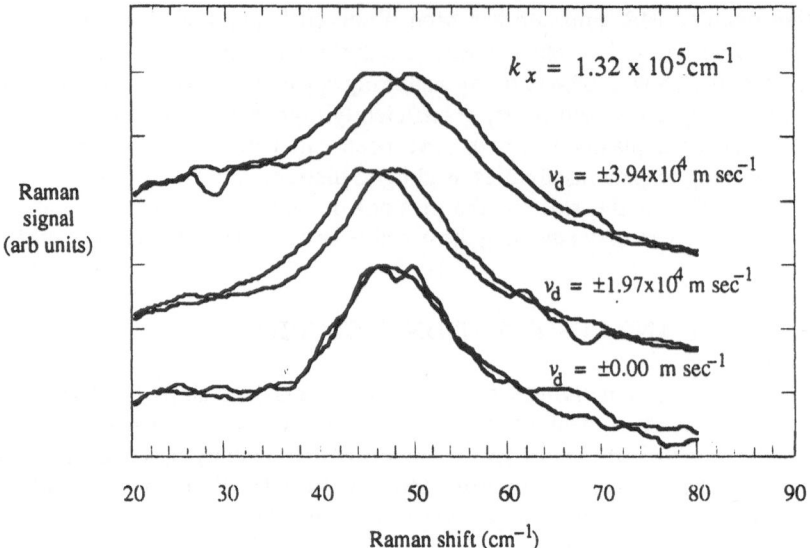

$k_x = 1.32 \times 10^5 \text{cm}^{-1}$

$v_d = \pm 3.94 \times 10^4 \text{ m sec}^{-1}$

$v_d = \pm 1.97 \times 10^4 \text{ m sec}^{-1}$

$v_d = \pm 0.00 \text{ m sec}^{-1}$

Raman shift (cm$^{-1}$)

**Figure 2** Raman spectra showing the upward and downward shifts of the plasmon frequency for several values of lateral drift velocity $v_d$, both parallel and anti-parallel to the lateral momentum transfer $k_x = 1.32 \times 10^5 \text{cm}^{-1}$. The curves have been displaced vertically for clarity.

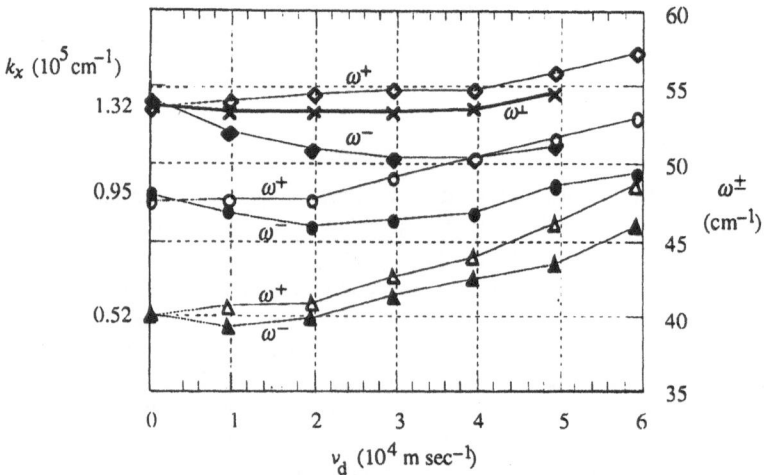

**Figure 3.** The plasmon frequencies $\omega^\pm$ measured by Raman spectroscopy as functions of lateral drift velocity $v_d$, for three values of lateral momentum transfer $k_x$. Also shown as $\omega^\perp$ is the shift for $k_x = 1.32 \times 10^5 \text{cm}^{-1}$ with the drift velocity perpendicular to the momentum transfer. The errors in energy are approximately given by the size of the symbols denoting the experimental points.

For $v_d = 0$, $\omega_0$ increases with $k_x$, according to Eqn (1); and $\omega^\perp$ (for which no Doppler shift is expected) is approximately constant, at least at low $v_d$, indicating that changes in $N_s$ for this sample are not significant. For $v_d \neq 0$, there is a splitting between the up- and down-stream plasmons $\omega^+$ and $\omega^-$, together with a shift in the mean plasmon energy, increasing with $v_d$. The Doppler splitting $(\omega^+ - \omega^-)$ is larger for larger $k_x$, and Figure 4 shows its magnitude as a function of $v_d$ for the three wavevectors $k_x$; superimposed are straight lines corresponding to Eqn (3); quite good agreement is obtained within the margin of error of the experimental splittings. Some deviation from the expected linear dependence might arise if the biasing of the sample did indeed produce some variation in

$N_s$, but such variations would have to be ~5% to account for the maximum deviations observed; moreover, under similar drift conditions, the deviations are also in opposite directions for $k_x = 0.52 \times 10^5$ and $k_x = 1.32 \times 10^5 \mathrm{cm}^{-1}$, so it appears more likely that they arise from errors in the determination of the precise plasmon peak positions. The shifts in the mean plasmon energy with drift observed in Figure 3 are not as yet understood.

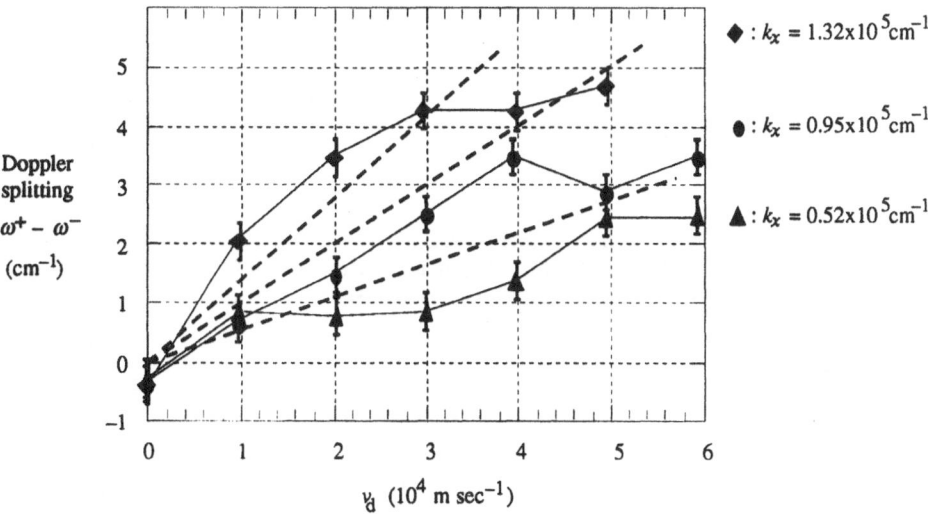

**Figure 4**. The Doppler splitting ($\omega^+ - \omega^-$) of the plasmon frequencies measured by Raman spectroscopy as functions of lateral drift velocity $v_d$, for three values of lateral momentum transfer $k_x$. The dotted lines show the splitting expected from Eqn (3).

## Far Infra-red Spectroscopy

Experiments have also been performed on GaAs/AlGaAs heterostructures (without p-type δ-doping) with superimposed aluminium gratings and with the drift current passed through the 2DEG perpendicular to the grating fingers. The gratings were fabricated by electron beam lithography and lift-off, and the period and mark fraction were measured using a scanning electron microscope. The number density and mobility of the 2DEG were $6.6 \times 10^{11} \mathrm{cm}^{-2}$ and $0.5 \times 10^6 \mathrm{cm}^2 \mathrm{V}^{-1}\mathrm{s}^{-1}$ respectively, both determined by Hall measurements, so the 2DEG channel had a resistance of about 25 Ω; the access resistance of the ohmic contacts was about 10 Ω. The area of the gratings were large, 2mm × 2mm, to obtain sufficient optical signal through the samples. The electric fields of the order of 500V/m resulted in drift currents of some tens of mA. The spectroscopy was performed on Bruker 113v Fourier transform spectrometer using a liquid helium cooled silicon bolometric detector. The samples were cooled, usually to 10K, in a superconducting magnet with optical access. Spectra were taken at a series of drift currents and ratioed with a reference spectrum to produce relative transmission spectra, eliminating instrumental and drift-independent features in the absolute spectra; the reference spectrum was taken with a magnetic field of a few Tesla to shift its (magneto-) plasmon to higher energy, away from that of the drifting plasmons, to achieve less complicated relative transmission spectra. The spectra were then fitted with a Lorentzian line-shape to obtain the plasmon peak energy. The spectra show a downward shift in plasmon frequency with increased drift current as shown in Figure 5; Figure 6 summarises data for two samples with grating periods 1.0μm and 0.75μm.

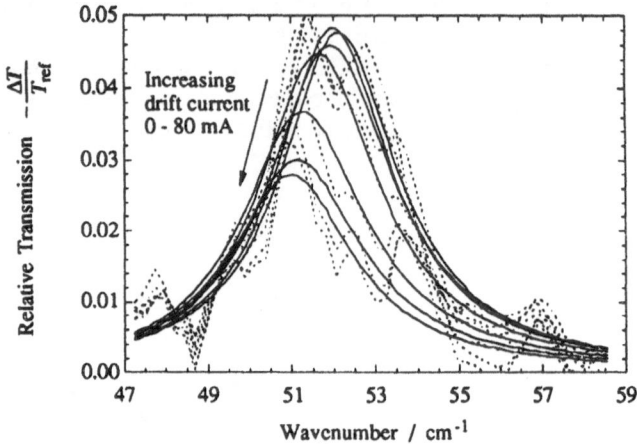

**Figure 5.** FIR relative transmission spectra (ratioed to a reference spectrum taken in a strong magnetic field) for a 2DEG plasmon under a grating of period 0.75μm, for various lateral drift currents. The solid lines are smoothed versions of the original data shown as background dots.

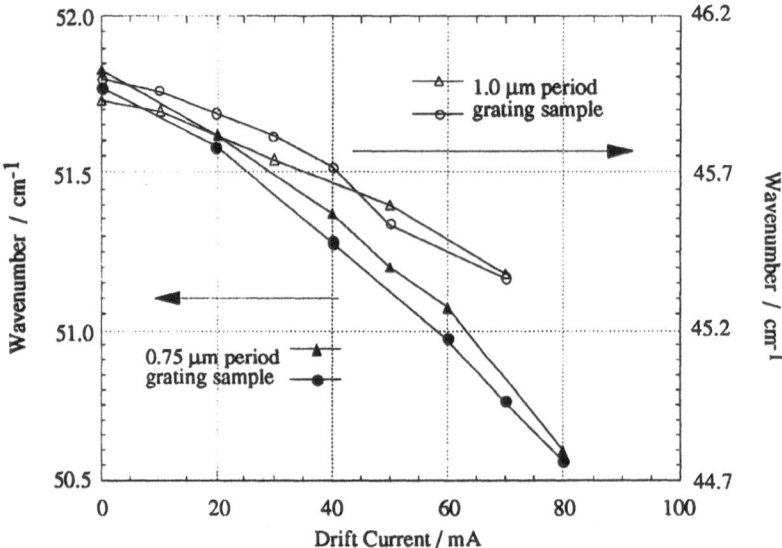

**Figure 6.** Plasmon resonance position plotted as a function of drift current for two experimental runs on two samples with grating periods 1.0μm and 0.75μm.

The shift is not linear with drift current, and may involve a decrease in $N_s$ caused by scattering at high currents; the shift is, however, greater for the smaller grating period, whereas one would expect any electrical change in $N_s$ to be independent of the overlaid grating.

## Computational Studies

The scattering matrix method[12] has been used to model the experimentally studied systems, with the drift velocity in the $x$-direction incorporated phenomenologically as a modification of conductivity tensor $\underline{\underline{\sigma}}$ of the 2DEG plasma sheet[13]:

$$\hat{\underline{\sigma}} = \sigma_0 \begin{pmatrix} \dfrac{i\omega}{\tau(\omega - k_x v_d)(\omega - k_x v_d + \frac{i}{\tau})} & 0 & 0 \\[3ex] 0 & \dfrac{i}{\tau(\omega + \frac{i}{\tau})} & 0 \\[3ex] 0 & 0 & 0 \end{pmatrix} \qquad \text{Eqn (5)}$$

where $\sigma_0$ ($= N_s e^2 \tau/m^*$) is the d.c. conductivity and $\tau$ is an appropriate scattering time. The expected shift of the plasmon with drift current for a range of values of $N_s$ are shown in Figure 7, together with the experimental data; it is apparent that the experimental shift in not consistent with the calculations for any fixed value of $N_s$, and we assume therefore that the experimental shift has contributions from two factors; the drift velocity induced shift, and a decrease in $N_s$ with drift current.

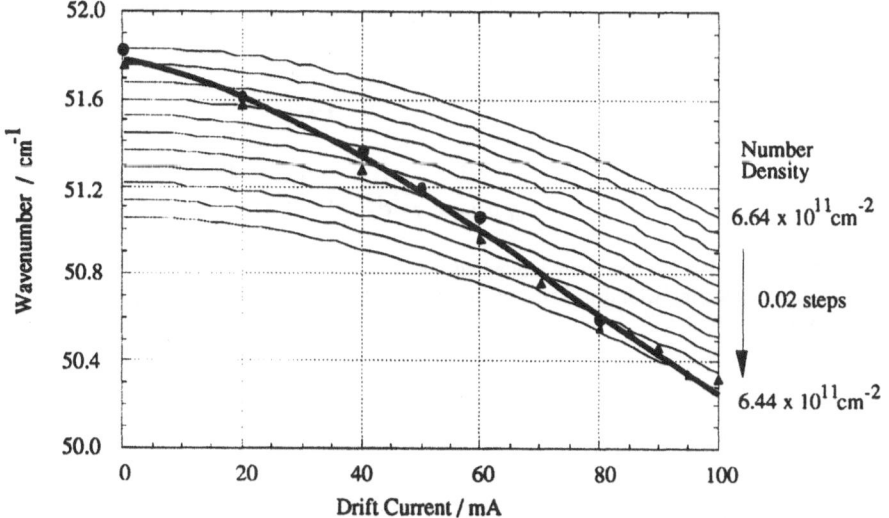

**Figure 7.** Scattering matrix calculation of the drifting plasmon frequency for a 0.75μm period grating and a range of 2DEG number densities (thin lines). Experimental results for a 0.75μm period grating sample are shown as points; the two symbols are for two experimental runs and the bold line is a guide for the eye.

From Figure 7 it is possible to deduce $N_s$ as a function of drift current from the intersections of the calculated and experimental curves, and the change at the largest current passed is about −3%; the decrease in $N_s$ is approximately linear with drift current, and is very similar for the two grating period samples, confirming the supposition that there is an electrically induced reduction in $N_s$ independent of other factors. Having deduced $N_s$ for a given drift current, the square root dependence of the plasmon frequency on $N_s$ allows the experimental value to be corrected. The current can also be converted to the carrier drift velocity since $J = N_s e v_d$. Figure 8 then shows the corrected plasmon frequency plotted as a function of drift velocity for the 0.75μm grating sample, together with the linear Doppler shift expected without a grating. Agreement between experiment and theory is good, and Figure 8 highlights the principal effects of the presence of the grating on the plasmon mode of the drifting 2DEG; the overall frequency is significantly reduced because

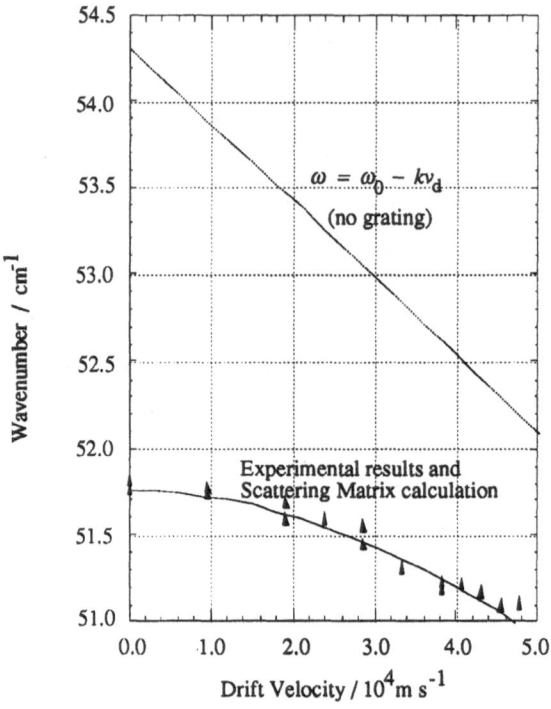

**Figure 8.** Corrected plasmon frequency plotted against drift velocity for a 0.75µm period grating sample, showing frequency lowering with drift velocity compared to the theoretical linear Doppler shift for a system without a grating.

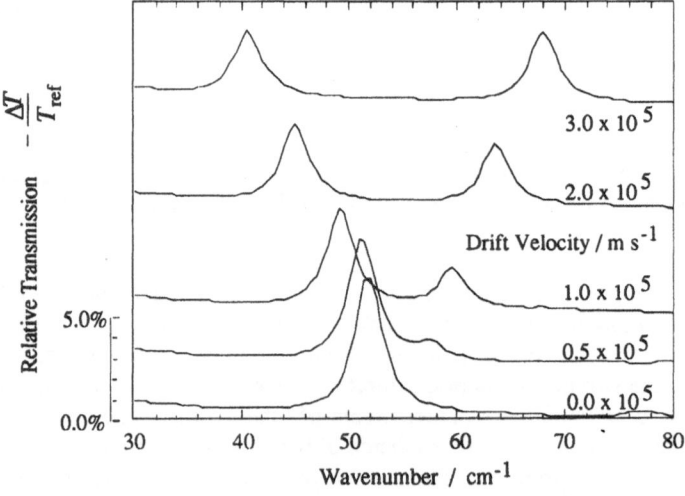

**Figure 9.** Scattering matrix calculations of relative transmission spectra for high drift velocities showing the upper mode coupling strength growing and the lower mode coupling strength decreasing with increasing drift velocity.

of the screening effect of the grating, and the drift induced shift is not a simple linear Doppler shift at low velocities.

To obtain a larger shift, one could use smaller period gratings to increase $k_x$, but the diffracted grating modes are evanescent, and the coupling strength thus falls as $e^{-Gh}$ where $h$ is the grating-2DEG separation; higher drift velocities could also increase the effect, but would cause problems with sample heating, though it may be possible to overcome this using a pulsed technique. However, scattering matrix calculations of the plasmon behaviour for the $0.75\mu m$ period grating sample have been made at higher values of $v_d$, and the resulting spectra are shown in Figure 9. The drift breaks the symmetry of the two modes at a band gap: the upper mode, which is non-radiative for $v_d = 0$ (for a symmetric grating profile), gradually increases in coupling strength, and shifts up in energy, as $v_d$ increases, while the lower mode decreases in strength and shifts downward. (Other researchers have reported observing both upper and lower plasmon modes at zero drift velocity[14,15] which implies that no centre of symmetry was present in the gratings used.) The upper and lower modes tend asymptotically to the linear Doppler shift, their separation being proportional to the drift velocity (Figure 10). The effect of asymmetric grating couplers can be quantified further, and investigations are in progress.

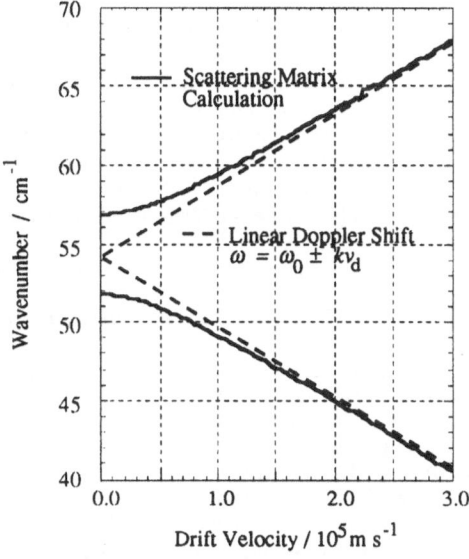

**Figure 10.** Calculated upper and lower plasmon mode frequencies as functions of increasing drift velocity. The calculated shifts are asymptotic to the linear Doppler behaviour at high drift velocities.

## CONCLUSIONS

For drifting 2DEGs without grating couplers, the drift induced shift is approximately linear as expected from simple phenomenological theory. The presence of a grating, required for direct coupling to light, substantially modifies the plasmon behaviour; there is an overall reduction in the frequency of the radiative plasmon mode at the grating wavevector, and its shift with drift velocity is not linear because of the periodic screening of the grating. The expected splitting of the plasmon in the presence of the grating is not observed at low drift velocities because the higher frequency mode is not coupled to radiation by a symmetric grating profile. There is good agreement between experimental FIR observations and scattering matrix calculations of the optical response of the drifting 2DEG under a grating, and comparison of the results allows us to infer that there is probably a small reduction in the 2DEG number density under drift conditions.

## ACKNOWLEDGEMENTS

The authors wish to express their gratitude to Dr J E F Frost, Dr D A Ritchie and Dr G A C Jones for supplying semiconductor wafers, to Dr C Shearwood and Dr D G Hasko for fabricating the grating structures, and to the U.K. Science and Engineering Research Council and the U.K. Ministry of Defence for providing funding for this project.

## REFERENCES

1.  F. Stern, Polarizability of a two-dimensional electron gas, *Phys. Rev. Lett.* 18:546 (1967).
2.  e.g. J. Cen, K. Kempa and P. Bakshi, Amplification of a new surface plasma mode in the type-I semiconductor superlattice, *Phys. Rev.* B38:10051 (1988).
3.  R. Gupta and B. K. Ridley, Two stream instability in two dimensional degenerate systems, *Phys. Rev.* B39:6208 (1989).
4.  S. J. Allen, F. Derosa and R. Bhat, Standing charge density waves driven by electron drift in patterned (Al, Ga)As/GaAs heterostructures, *Physica* 134B:332 (1985).
5.  G. Fasol, H. P. Hughes and K. Ploog, Resonance Raman scattering by intrasubband excitations of GaAs multi-quantum-well structures, *Surface Science* 170:497 (1986)
6.  S. J. Allen, D. C. Tsui and F. Derosa, Frequency dependence of the electron conductivity in the silicon inversion layer in the metallic and localised regimes, *Phys. Rev. Lett.* 35:1359 (1975).
7.  T. N. Theis, Plasmons in inversion layers, *Surface Science* 98:515 (1980).
8.  C. D. Ager, R. J. Wilkinson and H. P. Hughes, Periodic grating gate screening of plasmons in heterojunction structures, *J. Appl. Phys.* 71:1322 (1992).
9.  D. Heitmann and U. Mackens, Extended and local plasmons in a lateral superlattice, *Superlattices and Microstructures* 4:503 (1988).
10. R. A. Höpfel and E. Gornick, Two dimensional plasmons and far-infrared emission, *Surface Science* 142:412 (1984).
11. A. V. Chaplik, Collisionless absorption and possible far-infrared amplification in microstructured two dimensional systems, *Solid State Comm.* 65:1589 (1988).
12. C. D. Ager and H. P. Hughes, Scattering matrix modelling of the optical properties of multi-layered systems including lamellar gratings, *Phys. Rev.* B44:13452 (1991).
13  R J Wilkinson, Ph.D. Thesis, Cambridge University, 1991.
14. E. Batke, D. Heitmann and C. W. Tu, Plasmon and magnetoplasmon excitation in two dimensional electron space charge layers on GaAs, *Phys. Rev.* B34:6951 (1986).
15. D. Heitmann, Two dimensional plasmons and laterally microtructured space charge layers, *Surface Science* 170:332 (1986).

# TEMPERATURE-DEPENDENT SCREENING CALCULATION OF HOT-ELECTRON SCATTERING IN HEAVILY DOPED SEMICONDUCTORS

Kjeld O. Jensen,[1,2] J.M Rorison[1], and Alison B. Walker[3]

[1]Sharp Laboratories of Europe Ltd.
Oxford Science Park, Oxford OX4 4GA, United Kingdom

[2]Department of Physics, University of Essex
Colchester CO4 3SQ, United Kingdom
(present address)

[3]School of Physics, University of East Anglia
Norwich NR4 7TJ, United Kingdom

## INTRODUCTION

The advances in semiconductor growth techniques are leading to an ever increasing range of structures and devices displaying new electronic and optical properties. Many of these new properties are the direct result of hot electrons being generated and moving through the structures,[1] which implies that a detailed understanding of the scattering mechanisms affecting the transport of hot carriers is essential in understanding the new semiconductor physics.

In heavily doped semiconductors or semiconductors subject to intense photo-excitation carrier-carrier interactions (electron-electron, hole-hole and/or electron-hole) become very important. In the analysis of transport in hot-electron transistors/spectrometers[2] and hetero-junction bipolar transistors[3] scattering by carriers and polar longitudinal-optical phonons are found to be the dominant scattering mechanism for hot carriers. It is therefore necessary to have a good knowledge of the details of the carrier-carrier and carrier-phonon interactions and their interplay.

The analysis of the interactions between a carrier and a medium involves the study of the frequency ($\omega$) and wave vector ($q$) dependent dynamical screening function

*Negative Differential Resistance and Instabilities in 2-D Semiconductors,*
Edited by N. Balkan *et al.*, Plenum Press, New York, 1993

$\epsilon(\omega, q)$. For heavily doped semiconductors and metals the Random Phase Approximation (RPA) provides an accurate description[4,5]. Most of the various treatments of carrier-carrier interaction involve this approximation. For high carrier density systems like metals the zero-temperature RPA works well with $\epsilon(\omega, q)$ being the Lindhard dielectric function.[6,7] In this regime it is possible to make a clear distinction between collective (plasmon) and single-particle excitations and one can with reasonable accuracy divide the carrier-carrier interaction into collective and single-particle parts through the introduction of a cut-off wave vector[8,9]. However, for the lower carrier densities and lower excitation energies found in doped or optically excited semiconductors temperature effects are significantly more important than for metals. A result of this is that the collective and single particle excitation modes become intricately coupled making separation of these modes very difficult.

The dynamical screening function can be used directly to evaluate carrier-carrier scattering and also describes how carrier-lattice scattering (including carrier-phonon and carrier-impurity scattering) is modified by screening. In polar semiconductors mutual screening of lattice and electronic modes leads to coupled plasmon-LO phonon modes which within the RPA formalism can be described in terms of a coupling between the carrier and lattice screening functions[10-12].

In this paper we study three different approximations to the full RPA expression for the dynamical screening function in detail: (i) the Lindhard zero-temperature RPA [6,7] which should be valid at low temperatures and high carrier densities, (ii) a temperature-dependent two-pole approximation incorporating non-degenerate statistics[13,14] which should be valid at low densities and a wide range of temperatures, and (iii) the plasmon-pole approximation[3]. Our primary goal is to establish the ranges of validity of the different models, in particular to establish the importance of temperature effects and to determine the density range above which the degeneracy of the sea of carriers becomes significant. We present results for differential scattering rates and mean-free-paths as functions of doping level, electron energy and temperature for a hot electron traversing a heavily n-doped region in GaAs. Since this is a polar semiconductor we also consider the coupling of electrons to the LO phonons and present results for this system within the two-pole approximation.

A fuller account of this work will be published elsewhere.[15]

## THEORY

### Formalism

In a homogeneous medium at temperature $T$ with dielectric function $\epsilon(q, \omega; T)$ the doubly differential scattering rate for a particle with charge $Ze$, mass $m^*$, momentum $\hbar k$ and energy $E = \hbar^2 k^2/(2m^*)$ derived from linear response theory is[5,16]

$$\frac{d^2\Gamma(\omega, q, E)}{d\omega\, dq} = \frac{2m^* Z^2 e^2}{\hbar^2 (4\pi\epsilon_0)} \frac{1}{\pi k q} \text{Im}\left(\frac{-\epsilon_0}{\epsilon(q, \omega; T)}\right)$$
$$\times \frac{1}{(\exp(\hbar\omega/k_B T) - 1)} \Theta(q - q_-(\omega, E))\Theta(q_+(\omega, E) - q)$$
$$\times [1 - f(E - \hbar\omega - \mu)] \quad . \tag{1}$$

where $q$ and $\omega$ represent the momentum and energy transfer in the scattering process, respectively, and $\epsilon_0$ is the vacuum permittivity. The factor $1 - f$, where $f$ is the Fermi-function, takes into account the availability of final states for the scattered particle, assumed to be a fermion with chemical potential $\mu$. The wave vectors $q_-$ and $q_+$ are the minimum and maximum allowed momentum transfers determined by energy and momentum conservation for given values of $k$ and $\omega$:

$$q_- = \begin{cases} k - \sqrt{k^2 + 2m^*\omega/\hbar} & , \quad \omega < 0 \\ -k + \sqrt{k^2 + 2m^*\omega/\hbar} & , \quad \omega > 0 \end{cases} \quad q_+ = k + \sqrt{k^2 + 2m^*\omega/\hbar} \ . \tag{2}$$

Equation (1) shows that the scattering rates can be obtained directly if the dynamical response function $\epsilon(q, \omega; T)$ is known. Below we describe three different approximations to this response function.

## Model dielectric functions

If the sea of carriers is assumed to be a free electron gas present in a medium with a background dielectric constant $\epsilon_0^*$ and we set $T = 0$, $\epsilon(q, \omega) \equiv \epsilon(q, \omega; T = 0)$ corresponds to the well-known *Lindhard dielectric function*.[6,7] Two distinct contributions to the scattering rates arise from the Lindhard dielectric function: single particle and plasmon excitations. The dominant excitation at large $q$ is single particle excitation and at small $q$ plasmon excitation.

In the *plasmon-pole approximation* the full range of excitations is replaced by a single mode. The dielectric function is written as[3]

$$\frac{\epsilon(q, \omega)}{\epsilon_0} = 1 - \frac{\omega_{pl}^2}{(\omega^2 - (E_q/\hbar)^2)} \tag{3}$$

where the plasmon frequency $\omega_{pl}$ is defined by $\omega_{pl}^2 = 4\pi n e^2/(m^*\epsilon_0^*)$ with $n$ being the electron density and $E_q = \hbar^2 q^2/(2m^*)$. Thus the dispersion of the mode, which is given by the roots of the equation $\epsilon(q, \omega) = 0$, is $\omega(q) = \pm\sqrt{\omega_{pl}^2 + (E_q/\hbar)^2}$ where the plus and minus signs represent absorption and emission, respectively. The plasmon-pole approximation, as described by Eq.(3), corresponds to the zero-temperature limit of the RPA for a non-degenerate electron gas.

The *two-pole approximation*[13,17] to the temperature dependent RPA provides a more detailed description of the excitations than the single plasmon-pole approximation since it takes into account the temperature dependence of the dielectric function. The dielectric function for a non-degenerate electron gas, i.e. a gas obeying Maxwell-Boltzman statistics, can be written as

$$\frac{\epsilon(q, \omega)}{\epsilon_0} = 1 + \left(\frac{k_D}{q}\right)^2 \left[\frac{Z(\xi - \frac{a}{2}) - Z(\xi + \frac{a}{2})}{2a}\right] \tag{4}$$

with $\xi = \left[\sqrt{m^*/(2k_B T)}\right]\omega/q$, $a = \hbar q/\sqrt{2m^* k_B T}$, $k_D^2 = 4\pi n e^2/(\epsilon_0^* k_B T)$ and the function $Z$ is the plasma dispersion function.[18] In the two-pole approximation Z is approximated by[17]

$$Z(s + i\delta) = \frac{i\sqrt{\pi} + (\pi - 2)s}{1 - i\sqrt{\pi} - (\pi - 2)s^2} \ . \tag{5}$$

Inserting Eq.(5) in Eq.(4) results in a dielectric function which is the ratio of two fourth order polynomials in $\omega$ with coefficients which depend on $q$. The four (complex) roots of $\epsilon$ for a given $q$ correspond to the emission and absorption of two branches of excitation modes. Although no formal distinction can be made between different types of excitations, the lowest energy mode can be associated with single particle excitations while the highest energy mode accounts for plasmon excitation at low $q$ and approaches the single particle branch at high $q$.

A particular strength of the two-pole approximation is the ease with which the coupling of the electron system with other excitation modes can be incorporated. As an illustration of this we here present results for scattering off coupled electron-phonon modes. The dielectric function for the coupled system can be written:[14,19]

$$\epsilon(q,\omega) = \epsilon_e(q,\omega) + \epsilon_{ph}(\omega) - \epsilon_0 \qquad (6)$$

where $\epsilon_e$ is the dielectric function for the electronic system within the two-pole approximation as described above and $\epsilon_{ph}$ is the dielectric function due to screening by optical phonons:

$$\epsilon_{ph} = \epsilon_\infty^* + \frac{\epsilon_0^* - \epsilon_\infty^*}{1 - \omega^2/\omega_{TO}^2} \qquad (7)$$

where $\omega_{TO}$ is the frequency of transverse optical phonons and $\epsilon_\infty^*$ is the high frequency dielectric constant.

## Calculated quantities

In the next section we present a number of results derived from the model dielectric functions described above. From the doubly differential scattering rate, defined by (1), we have evaluated the (singly) differential scattering rate:

$$\frac{d\Gamma(\omega, E)}{d\omega} = \int_0^\infty dq \frac{d^2\Gamma(q, \omega, E)}{dq d\omega} \qquad . \qquad (8)$$

The total scattering rate is found as:

$$\Gamma(E) = \int_{-\infty}^\infty d\omega \int_0^\infty dq \frac{d^2\Gamma(q, \omega, E)}{dq d\omega} \qquad , \qquad (9)$$

which is related to the mean-free-path $\lambda$ as $\lambda(E) = (\hbar k/m^*)/\Gamma(E)$ .

## RESULTS

All results presented in this section correspond to $n$-doped GaAs. We have used the following parameters: $m^*=0.067m_0$, where $m_0$ is the free electron mass, $\epsilon_0^*=12.53\epsilon_0$, $\epsilon_\infty^*=10.9\epsilon_0$, and $\hbar\omega_{TO}=33.6$ meV.

The mean-free-path for hot electrons in $n$-doped GaAs at low (zero or near-zero) temperature as function of the primary electron energy is shown in Fig. 1. The figure gives results for all three models for the dielectric response and for two different doping levels. The rapid decrease in $\lambda$ at low energy for $n = 10^{18}$cm$-3$ signifies the on-set of plasmon emission (a similar behaviour occurs for $n = 10^{17}$cm$-3$ below the energy region for which data is shown in the figure). It is apparent from Fig. 1 that the

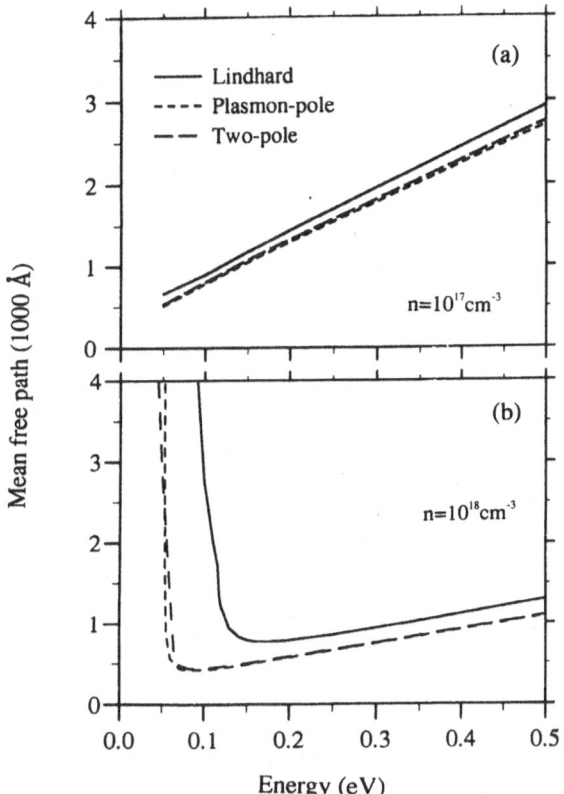

**Figure 1.** The mean-free-path $\lambda$ due to carrier-carrier scattering as a function of hot electron energy in $n$-doped GaAs for doping levels $10^{17}$ and $10^{18}$ cm$^{-3}$. The long-dashed lines show the results from the two-pole approximation at temperature 10 K, the short-dashed lines the results from the plasmon-pole approximation at zero temperature and the full lines the results from the zero temperature RPA.

two-pole approximation and the full zero-temperature RPA (Lindhard function) agree well when the primary energy of the hot carrier is much greater than the Fermi energy but the neglect of degeneracy in the two-pole approximation makes it inaccurate for energies comparable to the Fermi energy. For a typical hot electron energy, *eg.* in the base of a hot-electron transistor,[2,21] of 0.25 eV, this means that for doping densities near $10^{18}$ cm$^{-3}$ and above the two-pole approximation becomes inaccurate while for lower densities the neglect of degeneracy is justified.

Fig. 2 shows the temperature dependence of the mean-free-paths calculated within the two-pole and plasmon-pole approximations. It is clear from Fig. 2 that the temperature-dependence of the scattering is substantial for $n$ less than $10^{18}$cm$^{-3}$. It is thus important to incorporate the temperature effects when modelling hot electron transport at non-zero temperatures even for moderately high doping densities. In contrast, for $n = 10^{18}$ and above, the full zero temperature RPA describes the scattering even at non-zero temperatures reasonably accurately.

Figure 1 demonstrates that the two-pole and plasmon-pole approximations agree well at low temperatures. However, the temperature dependence of the scattering

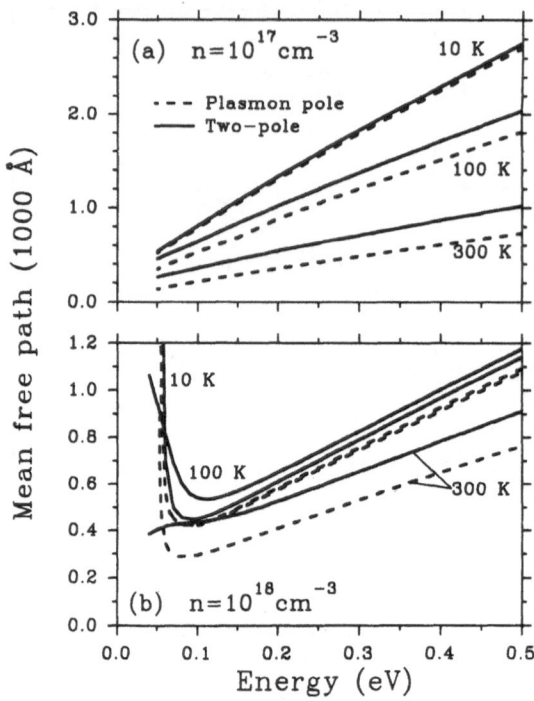

**Figure 2.** The mean-free paths of an electron in $n$-doped GaAs at three different temperatures at doping levels $10^{17}$ and $10^{18}$ cm$^{-3}$. Full and dashed curves were obtained with the two-pole and plasmon-pole approximations, respectively.

in the plasmon pole approximation is much stronger than in the two pole approximation as is demonstrated by Fig. 2. This is to be expected since the only temperature dependence included in the plasmon pole approximation is the Bose factor $[1 - exp(\hbar\omega/kT)]^{-1}$ in Eq.(1) while the temperature dependence of the screening, ie. $\epsilon(q,\omega)$, is ignored, unlike in the two-pole approximation.

A more detailed comparison between the two-pole and Lindhard approximations is provided by Fig.3 which shows the differential scattering rate for $n = 10^{17}$ cm$^{-3}$ and $E$=0.3 eV at low temperature. The most prominent feature is the plasmon emission which is represented by a sharp peak around -15 meV in all three models. However, the total contributions from plasmons and single particle scattering in the zero-temperature RPA are comparable in magnitude. The two approximations are seen to agree well at all $\omega$, including the regions where the scattering is dominated by single particle excitations but only the two-pole approximation describes the broadening of the on-set of the plasmon excitations due to plasmon damping.

The temperature dependence of the differential scattering rate within the two-pole approximation is given in Fig.4 for $n$=10$^{17}$ cm$^{-3}$. The figure shows that as the temperature is raised, absorption processes become increasingly important, the plasmon emission and absorption lines are broadened, and the scattering off low-energy single-particle excitations increases. The latter two features are not included in the simpler plasmon-pole approximation.

Finally we show in Fig. 5 mean-free-paths for scattering off a coupled system of

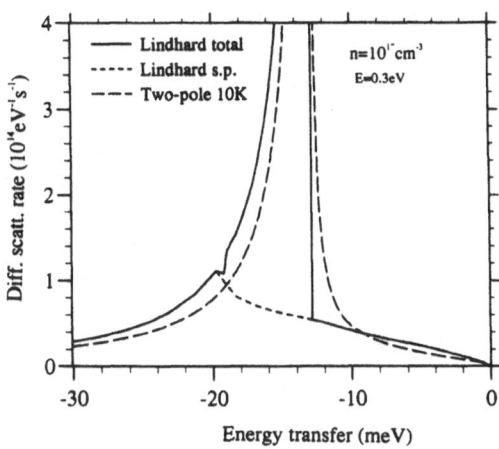

**Figure 3.** The differential scattering rate as function of energy transfer $\hbar\omega$ for $n=10^{17}$ cm$^{-3}$ and $E=0.3$ eV. The full curve shows the two-pole approximation result for temperature 10 K and the dot-dashed curve the zero temperature RPA (Lindhard) result. The contribution from single particle excitations within the zero temperature RPA is shown separately as the short-dashed line. The plasmon peak near -15 meV reaches in both cases values of $d\Gamma/d\omega$ above $50 \times 10^{14} eV^{-1}s^{-1}$.

phonons and electrons within the two-pole approximation for $n=5\times10^{17}$ cm$^{-3}$ where the coupling between optical phonons and plasmons is strong. The figure demonstrates the importance of the temperature dependence also when the electron system is coupled to phonons.

## DISCUSSION

An important conclusion which can be drawn from the results is that the zero-temperature RPA (Lindhard dielectric function) and the two-pole approximation together provide an adequate description of carrier- carrier scattering over a wide range of doping densities and temperatures. At high densities the degeneracy of the electron system is important but temperature effects are small, which means that the zero temperature RPA provide an accurate description at these densities. At low densities temperature effects are more pronounced but the use of Maxwell-Boltzmann statistics, *ie* assuming a non-degenerate electron gas, is justified for typical hot electron energies in devices.

In $n$-type GaAs the cross-over between the validity of the two models lies for typical hot-electron energies around $10^{18}$ cm$^{-3}$. This cross-over density will depend on the effective mass of the carriers. For higher masses the cross-over density will be higher and the two-pole approximation will be valid at substantially higher densities in Si hot electron transistors and in $p$-doped bases of hetero-junction bipolar transistors[2] where the hole masses are larger than the electron masses in GaAs.

Several authors have compared low-temperature results for hot-electron spectrometers or experiments on hot electrons in optically excited semiconductors with theories based on the Lindhard dielectric function coupled to LO phonons.[20–22] In all cases the agreement was found to be good. Unfortunately, we are not aware of experimental results taken at higher temperatures which would allow direct comparison to the present results and confirm the importance of the temperature-dependent screening as predicted by the two-pole approximation. However, for hot-carrier devices operating at room temperature these temperature effects would be important and should be included in modelling of these devices, e.g. by Monte Carlo simulation.

In addition to the results presented in the figures we performed calculations at the same electron densities as Hu and Das Sarma[23] who used the temperature-dependent degenerate RPA to calculate scattering rates for hot electrons in $n$-GaAs. We found,

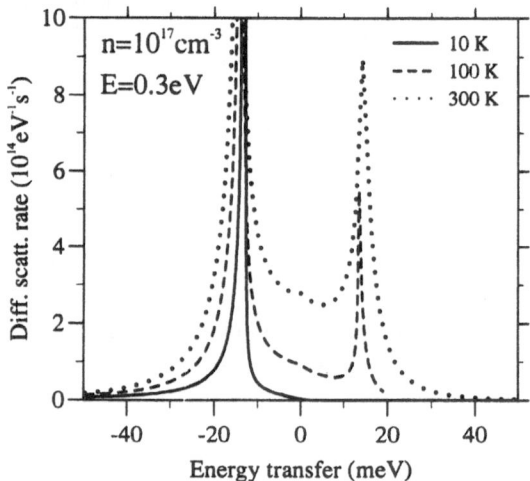

**Figure 4.** The differential scattering rate as function of energy transfer $\hbar\omega$ for $n=10^{17}$ cm$^{-3}$, $E=0.3$ eV and three different temperatures as indicated in the figure. All results were obtained with the two-pole approximation.

as expected, that the two calculations agree well (within 10%) when the hot-electron energy is substantially larger than the Fermi energy (in practice larger than 5 times $E_F$).

We have in this paper restricted ourselves to discussing the scattering of a single hot electron off a sea of conduction electrons in thermal equilibrium appropriate for hot-electron transistors. However, the present models, in particular the two-pole approximation, should also be valuable in modelling the contribution to the screening from electrons and holes in optically excited semiconductors, although in these cases complications arise because carriers are scattered by both electrons and holes in electron-hole plasmas [24,25,26] and because the scatterers (electrons, holes, phonons)

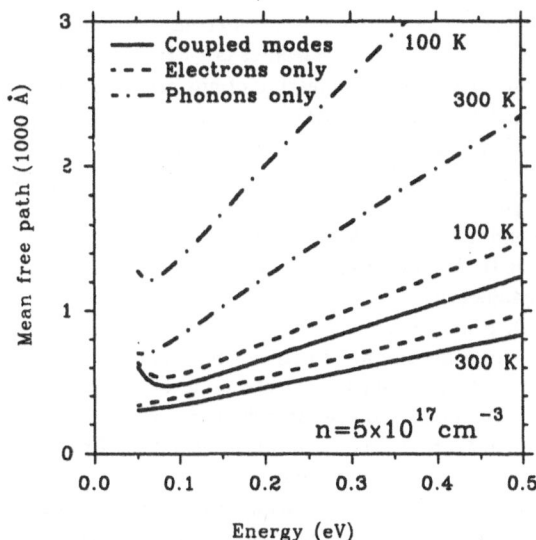

**Figure 5.** The mean-free-path for hot electrons in a system of coupled phonons and electrons with density $n=5\times10^{17}$ cm$^{-3}$ at temperatures 100 and 300 K calculated using the two-pole approximation for the electronic part of the dielectric function (full lines). The mean-free paths for systems of phonons and electrons only are shown as the dot-dashed and dashed lines, respectively, the latter calculated with the two-pole approximation.

are not necessarily in thermal equilibrium.[27,28] The most immediate extensions of our theory will be the calculation of scattering of carriers by coupled light and heavy hole plasmas including coupling with LO phonons and generalisation of the present 3-D theory to two-dimensional systems.[29]

## CONCLUSION

We have discussed a number of different descriptions of the scattering of hot carriers off high densities of carriers in semiconductors and examined the range of validity for each model. In particular, we have shown that the two-pole approximation to the temperature dependent RPA[13,14] is expected to be valid for the typical ranges of doping densities, temperatures and hot electron energies.

We have also emphasised the importance of including the temperature dependence of the dielectric response and, in particular, shown that the two-pole approximation provides a convenient way of including this effect in practical calculations.

## ACKNOWLEDGEMENTS

We would like to thank David Herbert (Defence Research Agency, Malvern) and David Neilson (University of New South Wales) for useful discussions and Robert Smith (Sharp Laboratories of Europe Ltd.) for computational assistance.

1.  B.K. Ridley, Hot electrons in low-dimensional structures, *Rep. Prog. Phys.* 54:169 (1991).

2. J.R. Hayes and A.F.J. Levi, Dynamics of extreme nonequilibrium electron transport in GaAs, *IEEE J. Quantum Electronics* QE-22:1744 (1986).

3. D.C Herbert, Band structure engineering of hot-carrier transport in SiGe heterojunction bipolar transistors, *Semicond.Sci.Technol.* 7:44 (1992).

4. G.D. Mahan, "Many-Particle Physics", Plenum, New York, (1981).

5. A.L. Fetter and J.D. Walecka, "Quantum Theory of Many-particle Systems", McGraw-Hill, New York, (1971).

6. J. Lindhard, *Kong. Dansk. Vidensk. Selsk. Mat.-Fys. Medd.* 28:1 (1954).

7. C.J. Tung and R.H. Ritchie, Electron slowing-down spectra in aluminium metal, *Phys. Rev. B* 16:4302 (1977).

8. D. Bohm and D. Pines, A collective description of electron interactions: III. Coulomb interactions in a degenerate electron gas, *Phys. Rev.* 92:609 (1953).

9. P. Lugli and D.K. Ferry, Effect of electron-electron scattering on monte-carlo studies of transport in sub-micron semiconductor-devices *Physica B* 117:251 (1983); Effect of electron-electron and electron-plasmon interactions on hot carrier transport in semiconductors *ibid* 129:532 (1985).

10. S.E. Kumekov and V.I. Perel', Relaxation of fast electrons in semiconductors at coupled plasmon-phonon oscillations, *Sov.Phys.Semicond.* 16:1291 (1983).

11. M.E. Kim, A. Das, and S.D. Senturia, Electron scattering interaction with coupled plasmon–polar-phonon modes in degenerate semiconductors, *Phys.Rev.B* 18:6890 (1978).

12. D. Neilson, J. Szymanski, and De-Xin Lu, Energy loss mechanism for hot electrons in GaAs, *J.Physique Coll.* 48:C5-263 (1987).

13. J.M. Rorison and D.C. Herbert, Electron-electron interaction in semiconductors with application to hot-electron transistors in silicon and gallium arsenide, *J. Phys. C* 19:3991 (1986).

14. J.M. Rorison and D.C. Herbert, Electorn-electron interaction in doped semiconductors including polar LO phonon effects: application to hot-electron spectrometers in GaAs, *J. Phys. C* 19:6357 (1986).

15. K.O. Jensen, J.M. Rorison and A.B. Walker, Temperature-dependent screening and carrier-carrier scattering in heavily doped semiconductors, submitted to *Phys.Rev. B*.

16. P.M. Platzman and P.A. Wolff, "Waves and Interactions in Solid State Plasmas" (Solid State Physics, Supplement 13), Academic Press, New York (1973).

17. D. Lowe and J.R. Barker, A model study of field-dependent dynamical screening due to mobile electrons in submicron semiconductor devices, *J. Phys. C* 18:2507 (1985).

18. B.D. Fried and S.D. Conte, "The Plasma Dispersion Function", Academic Press, London, (1961).

19. B.B. Varga, Coupling of plasmons to polar phonons in degenerate semiconductors, *Phys. Rev.* 137:A1896 (1965).

20. A.F.J. Levi, J.R. Hayes, P.M. Platzman and W. Weigmann, Injected-hot-electron transport in GaAs, *Phys.Rev.Lett.* 55:2071 (1985).

21. A.P. Long, P.H. Beton, and M.J. Kelly, Hot electron transport in heavily doped GaAs, *Semicond. Sci.Technol.* 1:63 (1986).

22. C.L. Peterson and S.A. Lyon, Observation of hot electron energy loss through the emission of phonon-plasmon coupled modes in GaAs, *Phys.Rev.Lett.* 65:760 (1990).

23. Ben Yu-Kuang Hu and S. Das Sarma, Finite temperature inelastic-scattering in a doped polar semiconductor , *Semicond. Sci. Technol.* 7: B305 (1992).

24. J.A. Kash, Carrier-carrier scattering GaAs: quantitavite measurements from hot $(e, A^0)$ luminescence, *Phys. Rev. B* 40:3455 (1989).

25. J.F. Young, N.L. Henry and P.J. Kelly, Full dynamic screening calculation of hot electron scattering rates in multicomponent semiconductor plasmas, *Solid-State Electronics*

32:1567 (1989); J.F. Young, P.J. Kelly, N.L. Henry and M.W.C. Dharma-wardana, Carrier density dependence of hot-electron scattering rates in quasi-equilibrium electron-hole plasmas, *Sol. State Commun.* 78:343 (1991).

26. K. Leo and J.H. Collet, Influence of electron-hole scattering on the plasma thermalization in doped GaAs, *Phys.Rev.B* 44:5535 (1991).

27. J.K. Jain, R. Jalabert, and S. Das Sarma, Many-body effects in a nonequilibrium electron-lattice system: coupling of quasiparticle excitations and LO phonons, *Phys.Rev.Lett.* 60:353 (1988); S. Das Sarma, J.K. Jain and R. Jalabert, Many-body theory of energy relaxation in an excited-electron gas via optical-phonon emission, *Phys. Rev. B* 41:3561 (1990).

28. M.W.C. Dharma-wardana, Nature of coupled-mode contributions to hot-electron relaxation in semiconductors, *Phys.Rev.Lett.* 66:197 (1991); S. Das Sarma and V. Korenman, Comment, *ibid* 67:2916 (1991); M.W.C. Dharma-wardana, Reply, *ibid* 66:2917 (1991).

29. V. Narayan, J.M. Rorison and K.O. Jensen (unpublished).

# OPTICAL MEASUREMENTS OF CARRIER MOBILITIES IN SEMICONDUCTORS USING ULTRAFAST PHOTOCONDUCTIVITY

Volkmar Brückner

Friedrich Schiller University Jena
Institute of Applied Optics
Max-Wien-Platz 1
O-6900 Jena, Germany

## INTRODUCTION

One of the most interesting parameters of semiconductors and devices is the carrier mobility in the valence and conduction band because the mobility can result in a limitation of the speed of the device. From the development of ultrafast measuring techniques new questions arise - because the carrier mobility is different at different energy states in the band or subband and the mobility can be influenced by external parameters like field strength applied, temperature, stress etc. Especially in low-dimensional semiconductor structures (quantum wells, quantum wires or quantum dots) we have a quantized energy structure. Furthermore, there are many physically different dynamical processes (intraband and interband transitions, different scattering mechanisms etc.). To perform mobility measurements it seems to be necessary to develop measuring techniques running on a time scale faster than the fastest dynamical process to be studied. Time-resolved photoconductivity seems to be a powerful tool to solve these problems down to a picosecond time scale.

In this paper we present results of the application of time-resolved photoconductivity to carrier mobility measurements. Furthermore, the experimental conditions and limitations will

*Negative Differential Resistance and Instabilities in 2-D Semiconductors*
Edited by N. Balkan *et al.*, Plenum Press, New York, 1993

373

be discussed in detail. The experimental results are compared with numerical calculations.

## METHODS OF TIME-RESOLVED PHOTOCONDUCTIVITY

The main idea of time-resolved photoconductivity developed by Auston[1] in the picosecond range is the generation of an electron-wave packet in the semiconductor, i.e. an electrical pulse, by an ultrashort light pulse. The switching device is an picosecond light illuminated semiconductor (bulk or layer) which is made by placing electrical contacts (mostly ohmic and low-resistance) onto the surface of the photoconductor (Fig. 1). The off-state resistance $R_{g,dark}$ of the gap between the contacts is determined by the dark conductivity and the device geometry (matched to the impedance Z of the whole system). Exciting the gap by a ps laser pulse of suitable frequency ($h\nu > E_g$, where $E_g$ is the energy gap of the semiconductor) creates charge carriers and, therefore, an electrical pulse is generated. The decay time is typically determined by the recombination process of the carriers.

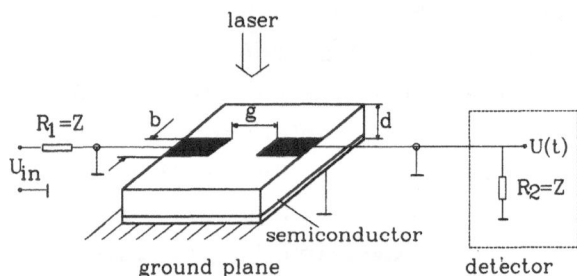

Fig. 1 Basic setup of time-resolved photoconductivity

To identify some scaling relations and to determine some limitations it is sufficient to consider the simple equivalent circuit in a quasi-stationary treatment. Taking into account the applied voltage $U_{in}$, the impedance of the whole system Z (typically 50 Ω) and the contact resistance $R_c$ one can derive a simple relation

$$U = U_{in} Z / (2Z + R_c + R_g) \qquad (1)$$

The gap resistance $R_g$ is determined by the conductivity $\sigma$ and by the gap geometry of the gap (Fig. 1)

$$R_g = \sigma^{-1} \ g/bd \qquad (2)$$

The conductivity is connected with the density N of the carriers created by the light pulse, and with their mobility $\mu_e$

$$\sigma = e\mu_e N \qquad (3)$$

For simplicity the corresponding values of holes are neglected here, because $\mu_h \ll \mu_e$.

Let us consider only the case of weak excitation so that $R_g > 2Z + R_c$ is satisfied. Then we find

$$U = ZeN\mu_e bd \cdot U_{in}/g \qquad (4)$$

$U_{in}/g$ is the field strength $E_{in}$ at the gap, and we find finally

$$U = Ze\mu_e Ndb \cdot E_{in} \qquad (5)$$

From this relation we can learn that

- we can measure the carrier relaxation N(t) from U(t) measurements[2] at constant mobility;
- we can calculate the carrier mobility from measurements of the maximum output voltage $U_0$ if the carrier density is known[3].

If carriers can be situated at different energy levels during the measuring time we have to modify formulae (5)

$$U = ZedbE_{in} \sum_i \mu_{e,i} N_i \qquad (6)$$

where $N_i$ and $\mu_i$ are the carrier density and mobility at the energy level $E_i$, respectively. In that case we are only able to measure the combined action of carrier density and mobility. To distinguish between them it is necessary to perform independent measurements.

By means of oscilloscopes or boxcar integrators it is possible to measure N(t) directly down to 50-100 ps. To measure faster recombination processes one can use a correlation or autocorrelation technique. However, mostly fast relaxation is connected with very low mobility. Therefore, in agreement with formulae (5) it is hard to measure the autocorrelation function due to the extremely low output voltage.

We developed a new technique[2] to measure the correlation function with two independent switches (Fig. 2a). Switch 1 is a semiconductor with high carrier mobility and relatively slow recombination (for example we used Cr:GaAs with $\mu_e = 5000$ cm$^2$/Vs and $\tau_R = 700$ ps). A part of the gap of switch 1 is covered by a light-tight layer acting as a high ohmic resitance. Exciting the other part of the gap by a ps pulse two electrical wave packets start to propagate in opposite dirctions. One of them reflects immediately at the covered part of the gap (with a phase jump). Interacting with the other electrical wave the reflected pulse extinguishes itself and the electrical pulse is "switched off". Therefore, the duration of the electrical pulse is no longer limited by the recombination time, but repeats the laser pulse duration.

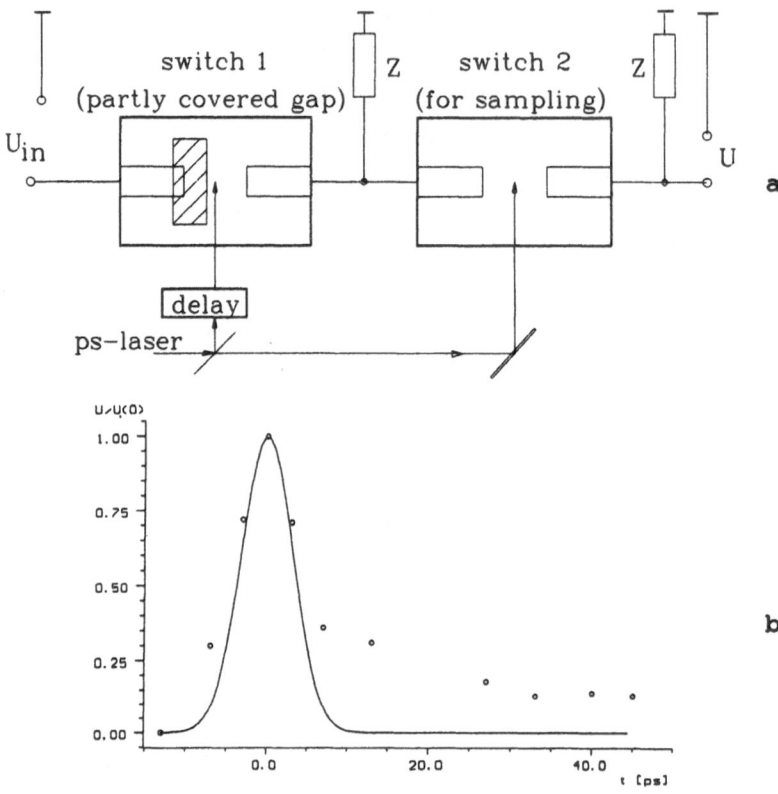

Fig. 2 Correlation measurements using a partly covered gap
Basic setup (a) and demonstration of the 6 ps time
resolution (b)

In order to measure the pulse duration of the electrical pulses we have used the well-known correlation technique, where switch 1 is a GaAs switch with a partly covered gap, and switch 2 is an a-Si switch with less than 10 ps recombination time (Fig. 2a).The response of the switching device to 6 ps pulses of the second harmonic of a glass laser is shown in Fig. 2b. Solid line depicts the convolution of two 6 ps pulses of Gaussian temporal shape. Therefore, the electrical pulse reflects the excitation laser pulse and its duration can be assumed to be about 6 ps. Some deviations at longer times (>20 ps) can be explained by a second 1-2 ns recombination process in the a-Si switch. Therefore, the time resolution of this correlator is about 6 ps limited by the exciting laser pulses. That means, this method can be used to measure recombination processes down to about 10 ps.

## EXPERIMENTS AND DISCUSSION

In accordance with the measuring methods we performed in general two types of experiments.

### (a) Carrier Mobilities at Different Energy Levels

It is well-known that carriers at different energy levels, i.e. with different excess energies above the band edge, have different relaxation rates (depending on the scattering mechanisms, e.g. electron-electron or electron-phonon scattering) and mobilities. From physical point of view the clearest way to perform these experiments would be to use a measuring technique faster than the fastest relaxation process and exactly matched to the energy level $E_i$ (this could be a single level as, for example, in 2D structures or a level in the conduction band as in bulk or amorphous semiconductors), i.e. to use ultrashort pulses of a certain frequency $h\nu = E_g + E_{exc}$, where the excess energy $E_{exc} \cong E_i - E_c$, $E_c$ - energy of the conduction band edge (Fig. 3a). In reality one can use tunable ultrashort laser pulses, but very often there are processes faster than the pulse duration (for example, intraband relaxation processes are typically runing on a femtosecond time scale).

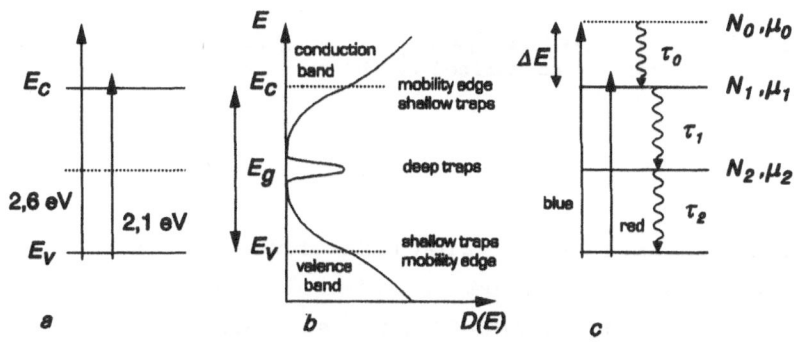

Fig. 3 To the model of carrier dynamics in H:a-Si

In our experiments we used the 25 ps pulses of a frequency-
doubled, Nd:YAG based parametric amplifier[4] together with the
correlation technique described above. The frequency ($h\nu$ = 2.1
eV) was chosen to excite the carriers either near to the
conduction band edge (more precisely the mobility edge - see
Fig. 3b) of the ion-beam damaged (amorphizised) and
hydrogenated (partly annealed) silicon or into the conduction
band ($h\nu$ = 2.6 eV, i.e. the excess energy is about 0.6 eV). The
results of the photoconductivity measurements are depicted in
Fig. 4 both in arbitrary units of the same scale (Fig. 4a) and
normalized to the maximum value of each measurement (Fig. 4b).
Our experimental results can be explained using a simple four-
level model of carrier relaxation in amorphous semiconductors
(Figs. 3b and c). The optical band gap in hydrogenated
amorphous silicon depends on the hydrogen concentration[5]. The
estimated value of the hydrogen concentration in our samples
corresponds to a band gap $E_g$ of about 2.0 eV. In the case of
red laser pulse excitation ($h\nu$ = 2.1 eV) the carriers will be
situated near the mobility edge of the conduction band.

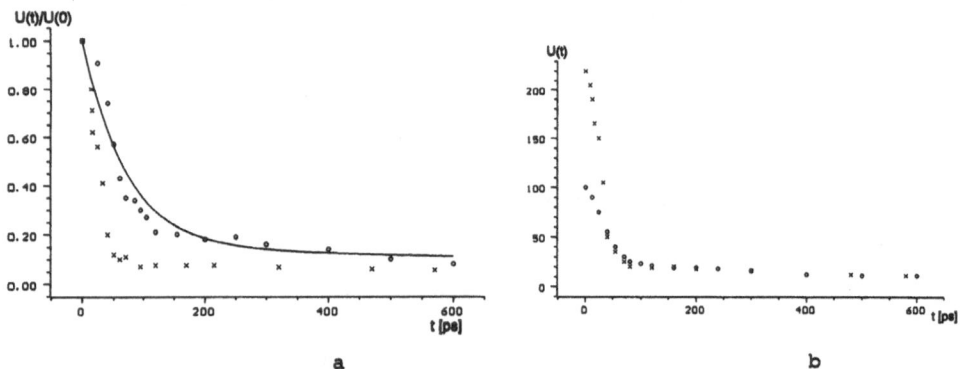

Fig. 4 Photoconductivity at "red" (open circles) and "blue" (crosses)
excitation; a - normalized to the maximum value, b - absolute values

The characteristic relaxation time of the photoexcited carriers
from these shallow traps (carrier density $N_1$) to deep traps
(density $N_2$) is $\tau_1$ (about 80 ps, see Fig. 4a, solid line) and
the relaxation time to lower levels (here probably to shallow

378

traps near the mobility edge of the valence band) is $\tau_2$ (about 1 ns). Using the blue excitation radiation (h$\nu$ = 2.6 eV) a high number of electrons with a great excess energy $E_{exc}$ is excited into the conduction band. During the thermalization process (intraband relaxation processes) the carriers lose their excess energy and are placed again near the band edge in shallow traps. This thermalization time $\tau_0$ in amorphous silicon was found to be less than 1 ps, that means, less than the laser pulse duration $\tau_p$ used in our experiments. So, there is only an effective carrier concentration $N_0 < N_1$ in the conduction band during the excitation process. Therefore, we have to expect the same time behaviour at 2.6 eV excitation if the mobility $\mu_0$ at higher excess energy is nearly the same as $\mu_1$. An increase in the absolute maximum output voltage (Fig. 4a) can be explained only if there is a significant influence of $\mu_0 N_0$ to the mobility $\sigma$. If $N_0 \ll N_1$; $\mu_0 > \mu_1$ should be true. Assuming rectangular excitation pulses $\tau_p$ = 25 ps and $\tau_0$ = 1ps roughly 24/25 of the total number of carriers relax to the band edge during the excitation and $N_0/N_1 \simeq 1/24$. Therefore, an increased photoconductivity signal can be explained only assuming $\mu_0 > 24\mu_1$. For more detailed measurements it seems to be necessary to tune the excitation wavelength in smaller steps.

## (b) Carrier Mobilities at Different Field Strength

It is well-known that the differential mobility of charged carriers in GaAs decreases at higher field strength applied and even acquires negative values. This is due to the specific band structure of GaAs; its conduction band has not only a central $\Gamma$ valley at zero wave vector, but also non-central valleys (L, X) where electrons have lower mobilities. The scattering from $\Gamma$ into X or L valleys causes well-known oscillations of the current through the semiconductor, the Gunn effect[6]. This behaviour can be studied by time-resolved photoconductivity experiments.

In the case of slow carrier relaxation the output voltage in equation (5) can be replaced by its maximum value $U_0$ and we get

$$U_0/ZedbN = \mu_e \, E_{in} = v = f(E_{in}) \qquad (7)$$

where v is the carrier drift velocity. The maximum carrier

density N can be calculated measuring the energy of the excitation laser pulse, taking into account reflectivity losses at the semiconductor surface and the penetration depth of the laser radiation. Thus, measuring the maximum output voltage $U_0$ the drift velocity v can be calculated. Performing measurements at different values $U_{in}$ (i.e. different $E_{in}$) we find the dependence $v(E_{in})$.

The experiments were performed using the 25 ps frequency-doubled pulses of a mode-locked Nd:YAG laser. The photon energy of 2.34 eV exceeds the band gap of the $\Gamma$ point of GaAs by about 1 eV. The electrical pulses generated in the switch were detected by a boxcar integrator with a time resolution of 220ps.

We investigated samples of chromium compensated GaAs ($N_{Cr}$ = 2 $10^{16}$ cm$^{-3}$, measured by SIMS). At all voltages applied we found a single exponential decay with a time constant of $\tau$ = (700+/-200) ps[3]. Therefore, we can assume that we can measure the true maximum value of the output voltage by the boxcar integrator. We studied three different GaAs surface qualities[3]:

(a) cut from a single crystal;

(b) polished by a combination of chemical and mechanical processes where a layer of about 100 $\mu$m has been removed;

(c) polished and additionally etched, which produce an optically clean surface by removing a layer of about 200 $\mu$m.

The measured v-$E_{in}$ characteristics at different excitation energies are shown in Fig. 5 for the case (c). At low excitation energy we find two different slopes, corresponding to two different mobilities. At low field strength (below 5 kV/cm) we found $\mu_c$ = 4000 cm$^2$/Vs, at $E_{in}$ > 12 kV/cm the slope is $\mu_e$ = 650 cm$^2$/Vs. In between we can assume scattering processes from the $\Gamma$ to the L or X valley and we find a negative differential mobility of about 1000cm$^2$/Vs. Assuming a constant scattering rate the comparison of the mobilities in the two valleys leads to the conclusion that the effective mass in the L valley is five times more than that in the $\Gamma$ valley.
Similar behaviour was also found in the other samples, the results in the case of low excitation are summarized in Tab. 1.

Differences in the values of the maximum mobility can be

explained by the decreasing number of vacancies with improved surface quality.

It is more difficult to explain the decreasing mobility at increasing initial carrier concentration.

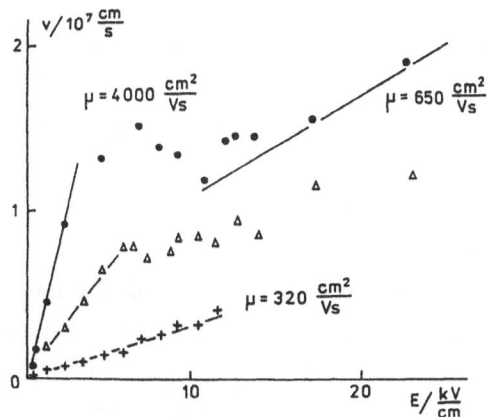

Fig. 5  Field-dependent mobility measurements in GaAs

Tab. 1 Carrier mobilities in different GaAs samples

| GaAs sample preparation | $\mu_\Gamma$ (cm²/Vs) | $\mu_L$ (cm²/Vs) | Excitation level $N_0$ (cm⁻³) |
|---|---|---|---|
| cut | 100 | ~ 30 | $4.0 \cdot 10^{15}$ |
| | 30 | | $1.0 \cdot 10^{17}$ |
| | 16 | | $4.5 \cdot 10^{17}$ |
| polished | 400 | 80 | $4.0 \cdot 10^{16}$ |
| | 20 | | $4.0 \cdot 10^{17}$ |
| etched | 4000 | 650 | $4.2 \cdot 10^{14}$ |
| | ~1000 | ~300 | $7.0 \cdot 10^{16}$ |
| | 320 | | $1.8 \cdot 10^{18}$ |

First we have to estimate the influence of the excited carriers on induced changes of the refractive index and/or the absorption coefficient. This we can do using the well-known Drude model for a free electron gas. The changes of the refractive index $\Delta n(t)$ and the free carrier absorption $\alpha_{fc}(t)$ can be calculated in the following manner:

$$\Delta n(t) = -\frac{N(t)e^2\lambda^2}{8\Pi^2\epsilon_0 c^2 m_{eff} n} \qquad (8)$$

$$\alpha_{fc}(t) = \frac{N(t)e^3\lambda^2}{4\Pi^2\epsilon_0 c^3 m_{eff}^2\mu_e n} \qquad (9)$$

With the standard values of GaAs (effective mass $m_{eff} = 0.07m_e$, refractive index $n = 3.4$, $\mu_e = 4000$ cm$^2$/Vs) together with the laser parameters ($\lambda = 0.53$ $\mu$m, absorption at this wavelength $\alpha \approx 10^4$ cm$^{-1}$) we find

$$\Delta n_{max} = -5.30 \; 10^{-22} \; N \qquad (8')$$
$$\alpha_{fc,max} = 7.0 \; 10^{-17} \; N \qquad (9')$$

where the carrier density $N$ is measured in cm$^{-3}$.

At carrier densities used in our experiments we cannot find any significant changes either in the refractive index or in the absorption. Therefore, our experimental results are not influenced by these parameters.

On the other hand, the stationary drift velocity in bulk GaAs can be calculated by Monte Carlo simulation including the standard scattering mechanisms[7]. The results should correspond to the low excitation data in Fig. 5. There is a reasonable agreement below 10 kV/cm but at higher field strength the calculated velocity increases much slower than the experimental values. May be these deviations can be explained by the photon energy (2.34 eV) which is high enough to populate the L valley directly, even at low field strength.

On the other hand, the arrival of charges at the contacts will cause the field to decrease depending on some regeneration time. As a consequence the effective field can be much lower, especially at high excitation density. We accounted for these effects in a second simulation, where we modeled the device as a capacitor - bulk GaAs as a dielectric separated from the striplines by thin (Schottky) barriers. Note that this introduces an asymmetry depending on the polarity of $U_{in}$. We also introduced a time constant $\tau$ representing the finite time for restoring the initial charges on the plates. We found that as a result of a finite $\tau$ the effective field decreases drastically. Taking our experimental conditions (pulse length, gap width) and assuming $\tau = 1$ ps, after averaging we indeed find a reduction of the deduced drift velocity by about a factor of 1.5-2 in good agreement with the experiment. Note that the simulations predict also some oscillations of the output voltage on a ps time scale, which we could not find in our experiments using the boxcar integrator due to its low time resolution.

Finally, we would like to estimate whether it is possible to perform similar experiments in 2D structures. Assuming a well thickness of 10 nm, a gap area of 0.5 mm$^2$, an absorption coefficient of $10^4$ cm$^{-1}$, and an excitation pulse energy of 1 mJ we should be able to generate carrier densities of about $5 \cdot 10^{21}$cm$^{-3}$. Therefore, it seems to be possible to perform similar experiments also with 2D structures. These experiments are in preparation.

## ACKNOWLEDGEMENT

I would like to thank H. Bergner and M. Lenzner (Jena) performing some of the experiments together with me. Furthermore, I am grateful to Piet van Hall (Eindhoven) for the help in Monte Carlo simulation.

## REFERENCES

1. D.H. Auston, Phys. Rev. Lett. 26:101(1975)
2. H. Bergner, V. Brückner, F. Kerstan, M. Lenzner, Scient. Instr. 3:41(1988)
3. H. Bergner, V. Brückner, M. Lenzner, Haoliang Li, R. Strobel, Int. J. of Optoelectr. 6:357(1991)
4. H. Bergner, V. Brückner, F. Kerstan, B. Schröder, M. Supianek, IEEE QE-21:1630(1985)
5. D.C. Allen, J.D. Joannopoulos, Phys. Rev. Lett. 44:43(1980)
6. J.B. Gunn, Sol.State Commun. 88:1(1963)
7. P.J. van Hall, P. Christianen, Paper presented at the Symposium on Ultrafast Processes in Semiconductors, Vilnius, Lithuania, Sept. 1992

# A TIME-DEPENDENT APPROACH FOR THE EVALUATION OF CONDUCTANCE IN TWO DIMENSIONS

K. Stratford and J.L. Beeby

Department of Physics and Astronomy
University of Leicester
Leicester LE1 7RH, U.K.

## 1   INTRODUCTION

A large growth in the experimental study of low-dimensional systems (LDS) in the past few years has revealed that systems such as quantum wires and dots have many interesting physical characteristics. In general, these devices are formed by the imposition of some artificial potential landscape on mobile electrons in, for example, high mobility GaAs–AlGaAs heterostructures. In the best samples at low temperatures the mean free path for inelastic scattering can be measured in tens of microns and transport is said to be *ballistic*. An important consequence of this desirable quality is that propagation of the electron wave is coherent on a length scale which is comparable to the device size.

The combination of these factors makes such mesoscopic systems highly amenable to simulation. The adoption of an effective mass approach to describe the interaction between electrons and the lattice potential means that the effect of the enforced potential landscape can be directly and easily studied. For a typical heterostructure we are able to make use of the GaAs effective mass $m^*=0.067$. In addition, the complication of inelastic scattering may be ignored when transport is ballistic. The usual approach to this kind of problem is to calculate the scattering matrix for the time-independent Schrödinger equation. This involves a numerical matrix inversion, the computational cost of which can become large (scaling as the cube of the number of basis states used in the description of the wave function). Many calculations of this type are also limited in the form of potential which may be addressed, and hence it may be of some interest to consider a slightly different angle of attack. In this paper an application of a time-*dependent* quantum-mechanical approach to the modelling of electronic transport in LDS is described. The method is most flexible is the choice of potential and can, in principle, be adapted to explore a wide range of phenomena[1,2]. It also has the advantage of requiring only modest computational effort.

The following section is concerned with the integration of the time-dependent Schrödinger equation (TDSE). For numerical purposes a scheme based on an expansion of

*Negative Differential Resistance and Instabilities in 2-D Semiconductors*
Edited by N. Balkan *et al.*, Plenum Press, New York, 1993

**385**

the time evolution operator as a truncated series of Chebyshev polynomials is utilised. A simple implementation is illustrated which describes some of the features of such a calculation including the use of a wave packet. Section 3 demonstrates how the time-dependent approach may be used to calculate the pertinent physical properties of a system: in this case the conductance of a quasi-one-dimensional quantum wire. This serves to highlight some of the advantages and disadvantages of the approach which are discussed briefly in the final section.

## 2 INTEGRATION OF THE TDSE

The motion of an electron of effective mass $m^*$ subject an external potential $V(\mathbf{r})$ is described by the time-dependent Schrödinger equation

$$i\hbar\frac{\partial}{\partial t}\Psi(\mathbf{r};t) = \mathcal{H}\Psi(\mathbf{r};t). \tag{1}$$

This is a first order partial differential equation with respect to time which may be formally integrated to yield

$$\Psi(\mathbf{r};t+\Delta t) = e^{-i\mathcal{H}\Delta t/\hbar}\Psi(\mathbf{r};t) \tag{2}$$

where the wave function at time $t + \Delta t$ is related to that at an earlier time $t$ by means of the time evolution operator $\exp[-i\mathcal{H}\Delta t/\hbar]$. This is an initial value type of problem: $\Psi(\mathbf{r};t)$ will determine the wave function for all time. The Hamiltonian for the electron can be written as

$$\mathcal{H} = \frac{1}{2m^*}(\mathbf{p} - e\mathbf{A})^2 + V(\mathbf{r}) \tag{3}$$

where $\mathbf{A}$ is the vector potential given by a suitable choice of gauge. This term is included for the sake of completeness but in the examples considered later no magnetic field is present. The fact that the exponent $\exp[-i\mathcal{H}\Delta t/\hbar]$ contains the Hamiltonian operator means that some manipulation is required to render it in a usable form. For this purpose a propagation scheme due to Tal-Ezer and Kosloff[3] is used, although other methods are available[2,4].

The time evolution operator may be expanded as an infinite sum in terms of the Chebyshev polynomials of imaginary argument $\Phi_n$. These are related to the more commonly encountered polynomials $T_n$ by $\Phi_n(z) = i^n T_n(-iz)$, where $z \in [-i, i]$. To this end we note that the function $\exp[-\mathcal{V}z]$ is expanded

$$e^{-\mathcal{V}z} = 2\sum_{n=0}^{\infty}{}'(-1)^n J_n(\mathcal{V})\Phi_n(z) \tag{4}$$

where the $J_n$ are ordinary Bessel functions of the first kind. The prime on the summation sign indicates that the first term should be multiplied by a factor of one half. In order to ensure that the argument of the Chebyshev polynomial lies on the correct interval $[-i,i]$ we write

$$e^{-i\mathcal{H}\Delta t/\hbar} = e^{-i\mathcal{V}}e^{-\mathcal{V}z} \tag{5}$$

where

$$z = \frac{i\mathcal{H}\Delta t}{\hbar\mathcal{V}} - i. \tag{6}$$

In these expressions the quantity $\mathcal{V}$ is proportional to the total time-energy phase space volume occupied by the system, that is, it is the highest relevant eigenvalue of the Hamiltonian multiplied by the quantity $\Delta t/2\hbar$. The full expansion of the time evolution operator is then given by combining equations (4), (5), and (6). An approximation is then made by truncating this infinite series so that

$$e^{-iHt/\hbar} \simeq 2e^{-i\mathcal{V}} \sum_{n=0}^{N}{}'(-1)^n J_n(\mathcal{V})\Phi_n(z). \tag{7}$$

Successive terms in the series are calculated by use of a recurrence relation for the Chebyshev polynomials $\Phi_n(z) = 2z\Phi_{n-1}(z) + \Phi_{n-2}(z)$ along with knowledge of the two lowest order polynomials $\Phi_0(z) = 1$ and $\Phi_1(z) = z$. The accuracy of the approximation is ensured by the fact that the value of the Bessel function $J_n(\mathcal{V})$ falls to zero exponentially quickly when $n$ exceeds $\mathcal{V}$. To guarantee that the series converges it is necessary to choose a value of $N$ which slightly exceeds $\mathcal{V}$.

In order to illustrate the use of the above approximation consider an example in which a Gaussian wave packet is propagated in two dimensions in the presence of a model potential representing a quantum wire in zero magnetic field. The required potential is set up on a rectangular grid along with the initial wave function

$$\Psi(x, y; t = 0) = [2\pi\delta_0^2]^{-\frac{1}{2}} e^{-(x-x_0)^2/4\delta_0^2} e^{-ik_x x} e^{-(y-y_0)^2/4\delta_0^2} e^{-ik_y y}. \tag{8}$$

The real space width of this circularly symmetric wave packet is given by the parameter $\delta_0$ while the initial position is $(x_0, y_0)$. It is well known that a wave packet of finite width in momentum space will spread in real space as a function of time and so it must be ensured that the initial width $\delta_0$ is large enough to prevent significant spreading in the time of interest ($\Delta t$).

The action of the time evolution operator on the initial wave function is then evaluated by means of the truncated series (7). A Fast Fourier Transform (FFT) algorithm is employed to calculate the spatial derivatives appearing in the Hamiltonian making use of the fact that a derivative in real space becomes a simple multiplication in momentum space. The value of $\Delta t$ is chosen so that the final transmitted and reflected portions of the wave packet are well separated. The transmitted intensity may be found by means of a numerical integration of $|\Psi(x, y; t = \Delta t)|^2$ over that part of the grid containing the transmitted packet.

The accuracy of the final result is dependent upon several factors. Firstly, it should be noted that the truncated expansion of $\exp[-i\mathcal{H}\Delta t/\hbar]$ is not unitary. However, provided that a sufficient number of terms is used in the expansion ($N$ is large enough) the integrated norm of the wave function is conserved to a high degree of accuracy. It can be seen that the size of the time step $\Delta t$ has no direct bearing on the accuracy of the propagation but it does affect the number of terms required to expand the time evolution operator. Secondly, the number of points used to define the grid must be sufficiently large that a smooth representation of both wave function and potential is provided. In addition, the dimensions of the grid must be large enough to accommodate the wave function in both real and momentum space throughout the time of interest. The use of an FFT algorithm to facilitate the calculation of spatial derivatives is accurate and efficient, the computation effort scaling as roughly $P \log P$ where $P$ is the total number of points used to make up the grid. The total computational effort for the simulation is then proportional to $NP \log P$. In general, with no magnetic field the number of terms, $N$, required in the expansion (7) is a few hundreds. The Chebyshev scheme does allow the inclusion of a magnetic field but can be expensive computationally as the contribution to $N$ from the magnetic field terms in the Hamiltonian scale roughly as $B^2$.

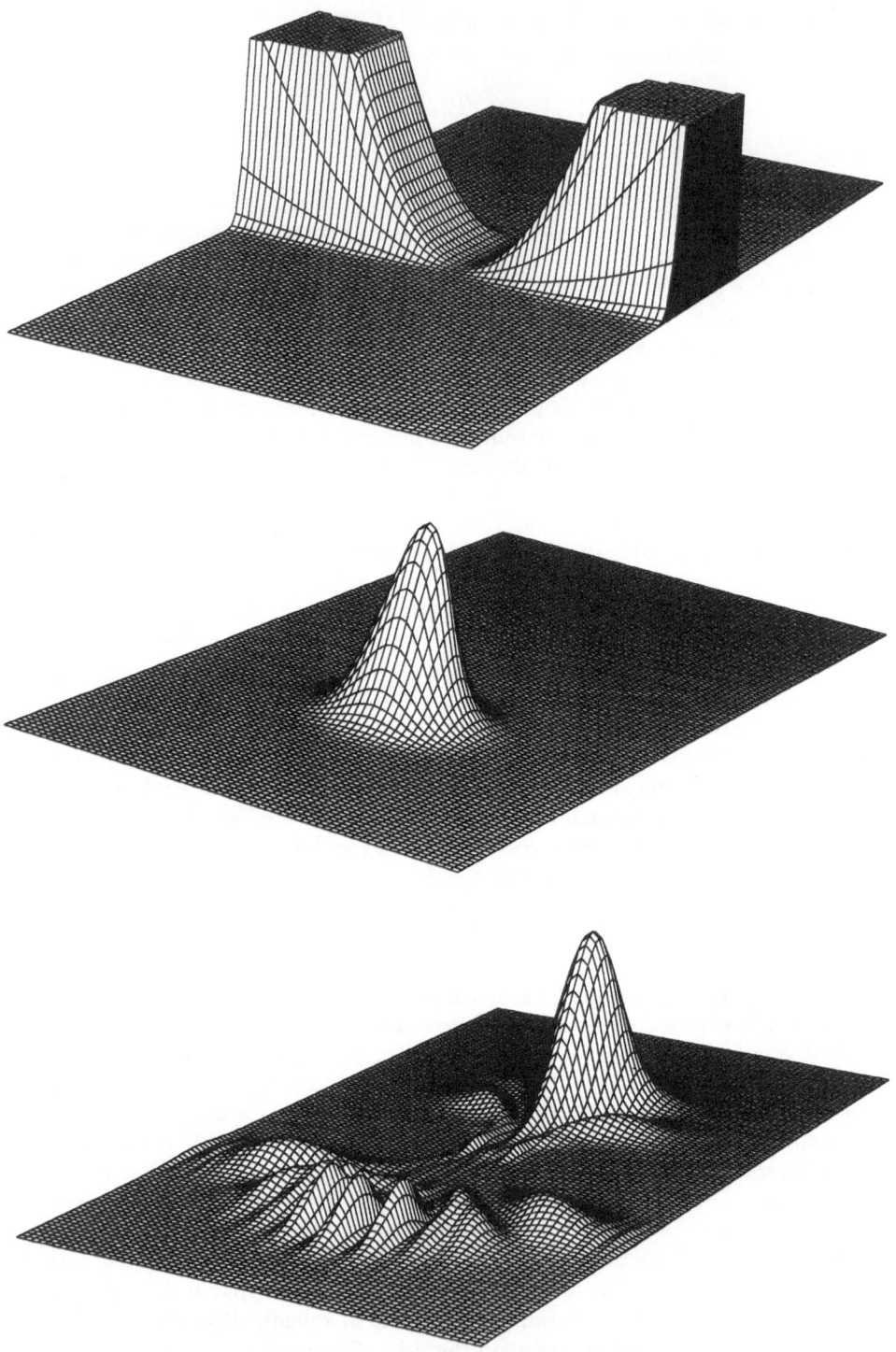

Figure 1. Contour plots of (top) the potential, representing a quasi-one-dimensional wire, $|\Psi(x,y)|^2$ for the initial Gaussian wave packet (centre), and the same quantity for the final packet after it has passed through the wire (bottom). The effects of diffraction can be seen clearly in the final wave function. It should be noted that the plot of the potential has been truncated as an aid to visualisation.

# 3  CONDUCTANCE

We turn now to the question of how the time-dependent approach may be utilised to evaluate a physical property of interest, for which we choose the conductance of a one-dimensional wire. Having seen in the previous section that it is a simple matter to calculate the transmission for a model potential it is necessary to find an appropriate expression for the conductance in terms of this quantity. In what follows it is assumed that the length of the wire runs in the $x$-direction.

The contribution to the current along the channel associated with the plane wave state $\psi_k$ with wavevector $k = (k_x, k_y)$ may be found by integrating the parallel current operator across the channel, viz

$$\langle \psi_k | j | \psi_k \rangle = \frac{i\hbar e}{2m^*} \int_{-\infty}^{\infty} \left[ \psi_k^* \frac{\partial \psi_k}{\partial x} - \psi_k \frac{\partial \psi_k^*}{\partial x} \right] dy. \tag{9}$$

We assume the application of a small potential difference, $V$, across the ends of the wire which will cause states at the Fermi circle to shift by an amount $\delta k_x \simeq eVm^*/\hbar^2 k_x$ in the $k_x$ direction. The total current, $I$, through the wire will be found by integrating the contributions (9) from all states within $\delta k_x$ of the Fermi wavevector. The conductance, $G = I/V$, can now be written down in terms of the matrix element for the current contributions

$$G = \frac{2em^*}{h^2} \int_{-k_F}^{k_F} \frac{1}{k_x} \langle \psi_k | j | \psi_k \rangle \, dk_y. \tag{10}$$

Equation (10) contains the current matrix element for the plane wave state $\psi_k$. Clearly, a useful expression for the conductance must contain quantities relevant to the use of the wave packet $\Psi(x,y)$. A consideration of the collision between a wave packet and a potential[5] shows that the matrix element $\langle \psi_k | j | \psi_k \rangle$ must be replaced by the quantity $eT(k)v/F$. This term introduces the group velocity of the wave packet $v$, the transmitted intensity (a function of incident direction) and a measure of incident flux in the direction of motion

$$F = \int |\Psi(x,0)|^2 \, dx, \tag{11}$$

$x$ being the direction of motion. It is possible to replace the group velocity $v$ by $\hbar k_F/m^*$ which gives, substituting for $\langle \psi_k | j | \psi_k \rangle$ in (10),

$$G = \frac{2e^2}{h} \frac{k_F}{2\pi F} \int_{-k_F}^{k_F} \frac{T(k)}{k_x} \, dk_y. \tag{12}$$

It can be seen that in this final expression for the conductance the properties related to the wave packet appear solely in the factor $F$ defined in equation (11).

By running several simulations for different incident directions (k vectors) the integral in equation (12) can be computed. For this purpose a three point Gaussian integration is employed, making use of the fact that the function $T(k)$ is even is this particular example. Indeed, this is the case for all potentials enjoying mirror symmetry down their length in conditions of zero magnetic field. In order to gain some idea of the accuracy of the result it was found that the three point integration produced the same result as a five point integration to roughly one part in $10^3$. The former was then considered sufficient for the purposes of this work.

A calculation of the conductance as a function of the wave packet energy has been performed for a potential similar to that seen in Fig. 1. The form of the one-dimensional confining potential is parabolic and so the nominal width is measured as twice the classical turning point of the harmonic oscillator ground state. In this example the width

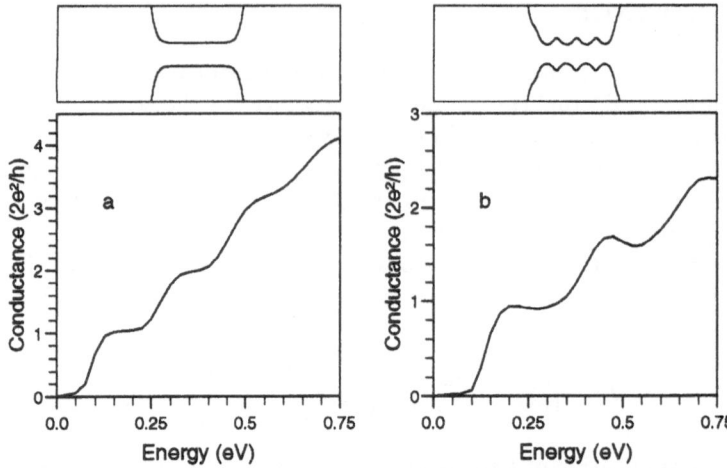

**Figure 2.** The conductance as a function of energy for two different model potentials. Quantisation is clearly exhibited in (a) for a simple straight wire while resonance structure arises when the width of the wire is modulated (b). A diagram representing an equipotential line is show in each case to give an idea of the wire geometry.

was chosen to be approximately 20Å while the length of the channel was 400Å. The ends of the channel were tapered smoothly to join with the wider two-dimensional region. In order to ensure that the wave packet should not pass straight through the channel in the fashion of a classical particle the real space width must be made larger than that of the aperture presented by the opening: in this case $\delta_0$=150Å. Fortunately, this requirement is compatible with with necessity of keeping wave packet spread to a minimum.

Fig. 2a shows the form of the conductance as a function of energy for the simple potential representing a wire with a smoothly varying confining potential. The familiar quantisation of conductance seen experimentally[6,7] is exhibited. It is understood that the spread in energy characteristic of a wave packet will broaden the edges of the steps. However, a deconvolution may be performed with the exact function for the energy spread to compensate and thus reveal the amount of broadening due to the potential geometry. Previous calculations using abrupt potentials[8-11] have shown strong oscillatory structure superimposed upon the conductance steps caused by interference between left- and right-going states in the channel arising from reflections at the ends of the wire. When a smoothly varying potential is used[12] these so-called "organ-pipe" resonances disappear, a fact consistent with the experimental situation. The present calculation supports this aspect of the findings in ref. 12.

An example of a result for a more complex potential is shown in Fig. 2b. The potential is similar to that studied in the first example but in this case is one having a smoothly modulated width along its length. The conductance of the resulting structure shows broad oscillations in the plateau of each step which form due to resonant reflections from the corrugations in the wire. This structure is not of quite the same origin as that seen in the typical double barrier resonant tunnelling problem. In addition, experiments on such a system usually measure the current as a function of the applied voltage. In this case it is possible to derive an appropriate expression for the current (as opposed to the conductance) which is

$$I = \frac{e\hbar}{\pi^2 m^*} \int_{-k_F}^{k_F} T(\mathrm{k}) k_x \sqrt{k_F^2 - k_x^2} \, dk_x. \tag{13}$$

In contrast to the conductance calculation a finite potential bias would be applied across the grid superimposed on the double barrier formation to drive the current. The integral over $k_x$ is evaluated as before.

## 4  CONCLUDING REMARKS

The TDSE provides a method by which transport processes may be studied to compliment use of the time-independent equation. The two examples given demonstrate its use for realistic potentials. While the spread in energy associated with a wave packet may be deemed undesirable for some applications, the flexibility of the approach and the relatively small computing resources required renders it worthy of note. This work has shown how the method can be tailored to the study of the conductance of one-dimensional wires and, as outlined, the double barrier resonant tunnelling problem could easily be addressed. The wave packet approach also offers the opportunity of excellent visualisation of the motion of an electron in an external potential, readily reconcilable with one's physical intuition of what the classical trajectory should be.

## Acknowledgement

One of us (K.S.) enjoyed financial support from the U.K. Science and Engineering Research Council while this work was in progress.

## REFERENCES

1. F. Ancilotto, A. Selloni, L.F. Xu, and E. Tosatti, Phys. Rev B39, 8322 (1989).

2. K. Michielsen and H. De Readt, J. Phys.: Condens. Matter 3, 8247 (1991).

3. H. Tal-Ezer and R. Kosloff, J. Chem. Phys. 81, 3967 (1984).

4. See, for example, J.A. Fleck, J.R. Morris, and M.D. Feit, Appl. Phys. 10, 129 (1976).

5. M.L. Goldberger and K.M. Watson, "Collision Theory", Chapter 3, Wiley, New York (1964).

6. B.J. van Wees, H. van Houten, C.W.J. Beenakker, J.G. Williamson, L.P. Kouwenhoven, D. van der Marel, and C.T. Foxon, Phys. Rev. Lett. 60, 848 (1988).

7. D.A. Wharam, T.J. Thornton, R. Newbury, M. Pepper, H. Ahmed, J.E.F. Frost, D.G. Hasko, D.C. Peacock, D.A. Ritchie, and G.A.C. Jones, J. Phys. C 21, L209 (1988).

8. G. Kirczenow, Sol. St. Commun. 68, 715 (1988).

9. L. Escapa and N. Garcia, J. Phys.: Condens. Matter 1, 2125 (1989).

10. D. van der Marel and E.G. Haanappel, Phys. Rev. B39, 7811 (1989).

11. E. Tekman and S. Ciraci, Phys. Rev. **B39**, 8772 (1989); Phys. Rev. **B40**, 8559 (1989).

12. J.A. Nixon, J.H. Davies, and H.U. Baranger, Phys. Rev. **B43**, 12 638 (1991).

# HIERARCHY OF CURRENT INSTABILITIES IN THE IMPACT
# IONIZATION AVALANCHE IN SEMICONDUCTORS

Kazunori Aoki

Department of Electrical Engineering
Faculty of Engineering
Kobe University
Rokkodai, Nada, Kobe 657, Japan

## INTRODUCTION

Lightly-doped or undoped semiconductors exhibit many different types of current instabilities and chaos related to the impact ionization of shallow impurities or low-temperature impurity avalanche breakdown.[1-15] In $n$-type GaAs,[1-9] we have recognized the existence of a hierarchy among the current instabilities ranging from low dimensional chaos to self-organized criticality (SOC[16-18]), which have been observed under different experimental parameter regimes. Among the instabilities, some observations (periodically-driven chaos[5,6] and crossover current instability[6,7]) agree well with the existing physical models,[5,8,9] while others[1-4,6,18] are much more complicated and even far from the low dimensional deterministic behaviors.[6,18] Among the current instabilities, there are many interesting problems to be solved.

In this paper, we shall discuss the physical processes and the dynamical behaviors for various types of current instabilities, from viewpoints of the experiments and the model simulations.The motivation of the present paper is to describe and classify the observed current instabilities and chaos in terms of the physical and mathematical models. For the chaotic behaviors, in order to stress the determinism, we have analyzed the statistical/thermodynamical quantities of correlation dimension,[9,19] Lyapunov spectrum,[20,21] $f(\alpha)$ spectrum[13,22] and *chaos temperature*.[23] For the periodically-driven chaos, we shall discuss the problems of the *central limit theorem and its breakdown*,[24,25] and the *deterministic noise amplification*.[23,26] For the latter problem, we have introduced chaos temperature in order to quantify the sensitive dependence on the external noise. Concerning the deterministic noise amplification, we propose a novel application by usage of period doubling bifurcation.

*Negative Differential Resistance and Instabilities in 2-D Semiconductors*
Edited by N. Balkan *et al.*, Plenum Press, New York, 1993

# HIERARCHY OF CURRENT INSTABILITIES

So far, we have observed in undoped $n$-type GaAs: (a) firing wave instability[1-4](FWI),(b) periodically-driven instability[5,9](PDI), (c) chaotic itinerancy[6](CI),(d) crossover current instability[7,8](COI), (e) cross talk coupling in the regime of COI[27,28] and (f) self-organized criticality.[6] Features of the dynamical behaviors are summarized in Table I. Among the instabilities, the PDI is the most fundamental and prototypical. Let us remember that the impact ionization avalanche of shallow impurities induces a simple S-shaped $J$-$E$ characteristic, where the process is essentially a first order phase transition at low temperatures. From the viewpoint of catastrophe theory,[29] the phase transition belongs to the cusp catastrophe in which the *perfect delay convention* by Thom[29] is a basic idea for the description of quasi-static regime or switching behavior (namely, critical slowing down). By applying a dc+ac bias of the form of $E_o + E_{ac}\sin(2\pi f_o t)$, the $J$-$E$ curve is perturbed periodically with the driving frequency $f_o$. When the driving period $1/f_o$ is shorter than the phase transition time (or switching time) $T_{tr}$, namely when $f_o T_{tr} > 1$, the perfect delay convention no more holds and the orbit deviates considerably from the static $J$-$E$ curve. The required driving frequency ranges from ~ 0.8 MHz to ~ 3 MHz. For the dynamical regime, the orbit experiences locally negative differential conductivity (NDC), $dJ/dE < 0$. Chaos appears if the control parameters meet with *local instability* condition. Correlation dimension of fully developed chaos is less than 3. Thus, the PDI exhibits low dimensional chaos.

## Table I. Hierarchy of current instabilities

| Types | PDI | COI | CI | FWI | SOC |
|---|---|---|---|---|---|
| Experimental parameter regime | dc+ac bias | dc bias + magnetic field | dc+ac bias + [1]$h\nu$ | pulsed voltage + [2]$h\nu$ | dc bias + [1]$h\nu$ |
| Frequencies | 0.8-3 MHz | 30-500 kHz | 0.8-3 MHz | 0.1-10 Hz | <100 kHz |
| Bifurcations | [3]Pd, [4]Int | [3]Pd, [5]Hopf [4]Int | ———— | [3]Pd | ———— |
| Correlation dimensions | < 3 | < 3 | ———— | 2 - 4.2 | ———— |
| Remarks | low dim. chaos | low dim. chaos | ———— | low & high dim. chaos | 1/f noise |

[1]$h\nu$: continuous photoexcitation near band gap energy (e.g, $\lambda$=812 nm).
[2]$h\nu$: Either continuous or light-chopped.
[3]Pd: period doubling bifurcation.
[4]Int: intermittency.
[5]Hopf: Hopf (torus) bifurcation route to chaos.

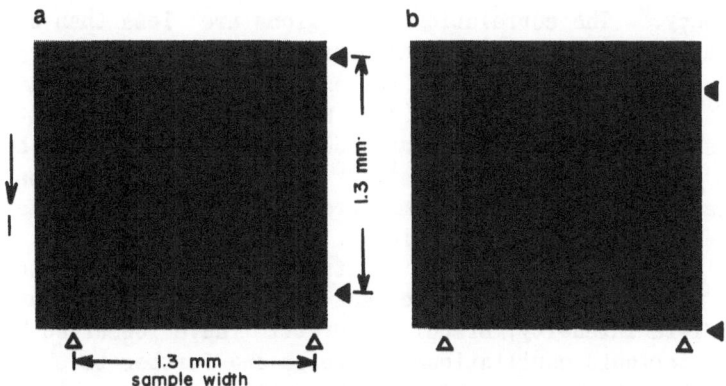

Figure 1. (a) Image of a stationary filament by LTSEM (see ref.32) and (b)image of turbulent pattern during the FWI induced by electron beam (see refs.32,33).

Under a continuous photoexcitation, however, the PDI exhibits different dynamics (CI): the long term behavior shows itinerancy among various periodic states and chaos.[6] The behavior is quite similar to the chaotic itinerancy as coined by Ikeda et al[30] and experimentally evidenced by Arecchi et al[31] in nonlinear laser optics. Under the continuous photoexcitation and the dc bias, the impact ionization avalanche induces $1/f$ noise in the frequency range less than 100 kHz. The physical origin of the $1/f$ noise in $n$-GaAs[6] and $p$-Ge[18] has been discussed in terms of the SOC model by Bak, Tang and Wiesenfeld.[16-18]

A quite different current instability has been observed under pulsed voltage and photoexcitation.[1-4] By applying a pulsed voltage with a repetition frequency of 10–40 kHz, stochastic firings of the filamentary currents can be seen on the pulses. In the low frequency range (0.1 Hz–10 Hz), the firings exhibit the density wave and the course-grained signal detected by a lock-in amplifier shows various types of peculiar oscillations and chaos.[1-4] The instability is termed "*firing wave instability*" (FWI). In order to confirm the deterministic nature, correlation dimension and Lyapunov spectrum are discussed.

Since the impact ionization avalanche induces a filamentary current between ohmic contacts (Fig.1(a)), one may imagine the irregular but deterministic (chaotic) switching between on/off states of the filament just like gaseous discharge, and further one might conjecture that the spatial pattern during the chaotic motion should not be so irregular if the correlation dimension of the current time series is not so high (and vice versa). But the problem is not so simple. A spatial pattern image of the filament during the electron-beam induced FWI is highly turbulent (Fig.1(b)),[32,33] while as we discuss later, the correlation dimension is relatively low.

Under the application of a longitudinal magnetic field with dc bias, the filamentary current is destabilized via the inhomogeneous magneto-resistance effect which acts upon the filament boundaries, resulting in spontaneous current oscillation (COI[6,7,8,27,28]). The oscillation frequencies range approximately from 30 kHz to 500 kHz. We can observe typically $2^k$ (k=0,1,2,..) period doubling bifurcation,[28] Hopf bifurcation and intermittency.[8] The correlation dimensions are less than 3, showing low dimensional chaos. With the three-terminal circuit geometry, cross talk coupling between two filamentary currents has been investigated with the COI regime. The physical process will be discussed in the text.

Not shown in Table I is spontaneous oscillations, which have been observed in $p$-Ge[10,11] and in $n$-GaAs[14,15] under dc bias condition and transverse magnetic fields. In these measurements,[10,14] the dc bias is applied with a heavily inclined load line (a large load resistor). Namely, the bias condition is almost "current controlled". The oscillation frequencies are a few kHz for $p$-Ge[10] and a few MHz for GaAs.[14,15] As a function of the magnetic field intensity, Brandl and Prettl[14] have observed Hopf bifurcation from periodic oscillations to torus and further to chaos. They have attributed the chaotic behavior to the Hall field instability and to the two-level impact ionization process as proposed by Schöll.[34]

## PERIODICALLY DRIVEN INSTABILITY

### Theoretical Descriptions

**Mathematical Consideration.** Let us consider a folded manifold $M$ which is an algebraic set:

$$M = \{x \in \mathbf{R}^3 \mid z = -\gamma_1 \partial V/\partial x\}, \tag{1}$$

where $x = (x, y, z)$ and $V = V(x, y)$ is the double well potential function with its derivative $\partial V/\partial x$ being a cubic function of $x$. We can interconnect the folded manifold with a dynamical behavior (the variables; $x, y$) by a couple of ODEs';

$$dx/dt = -\gamma_1 \{(x-\alpha)(x-\beta)(x-\gamma) + (y-a)[1+c(x-\beta)]\}, \tag{2}$$

$$dy/dt = -\gamma_2 \{y - y_o + cy(x-\beta) + y_{ac}\sin(2\pi f_o t)\}, \tag{3}$$

where $\gamma_1, \gamma_2, \alpha, \beta, \gamma, a$ and $c$ are the constants, $y_o$ is the control parameter and $y_{ac}$ is the modulation amplitude. Under the steady state condition ($\dot{x} = \dot{y} = 0$) with $c = 0$ and $y_{ac} = 0$, the solutions give an S-shaped $x$-$y$ characteristic which is a section of the manifold $M$ at $z = 0$, as shown in Fig.2(a). For any given $f_o$, the orbit lies exactly on $M$. Now the essential feature is the following. If $f_o T_{tr} \ll 1$, the orbit never crosses the unstable middle branch of the static $x$-$y$ curve (Fig.2(b)), showing a simply closed orbit. Namely, the orbit overshoots above $y_{2c}$ (below $y_{1c}$), switching on (off) from lower (upper) to upper (lower) branches within the time much faster than the driving period $1/f_o$. While, if $f_o > 1/T_{tr}$, incomplete switching occurs since there is insufficient time to complete the switching during one period. Nearby trajectories $x$ and $x'$ which have tangent unit vecotors $u_t$ and $u_t'$ diverges exponentially[9] (sensitive dependence on the initial conditions), since the local dynamics on the manifold are ruled by $|\dot{\xi}| = \varepsilon|\xi|$, where $\xi = x - x'$ and $\varepsilon$ is positive on the local plane of a large part of the manifold ($\varepsilon \infty |u_t \cdot u_t'|$) and is negative on the local plane of the manifold near the lower and upper branches (Fig.2(c)). As shown in Fig.2(c), during the chaotic behavior, the orbit crosses the unstable middle branch many times.

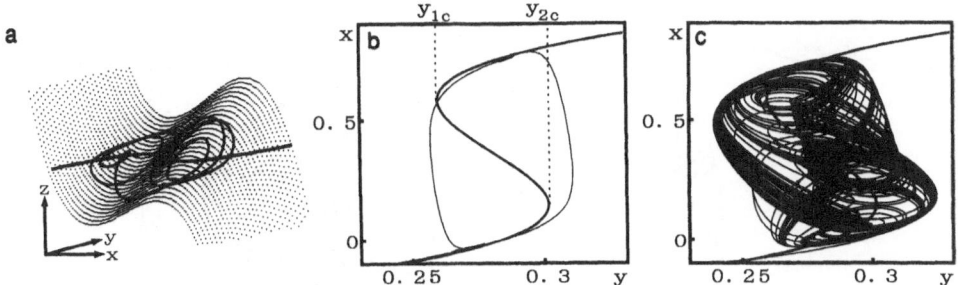

**Figure 2.** (a) Manifold $M$, (b) limit cycle and (c) chaos. $\gamma_1=1$, $\gamma_2=0.035$, $\alpha=0.1$, $\beta=0.2$, $\gamma=0.8$, $s=0.3$, $c=0.405$ and $y_0=0.2965$. $f_0=1/50$ and $y_{ac}=0.06$ in (b); $f_0=10/34$ and $y_{ac}=0.105$ in (c).

The argument holds for problems of the impact ionization avalanche, regarding $x$ and $y$ as the carrier density and the electric field, respectively.

**Impact Ionization Process - The Physical Model.** For the observed impact ionization avalanche and chaos[1-15] in semiconductors, extensive theoretical studies have been reported:(1) dielectric relaxation instability (*multi-level impact ionization model*) by Schöll,[34] (2) *breathing instability* of a current filament by Schöll,[35] (3) *Hall-field instability* by Hüpper and Schöll,[36] (4) dielectric relaxation instability (*one-level impact ionization model*) by Aoki and Yamamoto,[9] Aoki and Mugibayashi,[9] and (5) *crossover current instability* by Aoki.[27,28] Schöll describes in his multi-level impact ionization model[34] that the impact ionization process of the impurity excited states as well as the ground state play an essential role for the dielectric relaxation instability and chaos. However, for shallow donors in $n$-GaAs[5,6] and InSb[9] whose Rydberg energies of the first excited state (n=2) are too shallow and comparable to the thermal energy at 4.2 K, we claim that one should consider alternatively the one-level impact ionization model[9] (only impact ionization of the donor ground state is considered).

Especially in high-purity $n$-GaAs which is partially compensated by shallow acceptors, screening of the ionized impurity scattering and the acoustic phonon scattering by the impact-ionized electrons increases the electron mobility[37] and thus plays an important role in the impact ionization avalanche. In the one-level impact ionization model, we include the screening effect in the approximated formula of the impact ionization coefficient;

$$A_I = A_I^o \, exp(-C/E) + \gamma n \,, \qquad\qquad (4)$$

where $A_I^o, \gamma$ and $C$ are the constants, $E$ is the electric field and $n$ is the electron density. Leaving the details of the model,[9] we emphasize here only important aspects of the model simulation. Figure 3 shows the 3-D phase portrait of a strange attractor constructed by the embedding method, where the Cartesian coordinates $x, y$ and $z$ correspond to $I(t)$, $I(t+\tau)$ and $I(t+2\tau)$ ($I(t)$:current; $\tau$:time lag).[9] The correlation dimension was approximately 2.2, of which value agrees well with the experimental value.[9] The strange attractor is very thin, as recognized from Fig.3. In Fig.4, the 2-D

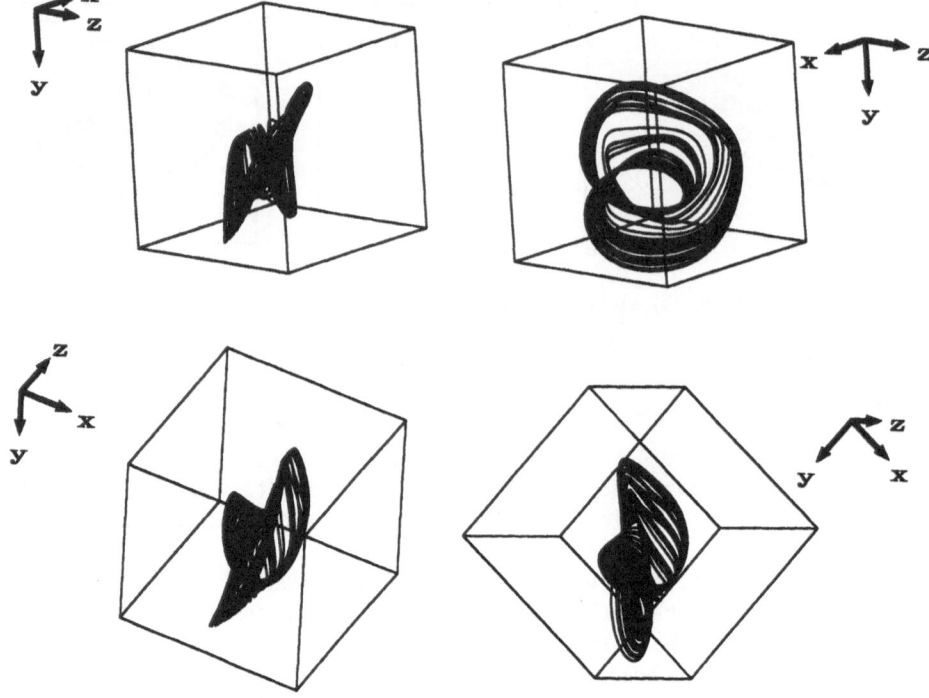

Figure 3. Zoological garden of chaos in the 3-D phase space, drawn by one-level impact ionization model.

phase portrait projected on the $x$–$y$ plane and the Poincaré surface of section are compared between the experiment and the model simulation. The Poincaré section indicates that the strange attractor is due to belt chaos. It is more clear if we enlarge the Poincaré section into finer and finer fractal structure, as shown in Fig.4(c).

One-level Impact Ionization Model Revised. In order to get a more precise physical picture, we have revised the model by taking the detailed energy balance equation into account.[28] The proposed model is given by,

$$\dot{\varepsilon} = -(\varepsilon - \varepsilon_L)/\tau(\varepsilon) + e\mu E^2, \qquad (5)$$

$$\dot{n} = n(N_D - N_A - n)A_I + (N_D - N_A - n)A_T - n(n + N_A)B_T, \qquad (6)$$

$$\dot{E} = -\gamma_d[E - E_o - E_{ac}\sin(2\pi f_o t) + cne\mu E], \qquad (7)$$

where $\varepsilon$ and $\varepsilon_L$ are the mean carrier energy and the thermal energy ($\varepsilon = 3k_B T_e/2$; $\varepsilon_L = 3k_B T_L/2$; $T_e$:electron temperature; $T_L$:lattice temperature), $\mu$ the electron mobility, $\tau(\varepsilon) \propto \varepsilon^{-1/2}$ the energy relaxation time due to acoustic deformation potential scattering, $n$ the electron density, $N_D$ and $N_A$ the donor and acceptor concentrations respectively (we assume an $n$-type semiconductor material with the effective doping concentration $N_D - N_A$), $A_I$ the impact ionization coefficient, $A_T$ the thermal ionization coefficient, $B_T$ the thermal recombination coefficient, $E$ the electric field, $\gamma_d$ the

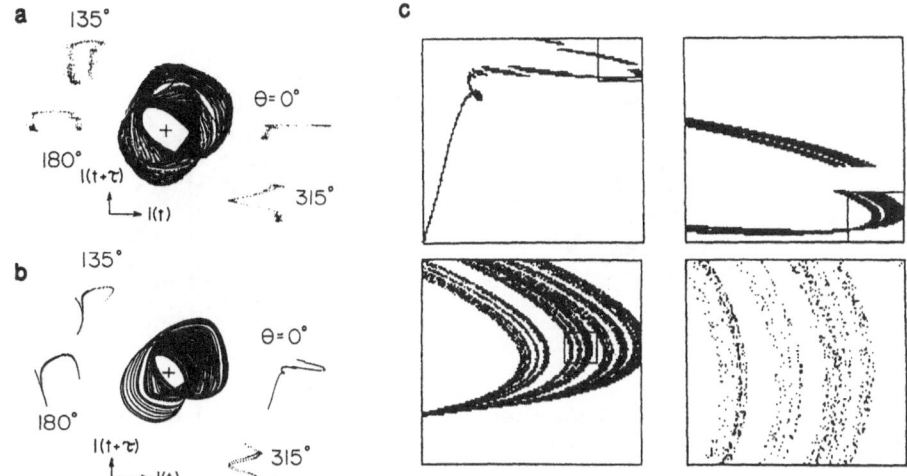

**Figure 4.** Projection of the 3D-phase portraits onto $x$-$y$ plane and the Poincaré sections; (a) experiments and (b) model simulation, and (c) successive magnifications of the Poincare section at $\theta=0$ in (b).

damping constant ($1/\gamma_d$:dielectric relaxation time), $c$ is a positive constant which specifies the slope of the load line, and the overdots in (5)-(7) denote the time derivatives. In (7), we drive the system periodically by a dc+ac field of the form of $E_o+E_{ac}\sin(2\pi f_o t)$. In the detailed energy balance equation (5), we use the Brooks-Herring (BH) formula[38] for the electron mobility;

$$\mu = a\varepsilon^{3/2}[(N_I+n)g(\beta_{BH})]^{-1} , \qquad (8)$$

$$g(\beta_{BH})=\ln(1+\beta_{BH}^2)-\beta_{BH}^2/(1+\beta_{BH}^2), \qquad (9)$$

where $a$ is a constant,[38] $N_I$ is the ionized impurity concentration given by $2N_A+n$, and $\beta_{BH}$ is the so-called BH coefficient given by $\beta_{BH}^2=16(m^*\varepsilon/\hbar^2)L_D^2$, $L_D=(2\varepsilon\kappa\kappa_o/3ne^2)^{1/2}$ ($L_D$ is the Debye length, $\kappa\kappa_o$ the dielectric constant, $m^*$ the effective electron mass, and $e$ the electron charge). The impact ionization coefficeient in (6) is approximated by,[27]

$$A_I \propto (\varepsilon_I/k_BT_e)^{1/2}[1+(2m^*\varepsilon_I/3)^{1/2}\mu E/k_BT_e]\exp(-\varepsilon_I/k_BT_e) , \qquad (10)$$

where $\varepsilon_I$ is the dissociation energy of the donor ground state. During the impact ionization process, the electron mobility plays an essential role for the occurence of the first order phase transition.[27] At low electric field, the electron mobility is initially ruled by the ionized acceptor scattering. At the breakdown field $E_b$, a slight increase in $n$ by the impact ionization of neutral donors (via $A_I$ in (6),(10)) leads to an increase in $\mu$ and an increase in the energy gain $e\mu E_b^2$ in (5), resulting in an increased electron temperature $T_e$. Such an incremental process is forwarded by the positive feedback via (5),(6), (8)-(10), followed by an electrical avalanche breakdown. During the transition, the Debye length

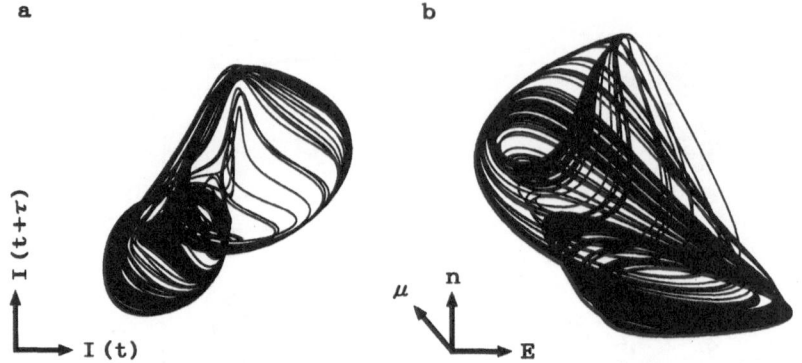

a       b

$I(t+\tau)$

$I(t)$

$\mu$   n

E

**Figure 5.** (a) Projection of the 3D–phase portrait onto $x$–$y$ plane and (b) the phase portrait in the $(n, E, \mu)$ phase space, simulated by the revised model.

contracts immediately due to the screening effect for the ionized impurity scattering. The reverse process occurs at the holding field $E_h$.

For the model simulations, a couple of the ODEs' ((5)-(7)) are reduced to dimensionless ones by using the following scale conversions:

$$\nu = n/(N_D-N_A); \quad \nu_I = N_I/(N_D-N_A); \quad \tilde{T}_e = T_e/T_L; \quad \tilde{\mu} = \mu/\mu_o; \quad \tilde{t} = t/\tau_r;$$

$$\tilde{E} = [(\mu_{ac}^o/\mu_o)^{1/2}/u_l]E; \quad \kappa_1 = A_I/\overline{A}_I; \quad \kappa_2 = A_T/(N_D-N_A)\overline{A}_I; \kappa_3 = B_T/\overline{A}_I,$$

where $\mu_{ac}^o$ is the zero-field electron mobility due to acoustic deformation potential scattering,[38] $u_l$ is the sound velocity, $\tau_r$ is the dielectric relaxation time ($\gamma_d = 1/\tau_r$ in (7)), and $\overline{A}_I$ is the impact ionization coeffi- at zero electric field. The parameter values have been given in ref.27, among which some important values are ; $T_L$=4.2 K, $\varepsilon_I$=6.8 meV, $\tau_r$=1 $\mu$s, $N_D$=1x10$^{14}$/cm$^3$, $N_D/N_A$=1.9, $\tilde{\mu}_o$=1x10$^4$ cm$^2$/Vs, $\overline{A}_I$=4.22x10$^{-7}$ cm$^3$/s, $A_T$=10$^6$/s, and $B_T$=1.097 x 10$^{-6}$ cm$^3$/s. Figure 5(a) shows 2-D phase portrait of the chaotic current ($I(t)$ vs $I(t+\tau)$), and Fig.5(b) the 3-D phase portrait in the $(n, E, \mu)$ phase space, respectively, where the current in Fig.5(a) includes the displacement current. Fig.5(a) is compared to Fig.4(b). The physical meaning of the chaotic orbit in Fig.5(b) is now apparent; the orbit forms the windings near the lower and upper branches and it shows a transition between the loops, whereby the electron mobility $\mu$ increases appreciably (several times the zero-field mobility $\mu_o$) when the orbit switches on from the lower to the upper branches. The electron temperature increases only by a factor of 1.2 in maximum. The chaotic orbit resembles somehow the Lorenz attractor.[39] The validity of the proposed model may be confirmed by an experimental fact (by Crandall[37]) that the Hall mobility of the ele- ctrons in high-purity n-GaAs increases sharply by impact ionization ava- lanche.

## Experimental New Findings

**Breakdown of the central limit theorem.** All the measurements were performed at 4.2 K. The samples used were high-purity epitaxial $n$-GaAs ($n\sim$ 2x10$^{14}$/cm$^3$) grown on a semi-insulating substrate. The thickness of the epitaxial layer was about 12 $\mu$m. The samples were cut into rectangular

400

shapes, with typical surface area dimensions of about 1 mm x 3 mm. For the experimental procedures, the author refers to refs.5,9. So far, we have carried out many experimental tests[5,6,9] for the chaotic behaviors in the PDI regime: (1) low dimensionality tests (correlation dimension[9], and $f(\alpha)$ spectrum[6]) and (2) bifurcation tests.[5] We vary one of the control parameters: dc bias $V_o$, amplitude of ac bias $V_{ac}$ and the driving frequency $f_o$. As a function of $V_{ac}^6$ or $V_{dc}$ (present study), we observe period doubling bifurcation. At the onset of chaos via the period doubling[6], the $f(\alpha)$ spectrum of the strange attractor[6] is found to be close to the ideal one of the Feigenbaum attractor. Periodic/chaotic bifurcation sequence has been described in ref.5.

Generally speaking, stationary fluctuations (stationary chaos) satisfy the central limit theorem. But, it has been shown by Mori et al[24,25] in the general chaos theory that the central limit theorem is broken by the intermittent behaviors. Recently, we have evidenced experimentally the breakdown of the theorem in our semiconductor experiments. Here, we shall note only the essential part of the experiment. We measure the current $I$ and sample voltage $V$ and calculate the instantaneous dissipation power: $U = (I-\langle I \rangle)(V-\langle V \rangle)$; and further calculate $U_m$ which is the averaged value over $m$ periods. Let $\tilde{U}_\infty$ be the average dissipation power $\langle U_m \rangle$ and $P(\tilde{U}:m)$ be the probability that $U_m$ takes the specified value $\tilde{U}$. By the generalized fluctuation theory of the dissipation energy by Mori et al[24,25], we have

$$U = -R_L(I-\langle I \rangle)^2 + V_{ac}(I-\langle I \rangle)\sin(2\pi f_o t), \qquad (11)$$

$$P(\tilde{U}:m) = \exp\{-m\,\psi(\tilde{U})\}\,P(\tilde{U}_\infty:m), \qquad (12)$$

where the $\psi(\tilde{U})$ spectrum is a parabolic (concave) function of $\tilde{U}$ when the long term fluctuation of $U_m$ satisfy the central limit theorem, otherwise it is linear at the lower energy side ($\tilde{U} < \tilde{U}_\infty$) and parabolic for $\tilde{U} \geq \tilde{U}_\infty$. We have evidenced (12) for the observed intermittency (boundary crisis[9]) in the prebreakdown regime (the details will appear elsewhere).

**Deterministic noise amplification by period doubling.** Any kinds of experimental systems suffer more or less from external noise, so that ob-

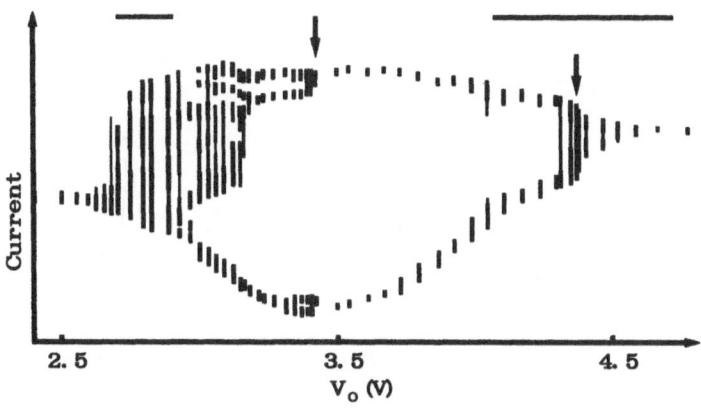

Figure 6. Bifurcation diagram observed in the PDI regime. $V_{ac}$=190 mV, $f_o$=1.83 MHz.

servation of the period doubling bifurcation is limited to several steps of the bifurcation sequences. This means that the chaotic systems are very sensitive to the external noise. Here, we show a novel application of the period doubling bifurcation in the chaos system by the *deterministic noise amplification*.[23,26] Periodic orbits near the bifurcation points exhibit local instability against external noise (microscopic information), and produce spontaneously the macroscopic information through the deterministic noise amplification. By the microscopic information, we mean, for instance, the radio waves from broadcasting stations or any other meaningful microscopic signals in general. Chaos is not necessarily required for the amplification.[26]

Figure 6 shows the bifurcation diagram, observed by varying the dc bias $V_O$, where $V_{ac}$=190 mV and $f_O$=1.83 MHz. By decreasing $V_O$ from above, we observe typically the period doubling bifurcation. Near the bifurcation points (see arrows in Fig.6), the periodic orbits are torus-modulated by the external radio frequencies $f_r$ (the modulation is quite significant[23]). By feeding the currents into the antenna terminal of radio, one can pick up clearly the amplified radio signals with the tuned frequencies at $f_O-f_r$. The horizontal bars in Fig.6 indicate the regions in which we could pick up the radio signals. In the chaotic regime, the radio signals were detected in the range of wide-band tuning frequencies near $1/2f_O$ or $1/4f_O$, where the radio signals were noisy mixed with the original chaos sound.

In order to quantify the deterministic noise amplification, we have introduced chaos temperature $T_{ch}$ by extending the Nyquist theorem. For the definition, the author referes to ref.26. In Fig.7, $T_{ch}$ is plotted as a function of $V_O$, correspondently to Fig.6. The chaos temperature ranges from 0.01 K to 5000 K. As seen in the figure, $T_{ch}$ increases sharply near the bifurcation points (see arrows). The maximum amplification factor $\beta$ of the noise power, which is the ratio of the maximum temperature to the minimum one in the periodic regime of the bifurcation diagram, was evaluated to be $10^3$-$10^5$, of which values, of course, depend upon the control parameter $V_{ac}$ and $f_O$ ($\beta$ is $\sim2\times10^3$ for the bifurcation diagram in Fig.6). It is found that the chaos temperature is a good indicator for the noise amplification and also for the attractor size of chaos. The chaos temperature is compared to the Lyapunov exponents $\lambda_1$ and $\lambda_2$, in Fig.8,

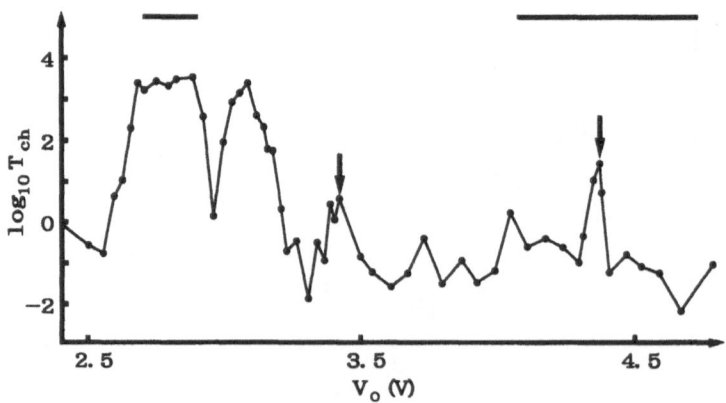

Figure 7. Chaos temperature as a function of $V_O$, correspondently to Fig.6.

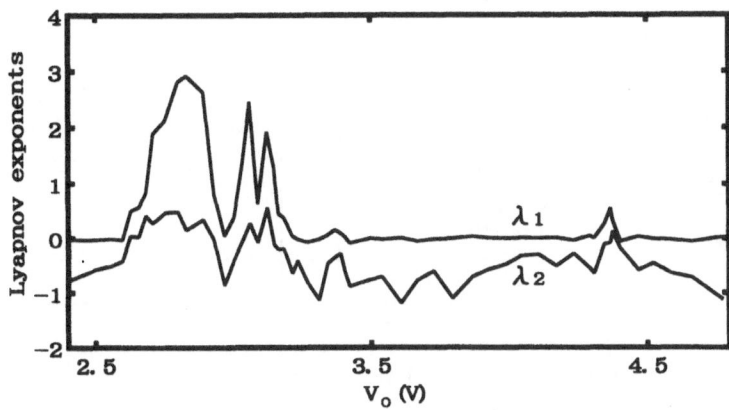

**Figure 8.** Lyapunov exponents $\lambda_1$ and $\lambda_2$ as a function of $V_o$, correspondetly to Fig.6.

where the exponents have been evaluated with the embedding dimension $m$=4 by using an algorithm used by Sano and Sawada[21]. The exponents show approximately the spectrum of 0,-,-,- in the periodic region and +,0,-,- in the chaotic region (precisely to say, $\lambda_2$ is somehow positive). It is interesting to note that $\lambda_2$ behaves similary with $T_{ch}$ as a function of $V_o$.

## FIRING WAVE INSTABILITY

The FWI of the filamentary currents has been modelled in analogy with the dynamical behaviors in the *neural network*.[1-4] Previously, we have reported some bifurcations[1,2,9] in the experiments. However, bifurcation routes from periodic states to chaos are not so fine, and the FWI system includes some unresolved problems on the precise description of the physical process and on the *determinism*. Firing of the filamentary current occurs in a stochastic manner under photoexcitation, but it forms the firing density wave in the macroscopic time scale.[1-4] Signals of the firing activity $S(t)$ is detected by a lock-in amplifier, where we use also the appied pulse voltage($\geq 8$ kHz) as a reference signal. Figure 9 (a)-(c) shows the output signals $S(t)$ and the phase portraits observed with

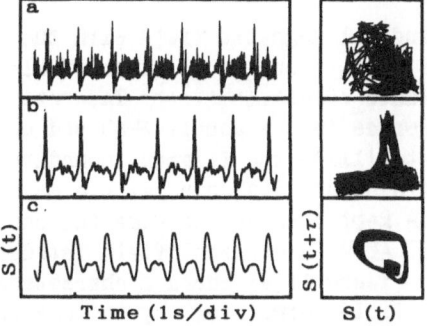

**Figure 9.** Waveforms $S(t)$ and the phase portraits in the FWI reegime. (a) $T$=1 ms, (b) $T$=10 ms and (c) $T$=100 ms.

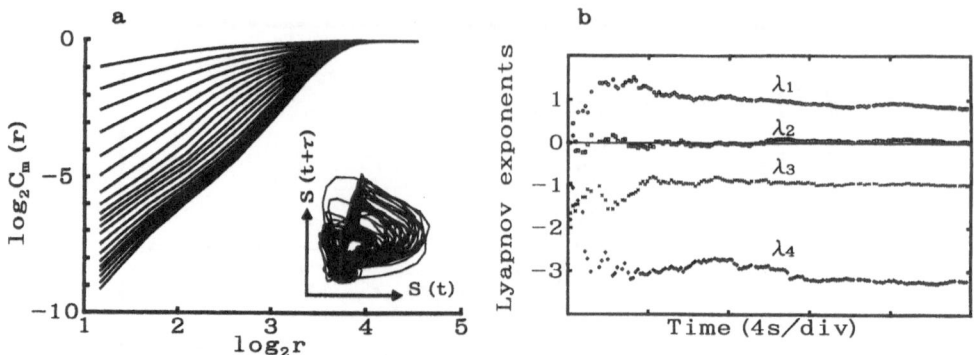

Figure 10. (a) Correlation integral $C_m(r)$ vs r in logarthmic scale, (b) Lyapunov spectrum.

different time constants of the lock-in amplifier, where the pulse voltage of 8 kHz is 0.952 V in height and the photoexcitation ($\lambda=812$ nm) is done using a light chopper (1.42 Hz). The oscillation is periodic in Fig.9(c), but additional small fluctuations are superimposed in Fig.9(a),(b), where the time constants of the lock-in amplifier are $T=1$ ms, $T=10$ ms and $T=100$ ms for Fig.9(a)-(c), respectively. The Lyapunov spectrum is 0,-,-,- for Fig.9(c), while +,-,-,- (the largest Lyapunov exponent $\lambda_1$ is nonnegligibly positive) for Fig.9(a),(b), where the spectra have been evaluated by a fixed condition of the algorithm. The latter is never expected from the chaos theory.[39] We attribute the spectrum +,-,-,- to the coexistence of different sizes of instabilities in space and time. By course-graining with $T=100$ ms (Fig.9(c)), only a large size instability is detected.

For the chaotic behavior, we show the correlation integral and Lyapunov spectrum $\lambda_i$ ($i=1,2,3,4$) in Fig.10 (a) and (b). The signal $S(t)$ was measured with $T=30$ ms. The correlation integral[9] $C_m(r)$ has been evaluated with the total data points $N=6144$. Slope $\tilde{\nu}$ of $\log_2 C_m(r)$ vs $\log_2 r$ increases with increasing the embedding dimension $m$ ($m=1 \sim 20$) and saturate above $m \geq 8$. The saturated value of $\tilde{\nu}=3.06 \pm 0.1$ gives the correlation dimension. From the Lyapunov spectrum +,0,-,- in Fig.10(b), chaos is relatively low dimensional, while the spatial pattern image of the filament during the electron beam-induced FWI is quite turbulent (Fig.1(b)).

## THE COI AND CROSS TALK COUPLING

Under the longitudinal magnetic field with the dc bias $V_o$, the filamentary current shows spontaneous oscillation (Table I) due to the inhomogeneous magnetoresistance effect.[7,8,27,28] Theoretically, an $S$-shaped $J$-$E$ characteristic is deformed into a doubly $S$-shaped curve. The inhomogeneous magnetoresistance destabilizes the filament boundary, so that the holding field $E_h$ shifts toward higher fields under the magnetic field, while the breakdown field $E_b$ is kept constant because the ohmic current is parallel to the magnetic field. Above a critical field intensity $B_c$, $E_h$ exceeds $E_b$, resulting in the COI. Figure 11(a) shows a characteristic of the carrier density vs electric field simulated by the one-level impact ionization model.[27] Since the middle branch is unstable, the current oscillates spontaneously when the electric field is biased in the range $E_b < E < E_h$. The ex-

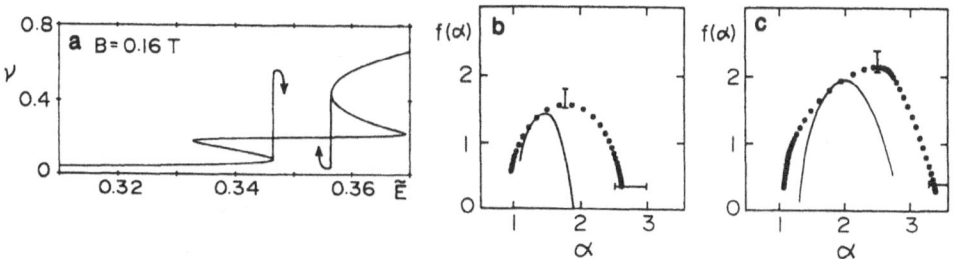

**Figure 11.** (a) Doubly S-shaped $\nu$-$E$ characteristic by one-level impact ionization model, and $f(\alpha)$ spectra in (b),(c) (see text).

perimental observations have been explained satisfactorily by the model simulations[8,27,28]. We have done quality test of $f(\alpha)$ spectrum[28] for the onset of chaos in Fig.11.(b) and for the developed chaos in Fig.11(c); solid circles are the experimental data and the solid lines are the model.

With a three-terminal circuit as shown in Fig.12, cross talk coupling between or among filamentary currents can be investigated[40]. We apply a source voltage (forward bias) $V_s$ by forming a Schottky-type point contact near the center position of the broad sample surface. Between the planar-type ohmic contacts, a balancer voltage $V_o$ is applied, where the current $I$ is measured by the voltage drop across a load resistor $R_L$(=500 $\Omega$). With $V_s$=1.78 V and $V_o$=0 V, filament 1 is turned on (see Fig.12). By applying a sawtooth bias instead of the dc bias $V_o$, an $I$-$V_o$ characteristic is measured. Figure 13 shows the $I$-$V_o$ curves under the longitudinal magnetic field with the intensity $B$. For $B$=3.6 mT, three types of the hysterisis are observed, where the branches denoted by a–d correspond to the flows of the filamentary currents schematically shown in Fig.12(a)-(d) respectively. We note that a small hysteresis observed at $V_o$ ~ 0.35 V in Fig.13 is reversed[40] in its shape, which is due to the cross talk current from filament 1(see Fig.12). In Fig.13, by increasing the magnetic field intensity, a hysteresis due to filament 2 shows narrowing and exhibit the COI with $B$ >96 mT. With further increase of $B$, the oscillatory region spreads into the wider range of $V_o$. With $B$> 222 mT, the current waveforms observed at a fixed bias show a clear evidence of cross talk coupling between the filaments 1 and 2 or between the filaments 1 and 3. However, the correlation

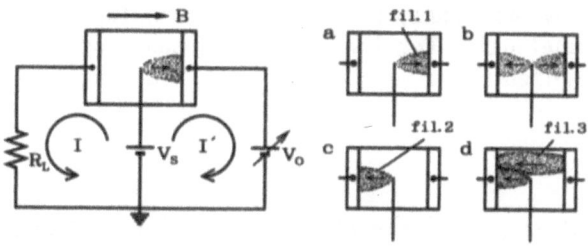

**Figure 12.** Three-terminal circuit. (a)-(d) correspond to the a-d branches of the $I$-$V_o$ curve in Fig.13.

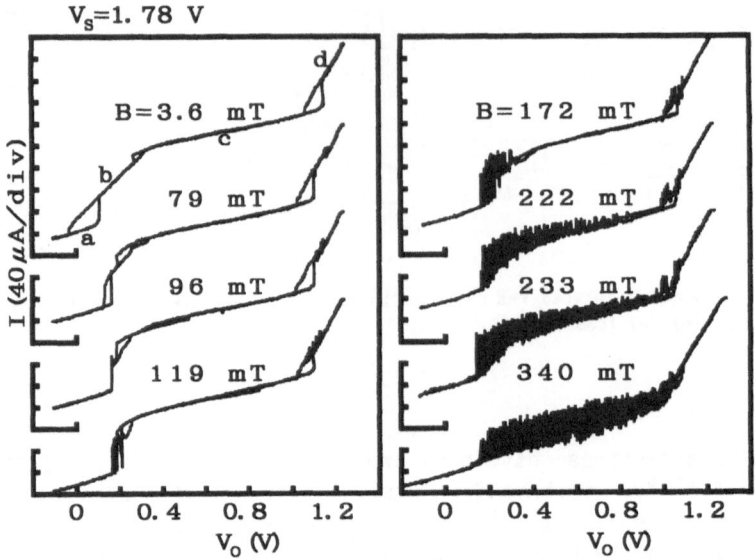

$V_S = 1.78$ V

B=3.6 mT    d    c    b    a

79 mT

96 mT

119 mT

B=172 mT

222 mT

233 mT

340 mT

I (40 μA/div)

$V_O$ (V)      $V_O$ (V)

0   0.4   0.8   1.2     0   0.4   0.8   1.2

**Figure 13.** $I$-$V_O$ characteristics under the various magnetic field intensities.

dimensions of the erratic current oscillations are considerably high (4 ~ 5) and no clear bifurcations have been observed.


## SUMMARY

We have reviewed various types of current instabilities in the impact ionization avalanche in semiconductor. The hierarchy of the instabilities appears by shifting the experimental parameter regime. A physical model of one-level impact ionization process has been discussed for the PDI, in a large part of the paper, since the process combined with the screening effect of the ionized impurity scattering is considered to be the true physical origin for impact ionization avalanche of shallow donors whose 1st excited Rydberg energy is not so large compared to the thermal energy, in the materials like GaAs and InSb. While the multi-level impact ioni- zation model by Schoell may give a true solution for the impact ioni- zation avalanche of shallow acceptors and deep impurities. Under the photoexcitation, the dynamical behaviors are rather more complex. For instance, analysis of Lyapunov spectrum suggests some coexistence of diff- erent sizes of instabilities in time and space. For the COI, the chaotic behaviors are low-dimensional, but crosstalk coupling in the COI regime is rather high-dimensional. For the crosstalk coupling, an extended model simulation is required.


## ACKNOWLEDGEMENT

The author thanks Y. Takahashi and A. Matsuura of Kobe University for their assistance and helpful discussions. He also thanks E. Schöll, J. Parisi, J. Peinke and U. Rau for fruitful discussions.

# REFERENCES

1. K. Aoki, T. Kobayashi and K. Yamamoto, J. Phys.Soc. Jpn. 51:2373 (1982).

2. K. Aoki and K. Yamamoto, Phys. Lett. 98A:72 (1983).

3. K. Aoki, K. Miyamae, T. Kobayashi and K. Yamamoto, Physica 117/118B+C:570 (1983).

4. K. Aoki, O.Ikezawa and K. Yamamoto, Phys. Lett. 106A:343 (1984).

5. K. Aoki, K. Yamamoto and N. Mugibayashi, J. Phys. Soc. Jpn. 57:26 (1988).

6. K. Aoki, Semicond. Sci. Technol. 7:B474 (1992).

7. K. Aoki, Y. Kawase, K. Yamamoto and N. Mugibayashi, J. Phys. Soc. Jpn. 59:20 (1990).

8. K. Aoki, Solid State Commun. 77:87 (1991); for the experiments see K. Aoki, T. Kondo and T. Watanabe, Solid State Commun. 77:91 (1991); K. Aoki, *Proc. 20th Int. Conf. on Phys. Semicond.*:2578 (World scientific, Singapore, 1990).

9. Y. Abe ed., Appl. Phys. A48:93-191 (1989); and references therein.

10. B. Röhricht, B. Wessely, J. Peinke, A. Mühlback, J. Parisi and P.P. Huebener, Physica 134B:281 (1985).

11. J. Peinke, B. Röhricht, A. Mühlbach, J. Parisi, Ch. Nöldeke, R.P. Huebener and O.E. Rossler, Z. Naturforsh 40a:562 (1985).

12. S.W. Teitsworth, R.M. Westervelt and E.E. Haller, Phys. Rev. Lett 51:825 (1983).

13. G. Gwinn and R.M. Westervelt, Phys. Rev. Lett. 59:157 (1987).

14. A. Brandl and W. Prettl, Festkörperprobleme 30:371 (1990).

15. J. Spangler, U. Margull and W. Prettl, Phys. Rev. B45:12137 (1992).

16. P. Bak, C. Tang and K. Wiesenfeld, Phys. Rev. Lett. 59:381 (1987).

17. P. Bak, C. Tang and K. Wiesenfeld, Phys. Rev. A38:364 (1988).

18. W. Clauss, A. Kittel, U. Rau, J. Parisi, J. Peinke and R.P. Huebener, Europhys. Lett. 12:423 (1990).

19. P. Grassberger and I.P. Procaccia, Physica 9D:189 (1983).

20. A. Wolf, J.B. Swift, H.L. Swinney and J.A. Vastano, Physica 16D:285 (1985).

21. M. Sano and Y. Sawada, Phys. Rev. Lett. 55:1082 (1985).

22. T.C. Halsey, M.H. Jensen, L.P. Kadanoff, I. Procaccia and B.I. Shraiman, Phys. Rev. A33:1141 (1986).

23. K. Aoki and K. Yamamoto, *Proc. 21th Int. Conf. on Phys. Semicond.* (Beijing,1992) in press.

24. H. Mori, H. Hata, T. Horita and T. Kobayashi, Prog.Theor.Phys. Supplement No.99:1(1989).

25. H. Mori, H. Okamoto and H. Tominaga, Prog. Theor. Phys. 85:6 (1991).

26. R.J.Deissler and J.D. Farmer, Physica D55:155 (1991).

27. K. Aoki, Phys. Lett. A152:485 (1991).

28. K. Aoki and T. Kondo, Phys. Rev. B45:3830 (1992).

29. R. Thom, *Structural Stability and Morphogenesis* (The Benjamin/Cummings, London, 1975).

30. K. Ikeda, K. Otsuka and K. Matsumoto, Prog. Theor. Phys. Supplement No.99:295 (1989).

31. F.T. Arechhi, G. Giacomelli, R.P. Ramazza and S. Residori, Phys.Rev.Lett. 65:2531 (1990).

32. K. Aoki, U. Rau, J. Peinke, J. Parisi and R.P. Huebener, J. Phys.Soc. Jpn. 59:420 (1990).

33. U. Rau, K. Aoki, J. Peinke, J. Parisi, W. Clauss and R.P. Huebener, Z. Phys.B-Condensed Matter 81:53 (1990).

34. E. Schöll, *Nonequilibrium Phase Transitions in Semiconductors* (Springer,New York, 1987).

35. E. Schöll, *Proc. 19th Int. Conf. on Phys. Semicond.* :1407 (Institute of Physics, Polish Academy of Sciences, 1988).

36. G. Hüpper and E. Schöll, *Proc. 20th Int. Conf. on Phys. Semicond.* :2463 (World Scientific, 1990).

37. R.S. Crandall, J. Phys. Chem. Solids 31:2069 (1970); Phys. Rev. B1:730 (1970).

38. K. Seeger, *Semiconductor Physics-An Introduction*, 3rd ed. :160 (Springer, Berlin, 1985).

39. H.G. Schuster, *Deterministic Chaos-An Introduction*, (VCH, 1988).

40. K. Aoki, T. Kondo and Y. Takahashi, Solid State Commun. 81:711 (1992).

# IMPACT IONIZATION OF EXCITONS AND DONORS IN CENTER-DOPED Al$_{0.3}$Ga$_{0.7}$As/GaAs QUANTUM WELLS

Helge Weman*

Department of Physics and Measurement Technology
Linköping University
S-581 83 Linköping
Sweden

## ABSTRACT

The effect on the low-temperature photoluminescence from AlGaAs/GaAs quantum wells of an electric field parallel to the quantum well plane has been investigated. The quantum well structures were grown by molecular-beam epitaxy, doped with Si in the center of the well, with well widths varying from 40 to 190 Å. Welldefined thresholds in the change of luminescence intensity from the free and bound excitons at fields near a few tens of V/cm have been observed, accompanied by a sharp increase in the current. The actual threshold is critically dependent on the well width of the sample, which is qualitatively explained by the differences in carrier mobility. Photoluminescence lifetime measurements of the excitons show that the recombination lifetime is unaffected during the quenching. The mechanism responsible for the observed effects is impact ionization of the excitons and shallow donors by the free carriers heated in the electric field.

## INTRODUCTION

Optical investigations on quantum wells (QWs) and superlattices have lead to a rather detailed understanding of quantum confined excitons in recent years.[1] In low-temperature photoluminescence (PL) experiments on such structures, both intrinsic (free exciton (FE)) and extrinsic (bound exciton (BE)) peaks have been found to be strongly dependent on the strength of the confinement.[2] The application of external perturbations such as electric, magnetic or strain fields in conjunction with PL is expected to yield additional information about the properties of these structures.

Méndez et al.[3] were the first to study the electric field induced effects on the low-temperature PL in undoped QWs. The electric field was applied perpendicular to the plane of the well and the excitonic features were found to shift to lower energies (Stark shift) and decrease in intensity, by increasing the electric field to some tens of kV/cm. Chemla et al.[4] were later able to observe the excitonic Stark shift even at room-temperature, and Miller et al.[5]

*Negative Differential Resistance and Instabilities in 2-D Semiconductors*
Edited by N. Balkan *et al.*, Plenum Press, New York, 1993

409

have demonstrated optical bistability, level shifting and modulation. Many other studies have been performed on the perpendicular field effect, and the observations are very well explained by the quantum-confined Stark effect, which is due to the field induced modifications of the electron and hole wave functions.[6,7] An applied electric field *parallel* to the QW plane is expected to produce large effects on the PL of excitons at low temperatures. In analogy to bulk semiconductors, carriers can be accelerated by the electric field to kinetic energies which are large enough to ionize the excitons by direct impact.[8] In bulk semiconductors a sharp threshold for impact ionization of excitons is usually observed; the threshold fields can be as low as a few V/cm in GaAs.[9-11] The first study on impact ionization in center-doped GaAs QWs were reported by the authors recently, where the threshold field was found to be critically dependent on the well width of the sample.[12] In this paper we review some of the results on the study of the effect of a parallel electric field on the free and bound excitons in AlGaAs/GaAs QWs.

## EXPERIMENTAL

The $Al_xGa_{1-x}As$/GaAs QW samples used in this study were grown by molecular-beam epitaxy (MBE) on semi-insulating (100) GaAs substrates. A buffer layer consisting of 500 Å GaAs, followed by a smoothing superlattice, and a 0.5 μm $Al_xGa_{1-x}As$ layer were grown on top of the substrate. The multiple-quantum-well (MQW) structure consisted of 50 periods of Si-doped GaAs wells between barriers of 150 Å thick $Al_xGa_{1-x}As$. The thickness of the QWs ($L_z$) was varied from 40 to 190 Å, between different wafers and the Al composition of the $Al_xGa_{1-x}As$ buffer layer and barriers was kept at $x = 0.3$. A schematic drawing of the sample structure and band diagram is shown in Fig. 1.

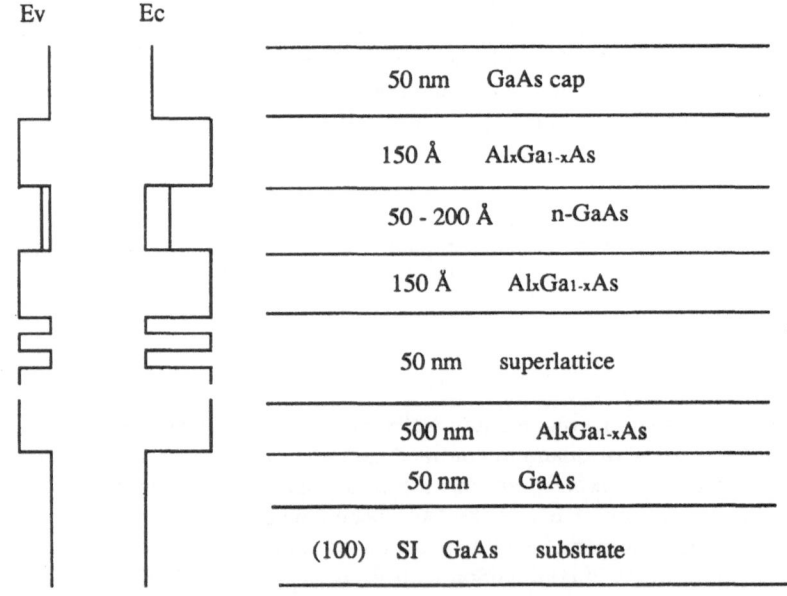

Figure 1. A schematic drawing of the layer-structure and band-diagram of the samples used.

The actual values of the well width and Al $x$ composition were determined by the quantum well FE and $Al_xGa_{1-x}As$ emission peak positions as observed by PL. The structures were doped with Si in the central 20% of the QW to a concentration of $5 \times 10^{16}$/cm$^3$. The nominal QW growth temperature was 690°C. Two ohmic contacts were made on the upper

(100) surface of the structures, separated by about 5 mm, by depositing In and strip annealing at 420°C for 2 min in a forming gas ambient.

The experimental setup used in this study is schematically drawn in Fig. 2. For above band gap excitation we used the 488 or 514 nm line from an Ar+ ion laser. A Ti:Sapphire laser pumped by an Argon laser (all lines), was used for selective excitation. The optical beam was mechanically chopped at around 1.5 kHz and focused to about a 0.5 mm diameter spot with an optical power of approximately 2 mW. The samples were mounted in a liquid helium cryostat and cooled to a temperature of 1.4 K. The luminescence was dispersed through a 1 m spectrometer and detected with a cooled GaAs-photomultiplier and a lock-in amplifier. PL decay-time measurements were performed with 5 ps pulses at 1.711 eV from a dye laser (Pyridine 2), synchronously pumped by a mode-locked argon laser, using a conventional time-correlated photon counting system to detect the decays. An electric field was applied parallel to the surface and the resulting photocurrent was measured with a curve tracer. The electric field quoted is approximated by dividing the applied voltage by the contact spacing.

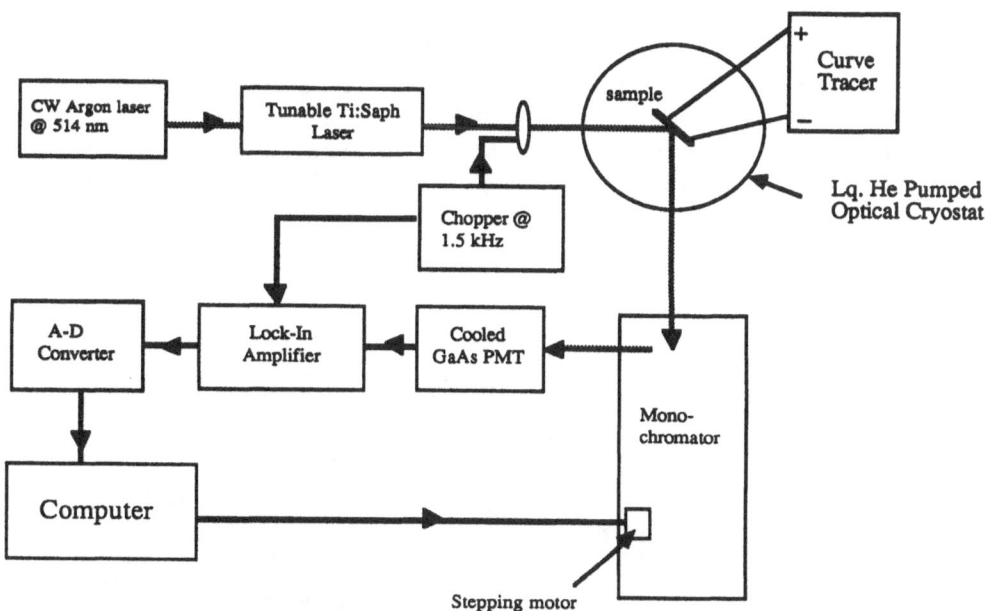

**Figure 2.** Schematic block diagram of the experimental setup.

## EXPERIMENTAL RESULTS

The solid line in Fig. 3 shows the PL spectrum at 1.4 K of a Si-doped 145 Å MQW. The laser energy was 1.746 eV, which is well above the expected FE and BE ground state transitions, but below the band gap energy of the barriers (1.88 eV). The FE ($n$=1 heavy-hole free exciton) and Si donor BE peaks overlap, with a separation in energy (the BE binding energy) of about 1.85 meV; this is close to the value of 2.0 meV obtained by Reynolds et al..[13] The application of an electric field reduces the intensity of the PL lines as can be seen in the dashed (24.5 V/cm) and dotted (49 V/cm) spectrum in Fig. 3 for the same sample. The PL quenching is accompanied by a small red shift of the BE energy peak of about 0.3 meV. This quenching occurs rapidly for fields above a threshold field $F_t$; below $F_t$ the luminescence appears unchanged. The decrease of the BE intensity, starting at $F_t$, is monotonic with a 40 % decrease around 60 V/cm. The FE on the other hand, decreasing to about 10 % around 30

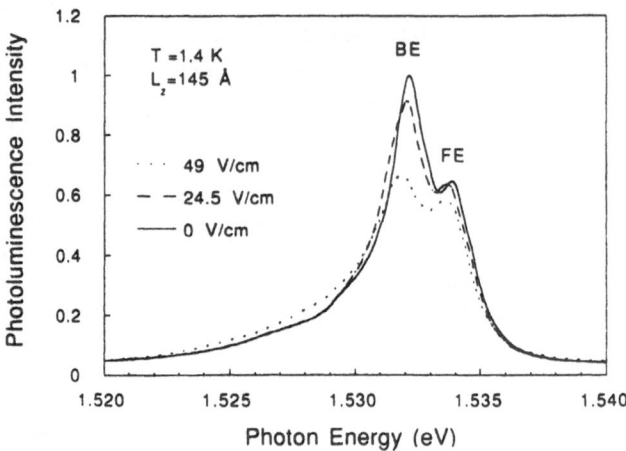

**Figure 3.** Photoluminescence (PL) spectrum (solid line) at 1.4 K of a 145 Å Si-doped $Al_xGa_{1-x}As/n$-GaAs MQW sample excited with a photon energy of 1.746 eV and a power of about 2 mW. The free exciton (FE) and bound exciton (BE) peaks are separated about 1.85 meV. The dashed (24.5 V/cm) and dotted curve (49 V/cm) show the corresponding results for the same sample with a parallel electric field applied.

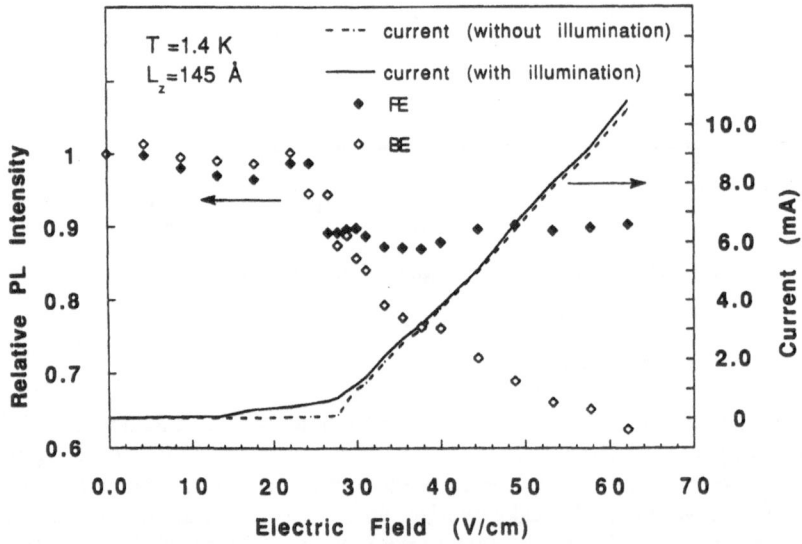

**Figure 4.** Relative PL peak intensity for the FE and BE (left scale) and current (right scale) vs applied electric field in the 145 Å Si-doped $Al_xGa_{1-x}As/n$-GaAs MQW sample. The current through the sample without illumination (dashed line) is due to donor electron impact ionization.

V/cm, stays more constant at higher fields. We have found that this effect is independent of the excitation photon energy in the range: 1.77-1.53 eV. In this range we are well below the band gap of the $Al_xGa_{1-x}As$ barrier (negligible absorption) so that we can exclude any electric field effects in this region. With excitation photon energies higher than the band gap of the barrier the PL quenching was very much reduced. With a fixed excitation photon energy it was found that the PL quenching rates depend on the excitation intensity.

To determine the actual threshold for the decrease in PL intensity we measured the relative FE and BE PL intensity as a function of field, together with the absolute value of the current flowing through the sample. The results are shown in Fig. 4 for the 145 Å MQW sample. It can be seen that the threshold for decrease in PL intensity and the threshold for increase of the current occur at approximately the same value of the applied field. At low fields the current is almost zero due to donor electron freeze out. Without illumination the

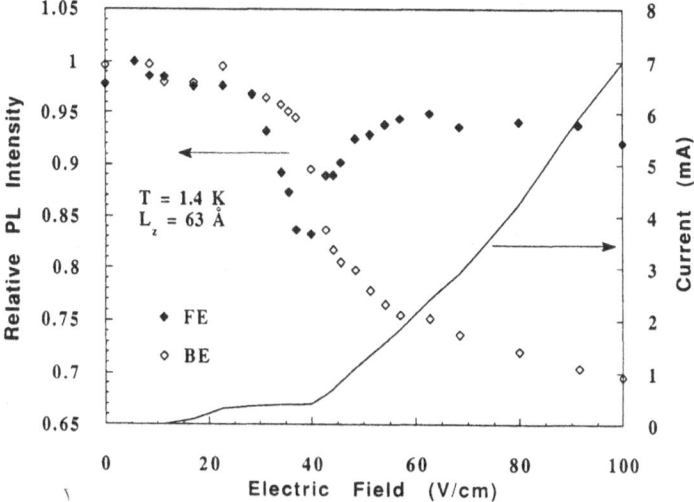

**Figure 5.** Relative PL peak intensity for the FE and BE (left scale) and current (right scale) vs applied electric field in the 63 Å Si-doped $Al_xGa_{1-x}As$/n-GaAs MQW sample.

current shows a sharp breakdown field due to impact ionization of the neutral donors. This field was always found to be slightly higher (a few V/cm) than the threshold field for the decrease in PL intensity. The slope of the *I-V* curve at higher fields is only slightly increased by photoexcitation. From these observations we conclude that the current measured with illumination is due primarily to an increase in the concentration of free carriers resulting from the PL quenching near the threshold (the non-linear part near the threshold), whereas it results primarily from the released donor electrons at higher fields (the linear part of the *I-V* curve). For the 145 Å MQW the threshold is found to be close to 22 V/cm for the BE and 25 V/cm for the FE. Qualitatively similar effects have also been observed at different threshold fields for samples with other well widths between 190 and 40 Å. The relative FE and BE PL intensity and the current as a function of field for a 63 Å MQW sample is shown in Fig. 5 . It can be seen that the BE intensity monotonically decreases in intensity from 37 V/cm. The FE decreases in intensity from approximately 32 V/cm, and then starts to increase again around 40 V/cm, before saturating at even higher fields.

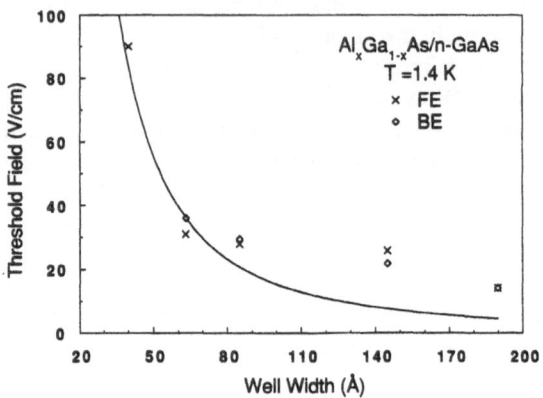

**Figure 6.** The threshold field ($F_t$) for the onset of the FE and BE PL quenching versus well width for the five Si-doped $Al_xGa_{1-x}As/n$-GaAs MQW samples studied in this work. The full line shows the $F_t \propto 1/L^{1.84}$ dependence that is expected for the 40 Å and 63 Å MQW sample if the carrier mobility, due to interface roughness, is dominating the threshold dependence.[21]

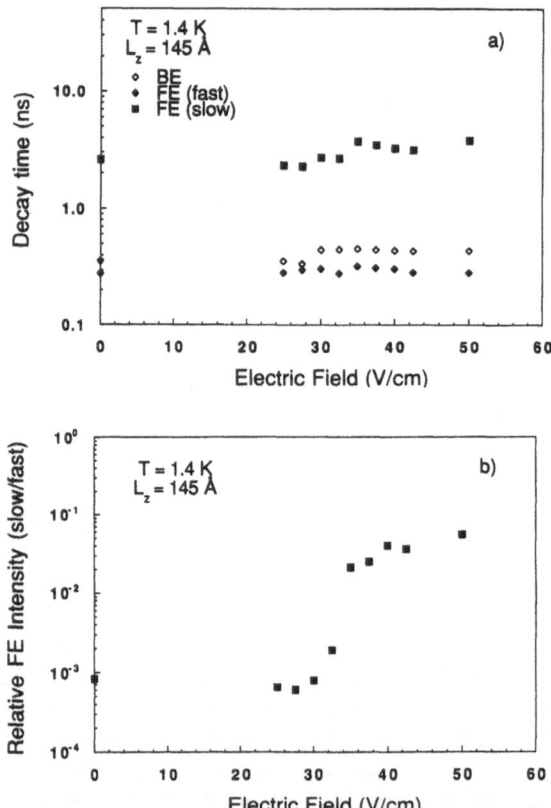

**Figure 7.** (a) The BE decay time and the fast and slow component of the FE decay time vs applied electric field. (b) The ratio dependence of the integrated FE intensity of the slow and fast decay component.

In Fig. 6 the threshold field dependence for the change in the FE and BE intensity on the quantum well thickness is summarized. These data show that the threshold field increases with decreasing well thickness. This consistent trend is exhibited both by the FE and BE threshold fields.

We have also measured the recombination lifetime at various electric fields for the FE and BE of the 145 Å MQW sample at 1.4 K. Without any electric field applied the FE and BE have a single exponential decay with a total lifetime of 280 ps and 360 ps. We have not seen any significant change in the lifetime when the electric field is applied, even when the PL intensity starts to be affected by the electric field, as shown in Fig. 7 (a) for the FE(fast) and BE. However near the threshold we observe a very weak component with a much longer decay time, which is gradually increasing in intensity as the field is increased above the threshold. Making a biexponential fit to the decay time we find that the longer lifetime is close to 3 ns and rather constant up to 60 V/cm, also shown in Fig. 7 (a), (FE(slow)). The ratio dependence of the integrated FE intensity of the slow to the fast component is shown in Fig. 7 (b).

## DISCUSSION

The parallel electric field induced PL quenching of excitons at fields as low as a few tens of V/cm, as shown here, is clearly very different from the perpendicular-field case (quantum confined Stark effect) where tens of kV/cm are needed for any PL effects.[3] The effect is also clearly different from the parallel electric field effects that are observed in absorption[7] where again very high electric fields are needed (tens of kV/cm) before the excitons were seen to be affected. The sharp onset for PL quenching, and the low electric fields required, strongly suggest impact ionization as the responsible mechanism. Impact ionization of excitons cannot be observed in an absorption experiment which studies only the creation of excitons in an electric field; on the other hand any electric field effect which affects the excitons before their annihilation will be observed in PL. Impact ionization has been discussed as a possible mechanism to explain some previous PL experiments carried out with a perpendicular field,[3,14,15] but the results are difficult to interpret due to the Stark effect.

Assuming that impact ionization is the mechanism which explains the data, we can outline the important ingredients for a theory. Impact ionization of a free or weakly bound exciton requires that the impacting carrier must have a kinetic energy which is comparable with the binding energy of the exciton.[16,17] These carriers are heated by the electric field. We assume the distribution function for hot carriers to be defined in terms of the carrier temperature $T_c$. In steady state the carrier temperature is defined by equating the power per carrier $P = \mu F^2$ gained from the electric field $F$ to the power per carrier $R(T_c)$ lost to processes such as phonon emission: $P = R(T_c)$. In this expression $\mu$ is the mobility which may be field dependent. The major contributions to the carrier energy relaxation rate $R$ are acoustic and optical phonon emission; for this case the carrier temperature for a given electric field and sample mobility can be readily computed. Given the carrier temperature, the impact ionization coefficient $A_{ii}$ has an Arrhenius dependence on the exciton binding energy $E_b$: $A_{ii} = A_0 \exp(-E_b/T_c)$.[17]

The factors which control the threshold field can be defined by a simple criterion. By analyzing the kinetic equations,[16] we find that the threshold occurs when the impact ionization rate becomes comparable with the exciton recombination rate. Assuming impact ionization by electrons with density $n$, the impact ionization rate is $nA_{ii}$. Also assuming that the recombination rate is dominated by the radiative recombination rate $R_x$, the threshold occurs when $nA_{ii} = R_x$. Furthermore, the energy relaxation rate is nearly linear at the onset of impact ionization. Thus we simply assume a linear energy relaxation rate, $R = c_0 T_c$, and obtain $T_c = \mu F^2/c_0$. Assembling these relationships, we find that the threshold field $F_t$ is defined by the expression:

$$F_t^2 = c_0 E_b/[\mu \ln(nA_0/R_x)] \qquad (1)$$

Using this expression, we can qualitatively analyze the physics controlling impact ionization to attempt an understanding of the well width dependence. By inspection, an increase in the threshold field can be caused by an increase in the exciton binding energy $E_b$, the radiative rate $R_x$, or the energy relaxation coefficient $c_0$. It can also be caused by a decrease in the free carrier mobility $\mu$, the free carrier concentration $n$, or the impact ionization coefficient $A_0$.

We first consider ionization energies for excitons and donors. These energies depend on the dissociation paths. FEs can dissociate into a free electron and a free hole with a dissociation energy which is approximately equal to the exciton binding energy. It should however be noted here that the "free" exciton observed in the PL spectrum in Fig. 3, is not really free, but localized in potential fluctuations due to roughness at the interfaces of the QW. This localized exciton (LE) is essential in treating the kinetics between the FE, LE and the BE correctly. The LE should be modeled as is the BE, that is as a trapped exciton with a specific binding energy of the same order as the BE.[18] Therefore when we look at the impact ionization kinetics for the "free" exciton, the situation is analogous to the BE ionization kinetics.

For a 145 Å QW, for example, the FE binding energy $E_b = 8$ meV.[19] The BEs we observe are likely trapped by neutral donors. These excitons have at least three dissociation paths. First, they can be released as FEs with a dissociation energy of 1.8 meV.[13] This low energy process is the only one which does not lead to release of charge carriers. Secondly they can be completely dissociated into an electron and a hole with a dissociation energy which is approximately 9.8 meV = 8.0 meV + 1.8 meV. Thirdly, they can remain bound while releasing an electron. The LE dissociation can lead to either a FE or a free electron and hole as a final state. Finally, a neutral donor can release an electron leaving a charged donor; the dissociation energy is approximately 11meV.[20] We relate the fast decay time (280 ps) to the LE recombination and the slow decay time (3 ns) to the FE recombination. This interpretation is supported by temperature dependent measurements and a blueshifted luminescence tail observed at the FE energy position at fields well above the threshold. The field independence of the decay times indicates that temperature effects are negligible during the impact ionization.

Next we consider the well-width dependence of these energies as an explanation for the data. Each of these dissociation energies has a weak dependence on well width with a maximum in binding energy at approximately 100 Å for the BE,[13] and approximately 35 Å for the FE binding energy.[19] This weak variation is not enough to explain the dependence of threshold field on well width.

A mobility which decreases with decreasing well width would, however, qualitatively explain the data. Such a decrease is expected if interface roughness were the dominant scattering mechanism.[21] This effect has been shown to dominate for modulation doped QWs with well widths below 60 Å, where the mobility was found to be proportional to about $L^{3.4}$. Using this relation with Eq. (1) we find that $F_t \propto 1/L^{1.84}$ for the 40 Å and 63 Å MQW sample (assuming that the other contributions are negligible between these two MQWs). This dependence is indicated by the full line in Fig. 6, showing a very good agreement with the actual data points for the 40 Å and 63 Å MQW sample. For the 145 Å and 190 Å MQW sample we believe that ionized impurity scattering is dominating, whereas the 85 Å MQW is in an intermediate range where both scattering events can contribute.

Since the impact ionization takes place under freeze out condition (no free carriers) we were unable to measure the Hall mobility in our samples under the right experimental conditions (1.4 K). However from Van der Pauw-Hall experiments at 10 K we have measured mobilities of close to $1 \times 10^4$ cm$^2$/Vs for the 85, 145 and 190 Å MQW samples, which is more than an order of magnitude lower (due to the intentional doping of the wells) than the mobilities in the samples used for the impact ionization studies of bulk GaAs (intentionally undoped).[9-11] This explains why we see threshold fields that are more than an order of magnitude larger than in the bulk GaAs study.[9] It was also recently shown by Masselink[22] that for center doped GaAs QWs, where ionized-impurity scattering is the dominating scattering mechanism, the mobility is even smaller than for an identical doping density in bulk GaAs.

The free carrier density $n$ dependence on well thickness is another possible explanation for the data. This quantity is also very difficult to measure, especially because it varies rapidly near the threshold. By inspection of Eq. (1), an increase in $n$ would reduce the threshold field. To test this relationship, we have varied the laser intensity while measuring the threshold voltage. At the highest intensities the photo-injected carrier density exceeds the background free carrier density and thus controls the threshold field; these data are in qualitative agreement with Eq. (1). However, since the threshold field has a weak, logarithmic dependence on the carrier density, we can rule this out as an explanation for the well width dependence.

Other effects seem to be weaker candidates to explain the data. For example, the energy relaxation rate could have a large effect because it enters Eq. (1) linearly. However, other work suggests that $c_0$ may decrease rather than increase as the quantum well width is decreased.[23] The recombination rate $R_x$ does not appear to explain the data; it has a weak and nonsystematic variation with well width.[12] A final candidate to explain the data is the impact ionization prefactor $A_0$. This quantity has a monotonically increasing dependence on increasing well width. Thus it has the qualitative behaviour needed to explain the data. However, the threshold field has a weak, logarithmic dependence on this quantity. Thus the variation of $A_0$ is not a strong candidate to explain the data.

The weak "FE" PL intensity dependence on applied electric field seems to be related to the donor electron impact ionization (onset around 28 V/cm). The dramatic free electron increase (several orders of magnitude) close to this threshold increases the equilibrium FE concentration, which seems to be canceled out by the "FE" impact ionization so that the PL intensity is close to constant. This effect is even more pronounced in other samples, like the 63 Å MQW sample (Fig. 5), where the "FE" PL intensity was found to increase above the threshold for the impact ionization of the donor electron. The donor BE however continues to decrease in intensity, since there are less available neutral donors for the excitons to be trapped at after the donor electron is impact ionized.


## CONCLUSIONS

In conclusion, we have shown that an electric field parallel to the plane of center-doped $n$-GaAs quantum wells, affects the free and bound exciton luminescence at fields of 10-50 V/cm. The luminescence intensity of the donor bound exciton is decreased beyond a critical threshold field, accompanied by a sharp increase in the current which is mainly due to the impact ionization of the shallow donor electron. The free exciton intensity dependence is more complex and seems to be related to the impact ionization of the donor electron. The critical threshold field is increasing with decreasing well width, which is qualitatively explained by the differences in carrier mobility. Photoluminescence lifetime measurements of the excitons show that the recombination lifetime is unaffected during the quenching. However near the threshold a weak component with a much longer decay time is observed, which is gradually increasing in intensity as the field is increased above the threshold. The fast decay is attributed to the recombination of excitons localized in potential fluctuations at the interfaces, and the slow decay to free exciton recombinaton. The mechanism responsible for the change in luminescence intensity is due to the impact ionization of the excitons as well as the shallow donor electrons, by free carriers that are heated in the electric field. The field independence of the decay times indicates that the lattice temperature is stable during impact ionization.

## ACKNOWLEDGMENTS

I am grateful to my collaborators on this project G.M. Treacy, H.P. Hjalmarson, K.K. Law, P. Bergman, J.L. Merz, and A.C. Gossard. I also want to thank B. Monemar for a critical reading of the manuscript. This work was carried out under the NSF Science and Technology Center for Quantized Electronic Structures (QUEST). The author was partially supported by a post doctoral scholarship from the Swedish Natural Science Research Council (NFR).

## REFERENCES

*Work performed while a post.doc at Center for Quantized Electronic Structures (QUEST), University of California, Santa Barbara, CA 93 106, USA.

1.  C. Weisbuch and B. Vinter, eds., "Quantum Semiconductor Structures", Chapter III, Academic Press, San Diego, (1991).
2.  M.D. Sturge and M.H. Meynadier, eds., "The optical properties of semiconducting quantum wells and superlattices", *J. Luminesc.*, 44: 199 (1989) and 45: 69 (1990).
3.  E.E. Méndez, G. Bastard, L.L. Chang, and L. Esaki, *Phys. Rev. B* 26 : 7101 (1982).
4.  D.S. Chemla, T.C. Damen, D.A.B. Miller, A.C. Gossard, and W. Wiegman, *Appl. Phys. Lett.*, 42: 864 (1983).
5.  D.A.B. Miller, J.S. Weiner, and D.S. Chemla, *IEEE J. Quantum Electron.*, QE-22: 1816 (1987).
6.  D.A.B. Miller, D.S. Chemla, T.C. Damen, A.C. Gossard, W. Wiegmann, T.H. Wood, and C.A. Burrus, *Phys. Rev. Lett.* 53 : 2173 (1984).
7.  D.A.B. Miller, D.S. Chemla, T.C. Damen, A.C. Gossard, W. Wiegmann, T.H. Wood, and C.A. Burrus, *Phys. Rev. B* 32 : 1043 (1985).
8.  V.M. Asnin, A.A. Rogachev and S.M. Ryvkin, *Fiz. Tekh. Polukrov.*, 1 : 1740 (1967) [*Sov. Phys.-Semicond.*, 1: 1445 (1968)].
9.  W. Bludau and E. Wagner, *Phys. Rev. B* 13 : 5410 (1976).
10. M. Yamawaki and C. Hamaguchi, *phys. stat. sol.* (b) 112 : 201 (1982).
11. Y. Horikoshi, A. Fischer, and K. Ploog, *Appl. Phys. A* 39 : 21 (1986).
12. H. Weman, G.M. Treacy, H.P. Hjalmarson, K.K. Law, J.L. Merz, and A.C. Gossard, *Phys. Rev. B* 45 : 6263 (1992).
13. D.C. Reynolds, C.E. Leak, K.K. Bajaj, C.E. Stutz, R.L. Jones, K.R. Evans, P.W. Yu, and W.M. Theis, *Phys. Rev. B* 40 : 6210 (1989).
14. R.C. Miller, and A. C. Gossard, *Appl. Phys. Lett.* 43 : 954 (1983).
15. P.W. Yu, D.C. Reynolds, K.K. Bajaj, C. W. Litton, W.T. Masselink, R. Fischer, and H. Morkoc, *J. Vac. Sci. Technol. B* 3 : 624 (1985).
16. D.L. Smith, D.S. Pan, and T.C. McGill, *Phys. Rev. B* 12 : 4360 (1975).
17. N. Sclar and E. Burstein, *J. Phys. Chem. Solids*, 2 : 1 (1957).
18. G. Bastard, C. Delalande, M.H. Meynadier, P.M. Frijlink, and M. Voos, *Phys. Rev. B.*, 29 : 7042 (1984).
19. R.L. Greene and K.K. Bajaj, D.E. Phelps, *Phys. Rev. B* 29 : 1807 (1984).
20. W.T. Masselink, Y-C. Chang, H. Morkoc, D. C. Reynolds, C.W. Litton, K.K. Bajaj, and P.W. Yu, *Sol. State. Commun.* 29 : 205 (1986).

21. R. Gottinger, A. Gold, G. Abstreiter, G. Weimann, and W. Schlapp, *Europhys. Lett.*, 6 : 183 (1988).
22. W.T. Masselink, *Phys. Rev. Lett.*, 66 : 1513 (1991).
23. D.M. C Smith, H.P. Hjalmarson, E.D. Jones, J.E. Schirber, T.J. Drummond, and L.R. Dawson, *Inst. Phys. Conf. Ser.* No 96: , IOP Publishing Ltd, Bristol , England, Ed. J.S. Harris, p 325, (1989).

# HOT EXCITON LUMINESCENCE IN QUANTUM WELLS AS A SPECTROSCOPIC TOOL

F. Calle[a], C. López[b], F. Meseguer, L. Viña, J.M. Calleja, C. Tejedor

Instituto de Ciencia de Materiales de Madrid (CSIC) and Facultad de Ciencias, Universidad Autónoma de Madrid, E-28049 Madrid, Spain

## INTRODUCTION

Hot carrier effects are the subject of much interest in the study of transport properties in semiconductor heterostructures because of their importance in device reliability. The present work is devoted to some aspects of their optical behaviour. Apart from the intrinsic interest of light emission from hot levels, this phenomenon may reveal new information on the electronic structure and be used as a powerful instrument from the spectroscopic point of view. Advantages of this technique arise from the state of non equilibrium of the recombining carriers which lead to clearer selection rules and, eventually cast new light on the phonon-exciton interaction.

It may seem that photoluminescence (PL) and absorption are complementary phenomena as they involve, respectively, the emission or absorption of a photon at the expense of the annihilation or creation of an electronic excitation. Commonly in intrinsic semiconductor structures Coulomb coupling produces a strong correlation between electron and hole so that one should speak of excitons rather than electrons and holes. In the process of absorption a photon is destroyed to create an excitonic state. The process of luminescence, on the other hand, can be viewed as the destruction of an exciton state, and the emission of a photon of the same energy. The exciton can be in any of its states, either fundamental or excited.

When an exciton is created from an electron-hole pair, it tends to disappear emitting its energy by two competing processes, radiative or non-radiative [1]. In polar intrinsic semiconductors the non-radiative thermalization of excess energy mainly takes place by the emission of phonons, which is a fast process (for LO phonons, of the order of $1/\omega_{LO} \simeq 10^{-13}$ s [2]). Other mechanisms are the acoustic interaction or the exciton-exciton collisions. The alternative channel is the direct recombination of

[a]Present address: Dpto. Ing. Electrónica. E.T.S.I.T. UPM. Ciudad Universitaria, E-28040 Madrid, Spain

[b]Present address: Dpto. Ingeniería E.P.S., Universidad Carlos III de Madrid, Leganés, E-28913 Madrid, Spain

*Negative Differential Resistance and Instabilities in 2-D Semiconductors*
Edited by N. Balkan *et al.*, Plenum Press, New York, 1993

421

the exciton. However, this process is, usually, much slower (in the range of $10^{-9}$s) and phonon emission takes place preferentially, until the lowest energy state of the exciton is reached, where a single phonon cannot carry the exciton energy and radiative recombination occurs. The lifetime ratio $\sim 10^4$ provides a rough estimate of the ratio of intensities between the main luminescence peak and phonon replicas.

Under these considerations, luminescence only takes place as a transition from the ground state exciton: a single peak will appear corresponding to the fundamental excitonic gap. Under certain conditions, however, electron-phonon interaction can be such that the intermediate states "live" long enough to be able to recombine and produce luminescence, thus a series of peaks, starting at the gap and separated by (roughly) one phonon energy, $\hbar\omega_{LO}$, appear in the luminescence spectrum. This is what is commonly called "hot luminescence" as it involves hot electrons and holes that recombine before thermalizing and was firstly reported in bulk by the Leningrad Group [3, 4]. For a review see reference [5]. This kind of process has been profusely used to characterize different semiconductor structures inasmuch as properties such as $\Gamma \rightarrow L$ valley scattering time, valence band anisotropy, LO phonon renormalization, valence band reconstruction, etc. [6] strongly influence luminescence.

A non equilibrium thermal distribution of carriers produced by an intense light excitation above the gap energy or a heavy doping may, likewise, lead to a hot luminescence process [7]. In this case, due to the large difference in effective mass between valence and conduction bands, a small distribution of holes may recombine with a comparatively large distribution of electrons; thus a high probability exists for hot electrons to recombine and develop a hot luminescence tail in the spectra.

Inter-Landau level transitions under magnetic field in bulk materials may be favoured, under resonant conditions, to produce luminescence from the Landau ladder. In this case, again, an enhancement of intermediate states life-time must take place to allow for the electronic recombination rather than phonon emission [8]. The same effect can be achieved by very high pumping rates ($\sim$ MW cm$^{-2}$), though in this case a dense electron-hole plasma is created which draws the phenomenon close to that of doping structures [9].

Hot luminescence, however, can take on a somewhat different character in layered semiconductors like quantum wells and superlattices under moderate excitation rates. In this case, luminescence from nonrelaxed states associated with excited excitonic levels may appear in the spectra. The combined effects of two dimensional quantization and magnetic field have proved to narrow the levels and enhance the nonradiative life time of excited states of both heavy and light hole excitons. Luminescence at excitonic energies far beyond the gap may be observed.

The purpose of this work is to analyze some aspects of this hot luminescence (HL), its polarized character, and its behaviour under magnetic field. We will also perform some excitation experiments on this HL, and compare the results with the usual excitation of the cold exciton. The analysis of HL and its excitation show very neat selection rules that permit to identify optical transitions, to determine Zeeman $g$-factors of excitons, and to study the lifetime of excitonic excited states in semiconductor quantum wells.

## EXPERIMENTAL

The samples used for the present study are MBE grown high quality GaAs-AlAs multiple quantum wells. The results reported come, mainly, from one of the samples that consists of 30 periods of 100 Å wells and barriers. A second sample, containing 30 periods with 90 Å wells and 50 Å barriers, was studied, yielding similar results. They were grown on GaAs (001) oriented substrates on top of a thick buffer, and were quality tested by X-ray diffraction.

The samples were placed in a superconducting magnet with optical access, at 2 K under magnetic fields up to 13.5 T, directed perpendicular to the growth axis. The optical measurements were realized in back scattering (Faraday configuration). Laser exciting light from an Kr$^+$ pumped LD/700 dye was set to circular polarization, with the help of an achromatic $\lambda/4$ plate, either to $\sigma^+$ or $\sigma^-$ helicity. Pumping rates of tens of W cm$^{-2}$ were used, and temperatures of 10 K were derived from the fitting of the high energy tail of the main luminescence band. The emitted light was analysed into its circular components with the help of a second achromatic $\lambda/4$ plate and a linear polarizer. With this set up four optical configurations are possible, namely: $\sigma^+\sigma^+$, $\sigma^+\sigma^-$, $\sigma^-\sigma^+$ and $\sigma^-\sigma^-$, where the first element refers to the polarization of light impinging on the sample and the second to that of the detected luminescence, all of them in the laboratory framework. Light was detected, after being dispersed by a 1 m double spectrometer, by a GaAs cooled photomultiplier in normal photon counting mode.

## MAGNETO-OPTICS OF MULTIQUANTUM WELLS

Magnetic field is a very powerful tool for the study of electronic properties of semiconductor structures and similar material systems. The effect of magnetic fields on semiconductor QWs is severe as it changes one of the most important and intimate properties: dimensionality.

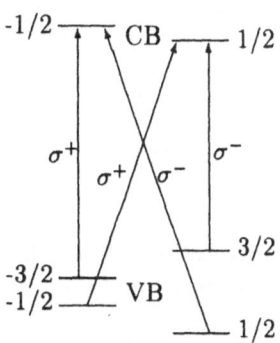

**Figure 1.** Valence and conduction levels in the $\mid j, m_j >$ representation. Arrows are labelled according to angular momentum conservation selection rules.

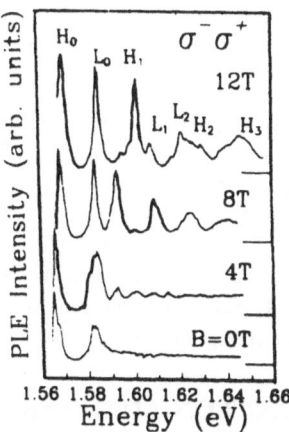

**Figure 2.** PLE spectra at several magnetic fields in the $\sigma^-\sigma^+$ configuration.

In terms of density of states, magnetic field quantization can be regarded as a squeezing of the, originally step shaped, DOS into a comb of sharp delta-like peaks, each corresponding to a Landau level. The main effect is, then, an accumulation of states in narrow, almost evenly spaced, regions. This localization in energy may, according to the uncertainty principle, lead to the enhancement of the levels nonradiative life time. In addition, optical transitions must obey new selection rules for angular momentum conservation that are summarized in Fig. 1, in the $\mid j, m_j >$ electron representation. The arrows represent photon absorption and $\sigma^+$ and $\sigma^-$ stand for the helicity of circularly polarized light.

Photoluminescence excitation (PLE) is an experiment in which detection energy is set to a position coincident with a certain emission, while exciting light is scanned. The emitted luminescence intensity is recorded, and yields information similar to that of absorption by states higher in energy than that whose PL is collected. Therefore, detecting at the lowest energy emission (heavy hole ground state), parallel polarization configuration ($\sigma^+\sigma^+$ or $\sigma^-\sigma^-$) PLE would show stronger heavy hole exciton transitions, while crossed polarizations ($\sigma^+\sigma^-$ or $\sigma^-\sigma^+$) would enhance light hole exciton transitions [10, 11]. For example, in Fig. 2, a series of PLE spectra are shown for different magnetic fields in the $\sigma^-\sigma^+$ configuration [12]. Here, $H_n$ and $L_n$ stand for the $n$th heavy or light magneto-excitonic state, respectively. The exciting light polarization selects which transitions are to be observed, while the emitted light polarization determines the relative intensities. Two sets of transition energies (ground and excited states of the $H_n$ and $L_n$ magneto-excitons) are illustrated in Figure 3 for different polarization configurations ($\sigma^+\sigma^+$ and $\sigma^-\sigma^+$, respectively), as a function of magnetic field. Anyway, rules are not always obeyed and both Stokes shifts and hole band mixing blur the picture.

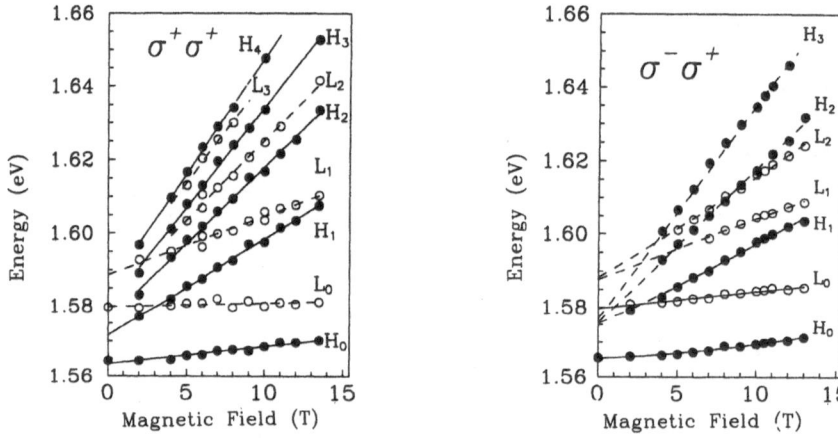

**Figure 3.** Energy positions of excitonic levels as obtained from PLE. The left panel shows results obtained in polarized configurations, whereas the right panel shows the same in depolarized configuration.

## HOT EXCITON LUMINESCENCE

The emission from the ground states of both $H_0$ and $L_0$ excitons, under magnetic field, in different polarization configurations, can be seen in Fig. 4, left. In contrast with the $H_0$ PL, and in spite of a much lower intensity, HL peaks from the light hole excitons are strongly polarized: both energy and intensity are configuration dependent according to the selection rules. The energy of the peaks do not depend on the exciting light polarization, however the exit polarization determines their position. Between outgoing $\sigma^+$ and $\sigma^-$, there is only a 0.5 meV splitting between the heavy hole exciton peaks, while the HL peaks are split by -3.5 meV. The analysis of HL allows to distinguish the third component of angular momentum, as they change with B according to $E_{spin} = g^* \mu B \sigma$. In particular, effective g-factors can be derived from the Zeeman splittings obtained from the comparison of $\sigma^+\sigma^+$ and $\sigma^-\sigma^+$ lines in Fig. 3, yielding $g_{hh} = 0.6$ and $g_{lh} = -4.3$ [13]. The same kind of measurement has been performed for a number of magnetic fields. In Fig. 4, right, three spectra at different magnetic fields are depicted in two polarization configurations. PL intensity increases monotonically with magnetic field, since it results from a long relaxation process. On the contrary, the intensity of polarized HL shows a striking dependence on B: for a fixed laser energy, magnetic field may locate different levels close to resonant absorption and enhance the emitted intensity. This is actually the case for $L_0$ at 9 T where the peak for $\sigma^+\sigma^-$ polarization is strengthened versus $\sigma^-\sigma^+$, as opposed to the 6 and 12 T cases. From these two figures one can deduce that it is very advantageous detecting HL rather than PL when obtaining excitation spectra: changing the polarization affects much more the light hole features than the heavy hole ones. This is an alternative way of obtaining the fan charts shown in Fig. 3, as will be explained later on.

Fitting of the exponential high energy tails of the luminescence peaks reveals a ground state exciton population in thermal equilibrium with the lattice, whereas hot carriers show higher temperatures as they depart from equilibrium: 20, 40 and 90 K respectively for $L_0$, $H_1$ and $L_1$.

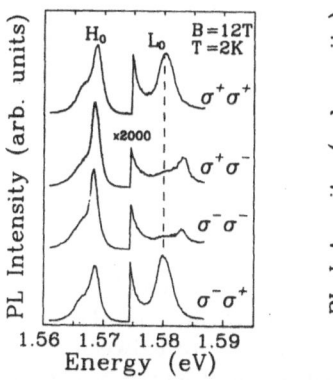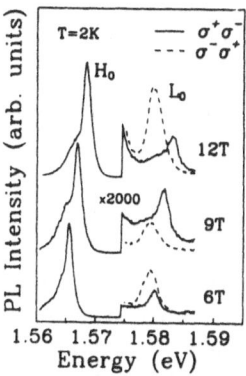

**Figure 4.** PL and HL in the four possible polarizations is shown. Left, at 12 T, in the four polarizations. Right, at different magnetic fields, in crossed and parallel configurations.

In order to study the dependence of HL on the exciting energy, several experiments were performed. It is sometimes not too easy to distinguish whether a spectral feature is a luminescent phenomenon or a Raman scattering process [14]. To check for this HL was excited, at a constant magnetic field, for different laser energies. The results can be seen in Fig. 5 (left), at 8T in the $\sigma^-\sigma^+$ polarization configuration. Three peaks can be observed: a main broad band corresponding to $L_0$ (fundamental light hole exciton) is accompanied by a sharp peak (LO phonon) with a satellite (interface phonon, IF) on the high energy side. When the laser energy is changed, the LO and interface phonon peaks move the same amount, whereas the HL peak remains in its position. For this configuration, $\sigma^-\sigma^+$, this band does not show any noticeable shift with field value.

Other HL lines can be shifted by the magnetic field. This can be clearly seen in Fig. 5 (right), where $H_1$ (first excited heavy hole exciton) energy is plotted for different

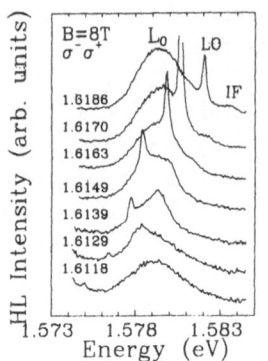 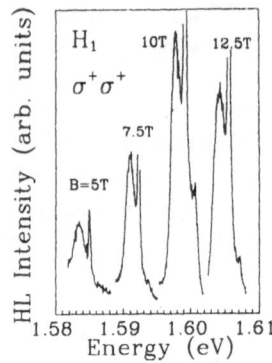

**Figure 5.** Left: $L_0$ HL is shown at a fixed magnetic field, and several laser energies. Right: $H_1$ HL at various magnetic fields.

fields. The accompanying peaks are, again, GaAs LO and interface phonons.

These results are summarized in Fig. 6 for $L_0$ and $H_1$ transitions. In PLE experiments, exciting light polarization determines the position of the peaks, whereas for HL measurements the observed energies depend on the detection polarization. Thus, results for $\sigma^+\sigma^+$ and $\sigma^-\sigma^+$ coincide, and correspond to those obtained in PLE with $\sigma^+$ incoming configuration (Fig. 3 left). One further point to stress is the fact that exciting light polarization is unimportant in HL, that is why $\sigma^+\sigma^+$ and $\sigma^-\sigma^+$ configurations cast the same results: it is sufficient to analyse the emitted light.

## HOT LUMINESCENCE EXCITATION

We can now turn to discuss luminescence excitation. It may be seen (left panel in Fig. 7) that in PLE the spectra are mainly determined by the incoming polarization: small differences appear between $\sigma^+\sigma^+$ and $\sigma^+\sigma^-$ or between $\sigma^-\sigma^-$ and $\sigma^-\sigma^+$. This is rather a problem as it hampers identification of levels: selection rules are obscure and one needs to study the behaviour under magnetic field in order to assign the

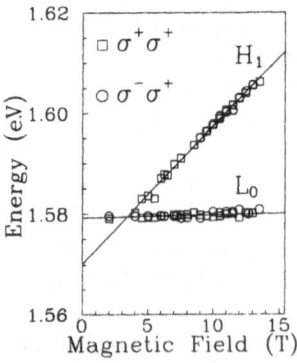

**Figure 6.** Summary of HL energies of both $L_0$ and $H_1$.

various peaks. In this case the assignment has been done, and can be seen in Fig. 3. However, when hot luminescence excitation (HLE) is performed (right panel in Fig. 7) sizeable differences appear between all four configurations.

Energies, as well as intensities, are much affected by polarization so that assignments become clearer and peaks are easily identifiable. Let us notice that, when detecting the emission of $L_0$, the selection rules are swapped: now, the $L_n$ excitonic peaks are more intense in the parallel configurations, and so are the $H_n$ peaks in the crossed ones.

There are two main reasons for so distinct behaviours of the selection rules in PLE and HLE. First, the Zeeman splitting of $H_0$ is much smaller than for $L_0$, which is a key point at high fields. Second, the Stokes shift affects the lowest energy optical transition, however, the hot states, due to a rapid recombination, are not affected by it. The difference is apparent in the spectra shown in Fig. 7 and 8 (left).

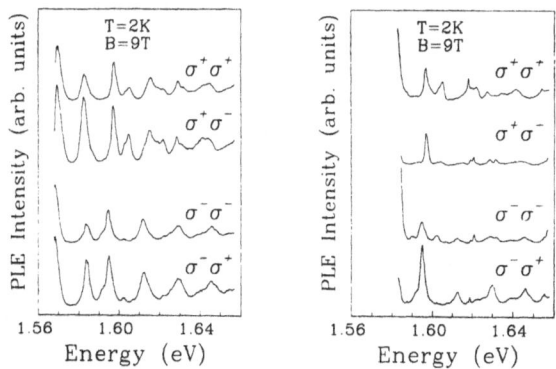

**Figure 7.** Left: PLE, detecting at the $H_0$ peak. Right: HLE, detecting at the $L_0$ peak. The four configurations are presented for B=9 T.

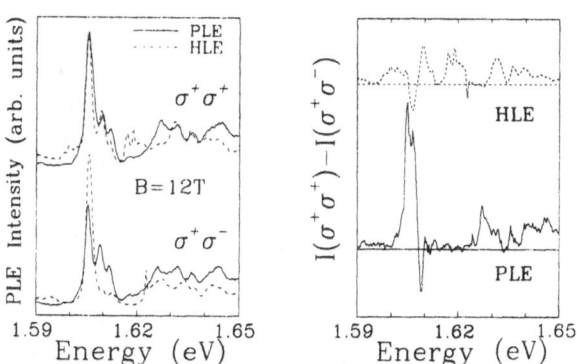

**Figure 8.** The left panel displays PLE (detecting at the $H_0$ peak) and HLE (detecting at the $L_0$ peak) at a fixed magnetic field (12T) in two configurations. In the right panel the different configurations have been subtracted to show the enhanced contrast of HLE.

Thus, one further advantage is that the lack of "contrast" between crossed and parallel configurations in the $H_0$ PLE is no longer a problem in the HLE. The increasing depolarization of the spectra with increasing magnetic field is much less felt when detecting HL: the selection rules for hot exciton recombination appear very neat and allow the unambiguous identification of the levels involved in the optical transitions. When subtracting spectra from conjugate configurations to emphasise selection rules, the augmented contrast in HLE becomes manifest. This has been done in the right part of Fig. 8 and one can see that PLE is much more featureless over large ranges of energy.

## ON LIFETIMES

Additional information on radiative lifetimes and their dependence on magnetic field can be obtained from a detailed analysis of the width and height of HL-peaks.

Let us assume a four-level system and concentrate on the recombination between the second and third levels with the ground level ($| 2\rangle \rightarrow | 1\rangle$ and $| 3\rangle \rightarrow | 1\rangle$). Denoting the pump rate ($| 1\rangle \rightarrow | 4\rangle$) by a characteristic time $n_0/\tau_p$, and radiative (nonradiative) lifetimes between levels $i$ and $j$ by $T_{ij}$ ($\tau_{ij}$), one gets the following dynamic equations [15]:

$$\frac{dn_3}{dt} = \frac{n_0}{\tau_p} - \frac{n_3}{\tau_{32}} - \frac{n_3}{T_{31}} \qquad \frac{dn_2}{dt} = \frac{n_3}{\tau_{32}} - \frac{n_2}{\tau_{21}} - \frac{n_2}{T_{21}}$$

In the steady state, where $dn_i/dt = 0$, bearing in mind that the intensities of recombination are proportional to the populations of the states, emission intensities result in

$$I_{31} \sim n_3 = \frac{n_0}{\tau_p} \frac{T_3}{1 + \frac{T_{31}}{\tau_{32}}} \qquad I_{21} \sim n_2 = \frac{n_0}{\tau_p} \frac{T_3}{1 + \frac{T_{31}}{\tau_{32}}} \frac{T_{21}}{\tau_{32}} \frac{1}{1 + \frac{T_{21}}{\tau_{21}}}$$

If we assume that $T_{21}/\tau_{21} \ll 1$, that is, all recombinations from the first state occur radiatively, one gets

$$\frac{I_{31}}{I_{21}} = \frac{\tau_{32}}{T_{21}},$$

which, in our case, has been found to be $\sim 10^{-3}$. Assuming a typical value of 500 ps as measured for this kind of structure [1],

$$T_{21} \sim 500\ ps \Rightarrow \tau_{32} \sim 500\ fs$$

This time is rather larger than those obtained for LO scattering in bulk GaAs [2], that are in the range of 150 fs. The fact that HL is observed implies very short recombination times and impedes non radiative scattering and, thereby, Stokes shifts.

In Fig. 9 we plotted the linewidth of the HL of $L_0$ peak at three selected magnetic fields, as a function of exciting energy. Although data dispersion is large, it may be seen that a decrease in linewidth of about 30% is attained at different exciting energies for each magnetic field. This decrease takes place at laser energies very near $L_0$ plus one LO phonon quantum (see Fig. 6). The laser energy at which this happens corresponds to the situation of resonant Raman scattering (RRS) [16], in which the scattered energy coincides with an actual electronic transition.

Most of the information gathered up to here could imply that HL supplies data not only overlapping with but, in some instances, complementary to that emerging from such sophisticated techniques as resonant Raman scattering. No Stokes shift affects the (fast) emission from non-relaxed states. This can be deduced from comparison between PL, HL, PLE and resonant Raman, illustrated in Fig. 10. PLE is determined by incoming polarization while PL, HL and outgoing RRS are set by the exit polarization. The $H_0$ PL peak should be located at higher energy (2 meV due to diamagnetic shift) than the first PLE peak if no Stokes shift were present (4 meV toward lower energy). Such a simple experiment as HL yields energy levels of higher excited states with the same accuracy as can be obtained in RRS.

It should be mentioned that HL, under appropriate magnetic field, can be used to determine the conditions for doubly resonant LO Raman scattering [16].

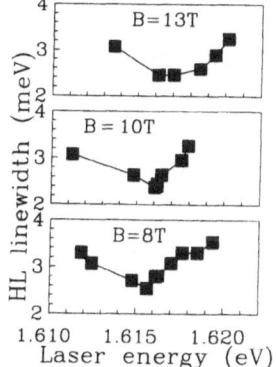

Figure 9. $L_0$ HL full width at half maximum, at three magnetic fields, as a function of exciting energy. It is important to notice that although laser energy is changing, HL energy stays the same.

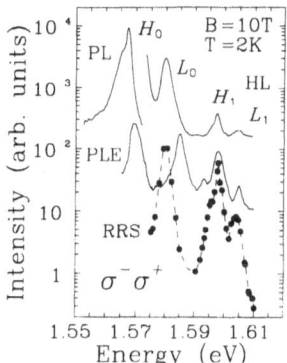

Figure 10. Full comparison between resonant Raman scattering intensity, PL and HL emissions, and PLE. Notice the direct comparison between the RRS and HL.

## SUMMARY

In conclusion, luminescence from hot excitons is a powerful tool for determining the electronic structure and properties of semiconductor heterostructures. The emission of HL, favoured in this case by the reduced dimensionality in the presence of a magnetic field, allows to determine the energies of several excitonic peaks. Hot luminescence excitation permits to identify more clearly the kind of excitons involved. On the other hand, hot luminescence can be considered as a sign of short lifetime for radiative recombination of excited states, that could be used as a test for semiconductor laser performance and a measure of scattering times.

## ACKNOWLEDGEMENTS

We are grateful to K. Ploog for providing the samples used in this experiment. This work has been partially financed by the Comisión Interministerial de Ciencia y Tecnología of Spain, Contract No. MAT-91-0201.

# REFERENCES

[1] T.C. Damen, J. Shah, D.Y. Oberly, D.S. Chemla, J.E. Cunningham, J.M. Kuo, Phys. Rev. **B 42**, 7434 (1990).

[2] G. Fasol, W. Hackemberg, H.P. Hughes, K. Ploog, E. Bauser, H. Kano, Phys. Rev. **B 41**, 1461 (1990).

[3] B.P. Zakharchenya, V.D. Dymnikov, I.Ya. Karlik, I.I. Reshina, J. Phys. Soc. Jpn. **49**, 573 (1980).

[4] D.N. Mirlin, I. Ya. Karlik, L.P. Nikitin, I.I. Reshina, V.F. Sapega, Solid State Comm. **37**, 757 (1981).

[5] S.A. Lyon, C.L. Petersen, "Ultrafast Laser Probe Phenomena in Bulk and Microstructure Semiconductors", II Proc. SPIE **942**, 264 (1988).

[6] G. Fasol, H.P. Hughes, Phys. Rev. **B 33**, 2953 (1986).

[7] B.J. Aitchison, N.M. Haegel, C.R. Abernathy, S.J. Pearton, Appl. Phys. Lett. **56**, 1154 (1990).

[8] F. Iikawa, T. Ruf, M. Cardona, Phys. Rev. **B 43**, 4849 (1991).

[9] R.Cingolani, G.C. La Rocca, H. Kalt, K. Ploog, M. Potemski, J.C. Maan, Phys. Rev. **B 43**, 9662 (1991).

[10] M. Potemski, J.C. Maan, A. Fasolino, K. Ploog, G. Weimann, Phys. Rev. Lett. **63**, 2409 (1989).

[11] L. Viña, F. Calle, C. López, J.M. Calleja, W.I. Wang, in Condensed Systems of Low Dimensionality, ed. J.L. Beeby. Plenum Press, New York (1991), p. 73.

[12] J.M. Calleja, F. Meseguer, F. Calle, C. López, L. Viña, C. Tejedor, K. Ploog, F. Briones, in Light Scattering in Semiconductor Structures and Superlattices, ed. D.J. Lockwood and J.F. Young. Plenum Press, New York (1991), p. 53.

[13] F. Meseguer, F. Calle, C. López, J.M. Calleja, L. Viña, C. Tejedor, K. Ploog, in The Physics of Semiconductors 1990, ed. E.M. Anastassakis and J.D. Joannopoulos. Vol. 21, World Scientific, Singapore (1990), p. 1461.

[14] A. Nakamura, C. Weisbuch, Solid State Electron. **21**, 1331 (1978).

[15] G. Bastard, Wave Mechanics Applied to Semiconductor Heterostructures, Les Editions de Physique, Paris (1988).

[16] F. Calle, J.M. Calleja, F. Meseguer, C. Tejedor, L. Viña, C. López, Phys. Rev. **B 44**, 1113 (1991).

COMBINED QUANTUM MECHANICAL-CLASSICAL MODELING
OF DOUBLE BARRIER RESONANT TUNNELING DIODES

T.G. Van de Roer

Department of Electrical Engineering,

Eindhoven University of Technology,

P.O. Box 513, NL-5600 MB Eindhoven, The Netherlands

## 1. INTRODUCTION

The Double Barrier Resonant Tunneling diode forms a member of a new class of electronic devices where thin-layer structures are used to manipulate the wavefunctions of electrons in order to obtain desired characteristics, in this case a current-voltage characteristic with a negative-resistance region [1].

From the point of view of applications it is important to have accurate models for the d.c. and a.c. characteristics of these devices. The actual thin-layer region where the shape of the wavefunction is important makes up only a small part of the device. It is packed between contact layers where electron transport is collision-dominated and where the classical billiard-ball model is adequate to describe what happens. Since these regions determine for a large part the I-V characteristics it is important to include them in the model. The problem then arises how to match the wavefunction solution to the particle description.

In this paper a model for the d.c. behaviour of DBRT diodes is presented which achieves this and is accurate enough for design purposes. In the model the diode is divided into three regions, an accumulation

*Negative Differential Resistance and Instabilities in 2-D Semiconductors*
Edited by N. Balkan *et al.*, Plenum Press, New York, 1993

431

region containing the injecting contact and spacer up to the first barrier, a barrier region consisting of the double barrier structure, and a depletion region between the second barrier and the collecting contact. Fig. 1 shows the layer stack of a typical DBRT diode and Fig. 2 gives a sketch of the conduction band profile at zero bias and at a high forward bias. The three regions can clearly be distinguished. The barrier region is where Schrödinger's equation has to be solved, the other two are treated as reservoirs [2] injecting and collecting electrons and to be described by classical models.

| | | | | |
|---|---|---|---|---|
| acc. reg. | GaAs | 0.5 $\mu$m | $2\times10^{18}$ | n |
| | GaAs | 50 nm | $2\times10^{16}$ | n |
| | GaAs | 2.5 nm | undoped | |
| bar. reg. | $Al_{0.4}Ga_{0.6}As$ | 5.6 nm | undoped | |
| | GaAs | 5 nm | undoped | |
| | $Al_{0.4}Ga_{0.6}As$ | 5.6 nm | undoped | |
| dep. reg. | GaAs | 2.5 nm | undoped | |
| | GaAs | 50 nm | $2\times10^{16}$ | n |
| | GaAs | 1 $\mu$m | $2\times10^{18}$ | n |
| | GaAs substrate | | $2\times10^{18}$ | n |

Fig. 1. Layer structure of a DBRT diode.

The connection between the barrier region and the contact regions is made by demanding continuity of the electric potential and field, as well as the transverse component of the wavevector at the interfaces. Also the position of the Fermi level in the accumulation region is assumed to determine the number of carriers that is available for tunneling.

## 2. TRANSPORT MODEL

The models used for the three regions shown in Figs. 1 and 2 will be described in the following.

### 2.1. The Accumulation Region

The region between the negatively polarized contact and the barriers is usually called accumulation region (see Fig. 2) because it is swamped by carriers injected from the contact which pile up against the first barrier. It provides a reservoir of carriers of which a small part tunnels through the barriers. The total tunnel current is directly proportional to the carrier density in this reservoir. Also this region imposes a boundary condition for the electric field on the barrier region and thus largely determines the position of the energy eigenstate of the well with respect to the fermi level in the accumulation region. At higher bias the electric field becomes very high and a triangular well is formed in the accumulation region as sketched in Fig. 1. Part of the electrons will now reside in subbands in this well and will tunnel from these two-dimensional states. This has a noticeable effect on the I-V characteristic as will be shown.

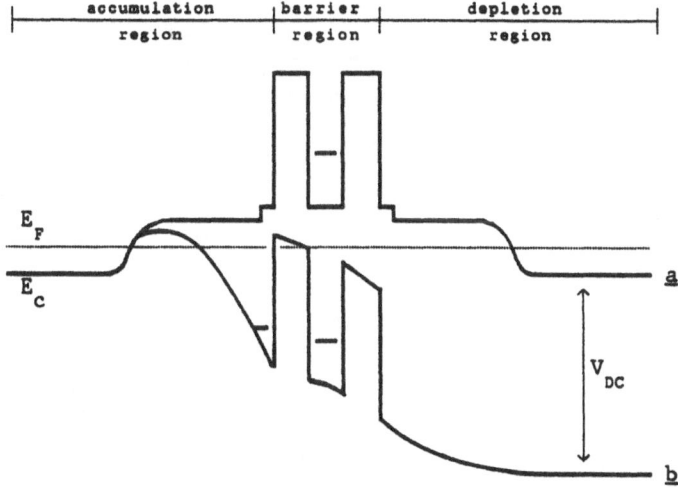

Fig. 2. Band diagram of DBRT diode at zero bias (a) and high bias (b).

The accumulation region is modeled by assuming that the Fermi level is constant throughout so that the carrier density is a function of the local potential only. This function, which is a well-known integral of the Fermi-Dirac distribution [3], can be inserted into the right hand side of Poisson's equation and the resulting non-linear equation is solved by successive overrelaxation. In this way values are obtained for the Fermi level and the electric field at the junction which serve as input values for the model of the barrier region. The potential profile calculated this way is used to calculate the positions of the subbands by numerical integration of Schrödinger's equation. This is not a fully self-consistent procedure but the error introduced is believed to be small.

Two subbands are taken into account and electrons at energy levels above the potential maximum of the accumulation region are assumed to be in 3-dimensional states.

## 2.2. The Barrier Region

In the barrier region the electric potential is calculated by selfconsistently solving Poisson's and Schrödinger's equations [5]. To simplify the calculations two assumptions have been made: in the solution of Poisson's equation the electron charge in the well is assumed to be concentrated in a sheet in the center of the well, and in Schrödinger's equation the trapezoidal barriers are replaced by rectangular ones so the solution can be written as a combination of plane waves. It has been shown [5] that the latter approximation makes very little difference compared to using the exact potential profiles.

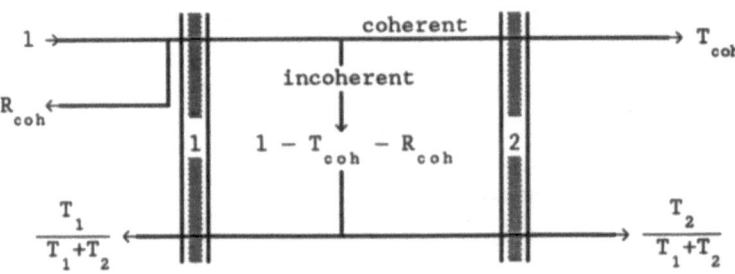

Fig. 3. *Contributions of coherent and incoherent tunneling electrons*

The effect of scattering is included by assuming that a damped wave function propagates in the quantum well. This is not a proper solution of Schrödinger's equation under the given circumstances, however, P. van Hall has shown [6] that such a solution can be obtained by introducing a complex potential in the well, a procedure which is often used in nuclear physics to represent scattering effects.

The transmission and reflection coefficients of the double-barrier system are given by

$$T_{coh} = \frac{\gamma T_1 T_2}{1 + \gamma^2 R_1 R_2 - 2\gamma\sqrt{R_1 R_2}\cos\phi} \tag{1a}$$

$$R_{coh} = \frac{R_1 + \gamma^2 R_2 - 2\gamma\sqrt{R_1 R_2}\cos\phi}{1 + \gamma^2 R_1 R_2 - 2\gamma\sqrt{R_1 R_2}\cos\phi} \tag{1b}$$

where $T_1$, resp. $T_2$ are the transmission coefficients of the left and right barrier, $R_1$ and $R_2$ are the corresponding reflection coefficients and $\phi$ is a phase factor representing the interference between the forward and backward traveling waves in the well.

Incoherent tunneling is represented by the factor $\gamma = \exp(-2\alpha L)$ where $\alpha$ is the damping coefficient of the wave function in the well and L is the width of the well [7].

A damped wave function means that we are losing coherently traveling electrons as the wave progresses and consequently the sum of $R_{coh}$ and $T_{coh}$ is not equal to one. These electrons are still in the well and have to get out through one of the two barriers. Following Büttiker [8] we assume that this goes in proportion to the transmission coefficients of the barriers. We thus get the scheme of Fig. 3 and for the total transmission coefficient:

$$T_{tot} = T_{coh} + (1 - T_{coh} - R_{coh})\frac{T_2}{T_1 + T_2} \qquad (2)$$

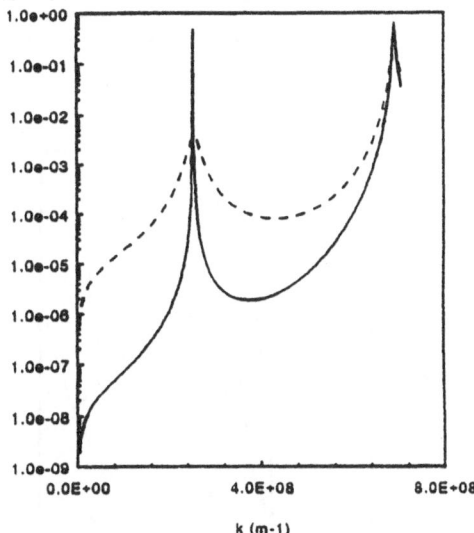

Fig. 4. Transmission coefficient against wavevector.
———— no scattering; - - - - - - with scattering.

In Fig. 4 $T_{tot}$ is sketched as a function of the transverse wavevector in the accumulation region and the influence of scattering is shown. To obtain the current density we have to multiply this by the current contributed by one electron and by the density of electrons having this wavevector which leads to the following expression

$$J_{out} = \frac{gem_1 kT}{4\pi^2\hbar^3}\int\limits_0^\infty T_{tot}\ln\left(1 + \exp\left(\frac{W_F - W_1}{kT}\right)\right)dW_1 \qquad (3)$$

where $W_1$ is the part of the electron energy resulting from the transverse wavevector, $W_F$ is the fermi level and the logarithmic term comes from the integration over the lateral coordinates.

## 2.3. The Depletion Region

This region presents a problem when accurate modeling is needed. A stream of almost mono-energetic high-speed carriers enters from the second barrier and is mixed with the carriers diffusing back from the contact, so a very peculiar distribution function results. However, the density of carriers injected from the barrier is much lower than the doping density so that the electric field profile in this region is mainly determined by the doping and the electrons diffusing back from the contact. Since these electrons are diffusing against the field they behave as if they were in an equilibrium situation. This enables us to apply the same model for this region as we have used for the accumulation region.

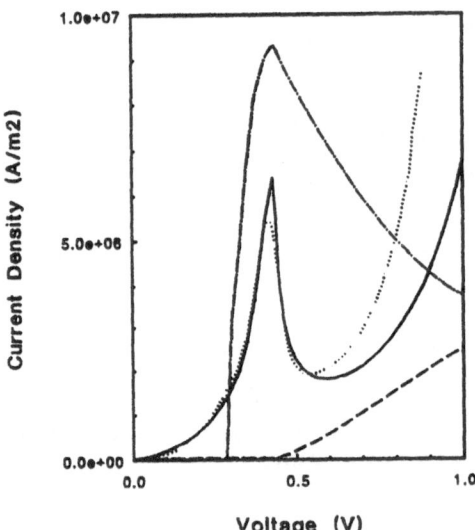

Fig. 5. J-V characteristics of the device of Fig. 1.
---------- Measured. ——————— Calculated.
— — — — Coherent/incoherent ratio (full scale=100%).
—.—.—.— Fraction of subband electrons (f. sc.=100%).

## 3. RESULTS

The I-V characteristic can now be calculated. It has been measured at room temperature on a device made at Nottingham University of which the technological parameters are those of Fig. 1.

The characteristic resulting from the model is given in Fig. 5 together with the measured characteristic. The only adjustable parameter in the model is the damping coefficient α. The best agreement was obtained with a value that is equivalent to a scattering time constant of about 30 fs, which is almost an order of magnitude shorter than the momentum relaxation time in bulk material. This indicates that additional scattering processes play an important role in these devices. The most likely causes are interface roughness scattering and non-uniform composition of the AlGaAs barriers.

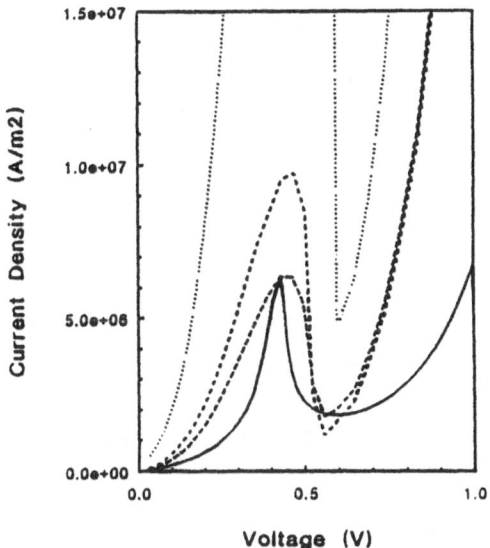

Fig. 6. Calculated J-V characteristics using different assumptions.
.......... Equal effective masses and coherent tunneling.
---------- + Including different effective masses.
— — — — + Including incoherent tunneling.
——————— + Including subbands in the accumulation region.

A third curve is added in Fig. 5 giving the ratio between coherent and incoherent current. It clearly indicates that over the bias range of interest incoherent tunneling is dominant. A fourth curve gives the fraction of electrons that tunnel from the first subband (tunneling from the second subband was insignificant over the whole range). Evidently the peak current comes mostly from the subband whereas at low and high bias the current is carried predominantly by 3-dimensional electrons.

In Fig. 6 different curves are displayed showing the effects of leaving out certain phenomena from the model, in particular subbands in the accumulation region, scattering and the difference in effective masses between the barriers and the well. It clearly shows the large influence of each of these phenomena.

It is a special problem to determine which mass should be used for the barriers since here the electrons are in the forbidden gap. We have used the conduction band effective mass of AlGaAs, calculated by a linear interpolation between those of GaAs and AlAs, on the assumption that due to the large bandgap of AlGaAs the electrons in the barriers are always much closer to the conduction band than to the valence band.

## 4. CONCLUSIONS

A model has been presented for DBRT diodes which includes incoherent tunneling and the effects of the contact regions. It has been found to yield good agreement with experiments and shows that it is possible to mix quantummechanical and classical approaches in one model.

## 5. ACKNOWLEDGEMENTS

Many people contributed in some form or another to this work. Thanks are due to H. Joosten and O. Abu-Zeid for contributions to the model, to M. Henini of Nottingham University for providing samples and to J.J.M. Kwaspen for measuring the I-V characteristics.

## 6. REFERENCES

1. L.L. Chang, L. Esaki, R. Tsu, Resonant Tunneling in Semiconductor Double Barriers, Appl. Phys. Lett. 24, (1974) 593.
2. M. Büttiker, Coherent and Sequential Tunneling in Series Barriers, IBM J. Res. Dev. 32, (1988), 63.
3. J.S. Blakemore, *Semiconductor Statistics*. Pergamon Press, London, 1962, pp 346-353.
4. T.G. van de Roer, H.P. Joosten, H. Noteborn, D. Lenstra, M. Henini: "Influence of Scattering on the I-V Characteristics of Double Barrier Resonant Tunneling Diodes", Physica B175, 301-306, (1991)
5. Joosten, H.P., Noteborn, H.J.M.F., Lenstra, D., Thin Solid Films 184, (1990) 199.
   Joosten, H.P., Numerical Study of Double Barrier Resonant Tunneling, AIO-2 Report, Dept. of Physics, Eindhoven University of Technology 1990.
6. P.J. van Hall, The Use of Imaginary Potential in DBRT Diodes, 5th Int. Conf. on the Physics of Electro-Optic Microstructures and Microdevices, Heraklion, Crete, July 30 - Aug. 3, 1990.
7. Jonson, M., Grincwaig, A.: "Effect of inelastic scattering on resonant and sequential tunneling in double barrier heterostructures", Appl. Phys. Lett. 1987, 51 1729-1731.

# INDEX

Acousto electric domains, 13, 127, 129, 130
Acousto electric effect, 127, 128, 138, 203, 212, 252
Acoustic phonon, 5, 13, 25, 68, 127, 130, 322
Acoustic phonon scattering, 5, 127, 130, 400, 421
Acoustic plasmon, 128, 136-138
Alenia MESFET, 224
Amorphous silicon, 378
Anderson-Crowell criteria, 32
Auger process, 27, 65
Autocorrelation technique, 375
Avalanche, 1, 239, 332, 395; *see also* Impact Ionisation
   and amplification, 130, 134, 135
   breakdown, 112, 113, 239, 262, 393, 399
   current, 314
   and instabilities, 283
   ionisation, 109-125
   multiplication, 221, 237, 292
   regime, 123

Balance equation, 25, 40-41; *see also* Energy balance equation
Ballistic transport, 66, 385
Barrier degradation, 308
Bethe's theory, 182
Bifurcation, 37-53, 186, 261-283, 393-409
Bilevel superlattice, 153-171
Birefrigence, 203-215
Bistability, 37-53, 195, 269, 275,
Bloch
   function, 102, 144
   oscillation, 8
   state, 100
   wave, 144
Boltzmann equation, 23-37, 72, 135
Born-Huang model, 142
Bragg Reflector, 8, 269-283
Breakdown, 40, 215-251, 261-269, 329; *see also* Avalanche
   voltage, 219-251
Breathing instability, 397
Bremsstrahlung, 219-251
Brooks-Herring formula, 393-399, 413
Capture
   of electrons, 2
   NDR, 2
   rate, 9, 204

Carrier
   depletion, 284
   injection, 119, 294
   leakage, 57, 58, 60, 310
   relaxation, 379
Carrier heating, 72, 180, 322
Catastrophe theory, 394
Central Limit theorem, 400
Chaos, 18, 37-53, 118, 261-269, 318, 393-409
   temperature, 393, 402
Chaotic attractor, 44
Chaotic itinerancy, 393
Chaotic orbit, 400
Chaotic oscillations, 44, 180, 210, 269, 275, 401
Chaotic repeller, 44
Channel
   pinch-off in FET, 305-319
   potential profile, 75
Charge injection transistor (CHINT), 53-83, 305-317
Chebyshev polynomial, 385-393
Chynoweth's law, 220, 225
Collision integral, 38, 39
Collisional broadening, 23, 24, 31, 32
Coherent transport, 435
   tunnelling, 434
Complemetary RST devices, 53-83
Continuity equation, 47, 181, 182
Continuum model 141-153, 335-351; *see also* Dielectric continuum model
Correlation technique, 376
   tunnelling, 434
Cross talk
   coupling, 394, 404
   current, 405
Current
   collapse, 109-127, 203
   filaments, 37-53, 109-127, 216-269, 251-261, 269-283, 331
   gain, 72
   instability, 37-53, 109-127, 127-141, 251-261, 294, 393, 395, 397
   oscillations, 11, 16, 37-53, 127-141, 179-189, 203-215, 265, 283-305, 393-409
   saturation, 1-23, 127-141, 283-305
   switching, 111
Current-controlled NDR, 1-23, 37-53, 109-127, 189, 251-261, 283-305, 261-269

Current-controlled NDR (*cont'd*)
189, 251-261, 283-305, 261-269

D-X centre formation, 88, 91, 219
Damped oscillations, 132
Dead-space effect, 243
Debye screening length, 129, 399
Deformation - potential, 14, 25, 27, 29, 34, 322
    constant, 23-27, 33-34
Deformation - potential coupling, 5
Deformation - potential scattering, 5, 327, 400
Delayed feedback control, 44
Delta ($\delta$) doping, 83-99, 317-335
Delta-funcion excitation, 7, 15
Density of states, 9, 25, 34
Depletion layer, 155, 184, 292
Deterministic noise amplification, 393, 401, 402
Detrapping, 117, 118
Dielectric
    breakdown, 1
    continuum model, 141-153, 335-351; *see
        also* Continuum model
    function, 90, 362, 363
    relaxation, 40-45, 180-187, 294
    relaxation instability, 397
    response, 369
Diode, 58-60
Differential optical comparator, 279
Diffusion coefficient, 40, 182
Dispersion relation, 149, 150, 338,343
Displacement current, 73
Dissipative structures, 261-269, 269-283
Domain boundary, 193-200
Domains, 49, 53, 54, 57, 72, 75, 190, 194, 199,
    269,310
Double barrier resonant tunnelling, 390, 431
Drift
    diffusion equation, 39, 47, 182
    instability, 40
    mobility, 121
    velocity, 6, 83-99, 283-305

Effective mass approximation, 32, 100
Einstein relation, 40, 182
Electrical instability, 3, 283, 285, 292
Electric field domains, 192-199, 262, 269, 280; *see also*
    Domains
    in superlattices, 199
Electroluminescence, 60-63, 131, 229, 251, 284-
    288
Electromechanical coupling
    constant, 129
Electron
    distribution, 25, 27, 29
    emission, 25, 99, 100
    mobility, 87, 93, 94, 112, 120, 127, 137, 138
    -hole plasma, 127, 136, 137
    impact ionisation coefficient, 246
    injection, 113, 116, 120, 154
    temperature, 23, 40-48, 57-76, 84, 132, 154,
        232, 287-305

Electron (*cont'd*)
    transfer, 2, 3, 7
Electron-electron scattering, 38, 87, 283, 321
Electron-heavy hole transition, 166, 167
Electron-hybrid phonon interaction, 144, 148, 150,
    347
Electron impurity scattering, 24
Electron-light hole transition, 167
Electron phonon scattering, 23-34, 141-153, 335-
    351
Electron screenings, 90
Electrooptic
    modulator, 276
    probing, 204, 207
    tensor, 205
Electrostatic boundary condition, 337
Energy balance equation, 25, 40, 41; *see also* Bal-
    ance equation
Energy-relaxation
    of hot electrons, 39, 56, 63, 66, 67, 110, 415
Exciton
    binding energy, 415
    -exciton collision, 421
    recombination rate, 415
    -Stark shift, 409

Fabry-Perot modulator, 276
Far Infra-red spectroscopy, 355
Fast Fourier transform, 387
Feigenbaum attractor, 401
Femto-second spectroscopy, 27
Field-controlled NDR, 5, 7, 318
Field-effect transistor (FET), 53-83, 215, 246,
    305-317
    accumulation mode, 76
Field induced capture, 2,9
Field-quenching, 3
Filament
    mobility, 121
    formation, 37-53, 109-127, 318, 395-405
    spike, 49
Fluctuations, 40
Fresnel loss, 65
Fringing field, 75, 79
    instability, 394, 403
Frustrated chaos, 331
Fuchs and Kliewer modes, 339
Full band
    model, 23, 34
    simulation, 24, 25, 29; *see also* Monte-Carlo

GaAs, 1-23, 23-37, 37-53, 99-109, 261-269, 373-
    385, 393-409
    Au doped, 1-23
    MESFET, 215-251
    patterned resonant structure, 171-179
    p-i-n, 269-283
GaAs/Ga$_{1-x}$Al$_x$As
    coupled quantum wires, 37-53
    delta-doped, 83-99, 409-421
    double quantum wells, 127-141, 283-305

GaAs/Ga$_{1-x}$Al$_x$As (cont'd )
 HBT, 215-251
 heterostructure, 1-23, 37-53, 99-109, 109-127,
   179-189, 385-393
 multiple quantum wells, 127-141, 203-215,
   251-261, 283-305
 NERFET, 305-317
 quantum state device, 153-171
 resonant tunnelling device, 431-438
 single quantum well, 37-53, 127-141, 141-153,
   283-305
 superlattice, 189-203
GaAs/AlAs, 99-109, 335-351, 421-431
GaAs/InGaAs, 215-251
GaInAs, 23-37
GaInAs/InAlAs, 53-83, 189-203
GaInAs/InP, 189-203
Gain factor, 137
Γ-L scattering, 27, 102,103,105, 251, 253, 325
Γ-X scattering, 27, 102, 103
Generation
 instability, 40
 lifetime, 39
 rate, 9, 224, 249
 recombination, 39, 40, 46, 328, 329
Germanium
 Au doped, 1-23
 p-type, 261-269
Global bifurcation, 43
Golden rule of Fermi , 23, 24, 148
Gunn effect, 3-23, 25, 28, 41, 153-171, 189-203,
   204, 212, 251, 269, 280, 283, 379

Hall field instability, 396, 397
HBT, 216, 246, 242
Heterostructure hot-electron diode, 47, 49, 50,
   305, 312, 313,
Heterostructure laser, 57, 284
Hierarchy of instabilities, 394
High-field domains, 2, 3, 11, 77, 78, 133, 187-
   199, 204, 251, 252, 283-290, 305-307
Hot electron, 1-24, 38, 53, 55, 63-65, 72, 111,
   137, 141, 179, 238-245, 289-294, 310,
   320-325, 368
 diode, 50
 domains, 54, 66, 67, 72-79
 emitter, 56
 fail, 66
 injection, 53, 54; see also Carrier injection
 instabilities see Instabilities
 luminescence, 245, 285, 422, 426
 percolation, 212, 283-305
 PL technique, 132
Hot exciton
 luminescence, 421, 424
 recombination, 428
HopF bifurcation, 40-47, 395, 396
Hot phonons, 121, 137, 251, 294
Huang and Zhu model, 150
Hybrid switching, 276
Hybridization, 18, 50, 144, 147, 341, 345, 346
Hydrodynamic model, 141, 338, 339, 340

Impact ionisation, 9-11, 23-32, 38-40, 63, 73,
   112-123, 215-240, 242-262, 270-285,
   307-394, 413
 avalanche diode, 11
 breakdown, 262
 coefficient, 28, 29, 112, 114, 115, 221, 222, 243,
   398, 416
 excitation, 10, 328
 of excitons, 409-415
 prefactor, 417
 rate, 27-29, 30, 32, 112, 113, 124, 224, 225,
   232,
 regime, 238
 threshold, 63, 243
Impurity barrier capture, 7, 9, 12,15, 283
Impurity scattering, 41, 83-97, 121, 180, 318,
   320, 399
Index ellipsoid, 204-206
InAs, 23-37
Injection
 and trapping, 113, 117
 current, 53, 76
 of minority carriers, 305
InP, 23-37, 251-261
InSb, 393-409
Instabilities, 2-23, 37-41, 49, 53, 54, 72, 79,
   125, 127, 161, 190, 212, 262, 267,
   287, 290-296, 397
Intervalley transfer, 2, 7 ,9 , 15, 25, 85, 87,
   154, 182, 189, 283
Interface
 optical modes, 141, 144, 150, 335, 337, 339,
   425, 426
 polaritons, 18, 342, 345
 potential, 41
 roughness, 89, 90, 137, 291, 416, 437
Inter-Landau level transition, 422
Internal radiative efficiency, 60, 67, 80
Intersubband scattering, 15, 16, 46, 145, 192
Intraband relaxation, 345, 346, 377, 379

Kane theory, 27, 28, 29, 32
Keldysh formula, 27, 28, 29, 32

Landau
 Büttiker formalism, 99
 level, 8
Lattice attenuation, 130, 132
Leakage current, 53-83, 308-317
Light emission, 53-83, 215-251, 284-305, 305-319
Light emitting logic devices, 53-83
Lindhard dielectric function, 361-373
Localised exciton, 416
Logic devices, 53-83
Longitudinal acoustic phonons, 25, 33, 362
Lorentz attractor, 400
Lucky electron model, 23, 29
Lyapunov spectrum, 393, 402, 404

Magnetically induced NDR, 8
Magneto-optic effect, 423-424
Mechanical interface mode, 338-351

MESFET, 84, 216-246
Microwave, 239, 255, 258, 269, 275, 283
Miniband transport, 189-200
Minority carrier injection, 306
Mixed scattering, 283
MODFET, 84
Momentum relaxation, 38, 39, 138, 181, 294
Momentum space transfer, *see* Intervalley transfer
Monte Carlo simulation, 23-29, 32, 72-84, 100, 112, 121, 182, 229, 242, 243, 368, 382
MOSFET, 29, 234
Multi-$\delta$ layers, 328
Multi-terminal logic, 61-70

Negative Differential Resistance (NDR), 1-23, 37, 41, 44, 56, 72-77, 109-127, 153, 171-179, 251-261, 261-269, 283-305
Negative mass, 2, 8
    amplifier, 2
Negative photoconductivity, 46
NERFET, 285, 306, 314
Neumann boundary, 72
Neural network, 403
Nl product, 13
Nipi structure, 284
Non-drifting hot phonons, 137, 294
Non-linear Bragg reflector, 276
Non-parabolicity, 6, 24, 283
Non-radiative lifetime, 46, 64, 421, 428
Non-resonant tunnelling, 38, 44-47, 192-200
NOR and NAND gate, 53, 68, 70, 71, 279
Nyquist theorem, 402

Optical beam induced current, 109, 120-124
Optical bistability, 270, 279, 280, 410
Optical comparator, 278
Optical hybrid phonon, 143-147
Optical non-linearity, 276
Optical phonons, 6, 8, 25, 65, 67, 121, 143, 147, 324, 335; *see also* Interface optical modes
Optical soliton, 279
Optically induced NDR, 8, 46, 124
Organ-pipe resonance, 390
ORNAND gate, 54-79
Oscillating filament, 270, 276, 278
Oscillation, 49, 50, 117, 131, 132, 251-261, 261-269, 283-305, 318, 330, 390, 396
    frequency, 210
    period, 134
    threshold, 134, 211, 257
Oscillatory instability, 38, 40-47, 50

PADRE, 72
Parallel conduction, 17, 50, 110, 111, 179
Parasitic effects, 59, 60, 61, 331
Patterned resonant structure, 171
Peak to valley ratio, 158
Percolation
    NDR, 78
    path, 8,9,204, 291
    channel, 291

Period doubling, 40, 43, 44, 210, 395, 401,402
Periodic instability, 30, 180, 393, 394, 396
Periodic repeller, 45
Periodic screening, 352, 359
Periodic switching mode, 267
Phonon induced NDR, 153, 154
Phonon, 32, 34, 38, 87, 97,182, 183
    exciton interaction, 421
    flux, 234, 324
    polar-LO, 5,6
    polarisation, 34
Photoconductivity, 157, 166, 378
    persistent, 120
Photothermal devices, 280
Piezo electric coupling, 14, 127
Pin structure, 196
Pinch-off, 154, 155, 228
Plasmon, 14, 18, 136, 351, 363, 366
    dispersion, 352
    emission, 364
    excitation, 363-366
    frequency, 351-357
    peak, 355
    resonance, 350
    shift, 353
    wavelength, 352
Pockels effect, 204
Poincaré surface, 398
Poisson equation, 112, 183, 242
Polariton, 18
Population inversion, 8
PL quenching, 413-415
Pseudomorphic HEMT, 236, 215, 216, 246

Quantum
    interference, 32
    wire, 50, 386
    transport equation, 24, 99, 105
    state transfer device, 153-171
    transmission coefficient, 183
    -confined Stark effect, 190, 196, 276, 410, 415
    efficiency, 218, 219
    well laser, 284
Quantum-mechanical tunnelling, 1; *see also* Tunnelling

Radiative efficiency, 56, 64, 65, 66, 68, 71, 80
Radiative recombination time
Radiative transitions, 9, 236, 422
Random phase approximation, 362
Raman
    scattering, 336-347, 428-429, 425
    spectroscopy, 335, 351-353
Real space transfer (RST), 15, 16, 37-53, 54-83, 85-98, 109-112, 131, 153-171, 179-189, 203, 204, 232, 236, 253, 283-305, 318-335
    current, 56-73
    of holes 63
    of hot electrons, 60, 80
    logic, 54, 68, 80

Real space transfer (RST) (*cont'd* )
    oscillator, 43, 44, 45
    structure, 64
    transistor, 53, 54, 57, 74, 79, 305, 307
    of hot electrons, 60, 80
    of holes
Recombination, 56, 64, 66, 328
    wave, 13
    current, 241
    radiation, 273
    lifetime, 415, 417
Resonant tunnelling, 38, 46, 171, 174-177, 189-200
Richardson's constant, 9
Ridley-Watkins-Hilsum mechanism, 189; *see also* Gunn effect
Runaway, 6

Scanning electron microscopy techniques, 265, 284, 290, 355
Scanning laser microscope, 121
Scattering-induced NDR, 5, 16, 87
Screening, 362, 368
SEED, 276, 278, 280
Self-organised criticality, 394
Shockley-Read-Hall model, 225, 270
Singular perturbation theory, 179, 187
Slab phonon modes, 336, 339
Slow domains, 3
Space charge waves, 15, 17
Spatio-temporal oscillation, 47-50, 183, 187, 264, 270
Spontaneous avalanche, 125
Spontaneous current oscillations, 180, 262-264, 298, 395
S-type NDR, 5, 10, 38, 47, 50, 270, 273, 278, 283, 313, 314
Stark shift, 197, 198, 475; *see also* Quantum confined Stark effect
Static domain, 180, 185, 199, 252, 257, 290 *see also* Stationary domains
Stationary domains, 3, 155, 159, 166, 395, 401
Streaming plasma instability, 353
Super threshold perturbation, 180-187

Temporal instability, 261, 266
Thermionic emission, 8, 37, 42, 46, 47, 48, 112, 180, 182, 187, 297
Thermal recombination coefficient, 398
Thomas-Fermi wavevector, 135
Three terminal CHINT, 61
Threshold discriminator, 218
Time-of-flight technique, 194, 325
Transient
    chaos, 44, 45
    current, 73
Transit-time
    NDR, 2, 10, 11
    mechanism, 5
Transverse optical phonon, 33
Trapping, 112, 115
Travelling domains, 43, 179, 184, 283

Triple hybrid, 143
Trobic, 121-125
Tunnelling, 10, 32, 37, 44, 46, 48, 57, 58,72, 73, 157, 166, 219, 229, 236
    current, 46, 47, 48, 195
    real space transfer, 15
    resonance, 191
Two-level impurity NDR, 10
Two-stream instability, 14, 127-141

Ultrafast photoconductivity, 373
Uniaxial stress, 7
Unipolar CHINT, 67, 72

Vertical hot electron transport, 37, 47, 284
Voltage controlled NDR, *see* NDR

Wannier -Stark ladder, 38
Well-to-well screening, 91
Well-width fluctuation, 138

Zeeman g-factors for excitons, 422
Zeeman splitting, 425
Zener diode, 1